NONPARAMETRIC METHODS IN GENERAL LINEAR MODELS

Probability and Mathematical Statistics (Continued)

MATTHES, KERSTAN, and MECKE • Infinitely Divisible Point Processes
MUIRHEAD • Aspects of Multivariate Statistical Theory
PARZEN • Modern Probability Theory and Its Applications
PURI and SEN • Nonparametric Methods in General Linear Models
PURI and SEN • Nonparametric Methods in Multivariate Analysis
RANDLES and WOLFE • Introduction to the Theory of Nonparametric Statistics
RAO • Linear Statistical Inference and Its Applications, *Second Edition*
RAO and SEDRANSK • W.G. Cochran's Impact on Statistics
ROHATGI • An Introduction to Probability Theory and Mathematical Statistics
ROHATGI • Statistical Inference
ROSS • Stochastic Processes
RUBINSTEIN • Simulation and The Monte Carlo Method
SCHEFFE • The Analysis of Variance
SEBER • Linear Regression Analysis
SEBER • Multivariate Observations
SEN • Sequential Nonparametrics: Invariance Principles and Statistical Inference
SERFLING • Approximation Theorems of Mathematical Statistics
TJUR • Probability Based on Radon Measures
WILLIAMS • Diffusions, Markov Processes, and Martingales, Volume I: Foundations
ZACKS • Theory of Statistical Inference

Applied Probability and Statistics

ABRAHAM and LEDOLTER • Statistical Methods for Forecasting
AGRESTI • Analysis of Ordinal Categorical Data
AICKIN • Linear Statistical Analysis of Discrete Data
ANDERSON, AUQUIER, HAUCK, OAKES, VANDAELE, and WEISBERG • Statistical Methods for Comparative Studies
ARTHANARI and DODGE • Mathematical Programming in Statistics
BAILEY • The Elements of Stochastic Processes with Applications to the Natural Sciences
BAILEY • Mathematics, Statistics and Systems for Health
BARNETT • Interpreting Multivariate Data
BARNETT and LEWIS • Outliers in Statistical Data, *Second Edition*
BARTHOLOMEW • Stochastic Models for Social Processes, *Third Edition*
BARTHOLOMEW and FORBES • Statistical Techniques for Manpower Planning
BECK and ARNOLD • Parameter Estimation in Engineering and Science
BELSLEY, KUH, and WELSCH • Regression Diagnostics: Identifying Influential Data and Sources of Collinearity
BHAT • Elements of Applied Stochastic Processes, *Second Edition*
BLOOMFIELD • Fourier Analysis of Time Series: An Introduction
BOX • R. A. Fisher, The Life of a Scientist
BOX and DRAPER • Evolutionary Operation: A Statistical Method for Process Improvement
BOX, HUNTER, and HUNTER • Statistics for Experimenters: An Introduction to Design, Data Analysis, and Model Building
BROWN and HOLLANDER • Statistics: A Biomedical Introduction
BUNKE and BUNKE • Statistical Inference in Linear Models, Volume I
CHAMBERS • Computational Methods for Data Analysis
CHATTERJEE and PRICE • Regression Analysis by Example
CHOW • Econometric Analysis by Control Methods
CLARKE and DISNEY • Probability and Random Processes: A First Course with Applications

Applied Probability and Statistics (Continued)

COCHRAN • Sampling Techniques, *Third Edition*
COCHRAN and COX • Experimental Designs, *Second Edition*
CONOVER • Practical Nonparametric Statistics, *Second Edition*
CONOVER and IMAN • Introduction to Modern Business Statistics
CORNELL • Experiments with Mixtures: Designs, Models and The Analysis of Mixture Data
COX • Planning of Experiments
DANIEL • Biostatistics: A Foundation for Analysis in the Health Sciences, *Third Edition*
DANIEL • Applications of Statistics to Industrial Experimentation
DANIEL and WOOD • Fitting Equations to Data: Computer Analysis of Multifactor Data, *Second Edition*
DAVID • Order Statistics, *Second Edition*
DAVISON • Multidimensional Scaling
DEMING • Sample Design in Business Research
DILLON and GOLDSTEIN • Multivariate Analysis: Methods and Applications
DODGE and ROMIG • Sampling Inspection Tables, *Second Edition*
DOWDY and WEARDEN • Statistics for Research
DRAPER and SMITH • Applied Regression Analysis, *Second Edition*
DUNN • Basic Statistics: A Primer for the Biomedical Sciences, *Second Edition*
DUNN and CLARK • Applied Statistics: Analysis of Variance and Regression
ELANDT-JOHNSON and JOHNSON • Survival Models and Data Analysis
FLEISS • Statistical Methods for Rates and Proportions, *Second Edition*
FOX • Linear Statistical Models and Related Methods
FRANKEN, KÖNIG, ARNDT, and SCHMIDT • Queues and Point Processes
GALAMBOS • The Asymptotic Theory of Extreme Order Statistics
GIBBONS, OLKIN, and SOBEL • Selecting and Ordering Populations: A New Statistical Methodology
GNANADESIKAN • Methods for Statistical Data Analysis of Multivariate Observations
GOLDBERGER • Econometric Theory
GOLDSTEIN and DILLON • Discrete Discriminant Analysis
GREENBERG and WEBSTER • Advanced Econometrics: A Bridge to the Literature
GROSS and CLARK • Survival Distributions: Reliability Applications in the Biomedical Sciences
GROSS and HARRIS • Fundamentals of Queueing Theory, *Second Edition*
GUPTA and PANCHAPAKESAN • Multiple Decision Procedures: Theory and Methodology of Selecting and Ranking Populations
GUTTMAN, WILKS, and HUNTER • Introductory Engineering Statistics, *Third Edition*
HAHN and SHAPIRO • Statistical Models in Engineering
HALD • Statistical Tables and Formulas
HALD • Statistical Theory with Engineering Applications
HAND • Discrimination and Classification
HILDEBRAND, LAING, and ROSENTHAL • Prediction Analysis of Cross Classifications
HOAGLIN, MOSTELLER and TUKEY • Exploring Data Tables, Trends and Shapes
HOAGLIN, MOSTELLER, and TUKEY • Understanding Robust and Exploratory Data Analysis
HOEL • Elementary Statistics, *Fourth Edition*

(*continued on back*)

Nonparametric Methods in General Linear Models

Madan Lal Puri
Indiana University, Bloomington

Pranab Kumar Sen
University of North Carolina, Chapel Hill

JOHN WILEY & SONS
New York Chichester Brisbane Toronto Singapore

Library of Congress Cataloging in Publication Data:

Puri, Madan Lal.
Nonparametric methods in general linear models.

Bibliography: p.
Includes index.
1. Linear models (Statistics) 2. Nonparametric statistics. I. Sen, Pranab Kumar, 1937– II. Title. III. Series.

QA276.P84 1985 519.5 84-25813
ISBN 0-471-70227-7

Printed in the United States of America

10 9 8 7 6 5 4 3 2 1

Preface

This book attempts to provide a systematic account of the developments in nonparametric statistical inference relating to general (multivariate) linear models. There has been a great upsurge of activity in this area during the past two decades, and this book presents some aspects of this work in a logically integrated and systematic form.

The book is written for specialists, teachers, and advanced graduate students with a background knowledge of parametric linear models (and multivariate statistical theory) as well as basic nonparametric inference. Familiarity with probability theory at the graduate level is also presumed.

The book consists of eight chapters and an Appendix. Chapter 1 provides a brief outline of the material covered in the book and also refers to some related topics not discussed in detail in later chapters. Chapters 2, 3, and 4 constitute Part 1 of the book, where the distribution theory of rank statistics is developed in a systematic manner. Chapter 2 is devoted to linear rank statistics in the univariate setup, Chapter 3 to signed rank statistics in the univariate setup, and Chapter 4 to their multivariate generalizations. Chapters 5, 6, 7, and 8 form Part 2 of the book and are devoted to the statistical inference procedures related to general linear models. Rank tests for simple univariate as well as multivariate linear models are considered in Chapter 5. Chapter 6 is mainly devoted to the development of the theory of rank estimates for such models. Tests for composite hypotheses in linear models, for unrestricted as well as restricted alternatives, are treated in Chapter 7. Miscellaneous rank tests for some other problems (including grouped data, stochastic predictors, censored data, growth-curve models, and analysis-of-covariance models) are considered in Chapter 8. The Appendix deals with the derivation of some of the results used in previous chapters. Centering of rank statistics and asymptotic linearity results on rank statistics are treated here.

Throughout the book, the main emphasis has been placed on rank-based procedures. For some of the developments based on U-statistics and von Mises' functionals, we refer to our earlier monograph *Nonparametric Methods in Multivariate Analysis* (Wiley, 1971). The main emphasis here is on the general linear models. For some simple (one-, two-, or several-sample) models as well as for some blocked designs, the reader is also referred to the above monograph. We may, however, mention that besides the broader framework of this book, the techniques employed are of more recent origin than the ones in our earlier monograph. In particular, the general approach of the late Professor Jaroslav Hajék has inspired the basic philosophy of this book.

Substantial parts of this book have been offered in courses at the University of North Carolina, Chapel Hill; Indiana University, Bloomington; Albert-Ludwigs Universität, Freiberg in Brisgau, West Germany; the University of Göttingen, West Germany; the University of Washington, Seattle; and other places. Also, during the past ten years, various parts were presented in professional meetings and colloquiums in the United States as well as other countries. We have greatly benefited from the valuable comments and criticisms of our colleagues and (former) students, and we regret that we cannot thank them individually here.

The writing of this book was supported in part by the following research grants and contracts (during the past 12 years): (1) National Institutes of Health, Grant number GM-12868 and Contract number NIH-HHLBI-71-2243L, (2) Aerospace Research Laboratories, U.S. Air Force Contracts F33615-70-C-1124 and F33615-71-C-1927, (3) Air Force Office of Scientific Research, AFSC, USAF, Contract numbers AFOSR 71-2009, 74-2736, and 76-2927, and (4) National Science Foundation, Grant number MCS76-00951. The support of these organizations is very much appreciated.

Generous support from both the University of North Carolina and Indiana University constituted a vital factor in this work from initiation to completion.

It is indeed a pleasure to acknowledge with profound thanks the invaluable cooperation and help we received from Ms. Beatrice Shube, Editor, Wiley-Interscience, in the accomplishment of this project. Finally, we would like to thank our wives Uma and Gauri for their generous support and enormous patience throughout the whole undertaking.

MADAN L. PURI
PRANAB K. SEN

Bloomington, Indiana
Chapel Hill, North Carolina
February 1985

Contents

NONPARAMETRIC METHODS IN GENERAL LINEAR MODELS

PART 1

Distribution Theory of Rank Statistics

CHAPTER 1

Preliminaries and General Objectives

1.1. INTRODUCTION

The past few decades have witnessed a fundamental growth of the theory of nonparametric statistical inference. Nonparametric procedures were developed originally for simple hypothesis-testing problems relating to randomness, location of a distribution, differences of two (or more) distributions, bivariate independence, and some goodness-of-fit models. However, the recent developments encompass a much wider field where the theory of estimation and that of hypothesis testing have mingled together in a very nice manner, covering a broad class of experimental designs, including sequential as well as multiresponse models. The primary objective of this book is to present a systematic and comprehensive account of the nonparametric statistical analysis of some general (multivariate) linear models.

Statistical analysis of linear models not only constitutes one of the most active areas of fruitful research, but also provides the basic tools for the planning and analysis of research schemes pertaining to various experimental setups, arising in various fields of experimental science and technology as well as socioeconomic, behavioral, and biomedical investigations. A variety of textbooks, ranging from the elementary to the advanced, have appeared during the past three decades. However, most of these deal exclusively with parametric setups, exploiting the classical least-squares estimation theory and likelihood-ratio-based testing theory requiring on certain distributional assumptions. Lately, some attention has been focused on the lack of *robustness* of these parametric procedures (against possible departures from the basic assumptions relating to the models), and in consequence there have been calls for alternative robust procedures. The need for such robust procedures is particularly felt in many linear models, where the normality of the distribution of the error variables may be open to question and for nonnormal distributions, the simplicity of the parametric procedures may be compromised. In multivariate theory, the need for procedures valid for nonnormal populations is felt even more strongly; there are not many well-behaved nonnormal multivariate distributions, and, for such as do exist, suitable inference procedures are either not available or too complicated to apply. The present book attempts to provide a broad coverage of some robust nonparametric procedures pertaining to a general class of (univariate as well as multivariate) linear models.

In the parametric linear models, apart from the basic additivity (linearity) of the various effects (attributed to various assignable factors), the distributions of the error components are assumed to be of some specified forms (mostly, normal). Statistical analysis for such models, based on such distributional assumptions, may be restricted in scope to where these assumptions remain valid. In a variety of practical applications, the linearity of the models may be broadly justified (by suitable transformations, if needed), although the distributional assumptions (on the error or other components) may remain questionable. Therefore, it is natural to seek solutions which remain insensitive to the choice of distributional assumptions, that is, which are robust for a broad class of underlying distributions. The theory of nonparametric inference to be systematically developed in this book pertains to this general objective.

For statistical analysis of linear models, whereas a variety of textbooks are available on the parametric case, there are not many books or monographs on the nonparametric case. In fact, the theory has not been developed in the nonparametric case in its full generality, though the need for nonparametric procedures has been felt strongly. For the univariate case and some simple linear models such as the one-, two-, or several-sample (one-way) analysis-of-variance models, two-way layouts, and some simple regression models, descriptions of certain nonparametric procedures have appeared in some books. However, for the general multivariate linear models, with the exception of the 1971 monograph *Nonparametric Methods in Multivariate Analysis* by the present authors, there is practically no detailed treatment of the nonparametric procedures in a systematic and integrated form.

Even the abovementioned monograph is devoted primarily to simple models, including one-sample and multisample problems, two-way layouts, and multivariate independence problems; it does however, provide the incentive to consider more general linear models in an up-to-date fashion, and that is done in this book. The results of Chapters 4, 5, and 6 of the earlier monograph may be characterized as special cases of parallel ones treated in this book, though the results of Chapters 7, 8, and 9 are not further generalized here. There is also a basic difference in the approaches of these two books, as will be made clear in Section 1.2. Thus, the current book in no way attempts to replace the earlier one. The comparative simplicity of the theory (for the specific problems) in the earlier book has been compensated here by increased generality and wider coverage, so that each one has its own purpose. In the next section, we present a more detailed description of the basic themes of this book.

1.2. THE ORGANIZATION OF THIS BOOK

This book is intended for graduate students as well as research scholars specializing in the areas of nonparametric and linear models. We take it for granted that the reader is familiar with the theory of probability and statistical inference (at the standard graduate level), with at least some exposure to nonparametric statistics, linear models, and multivariate analysis. Familiarity with advanced calculus, real analysis, and matrix algebra is also presumed.

The main objective of the book is the development of statistical procedures based on ranks for some general linear models. In this context, the distribution theory of rank statistics plays a fundamental role and provides the basis for the statistical analysis. For this reason, the distribution theory is presented separately in Chapters 2, 3, and 4 (Part 1) and then incorporated in Chapters 5, 6, 7, and 8 (Part 2) in the formulation of the inference procedures. Some of the technical derivations are postponed to the Appendix for convenience of reading and presentation.

The distribution theory of rank statistics has been developed systematically and in increasing generality by a host of workers. The permutational central limit theorems—due to Wald and Wolfowitz (1944), Noether, (1949), Hoeffding (1951a), Motoo (1957), and Hájek, (1961), among others—are the forerunners in this direction and pertain to the asymptotic distribution of rank statistics under suitable hypotheses of invariance. A great many of the rank statistics may also be expressed in terms of U-statistics; Hoeffding (1948) was the first to utilize a projection technique in the simplification of the asymptotic distribution theory of U-statistics, and he paved the way for active research in asymptotic nonparametrics. For the two-sample problem, another ingenious approach to the study of the asymptotic distribution theory of rank statistics is due to Chernoff and Savage (1958). An alternative "weak-convergence approach" due to Pyke and Shorack (1968a, b) works very elegantly for single as well as several-sample problems, but may be operationally difficult for a general linear model. Most of these results have been discussed in detail in Puri and Sen (1971).

The main emphasis of the present book is on the later approach of Hájek (1968), which is characterized by the relaxation of the basic regularity conditions (on the underlying distributions and the score functions) and by adaptability to general linear models (containing the single- and several-sample problems as special cases). In Part 1 the distribution theory of rank statistics is presented in a systematic and logically integrated form. For the univariate setup, the distribution theory of linear rank statistics is systematically explored in Chapter 2. The theory, mainly adapted from Hájek (1968), rests on a powerful variance inequality and a polynomial approximation scheme for linear rank statistics, where the basic regularity conditions are trimmed to a minimum and the theory is not necessarily confined to local alternatives. For some local alternatives, an alternative (*contiguity*) approach of Hájek (1962) works well under somewhat different sets of regularity conditions, and this is also discussed briefly in the same Chapter. Following Hoeffding (1973), the centering of linear rank statistics (with respect to the asymptotic normal distribution) is also considered in Chapter 2. Chapter 3 deals with the distribution theory of signed rank statistics, in the univariate setup. The results are mainly adapted from Hušková (1970), and they run parallel to the ones in the preceding chapter. Chapter 4 is devoted to the multivariate generalizations of the theory developed in Chapters 2 and 3, and these results are mainly adapted from Puri and Sen (1969a) and Hušková (1971). In all these chapters, though the asymptotic distribution theory has been developed under the full generality, no attempt has been made to consider problems related to asymptotic expansions (e. g. Edgeworth or Gram–Charlier series), or rate of convergence (e. g. Berry–Esséen or Prohorov distance estimates), or large-deviation probabilities. The study of these problems (which in general provides deeper analyses than the statistical results of asymptotic normality) belongs to a different domain and is outside the scope of the present book. However, for the reader interested in these topics, we refer to the books by Petrov (1975), Bhattacharya and Ranga Rao (1976), and Serfling (1980), and the papers by Jurečková and Puri (1975), Bergström and Puri (1977), Hušková (1977a, b; 1979), Does (1982), Puri and Wu (1985b), Puri and Seoh (1984a–d, 1985a–c) Denker, Puri, and Rösler (1985), Puri and Ralescu (1982b), Ralescu and Puri (1985) Kallenberg (1982), Puri, Ralescu, and Seoh (1983), Bhattacharya and Puri (1983), and Seoh, Ralescu, and Puri (1985), and the references cited therein. Similarly, no attempt has been made to strengthen the asymptotic normality results to suitable (weak or strong) invariance principles for rank statistics. Such results are available in integrated form elsewhere (viz. Sen (1981a)). In practical applications, for nonparametric statistics, the approximations work out well even for moderately large sample sizes, and for finite sample sizes the procedures are generally conservative (insuring validity). Thus, this sacrifice of refinement need not be of great concern.

Genuinely distribution-free rank-order tests for some univariate as well as multivariate linear models are considered in Chapter 5. For such models, under the postulated null hypotheses, the joint distribution of the observable random variables remains invariant under appropriate groups of transformations, and this invariance generates distribution-free tests; for the multivariate case, one may have to have recourse to conditionally distribution-free tests. For these tests, both linear rank statistics and signed linear rank statistics are employed, and the distribution theory developed in Part 1 is incorporated in the study of the (asymptotic) properties of these tests. Asymptotic power properties and efficiency and optimality results for the rank tests are presented along with the parallel results for the classical normal-theory analysis-of-variance (and multivariate analysis-of-variance) tests. Simplifications for some specific models (for which the theory developed in Puri and Sen, 1971, applies) are also discussed. The results of this chapter are mainly adapted from Hájek (1962) and Puri and Sen (1969a, b).

Chapter 6 deals with the estimation of parameters in a linear model based on suitable rank statistics. Estimates of the slope and intercept in a simple regression model (univariate case), due to Adichie (1967b), are considered and subsequently generalized to the multiple-regression model and/or the multivariate case. The results are adapted from Jurečková (1969, 1971a, b), Sen and Puri (1969), and others. The asymptotic linearity of linear rank statistics (and signed rank statistics) in regression parameters, studied by Jurečková (1969), 1971a, b) and Van Eeden (1972) provides a convenient tool for the study of the asymptotic properties of these rank-order estimators. Results on asymptotic (multi)normality and asymptotic relative efficiency for these rank estimators are studied in detail, along with parallel results on the classical least-squares estimators and the maximum-likelihood estimators. The interval estimation problem for rank-based estimators has also been studied.

Chapter 7 is devoted to the problem of subhypothesis testing in linear models based on aligned rank statistics. The tests of Chapter 5 rest on some invariance structures which may not hold for a subhypothesis relating a general linear model, and hence, genuinely distribution-free rank tests for subhypotheses testing may not exist. Nevertheless, using the Jurečková linearity results of Chapter 6, rank statistics, constructed from the residuals, lead to some aligned rank tests which are at least asymptotically distribution-free and share good asymptotic properties. Tests for general linear hypotheses based on aligned rank statistics have been considered by Sen (1969a), Puri and Sen (1973), Sen and Puri (1977), and Adichie (1978), and their results are presented here in a unified manner. Efficiency and optimality results are also presented along with simultaneous inference procedures. In various multiparameter models, the null hypothesis is often tested against some *restricted alternatives* defined either in terms of some inequality restraints on the parameters (under testing) or in terms of a restricted subspace of the parameter space. Testing the equality of the treatment effects in a multisample or randomized block design against ordered alternatives, and testing against one-sided alternatives in the multivariate one-sample location or general regression problems, are notable examples of this type. The theory of rank-order tests developed earlier in Chapter 5 for simple hypothesis problems and in earlier sections of this chapter for composite hypothesis problems, for global alternatives, is extended to cover restricted alternatives and presented in section 7.6 The *union–intersection* (UI) principle, due to Roy (1953), provides a convenient way of dealing with these problems, and some basic results in the *nonlinear-programming* theory provide the desired solutions. These results are mainly due to Chinchilli and Sen (1981a, b). As a general rule, increasing the number of parameters in a model, parametric or nonparametric, leads to some loss of efficiency of the inference procedures. In many problems of statistical inference, based on extraneous information, one may attempt to reduce the number of parameters, though the possibility of such a reduction may not altogether be taken for granted. Thus, a *preliminary test* may be performed to justify the validity of such dimension reductions, and the actual inference procedures may be formulated following this preliminary test. Nonparametric inference procedures following a preliminary test, for various simple (univariate as well as multivariate) linear models, developed mostly by Saleh and Sen (1978, 1982, 1984a, b) and Sen and Saleh (1979), are discussed in Section 7.7. An important combination of the methods developed in the last two sections of Chapter 7 relates to the *profile analysis*, and rank-order tests for this problem, adapted from Chinchilli and Sen (1982), are considered in the last section of Chapter 7.

Rank tests for some miscellaneous problems in linear models are considered in Chapter 8. These include the case of *stochastic predictors*, for which some conditionally distribution-free tests (based on either *pure* or *mixed-rank statistics*) are due to Ghosh and Sen (1971b); rank tests for *mixed models* involving both stochastic and nonstochastic predictors, due to Sen and Puri (1970); rank tests for some *growth-curve models*, due to Ghosh, Grizzle, and Sen (1983), Sen (1973b), Woolson and Sen (1974), and Koziol et al. (1981), among others; rank tests for *grouped data* (relating to ordered categories), due to Sen (1967) and Ghosh (1973a, b); and rank tests under *censoring* of various types. The last topic relates to some useful statistical procedures for survival-analysis and life-testing problems, where because of practical limitations on time and cost of experimentation, one has to adapt the technique to an incomplete experiment, curtailed either at a fixed point of time or after a given number of responses have been obtained. There are other types of censoring too. In some of these problems, genuinely distribution-free tests are available; in most of the others one may obtain suitable conditionally distribution-free tests, and these tests are studied in Chapter 8 along with the allied (asymptotic) power properties.

The basic results in Chapters 2, 3, 4, 5, 6, and 8 are considered without proofs for convenience of reading; some proofs are provided in the Appendix. In particular, the results on the centering of linear rank statistics, multivariate permutational central limit theorems, and asymptotic linearity results (in shift or regression parameters) on linear and signed rank-order statistics (mostly due to Jurečková, 1969, 1971a, b) are discussed there and are also of independent interest.

CHAPTER 2

Distribution Theory of Linear Rank-Order Statistics

2.1. INTRODUCTION

For independent and real-valued random variables, the distribution theory of simple linear (or regression-type) rank statistics is considered in this chapter. The case of signed rank statistics will be taken up in Chapter 3, and the multivariate generalizations of these results will be presented in Chapter 4. Most of the results in this chapter are based on the fundamental work of Hájek (1968), with further extensions or generalizations by Dupač (1970), Dupač and Hájek (1969), Hoeffding (1973), Koul and Staudte (1972), Puri and Ralescu (1982a, b, 1984a), and Vorlíčková (1970, 1972) among others. We mainly consider the case of continuous distribution functions (d.f.'s) as well as continuous score functions. A few problems at the end of the chapter illustrates the modifications necessary when the score functions are not continuous everywhere. Also, we consider mainly the general setup, without restricting ourselves to contiguous alternatives. For contiguous alternatives, a very detailed treatment of linear rank statistics is given by Hájek and Šidák (1967, Chapter VI). However, for the sake of completeness, a brief review of it is given in Section 2.7.

2.2. LINEAR RANK STATISTICS

Let $X_1, \ldots, X_N$ be independent random variables with continuous d.f.'s $F_1(x), \ldots, F_N(x)$, respectively, all defined on the real line $(-\infty, \infty)$. For every $N \geq 1$, we define

$$R_{Ni} = \sum_{j=1}^{N} u(X_i - X_j), \qquad i = 1, \ldots, N \tag{2.2.1}$$

[where $u(t) = 1$ or 0 according as $t \geq$ or < 0], so that R_{Ni} is the rank of X_i among $X_1, \ldots, X_N$ for $1 \leq i \leq N$. By virtue of the assumed continuity of $F_1, \ldots, F_N$, ties among $X_1, \ldots, X_N$ may be neglected, in probability, and hence $(R_{N1}, \ldots, R_{NN})$ is some permutation of $(1, \ldots, N)$.

A simple linear rank statistic $L_N = L_N(X_1, \ldots, X_N)$ depends on $X_1, \ldots, X_N$ only through their ranks $R_{N1}, \ldots, R_{NN}$, and is characterized by a *score function* $\phi(u)$ $(0 < u < 1)$ and a set of known (regression) constants $\{c_1, \ldots, c_N\}$, where we assume that $c_1, \ldots, c_N$ are not all equal or null. To define L_N, we consider a set of scores, defined by

$$a_N(i) = E\phi\left(U_N^{(i)}\right) \quad \text{or} \quad \phi\left(\frac{i}{N+1}\right), \qquad 1 \leq i \leq N, \tag{2.2.2}$$

where $U_N^{(1)} \leq \cdots \leq U_N^{(N)}$ are the ordered random variables of a sample of size N from the rectangular (0, 1) d.f., so that $EU_{Ni} = i(N+1)^{-1}, 1 \leq i \leq N$. Then L_N is defined by

$$L_N = \sum_{i=1}^{N} c_i a_N(R_{Ni}), \qquad N \geq 1. \tag{2.2.3}$$

In passing, we may remark that if we let $\phi(u) = u : 0 < u < 1$ or $\phi(u) = \Phi^{-1}(u)$ [$\Phi(x)$ being the standard normal d.f.], the scores in (2.2.2) are termed the Wilcoxon or the normal scores, and the corresponding L_N the (generalized) Wilcoxon or the normal score statistic.

Statistics of the type (2.2.3) play a fundamental role in the theory of nonparametric statistical inference. For example, in the two-sample problem, where $F_1 = \cdots = F_m = F$, $F_{m+1} = \cdots = F_N = G$, $N = m + n$, for testing the hypothesis that $F \equiv G$ (against location and/or scale alternatives), rank tests are usually based on $\sum_{i=1}^{m} a_N(R_{Ni})$ for suitable score functions, and this is a particular case of (2.2.3) where $c_1 = \cdots = c_m = 1$, $c_{m+1} = \cdots = c_N = 0$. Hoeffding (1951b) and, later on, Terry (1952) arrived at L_N from the consideration of maximum local power of rank tests for regression alternatives. All these are explained in detail in Chapter 3 of Puri and Sen (1971). Throughout this book, rank statistics L_N of the form (2.2.3) will be used for testing and estimating parameters in general multivariate linear models. For this reason, we proceed to study first the distribution theory of L_N.

Since $\mathbf{R}_N = (R_{N1}, \ldots, R_{NN})$ is a permutation of $(1, \ldots, N)$, it can only assume $N!$ possible realizations $\{(i_1, \ldots, i_N) : 1 \leq i_1 \neq \cdots \neq i_N \leq N\}$. The probability distribution of $\mathbf{R}_N$ over the $N!$ possible realizations, in turn, determines the probability distribution of L_N. In the particular case of $F_1 = \cdots = F_N = F$, all possible $N!$ realizations of $\mathbf{R}_N$ are equally likely, so that the distribution of $\mathbf{R}_N$ is independent of F, and hence L_N is a distribution-free statistic. On the other hand, if the F_i are not all identical,

$$P\{R_{Ni}=j\} = \sum_j \int \prod_{\alpha=1}^{j-1} F_{s_\alpha}(x) \prod_{\beta=j+1}^{N} \left[1 - F_{s_\beta}(x)\right] dF_i(x), \quad (2.2.4)$$

for $1 \le j \le N$, where the summation Σ_j extends over all possible choices of distinct $(s_1, \dots, s_{j-1})$ from $(1, \dots, j-1, j+1, \dots, N)$, with $s_j = i$, and $(s_{j+1}, \dots, s_N)$ as the complementary set. Thus, (2.2.4) depends on $F_1, \dots, F_N$, and in general the joint probability of $R_{Ni_1} = j_1, \dots, R_{Ni_k} = j_k$, for every $k \ge 1$, depends on $F_1, \dots, F_N$. Consequently, $\mathbf{R}_N$ no longer has all $N!$ realizations equally likely, and hence, the distribution of L_N depends on $F_1, \dots, F_N$ through the dependence of the distribution of $\mathbf{R}_N$ on $F_1, \dots, F_N$. The probability that $\{R_{N1} = j_1, \dots, R_{NN} = j_N\}$ (where $j_1, \dots, j_N$ is any permutation of $1, \dots, N$) is given by

$$\int \cdots \int_{A_N} \prod_{\alpha=1}^{N} dF_\alpha(x_{j_\alpha}), \qquad \mathbf{j} = (j_1, \dots, j_N), \quad (2.2.5)$$

where $A_N = \{(x_1, \dots, x_n) : x_1 \le x_2 \le \cdots \le x_N\}$. Thus, the computation of the exact distribution of L_N with the aid of (2.2.5) involves tedious quadrature formulae and depends heavily on $\{F_1, \dots, F_N\}$ and $\{c_1, \dots, c_N\}$. Moreover, the ranks $R_{N1}, \dots, R_{NN}$ are not independent, and hence, the usual central limit theorems may not readily apply to L_N. The study of the large-sample distribution of L_N requires different tools. In the subsequent sections, we present the asymptotic distribution theory of L_N when $F_1, \dots, F_N$ are not necessarily the same. We mainly follow the ingenious treatment of Hájek (1968). Basically, we assume that $\phi(u)$ is expressible as the difference of two monotonically nondecreasing, absolutely continuous, and square-integrable score functions. For such score functions, one can bound the variance of L_N and also apply the polynomial approximation due to Weierstrass and Bernoulli. With that end, we first study the variance inequality for L_N and then consider the distribution theory of L_N: first, for score functions with bounded second derivatives, and finally, by the polynomial approximation, for the general case of square-integrable functions.

2.3. THE HÁJEK VARIANCE INEQUALITY

First, we assume that $\phi(u)$ is nondecreasing in $u: 0 < u < 1$, so that

$$a_N(1) \le \cdots \le a_N(N) \qquad \text{for every} \quad N \ge 1. \quad (2.3.1)$$

Theorem 2.3.1 (Hájek, 1968). *Whenever $F_1, \dots, F_N$ are all continuous and* (2.3.1) *holds,*

$$\operatorname{Var} L_N \le 21\left\{ \max_{1 \le i \le N} (c_i - \bar{c}_N)^2 \sum_{i=1}^{N} [a_N(i) - \bar{a}_N]^2 \right\}, \qquad (2.3.2)$$

where

$$\bar{c}_N = N^{-1} \sum_{i=1}^{N} c_i \quad \textit{and} \quad \bar{a}_N = N^{-1} \sum_{i=1}^{N} a_N(i).$$

The proof of the theorem is based on the following lemmas which are also of independent interest.

Lemma 2.3.2 (Hájek, 1968). Let $\alpha(x_1, \dots, x_N)$ and $\beta(x_1, \dots, x_N)$ be real functions, nondecreasing in each x_i, $1 \le i \le N$. Let $X_1, \dots, X_N$ be independent random variables such that $E|\alpha(X_1, \dots, X_N)| < \infty$. Then

$$\bar{\alpha}(x_1, \dots, x_k) = E\left[\alpha(X_1, \dots, X_N) | X_1 = x_1, \dots, X_k = x_k\right] \qquad (2.3.3)$$

is a nondecreasing function in each x_i, $1 \le i \le k$, $\forall k \le N$, and

$$\operatorname{Cov}[\alpha(X_1, \dots, X_N), \beta(X_1, \dots, X_N)] \ge 0, \qquad (2.3.4)$$

provided the covariance exists.

Lemma 2.3.3 (Hájek, 1968). For arbitrary $a_N(1), \dots, a_N(N)$, and continuous $F_1, \dots, F_N$, whenever $y < x$,

$$E\left[a_N(R_{Ni}) | X_i = x,\ X_j = y\right] - E\left[a_N(R_{Ni}) | X_i = x\right]$$

$$= \left[u(x-y) - F_j(x)\right] \sum_{k=2}^{N} [a_N(k) - a_N(k-1)]$$

$$\times P\{R_{Ni} = k | X_i = x,\ X_j = y\}, \qquad (2.3.5)$$

for $i \neq j$, where $u(t) = 1$ or 0 according as $t \ge 0$ or < 0, and where $X_j = y$ may also be replaced by $X_j \le x - 1$.

Lemma 2.3.4 (Hájek, 1968). Under the setup of Lemma 2.3.3,

$$\operatorname{Cov}\left[a_N(R_{Ni}), a_N(R_{Nj})\right]$$

$$= E\left\{\operatorname{Cov}\left[a_N(R_{Ni}), a_N(R_{Nj})|X_i, X_j\right]\right.$$

$$\left.+\operatorname{Cov}\left\{E\left[a_N(R_{Ni})|X_i, X_j\right], E\left[a_N(R_{Nj})|X_i, X_j\right]\right\}\right\}. \quad (2.3.6)$$

Lemma 2.3.5 (Hájek, 1968). If (2.3.1) holds and $F_1, \ldots, F_N$ are all continuous, then for every $1 \le i < j \le N$,

$$\operatorname{Cov}\left\{E\left[a_N(R_{Ni})|X_i, X_j\right], E\left[a_N(R_{Nj})|X_i, X_j\right]\right\} \le 0, \quad (2.3.7)$$

$$\operatorname{Cov}\left\{a_N(R_{Ni}), a_N(R_{Nj})|X_i = x, X_j = y\right\} \ge 0, \quad (2.3.8)$$

Lemma 2.3.6 (Hájek, 1968). For arbitrary $a_N(1), \ldots, a_N(N)$ and continuous $F_1, \ldots, F_N$,

$$\operatorname{Cov}\left[E\left(a_N(R_{Ni})|X_i, X_j\right), E\left(a_N(R_{Nj})|X_i, X_j\right)\right]$$

$$= \operatorname{Cov}\left[E\left(a_N(R_{Ni})|X_i\right), E\left(a_N(R_{Nj})|X_i\right)\right]$$

$$+\operatorname{Cov}\left[E\left(a_N(R_{Ni})|X_j\right), E\left(a_N(R_{Nj})|X_j\right)\right]$$

$$+E\left[\left\{E\left(a_N(R_{Ni}|X_i, X_j\right) - E\left(a_N(R_{Ni})|X_i\right)\right\}\right.$$

$$\left.\times\left\{E\left(a_N(R_{Nj}|X_i, X_j\right) - E\left(a_N(R_{Nj})|X_j\right)\right\}\right]. \quad (2.3.9)$$

Lemma 2.3.7 (Hájek, 1968). For every k $(1 \le k \le N)$ and continuous $F_1, \ldots, F_N$,

$$P\{R_{Ni} = k|X_i = x, X_j = x - 1\}$$

$$\le P\{R_{Ni} = k|X_i = x\} + P\{R_{Ni} = k - 1|X_i = x\}. \quad (2.3.10)$$

In the sequel when we prove the theorem and the lemmas, we write $R_{Ni} = R_i$ and $a_N(i) = a_i$, $1 \le i \le N$.

Proof of Lemma 2.3.2. Let $x_i' > x_i$. Then

$$\bar{\alpha}(x_1,\ldots,x_{i-1},x_i',x_{i+1},\ldots,x_k) - \bar{\alpha}(x_1,\ldots,x_{i-1},x_i,x_{i+1},\ldots,x_k)$$

$$= \int\cdots\int\big[\alpha(x_1,\ldots,x_{i-1},x_i',x_{i+1},\ldots,x_N)$$

$$-\alpha(x_1,\ldots,x_{i-1},x_i,x_{i+1},\ldots,x_N)\big]$$

$$\times dF_{k+1}(x_{k+1})\cdots dF_N(x_N) \geq 0,$$

since the integrand is nonnegative by hypothesis. This proves (2.3.3).

We shall prove (2.3.4) by induction. Let $N = 1$, and let Y_1 be an independent copy of X_1. Then, it is easy to check that $\mathrm{Cov}[\alpha(X_1), \beta(X_1)] = \frac{1}{2}E\{[\alpha(X_1) - \alpha(Y_1)][\beta(X_1) - \beta(Y_1)]\} \geq 0$, since $\alpha(x)$ and $\beta(x)$ are nondecreasing functions. Thus (2.3.4) holds for $N = 1$. To prove the general case, first note (see Exercise 2.3.1) that

$$\mathrm{Cov}\big[\alpha(X_1,\ldots,X_N),\beta(X_1,\ldots,X_N)\big]$$

$$= \mathrm{Cov}\big[\bar{\alpha}(X_1),\bar{\beta}(X_1)\big]$$

$$+\int \mathrm{Cov}\big[\alpha(x,X_2,\ldots,X_N),\beta(x,X_2,\ldots,X_N)\big]\,dF_1(x). \quad (2.3.11)$$

Now assume that (2.3.4) holds for $N - 1$ variables $X_2,\ldots,X_N$. Then $\mathrm{Cov}[\alpha(x, X_2,\ldots,X_N), \beta(x, X_2,\ldots,X_N)] \geq 0$, and hence the second term on the right-hand side of (2.3.11) is ≥ 0. Also, the first term is ≥ 0. Hence, $\mathrm{Cov}[\alpha(X_1,\ldots,X_N), \beta(X_1,\ldots,X_N)] \geq 0$. The proof follows by induction.

For another proof of (2.3.4), see Exercise 2.3.2.

Before we prove Lemma 2.3.3, we note that Lemma 2.3.4 follows by using the fact that the unconditional expectation is the expectation of the conditional expectation. Also, by (2.2.1), R_i is a nondecreasing function of X_i and $X_j' = (-1)X_j$, $j \neq i$. Thus, on letting $X_i = x$, $X_j = y$, where x and y are fixed, note that R_i is nondecreasing in X_k and R_j is nondecreasing in X_k, for $k(\neq i \neq j) = 1,\ldots,N$. Thus, both $a(R_i)$ and $a(R_j)$ are nondecreasing in X_j, $k \neq j \neq i$, when (2.3.1) holds. Hence, (2.3.8) follows directly from Lemma 2.3.2. The proof of (2.3.7) is left as an exercise.

Proof of Lemma 2.3.3. To simplify notation, we take $i = 1$ and $j = 2$. Let $B(k|p_1,\ldots,p_N)$ denote the probability of k successes in N independent trials when the probability of success in the ith trial is p_i, $i = 1,\ldots,N$.

Then, since $R_i = \sum_{i=1}^{N} u(X_i - X_j)$, $1 \le i \le N$, we have

$$P(R_1 = k | X_1 = x, X_2 = y) = B(k|1, u(x-y), F_3(x), \dots, F_N(x)) \tag{2.3.12}$$

and

$$P(R_1 = k | X_1 = x) = B(k|1, F_2(x), \dots, F_N(x)). \tag{2.3.13}$$

Next, since

$$\begin{aligned} B(k|1, F_2(x), \dots, F_N(x)) &= F_2(x) B(k|1,1, F_3(x), \dots, F_N(x)) \\ &\quad + [1 - F_2(x)] B(k+1|1,1, F_3(x), \dots, F_N(x)), \end{aligned} \tag{2.3.14}$$

we obtain, using (2.3.12), (2.3.13), and (2.3.14), that

$$\begin{aligned} P(R_1 = k &| X_1 = x, X_2 = y) - P(R_1 = k | X_1 = x) \\ &= B(k|1, u(x-y), F_3(u), \dots, F_N(x)) \\ &\quad - B(k|1, F_2(x), \dots, F_N(x)) \\ &= u(x-y) B(k|1,1, F_3(x), \dots, F_N(x)) \\ &\quad + [1 - u(x-y)] B(k|1,0, F_3(x), \dots, F_N(x)) \\ &\quad - F_2(x) B(k|1,1, F_3(x), \dots, F_N(x)) \\ &\quad - [1 - F_2(x)] B(k|1,0, F_3(x), \dots, F_N(x)) \\ &= [u(x-y) - F_2(x)] \\ &\quad \times [B(k|1,1, F_3(x), \dots, F_N(x)) \\ &\quad - B(k+1|1,1, F_3(x), \dots, F_N(x))] \\ &= [u(x-y) - F_2(x)] \\ &\quad \times [P(R_1 = k | X_1 = x, X_2 = x-1) \\ &\quad - P(R_1 = k+1 | X_1 = x, X_2 = x-1)]. \end{aligned} \tag{2.3.15}$$

Note that in the above expression $x - 1$ can be replaced by any number smaller than x.

Now, by definition,

$$E\left[a(R_1)|X_1 = x,\ X_2 = y\right] - E\left[a(R_1)|X_1 = x\right]$$
$$= \sum_{k=1}^{N} a(k)\left[P(R_1 = k|X_1 = x,\ X_2 = y) - P(R_1 = k|X_1 = x)\right]. \tag{2.3.16}$$

The result now follows by using (2.3.15) in (2.3.16) and the fact that $P(R_1 = 1|X_1 = x,\ X_2 = x - 1) = 0$.

The proof of Lemma 2.3.4 follows by standard computations, and hence is left as an exercise.

Proof of Lemma 2.3.7. For simplicity take $i = 1,\ j = 2$. Then,

$$\begin{aligned}
P(R_i = k|X_i = x) &+ P(R_i = k - 1|X_i = x) \\
&= F_2(x)B(k|1,1,F_3(x),\dots,F_N(x)) \\
&\quad + [1 - F_2(x)]B(k|1,0,F_3(x),\dots,F_N(x)) \\
&\quad + F_2(x)B(k-1|1,1,F_3(x),\dots,F_N(x) \\
&\quad + [1 - F_2(x)]B(k|1,1,F_3(x),\dots,F_N(x)) \\
&\geq B(k|1,1,F_3(x),\dots,F_N(x)) \\
&= P(R_1 = k|X_1 = x,\ X_2 = x - 1).
\end{aligned}$$

The result follows.

Proof of Theorem 2.3.1. Without any loss of generality, we may assume that $\bar{a}_N = \bar{c}_N = 0$. Then, we have to prove that

$$\operatorname{Var} L_N \leq 21\left[\max_{1 \leq k \leq N} c_k^2\right] \sum_{i=1}^{N} a^2(i). \tag{2.3.17}$$

Now, writing $a(i) = a_i$, $1 \leq i \leq N$, noting that by (2.2.3)

$$\operatorname{Var} L_N = \sum_{i=1}^{N} c_i^2 \operatorname{Var}(a(R_i)) + \sum_{i \neq j = 1}^{N} c_i c_j \operatorname{Cov}\left[a(R_i), a(R_j)\right], \tag{2.3.18}$$

and noting that $\mathrm{Var}(X) \le EX^2$, it follows that

$$\sum_{i=1}^{N} c_i^2 \mathrm{Var}[a(R_i)] \le \sum_{i=1}^{N} c_i^2 E[a(R_i)]^2 \le \max_{1\le i\le N} c_i^2 \sum_{i=1}^{N} E[a(R_i)]^2$$
$$= \max_{1\le i\le N} c_i^2 E\left\{ \sum_{i=1}^{N} [a(R_i)]^2 \right\} = \max_{1\le i\le N} c_i^2 \sum_{i=1}^{N} a_i^2. \tag{2.3.19}$$

Now consider the second term on the right side of (2.3.18). Since R_i is an increasing function of X_i and $-X_j$, $j \ne i$, and since $a_1 \le \cdots \le a_N$, by (2.3.4) we have $\mathrm{Cov}[a(R_i), a(R_j)|X_i, X_j] \ge 0$ $\forall i \ne j, \dots, N$. Hence, using (2.3.6), (2.3.7), (2.3.8) and replacing c_i, c_j by $\max_{1\le i\le N} c_i$, we find that

$$\sum_{\substack{i=1 \\ i\ne j}}^{N} \sum_{j=1}^{N} c_i c_j \mathrm{Cov}\big[a(R_i), a(R_j)\big]$$
$$\le \max_{1\le i\le N} c_i^2 \left[\sum_{\substack{i=1 \\ i\ne j}}^{N} \sum_{j=1}^{N} E\Big\{\mathrm{Cov}\big[a(R_i), a(R_j)|X_i, X_j\big]\Big\} \right.$$
$$\left. - \sum_{\substack{i=1 \\ i\ne j}}^{N} \sum_{j=1}^{N} \mathrm{Cov}\Big\{ E\big[a(R_i)|X_i, X_j\big], E\big[a(R_j)|X_i, X_j\big]\Big\} \right]. \tag{2.3.20}$$

Now since $\mathrm{Var}[\Sigma_{i=1}^{N} a(R_i)] = 0$, it follows that

$$\sum_{\substack{i=1 \\ i\ne j}}^{N} \sum_{j=1}^{N} \mathrm{Cov}\big[a(R_i), a(R_j)\big] = -\sum_{i=1}^{N} \mathrm{Var}\big[a(R_i)\big] \le 0. \tag{2.3.21}$$

Using (2.3.21) and Lemma 2.3.4, we obtain

$$\sum_{\substack{i=1 \\ i\ne j}}^{N} \sum_{j=1}^{N} E\Big\{\mathrm{Cov}\,\big[a(R_i), a(R_j)\big]\big|X_i, X_j\Big\}$$
$$\le -\sum_{\substack{i=1 \\ i\ne j}}^{N} \sum_{j=1}^{N} \mathrm{Cov}\Big\{ E\big[a(R_i)|X_i, X_j\big], E\big[a(R_j)|X_i, X_j\big]\Big\}. \tag{2.3.22}$$

Using (2.3.22), (2.3.7), and (2.3.8) in (2.3.20), it follows that

$$\sum_{\substack{i=1 \\ i \neq j}}^{N} \sum_{j=1}^{N} c_i c_j \operatorname{Cov}\left[a(R_i), a(R_j)\right]$$

$$\leq -2 \max_{1 \leq i \leq N} c_i^2 \sum_{\substack{i=1 \\ i \neq j}}^{N} \sum_{j=1}^{N} \operatorname{Cov}\left\{E\left[a(R_i)|X_i, X_j\right], E\left[a(R_j)|X_i, X_j\right]\right\}. \tag{2.3.23}$$

Looking at (2.3.18), (2.3.19), and (2.3.23), we find that to prove the theorem, we have to show

$$-\sum_{\substack{i=1 \\ i \neq j}}^{N} \sum_{j=1}^{N} \operatorname{Cov}\left\{E\left[a(R_i)|X_i, X_j\right], E\left[a(R_j)|X_i, X_j\right]\right\} \leq 10 \sum_{i=1}^{N} a_i^2. \tag{2.3.24}$$

Using Lemma 2.3.6, we can rewrite the left side of (2.3.24) as

$$-\sum_{\substack{i=1 \\ i \neq j}}^{N} \sum_{j=1}^{N} \operatorname{Cov}\left\{E\left[a(R_i)|X_i, X_j\right], E\left[a(R_j)|X_i, X_j\right]\right\}$$

$$= \sum_{\substack{i=1 \\ i \neq j}}^{N} \sum_{j=1}^{N} (A_{ij} + B_{ij} + C_{ij}), \tag{2.3.25}$$

where

$$A_{ij} = -\operatorname{Cov}\left\{E\left[a(R_i)|X_i\right], E\left[a(R_j)|X_i\right]\right\}, \tag{2.3.26}$$

$$B_{ij} = -\operatorname{Cov}\left\{E\left[a(R_i)|X_j\right], E\left[a(R_j)|X_j\right]\right\}, \tag{2.3.27}$$

and

$$C_{ij} = -E\Big\{\Big(E\left[a(R_i)|X_i, X_j\right] - E\left[a(R_i)|X_i\right]\Big)\Big(E\left[a(R_j)|X_i, X_j\right] - E\left[a(R_j)|X_j\right]\Big)\Big\}. \tag{2.3.28}$$

We now find the bounds of each of the sums on the right side of (2.3.25). First consider $\sum\sum_{i \neq j} A_{ij}$. Routine computations yield

$$\sum_{i=1}^{N} \sum_{\substack{j=1 \\ i \neq j}}^{N} A_{ij} = \sum_{i=1}^{N} \operatorname{Cov}\left\{ E\left[a(R_i)|X_i\right], E\left[\left(a(R_i) - \sum_{j=1}^{N} a(R_j)\right)\Big|X_i\right]\right\}$$

$$= \sum_{i=1}^{N} \operatorname{Cov}\left\{ E\left[a(R_i)|X_i\right], E\left[a(R_i)|X_i\right]\right\}$$

$$= \sum_{i=1}^{N} \operatorname{Var}\left\{ E\left[a(R_i)|X_i\right]\right\} \leq \sum_{i=1}^{N} \operatorname{Var}\left[a(R_i)\right]$$

$$= \sum_{i=1}^{N} E\left[a(R_i)\right]^2 = E \sum_{i=1}^{N} \left[a(R_i)\right]^2 = \sum_{i=1}^{N} a_i^2 \tag{2.3.29}$$

as

$$\sum_{j=1}^{N} a(R_j) = \text{const}.$$

By symmetry

$$\sum_{i=1}^{N} \sum_{\substack{j=1 \\ i \neq j}}^{N} B_{ij} \leq \sum_{i=1}^{N} a_i^2. \tag{2.3.30}$$

The proof will be complete if we show that

$$\sum_{i=1}^{N} \sum_{\substack{j=1 \\ i \neq j}}^{N} C_{ij} \leq 8 \sum_{i=1}^{N} a_i^2. \tag{2.3.31}$$

Using Lemma 2.3.3, we obtain

$$C_{ij} = -E\left\{\left[u(X_i - X_j) - F_j(X_i)\right] \sum_{k=2}^{N} (a(k) - a(k-1)) \times P(R_i = K|X_i = x_i, X_j = x_i - 1)\right\}$$

$$\times \left\{\left[u(X_j - X_i) - F_i(X_j)\right] \sum_{h=2}^{N} (a(h) - a(h-1)) \times P(R_j = h|X_j\text{-}x_j, X_i = x_j - 1)\right\}. \tag{2.3.32}$$

Using Lemma 2.3.7 and the fact that

$$\left|\left[u(X_i - X_j) - F_j(X_i)\right]\left[u(X_j - X_i) - F_i(X_j)\right]\right| \le 1, \tag{2.3.33}$$

we find that

$$\begin{aligned}
C_{ij} &\le E \sum_{k=2}^{N} \sum_{h=2}^{N} [a(k) - a(k-1)][a(h) - a(h-1)] \\
&\quad \times \left[\{P(R_i = k|X_i) + P(R_i = k-1|X_i)\}\right. \\
&\qquad \left.\times \{P(R_j = h|X_j) + P(R_j = h-1|X_j)\}\right] \\
&= \sum_{k=2}^{N} \sum_{h=2}^{N} [a(k) - a(k-1)][a(h) - a(h-1)] \\
&\quad \times \left[P(R_i = k)P(R_j = h) + P(R_i = k)P(R_j = h-1)\right. \\
&\qquad \left. + P(R_i = k-1)P(R_j = h) + P(R_i = k-1)P(R_j = h-1)\right].
\end{aligned} \tag{2.3.34}$$

This implies that

$$\begin{aligned}
\sum_{\substack{i=1\\ i\neq j}}^{N} \sum_{j=1}^{N} C_{ij} \le \sum_{k=2}^{N} \sum_{h=2}^{N} & [a(k) - a(k-1)][a(h) - a(h-1)] \\
& \times \left[\sum\sum_{i\neq j} P(R_i = k)P(R_j = h)\right. \\
& \quad + \sum\sum_{i\neq j} P(R_i = k)P(R_j = h-1) \\
& \quad + \sum\sum_{i\neq j} P(R_i = k-1)P(R_j = h) \\
& \quad \left. + \sum\sum_{i\neq j} P(R_i = k-1)P(R_j = h-1)\right].
\end{aligned} \tag{2.3.35}$$

Now

$$\sum_{i \neq j}\sum P(R_i = k)P(R_j = h)$$

$$= \sum_{i=1}^{N} P(R_i = k) \sum_{j=1}^{N} P(R_j = h) - \sum_{i=1}^{N} P(R_i = k)P(R_i = h) \leq 1.$$

Similarly, each of the other sums in the bracket is ≤ 1. Hence

$$\sum_{\substack{i=1 \\ i \neq j}}^{N} \sum_{j=1}^{N} C_{ij} \leq 4 \sum_{k=2}^{N} \sum_{h=2}^{N} [a(k) - a(k-1)][a(h) - a(h-1)]$$

$$= 4(a_N - a_1)^2 \leq 8 \sum_{i=1}^{N} a_i^2.$$

The proof follows.

2.4. PROJECTION APPROXIMATION FOR S_N FOR BOUNDED SCORES

For statistics not expressible as sums of independent random variables, the projection technique often provides a fruitful way of deriving asymptotic normality. For example, for U-statistics, Hoeffding (1948) used the projection technique to reduce the statistic to the sum of two components of which the first involves independent random variables, while the other converges to zero (in probability) at a faster rate. Similar projection techniques are used in the derivation of the asymptotic normality of one and several sample rank-order statistics, discussed in detail in Chapter 3 of Puri and Sen (1971). In the present context, the projection technique consists in approximating a statistic $S(X_1, \ldots, X_N)$ by $\Sigma_{i=1}^{N} l_{Ni}(X_i)$ where the $l_{Ni}(X_i)$ are square-integrable, and then using the result on S_N when the score function $\phi(u)$ has a bounded second derivative inside $(0, 1)$. The basic projection result is contained in the following lemma.

Lemma 2.4.1 (Hájek, 1968). Let $X_1, \ldots, X_N$ be independent r.v.'s. Let $S_N = s(X_1, \ldots, X_N)$ and $L_N = \Sigma_{i=1}^{N} l_i(X_i)$ be two statistics such that $ES_N^2 <$

∞ and $El_i^2(X_i) < \infty$, $i = 1, \ldots, N$. Denote

$$\hat{S}_N = \sum_{i=1}^{N} E(S_N | X_i) - (N-1) E S_N. \tag{2.4.1}$$

Then

$$E\hat{S}_N = ES_N, \tag{2.4.2}$$

$$E(S_N - \hat{S}_N)^2 = \operatorname{Var} S_N - \operatorname{Var} \hat{S}_N, \tag{2.4.3}$$

and

$$E(S_N - L_N)^2 = E(S_N - \hat{S}_N)^2 + E(\hat{S}_N - L_N)^2. \tag{2.4.4}$$

Proof. Equation (2.4.2) is trivial. Equation (2.4.3) follows from (2.4.4) by taking $L_N = ES_N = E\hat{S}_N$. We therefore prove (2.4.3). Without any loss of generality, we may assume $ES_N = E\hat{S}_N = 0$. Now

$$E(S_N - L_N)^2 = E(S_N - \hat{S}_N)^2 + E(\hat{S}_N - L_N)^2$$
$$+ 2E(S_N - \hat{S}_N)(\hat{S}_N - L_N). \tag{2.4.5}$$

But

$$E(S_N - \hat{S}_N)(\hat{S}_N - L_N)$$
$$= E\left[(S_N - \hat{S}_N) \sum_{i=1}^{N} \{E(S_N | X_i) - l_i(X_i)\}\right]$$
$$= \sum_{i=1}^{N} E[(S_N - \hat{S}_N)\{E(S_N | X_i) - l_i(X_i)\}]$$
$$= \sum_{i=1}^{N} E\left[E\{(S_N - \hat{S}_N) | X_i\}\{E(S_N | X_i) - l_i(X_i)\}\right]$$
$$= 0, \quad \text{since} \quad E\{(S_N - \hat{S}_N) | X_i\} = 0. \tag{2.4.6}$$

The proof follows.

The following lemma is useful in proving that the remainder term $S_N - \hat{S}_N$ may be treated in terms of mean-square convergence (which incidentally is easier to handle than the convergence in probability). We shall prove this lemma for the special case when S_N can be represented as $S_N = \sum_{i=1}^N S_{Ni}$.

Lemma 2.4.2 (Hájek, 1968). If $S_N = \sum_{i=1}^N S_{Ni}$, then

$$E(S_N - \hat{S}_N)^2 \le \sum_{i=1}^N E[S_{Ni} - E(S_{Ni}|X_i)]^2$$
$$+ \sum\sum_{i \ne j} \Big\{ E\big([S_{Ni} - E(S_{Ni}|X_i)][S_{Nj} - E(S_{Nj}|X_j)]\big)$$
$$- \sum_{k \ne (i,j)} \operatorname{Cov}[E(S_{Ni}|X_k), E(S_{Nj}|X_k)] \Big\}, \quad (2.4.7)$$

where $\hat{S}_N$ is given by (2.4.1).

Proof. (Outline) We first express $E(S_N - \hat{S}_N)^2$ as

$$E(S_N - \hat{S}_N)^2 = \operatorname{Var} S_N - \operatorname{Var} \hat{S}_N = \left[\sum_{i=1} A_i + \sum\sum_{i \ne j} B_{ij} \right], \quad (2.4.8)$$

where

$$A_i = \operatorname{Var} S_{Ni} - \operatorname{Var} \sum_{k=1}^N E(S_{Ni}|X_k), \quad (2.4.9)$$

and

$$B_{ij} = \operatorname{Cov}(S_{Ni}, S_{Nj}) - \sum_{k=1}^N \operatorname{Cov}[E(S_{Ni}|X_k), E(S_{Nj}|X_k)]. \quad (2.4.10)$$

It is easy to show that [see problem 2.4.1]

$$A_i \le \operatorname{Var} S_{Ni} - \operatorname{Var} E(S_{Ni}|X_i) = E[S_{Ni} - E(S_{Ni}|X_i)]^2 \quad (2.4.11)$$

and

$$\operatorname{Cov}(S_{Ni}, S_{Nj}) = \operatorname{Cov}[E(S_{Ni}|X_i), E(S_{Nj}|X_i)]$$
$$+ \operatorname{Cov}[E(S_{Ni}|X_j), E(S_{Nj}|X_j)]$$
$$+ E\{[S_{Ni} - E(S_{Ni}|X_i)][S_{Nj} - E(S_{Nj}|X_j)]\}. \quad (2.4.12)$$

Using (2.4.9) through (2.4.12) in (2.4.8), we get the desired result.

Theorem 2.4.3. *If ϕ possesses a bounded second derivative inside* $(0,1)$ *and* $S_N = \sum_{i=1}^N c_i \phi(R_{Ni}/(N+1))$, *then there exists a constant* $K(\phi)$ *such that*

$$E(S_N - \hat{S}_N)^2 \le K(\phi) N^{-1} \sum_{i=1}^N (c_i - \bar{c})^2, \tag{2.4.13}$$

$K(\phi) < \infty$, *where* $\hat{S}_N$ *is given by* (2.4.1) *and* $c_1, \ldots, c_N$ *are any constants.*

Proof. Without any loss of generality, we let $\bar{c} = 0$. Let then

$$\rho_i = \frac{R_{Ni}}{N+1}, \qquad 1 \le i \le N. \tag{2.4.14}$$

Then $S_N = \sum_{i=1}^N c_i \phi(\rho_i)$, so that by Taylor's expansion of $\phi(\rho_i)$ about $E(\rho_i | X_i)$, we obtain that

$$\phi(\rho_i) = \phi(E(\rho_i|X_i)) + [\rho_i - E(\rho_i|X_i)]\phi'(E(\rho_i|X_i))$$
$$+ [\rho_i - E(\rho_i|X_i)]^2 \lambda_i(X_i), \tag{2.4.15}$$

where $\lambda_i^2(x) < K_2$ (say), $i = 1, \ldots, N$; $-\infty < x < \infty$.

Using (2.4.14) and (2.4.15), we can rewrite S_N as

$$S_N = S_{N1} + S_{N2} + S_{N3}, \tag{2.4.16}$$

where

$$S_{N1} = \sum_{i=1}^N c_i \rho_i \phi'[E(\rho_i|X_i)], \tag{2.4.17}$$

$$S_{N2} = \sum_{i=1}^N c_i \{\phi[E(\rho_i|X_i)] - E(\rho_i|X_i)\phi'[E(\rho_i|X_i)]\}, \tag{2.4.18}$$

and

$$S_{N3} = \sum_{i=1}^N c_i [\rho_i - E(\rho_i|X_i)]^2 \lambda_i(X_i). \tag{2.4.19}$$

Now denote the projection of S_{Ni} by $\hat{S}_{Ni}$, $i = 1, 2, 3$. This means

$$\hat{S}_{Ni} = \sum_{k=1}^N E(S_{Ni}|X_{Nk}) - (N-1) E S_{Ni}, \qquad i = 1, 2, 3. \tag{2.4.20}$$

Writing $\tau_i(X_i) = c_i\{\phi[E(\rho_i|X_i)] - E(\rho_i|X_i)\phi'[E(\rho_i|X_i)]\}$, we note that $S_{N2} = \sum_{i=1}^{N}\tau_i(X_i)$, and

$$\hat{S}_{N2} = \sum_{k=1}^{N}\sum_{i=1}^{N} E\left[\tau_i(X_i)|X_k\right] - (N-1)ES_{N2} = S_{N2}, \quad (2.4.21)$$

after routine computations. Thus

$$\begin{aligned} E(S_N - \hat{S}_N) &= E(S_{N1} - \hat{S}_{N1} + S_{N3} - \hat{S}_{N3})^2 \\ &\le 2E(S_{N1} - \hat{S}_{N1})^2 + 2E(S_{N3} - \hat{S}_{N3})^2 \\ &\le 2E(S_{N1} - \hat{S}_{N1})^2 + \operatorname{Var} S_{N3} \qquad [\text{using } (2.4.3)] \\ &\le 2E(S_{N1} - \hat{S}_{N1}) + 2ES_{N3}^2. \end{aligned} \quad (2.4.22)$$

The theorem will follow if we prove

Lemma 2.4.4. Under the condition of Theorem 2.4.3,

$$ES_{N3}^2 \le K_3 \cdot N^{-1}\sum_{i=1}^{N} c_i^2, \quad (2.4.23)$$

$$E(S_{N1} - \hat{S}_{N1})^2 \le K_4 N^{-1}\sum_{i=1}^{N} c_i^2. \quad (2.4.24)$$

Proof of (2.4.23). By the Cauchy–Schwartz inequality, we have

$$ES_{N3}^2 \le K_2\sum_{i=1}^{N} c_i^2 \cdot E\left[\sum_{i=1}^{N}\{\rho_i - E(\rho_i|X_i)\}^4\right]. \quad (2.4.25)$$

Now

$$E\{\rho_i - E(\rho_i|X_i)\}^4 = (N+1)^{-4}E\{R_i - E(R_i|X_i)\}^4. \quad (2.4.26)$$

Writing $p_j = E[u(X_i - X_j)|X_i = x]$, it is easy to check that

$$\begin{aligned} E\{(R_i - E(R_i|X_i))^4|X_i\} &= \sum_{i=1}^{N}\{(1-p_j)^4 p_j + p_j^4(1-p_j)\} \\ &\quad + 6\sum_{j<k}\sum p_j p_k(1-p_j)(1-p_k) \\ &\le \frac{(N+1)^2}{4}. \end{aligned} \quad (2.4.27)$$

Using (2.4.26) and (2.4.27) in (2.4.25), we see the desired result.

Proof of (2.4.24). Denoting

$$S_{1i}^{(N)} = c_i \rho_i \phi'[E(\rho_i | X_i)], \tag{2.4.28}$$

we can rewrite S_{N1} as

$$S_{N1} = \sum_{i=1}^{N} S_{1i}^{(N)}. \tag{2.4.29}$$

Applying Lemma 2.4.2, we obtain

$$E(S_{N1} - \hat{S}_{N1})^2 \le \sum_{i=1}^{N} C_i + \sum\sum_{i \ne j} D_{ij} - \sum\sum_{i \ne j} \sum_{k \ne (i,j)} E_{ij,k}, \tag{2.4.30}$$

where

$$C_i = E\left[S_{1i}^{(N)} - E\left(S_{1i}^{(N)} | X_i\right)\right]^2, \tag{2.4.31}$$

$$D_{ij} = E\left[S_{1i}^{(N)} - E\left(S_{1i}^{(N)} | X_i\right)\right]\left[S_{1j}^{(N)} - E\left(S_{ij}^{(N)} | X_j\right)\right]. \tag{2.4.32}$$

and

$$E_{ij,k} = \mathrm{Cov}\left[E\left(S_{1i}^{(N)} | X_k\right), E\left(S_{1j}^{(N)} | X_k\right)\right]. \tag{2.4.33}$$

We bound each term on the right side of (2.4.30) separately. First consider C_i.

Let $[\phi'(t)]^2 \le K_1$, $0 < t < 1$. Then

$$\begin{aligned} C_i &= c_i^2 (N+1)^{-2} E\left\{R_i \phi'[E(\rho_i | X_i)] - E\left[\{R_i \phi'[E(\rho_i | X_i)]\} | X_i\right]\right\}^2 \\ &\le c_i^2 K_i (N+1)^{-2} E[R_i - E(R_i | X_i)]^2 \\ &= c_i^2 K_1 (N+1)^{-2} E[\mathrm{Var}(R_i | X_i)]^2 \\ &\le c_i^2 K_1 (N+1)^{-2} \frac{N}{4} \le \frac{c_i^2 K_1 (N+1)^{-1}}{4}. \end{aligned} \tag{2.4.34}$$

Next, consider D_{ij}. We express it as

$$\begin{aligned} D_{ij} = E&\left[\mathrm{Cov}\left(S_{1i}^{(N)}, S_{1j}^{(N)} | X_i, X_j\right)\right] \\ &+ E\left[\left\{\left(S_{1i}^{(N)} | X_i, X_j\right) - E\left(S_{1i}^{(N)} | X_i\right)\right\}\right. \\ &\quad \left. \times \left\{E\left(S_{1j}^{(N)} | X_i, X_j\right) - E\left(S_{1j}^{(N)} | X_j\right)\right\}\right]. \end{aligned} \tag{2.4.35}$$

Let us consider $E\,\mathrm{Cov}[(S_{1i}^{(N)}, S_{1j}^{(N)})|X_i, X_j]$:

$$\begin{aligned}
&\mathrm{Cov}\left[\left(S_{1i}^{(N)}, S_{1j}^{(N)}\right)\middle|X_i, X_j\right] \\
&= c_i c_j (N+1)^{-2} \sum_{\substack{k=1 \\ k\neq(i,j)}}^{N} \mathrm{Cov}\Big[\big(u(X_i - X_k)\phi'[E(\rho_i|X_i)], \\
&\qquad\qquad \times u(X_j - X_k)\phi'[E(\rho_j|X_j)]\big)\Big|X_i, X_j\Big] \\
&= c_i c_j (N+1)^{-2} \sum_{\substack{k=1 \\ k\neq(i,j)}}^{N} \phi'[E(\rho_i|X_i)]\phi'[E(\rho_j|X_j)] \\
&\qquad \times\left[E\left\{u(X_i - X_k)u(X_j - X_k)\middle|X_i, X_j\right\} - F_k(X_i)F_k(X_j)\right] \\
&= c_i c_j (N+1)^{-2} \sum_{\substack{k=1 \\ k\neq(i,j)}}^{N} \left[\min\{F_k(X_i), F_k(X_j)\} - F_k(X_i)F_k(X_j)\right] \\
&\qquad \times\phi'[E(\rho_i|X_i)]\phi'[E(\rho_j|X_j)]. \qquad (2.4.36)
\end{aligned}$$

Thus

$$\begin{aligned}
&E\,\mathrm{Cov}\left[\left(S_{1i}^{(N)}, S_{1j}^{(N)}\right)\middle|X_i, X_j\right] \\
&= c_i c_j (N+1)^{-2} \sum_{\substack{k=1 \\ k\neq i,j}}^{N} \int\!\!\int \{\min[F_k(x), F_k(y)] - F_k(x)F_k(y)\} \\
&\qquad \times\phi'[E(\rho_i|X_i = x)]\phi'[E(\rho_j|X_j = y)]\,dF_i(x)\,dF_j(y). \qquad (2.4.37)
\end{aligned}$$

The right side can also be expressed differently as follows. Denote

$$l_{ik}(s) = \int [u(x - s) - F_k(x)]\phi'[E(\rho_i|X_i = x)]\,dF_i(x). \qquad (2.4.38)$$

Then, it is easy to check that

$$\begin{aligned}
E\,\mathrm{Cov}\left[\left(S_{1i}^{(N)}, S_{1j}^{(N)}\right)\middle|X_i, X_j\right] &= c_i c_j (N+1)^{-2} \\
&\quad \times \sum_{k\neq(i,j)} \mathrm{Cov}\left[l_{ik}(X_k), l_{jk}(X_k)\right].
\end{aligned} \qquad (2.4.39)$$

We now consider $E_{ij,k}$ given by (2.4.33). By definition,

$$E_{ij,k} = E\left[\left\{E\left(S_{1i}^{(N)}|X_k\right) - E\left(S_{1i}^{(N)}\right)\right\}\left\{E\left(S_{1j}^{(N)}|X_k\right) - E\left(S_{1j}^{(N)}\right)\right\}\right]. \tag{2.4.40}$$

Using Lemma 2.3.3, we obtain

$$\begin{aligned} E\left(S_{1i}^{(N)}|X_k\right) - E\left(S_{1i}^{(N)}\right) &= c_i E_{X_i}\left[\phi'[E(\rho_i|X_i)]\{E(\rho_i|X_i, X_k) - E(\rho_i|X_i)\}\right. \\ &= c_i(N+1)^{-1}\int[u(x - X_k) - F_k(x)] \\ &\quad \times \sum_{r=2}^{N} P\left(R_i = r|\, X_i = x,\, \underset{i \neq k}{X_r = x - 1}\right) \\ &\quad \times \phi'[E(\rho_i|X_i = x)]\, dF_i(x) \\ &= c_i(N+1)^{-1}\int[u(x - X_k) - F_k(x)] \\ &\quad \times \phi'[E(\rho_i|X_i = x)]\, dF_i(x), \end{aligned} \tag{2.4.41}$$

since $\sum_{r=2}^{N} P(R_i = r|X_i = x,\ X_k = x - 1,\ i \neq k) = 1$. Using (2.4.41) in (2.4.40), we obtain for every $k \neq i \neq j$

$$E_{ij,k} = c_i c_j (N+1)^{-2}\operatorname{Cov}\left[l_{ik}(X_k), l_{jk}(X_k)\right]. \tag{2.4.42}$$

Finally, we consider the second term on the right side of (2.4.35). First note that

$$\begin{aligned} &E\left(S_{1i}^{(N)}|X_i, X_j\right) - E\left(S_{1i}^{(N)}|X_i\right) \\ &\quad = c_i(N+1)^{-1}\phi'[E(\rho_i|X_i)]\left[E(R_i|X_i, X_j) - E(R_i|X_i)\right] \\ &\quad = c_i(N+1)^{-1}\phi'[E(\rho_i|X_i)]\left[u(X_i - X_j) - F_j(X_i)\right], \end{aligned} \tag{2.4.43}$$

using Lemma 2.3.3 and proceeding as previously noted. Similarly,

$$\begin{aligned} &E\left(S_{1j}^{(N)}|X_i, X_j\right) - E\left(S_{1j}^{(N)}|X_j\right) \\ &\quad = c_j(N+1)^{-1}\phi'(E\rho_j|\, X_j))\left[u(X_j - X_i) - F_i(X_j)\right]. \end{aligned} \tag{2.4.44}$$

Thus

$$\left|E\left[\left\{E\left(S_{1i}^{(N)}|X_i, X_j\right) - E\left(S_{1i}^{(N)}|X_i\right)\right\}\left\{E\left(S_{1j}|X_i, X_j\right) - E\left(S_{1j}|X_j\right)\right\}\right]\right|$$

$$\leq K_1|c_i c_j|(N+1)^{-2}E\left|\left[u(X_i - X_j) - F_j(X_i)\right]\right.$$

$$\left.\times\left[u(X_j - X_i) - F_i(X_j)\right]\right|$$

$$\leq \frac{K_1}{4}|c_i c_j|(N+1)^{-2}; \tag{2.4.45}$$

the last inequality follows by using the Cauchy–Schwartz inequality and the fact that $E[u(X_i - X_j) - F_j(X_i)]^2 \leq \frac{1}{4}$.

Using (2.4.39) and (2.4.45) in (2.4.35), we obtain

$$D_{ij} \leq |c_i c_j|(N+1)^{-2}\left[\sum_{k \neq i, j} \text{Cov}\left[l_{ik}(X_k), l_{jk}(X_k)\right] + \frac{K_1}{4}\right]. \tag{2.4.46}$$

The desired result is obtained by using (2.4.34), (2.4.46), and (2.4.42) in (2.4.30).

Theorem 2.4.5 (Hájek, 1968). *Under the assumption of Theorem* 2.4.3, *there exists a constant* $M = M(\phi)$ *such that for* L_N *defined by* (2.2.3),

$$E\left(L_N - EL_N - \sum_{i=1}^{N} Z_i\right)^2 \leq MN^{-1}\sum_{i=1}^{N}(c_i - \bar{c})^2 \tag{2.4.47}$$

and

$$(EL_N - \mu)^2 \leq MN^{-1}\sum_{i=1}^{N} c_i^2, \tag{2.4.48}$$

where for each i $(= 1, \ldots, N)$,

$$Z_i = l_i(X_i)$$

$$= (N+1)^{-1}\sum_{j=1}^{N}(c_j - c_i)\int\left[u(x - X_i) - F_i(x)\right]\phi'[H(x)]\,dF_j(x) \tag{2.4.49}$$

and

$$\mu = \sum_{i=1}^{N} c_i \int \phi[H(x)]\, dF_i(x), \qquad H(x) = N^{-1} \sum_{j=1}^{N} F_j(x). \quad (2.4.50)$$

Proof. We have for $L_N = S_N$, defined in Theorem 2.4.3,

$$\hat{S}_N - E\hat{S}_N = \hat{S}_N - ES_N = \sum_{i=1}^{N} E(S_N | X_i) - NES_N$$

$$= \sum_{i=1}^{N} \sum_{j=1}^{N} c_j \left\{ E\left[\phi(\rho_j) | X_i \right] - E\left[\phi(\rho_j) \right] \right\}. \quad (2.4.51)$$

Furthermore, since

$$\sum_{j=1}^{N} \phi(\rho_j) = \sum_{j=1}^{N} \phi\left(\frac{R_j}{N+1} \right) = \sum_{j=1}^{N} \phi\left(\frac{j}{N+1} \right) = \text{const.}, \quad (2.4.52)$$

we have

$$E\left[\sum_{j=1}^{N} \phi(\rho_j) \right] = \sum_{j=1}^{N} \phi(\rho_j) = E\left[\sum_{j=1}^{N} \phi(\rho_j) \middle| X_i \right] = \text{const.}$$

Therefore,

$$\sum_{j=1}^{N} \phi(\rho_j) = \sum_{j=1}^{N} E\left[\phi(\rho_j) | X_i \right] = \sum_{j=1}^{N} E\phi(\rho_j) = \sum_{j=1}^{N} \phi\left(\frac{j}{N+1} \right).$$

(2.4.53)

Thus

$$\hat{S}_N - E\hat{S}_N = \sum_{i=1}^{N} \sum_{j=1}^{N} (c_j - c_i)\left[E\left\{ \phi\,(\rho_j) | X_i \right\} - E(\rho_j) \right]. \quad (2.4.54)$$

We now evaluate $E\{\phi(\rho_j)|X_i\} - E\phi(\rho_j)$. Using Lemma 2.3.3, we obtain

$$E\left[\phi(\rho_j)|X_i\right] - E\phi(\rho_j) = E_{X_j}\left[E\{\phi(\rho_j)|X_i, X_j\right] - E\left[\phi(\rho_j)|X_j\right]\right]$$

$$= E_{X_j}\left[\{u(X_j - X_i) - F_i(X_j)\}\sum_{k=2}^{N}\left[\phi\left(\frac{k}{N+1}\right) - \phi\left(\frac{k-1}{N}\right)\right]\right.$$

$$\left.\times P\left(\rho_j = \frac{k}{N+1}\middle| X_j, X_i = X_j - 1, i \neq j\right)\right]$$

$$= \int\{u(x - X_i) - F_i(x)\}$$

$$\times E\left\{\left[\phi(\rho_j) - \phi\left(\rho_j - \frac{1}{N+1}\right)\right]\middle| X_j = x, X_i = x - 1\right\} dF_j(x).$$

(2.4.55)

By Taylor's theorem,

$$\phi(\rho_j) - \phi\left(\rho_j - \frac{1}{N+1}\right) = (N+1)^{-1}\phi'(\rho_j)$$

$$- \frac{(N+1)^{-2}}{2}\phi''\left(\rho_j - \frac{\theta}{N+1}\right), \qquad 0 < \theta < 1.$$

Using this relation in (2.4.55), we obtain

$$E\left[\phi(\rho_j)|X_i\right] - E\phi(\rho_j) = (N+1)^{-1}\int\{u(x - X_i) - F_i(x)\}$$

$$\times E\left\{\phi'(\rho_j)|X_j = x, X_i = x - 1\right\} dF_j(x)$$

$$- \tfrac{1}{2}(N+1)^{-2}\int\{u(x - X_i) - F_i(x)\}$$

$$\times E\left\{\phi''\left(\rho_j - \frac{\theta}{N+1}\right)\middle| X_j = x,\right.$$

$$\left.\times\ X_i = x - 1\right\} dF_j(x). \qquad (2.4.56)$$

Since

$$\left|E\left\{\phi''\left(\rho_j - \frac{\theta}{N+1}\right)\middle|X_j = x,\ X_i = x-1\right\}\right| \le \sqrt{K_2},$$

we can write

$$\left|E\left\{\phi''\left(\rho_j - \frac{\theta}{N+1}\right)\middle|X_j = x,\ X_i = x-1\right\}\right|$$

$$= \sqrt{K_2}\cdot\theta(x), \qquad |\theta(x)| \le 1.$$

Also $|u(x - X_i) - F_i(x)| = |\alpha(x, X_i)| \le 1$. Thus

$$\int\{u(x-X_i) - F_i(x)\}E\left|\left\{\phi''\left(\rho_j - \frac{\theta}{N+1}\right)\middle|X_j\right.\right.$$

$$\left.\left.= x,\ X_i = x-1\right\}dF_j(x)\right| = \sqrt{K_2}\,\alpha_i, \quad (2.4.57)$$

where $|\alpha_i| = |\alpha(X_i)| \le 1$. Hence

$$E\left[\phi\left(\rho_j\right)\middle|X_i\right] - E\phi(\rho_j) = (N+1)^{-1}\int\{u(x-X_i) - F_i(x)\}$$

$$\times E\left\{\phi'\left(\rho_j\right)\middle|X_j = x,\ X_i = x-1\right\}dF_j(x)$$

$$+(N+1)^{-2}\sqrt{K_2}\,\alpha_i.$$

Now let us consider $E\{\phi'(\rho_j)|X_j = x,\ X_i = x-1\}$. By the Taylor theorem,

$$\phi'(\rho_j) = \phi'[H(x)] + \left[\rho_j - H(x)\right]\phi''\left[H(x) + \theta\left(\rho_j - H(x)\right)\right],$$

$$\text{where } 0 \le \theta \le 1. \quad (2.4.58)$$

Therefore,

$$E\left[\phi'\left(\rho_j\right)\middle|X_j = x,\ X_i = x-1\right] = \phi'[H(x)] + 2N^{-1/2}\sqrt{K_2}\,\beta_i,$$

$$\text{where } \quad |\beta_i| \le 1. \quad (2.4.59)$$

Using (2.4.58) and (2.4.59) in (2.4.55), we obtain

$$E\left[\phi(\rho_j)|X_i\right] - E\phi(\rho_j)$$

$$= (N+1)^{-1}\int\{u(x-X_i) - F_i(x)\}\phi'[H(x)]\,dF_j(x)$$

$$+2N^{-1/2}(N+1)^{-1}\sqrt{K_2}\,\beta_i\delta_i(X_i) + (N+1)^{-2}\sqrt{K_2}\,\alpha_i(X_i)$$

$$= (N+1)^{-1}\int\{u(x-X_i) - F_i(x)\}\phi'[H(x)]\,dF_j(x)$$

$$+3N^{-3/2}\sqrt{K_2}\,Y_i(X_i), \qquad |Y_i(X_i)| \le 1.$$

Now set

$$Y_i = \sum_{j=1}^{N}(c_j - c_i)\Big\{E\left[\phi(\rho_j)|X_i\right] - E\left[\phi(\rho_j)\right]$$

$$-\frac{1}{N+1}\int\{u(x-X_i) - F_i(x)\}\phi'[H(x)]\,dF_j(x) \qquad (2.4.60)$$

and note that $Y_i = Y(X_i)$ are independent r.v.'s and $EY_i = 0$, $i = 1,\dots,N$. Thus

$$E\left(\hat{S}_N - E\hat{S}_N - \sum_{i=1}^{N} Z_i\right)^2 = E\left(\sum_{i=1}^{N} Y_i\right)^2 = \sum_{i=1}^{N} EY_i^2$$

$$= 9N^{-3}K_2\sum_{i=1}^{N} E\left\{\sum_{j=1}^{N}(c_j - c_i)Y_i\right\}^2$$

$$\le 9N^{-3}K_2\sum_{i=1}^{N}\left(\sum_{j=1}^{N}|c_j - c_i|\right)^2 \qquad (|Y_i| \le 1)$$

$$\le 18N^{-1}K_2\sum_{i=1}^{N}(c_i - \bar{c})^2. \qquad (2.4.61)$$

Now

$$E\left(S_N - ES_N - \sum_{i=1}^{N} Z_i\right)^2 = E\left(S_N - \hat{S}_N + \hat{S}_N - E\hat{S}_N - \sum_{i=1}^{N} Z_i\right)^2$$

$$= E(S_N - \hat{S}_N)^2 + E\left(\hat{S}_N - E\hat{S}_N - \sum_{i=1}^{N} Z_i\right)^2$$

$$\leq MN^{-1} \sum_{i=1}^{N} (c_i - \bar{c})^2, \qquad M < \infty, \qquad (2.4.62)$$

by (2.4.13) and (2.4.61), where we let $M = 2K + 36K_2^2$. This proves (2.4.47). To prove (2.4.48), first note that

$$\phi(\rho_i) = \phi[H(X_i)] + [\rho_i - H(X_i)]\phi'[H(X_i)] + \frac{[\rho_i - H(X_i)]^2}{2!} k_i(\rho_i).$$

And, then proceeding as above, we obtain

$$|E\phi(\rho_i) - E\phi[H(X_i)]| \leq K_4 N^{-1}. \qquad (2.4.63)$$

Hence,

$$(ES_N - \mu)^2 = \left(\sum_{i=1}^{N} c_i[E\phi(\rho_i) - E\phi(H(X_i))]\right)^2$$

$$\leq \sum_{i=1}^{N} c_i^2 \sum_{i=1}^{N} [E\phi(\rho_i) - E\phi(H(X_i))]^2$$

$$\leq MN^{-1} \sum_{i=1}^{N} c_i^2, \qquad \text{where} \quad M = K_4^2. \qquad (2.4.64)$$

2.5. ASYMPTOTIC NORMALITY OF L_N

In this section, we study the asymptotic normality of L_N defined by (2.2.3) when the scores $a_N(i)$, $1 \leq i \leq N$, are defined by (2.2.2). W assume that ϕ is known and fixed, whereas N, $(c_1, \ldots, c_N)$, and $(F_1, \ldots, F_N)$ could vary.

We shall first prove the asymptotic normality of L_N for the case when the score generating function ϕ has a bounded second derivative. Later on we shall relax this assumption.

Theorem 2.5.1 (Hájek, 1968). *For $a_N(i)$ defined by (2.2.2) and $\phi(t)$ having a bounded second derivative inside* (0,1), *whenever, for every $\varepsilon > 0$, there exists a K_ε ($< \infty$) such that*

$$\operatorname{Var} L_N > K_\varepsilon \Big[\max_{1 \le i \le N} (c_i - \bar{c}_N)^2 \Big], \qquad \bar{c}_N = N^{-1} \sum_{i=1}^{N} c_i, \tag{2.5.1}$$

then

$$\sup_x \left| P\{ L_N - EL_N < x (\operatorname{Var} L_N)^{1/2} \} - \Phi(x) \right| < \varepsilon, \tag{2.5.2}$$

where $\Phi(x) = (2\pi)^{-1/2} \int_{-\infty}^{x} e^{-t^2/2}\, dt$. The assertion remains true if we replace $\operatorname{Var} L_N$ *in* (2.5.1) *and* (2.5.2) *by*

$$\sigma_N^2 = \sum_{i=1}^{N} \operatorname{Var} Z_i, \tag{2.5.3}$$

where the Z_i are defined by (2.4.49).

Proof. From (2.4.49), we have $EZ_i = 0$. Let $G_i(x)$ be the c.d.f. of Z_i. Then by the Lindeberg version of the central limit theorem, for every $\varepsilon > 0$ there exists a $\delta > 0$ such that

$$\frac{1}{\sigma_N^2} \sum_{i=1}^{N} \int_{|x| > \delta\sigma} x^2\, dG_i(x) < \delta \tag{2.5.4}$$

implies

$$\sup_x \left| P\left(\sum_{i=1}^{N} Z_i < x\sigma_N \right) - \Phi(x) \right| < \frac{\varepsilon}{4}. \tag{2.5.5}$$

Next, since $\Phi(x)$ is a continuous function of x, there exists a $\beta > 0$ such that

$$|\Phi(x) - \Phi(x \pm \beta)| < \frac{\varepsilon}{4} \qquad \forall x \in (-\infty, \infty). \tag{2.5.6}$$

Thus, using (2.5.5) and (2.5.6), we have

$$\sup_x \left| P\left(\sum_{i-1}^{N} \frac{Z_i}{\sigma} \le x \pm \beta \right) - \Phi(x) \right| < \varepsilon/4 + \varepsilon/4 = \varepsilon/2. \tag{2.5.7}$$

Set

$$K_\varepsilon = \left[2\delta^{-1} \sup_{t\in(0,1)} |\phi'(t)| + (2\varepsilon^{-1}\beta^{-1} + 1)M^{1/2}\right]^2, \tag{2.5.8}$$

where M is a constant appearing in (2.4.47). We shall show that (2.5.1) with K_ε given by (2.5.8) implies (2.5.2).

First let us consider the case when $a_N(i) = \phi(i/(N+1))$. To prove the theorem, we need the following inequalities, which we ask the reader to verify (see Exercise 2.5.1):

$$\left|\sigma_N - \sqrt{\operatorname{Var} L_N}\right| \le M^{1/2} \max_i |c_i - \bar{c}|, \tag{2.5.9}$$

$$\delta\sigma_N \ge \max_i |c_i - \bar{c}|\{\sup|\phi'(t)|\}, \tag{2.5.10}$$

$$\beta^2\sigma_N^2 \ge \frac{4M}{\varepsilon} \max_i (c_i - \bar{c})^2, \tag{2.5.11}$$

$$|Z_i| \le 2\max_i |c_i - \bar{c}| \sup|\phi'(t)|. \tag{2.5.12}$$

From (2.5.10) and (2.5.12), we obtain that $\delta\sigma_N \ge |Z_i|$ with probability 1 for $i = 1, \ldots, N$. This implies

$$\frac{1}{\sigma_N^2} \sum_{i=1}^N \int_{|x|>\delta\sigma_N} x^2 \, dG_i(x) = 0. \tag{2.5.13}$$

Thus the Lindeberg condition (2.5.4) is satisfied. Hence (2.5.5) holds. Now,

$$\begin{aligned}
&P\left(\frac{L_N - EL_N}{\sigma_N} < x\right) \\
&\le P\left(\sum_{i=1}^N \frac{Z_i}{\sigma_N} < x + \beta\right) + P\left(\left|L_N - EL_N - \sum_{i=1}^N Z_i\right| > \sigma_N\beta\right) \\
&\le \Phi(x) + \frac{\varepsilon}{2} + P\left(\left|L_N - EL_N - \sum_{i=1}^N Z_i\right| > \sigma_N\beta\right) \\
&\le \Phi(x) + \frac{\varepsilon}{2} + \frac{E\left(L_N - EL_N - \sum_{i=1}^N Z_i\right)^2}{\sigma_N^2\beta^2} \\
&\le \Phi(x) + \frac{\varepsilon}{2} + \frac{\varepsilon}{4} = \Phi(x) + \frac{3\varepsilon}{4}.
\end{aligned} \tag{2.5.14}$$

The last inequality is obtained using Theorem 2.4.3 and (2.5.11). Proceeding analogously, we obtain

$$P\left(\frac{L_N - EL_N}{\sigma_N} < x\right) \geq \Phi(x) - \frac{3\varepsilon}{4}. \tag{2.5.15}$$

Thus, from (2.5.14) and (2.5.15),

$$\sup_x \left| P\left(\frac{L_N - EL_N}{\sigma_N} < x\right) - \Phi(x)\right| < \frac{3\varepsilon}{4}. \tag{2.5.16}$$

This proves (2.5.2) with $\operatorname{Var} L_N$ replaced by σ_N^2.

Now let $\varepsilon < 1$. Then from (2.5.11),

$$\beta\sigma_N \geq \frac{2M^{1/2}}{\varepsilon^{1/2}} \max_i |c_i - \bar{c}| > \left|\sigma_N - \sqrt{\operatorname{Var} L_N}\right|. \tag{2.5.17}$$

Using (2.5.6), (2.5.16), and (2.5.17), we obtain (2.5.2).

Now let us consider the case when $a_N(i) = E\phi(U_N^{(i)})$, where $U_N^{(i)}$ is the ith order statistic in a sample of size N from the $R(0,1)$ distribution. By the Taylor theorem,

$$\phi(U_N^{(i)}) = \phi\left(\frac{i}{N+1}\right) + \left(U_N^{(i)} - \frac{i}{N+1}\right)\phi'\left(\frac{i}{N+1}\right)$$
$$+ \left(U_N^{(i)} - \frac{i}{N+1}\right)^2 \phi''(\zeta_i),$$

so that

$$E\phi(U_N^{(i)}) = \phi\left(\frac{i}{N+1}\right) + k_{iN}, \tag{2.5.18}$$

$$\text{where } |k_{iN}| \leqq KN^{-1} \qquad \left[\text{as } \operatorname{Var} U_N^{(i)} = O(N^{-1})\right].$$

Now writing

$$L_N = \sum_{i=1}^{N} (c_i - \bar{c}_N)\phi\left(\frac{R_{Ni}}{N+1}\right) + \bar{c}_N \sum_{i=1}^{N} \phi\left(\frac{i}{N+1}\right);$$

$$L'_N = \sum_{i=1}^{N} (c_i - \bar{c}_N) E\phi\left(U_N^{(R_{Ni})}\right) + \bar{c}_N \sum_{i=1}^{N} E\phi\left(U_N^{(i)}\right),$$

and noting that

$$E\phi\left(U_N^{(R_{Ni})}\right) = \phi\left(\frac{R_{Ni}}{N+1}\right) + |k_{R_{Ni},N}|, \qquad |k_{R_{Ni},N}| \le KN^{-1},$$

we obtain, after simple computations, that

$$E\left(L_N - EL_N - L'_N + EL'_N\right)^2 \le K^2N^{-1}\sum_{i=1}^{N}(c_i - \bar{c}_N)^2.$$

Thus, $L_N - EL_N$ is asymptotically equivalent to $L'_N - EL'_N$. The theorem follows.

In most of the applications, as we shall see in later chapters, we need to assume that the d.f.'s $F_1, \ldots, F_N$ are close to each other in a certain sense. For example, these d.f.'s may differ only in certain shifts of small variation. In such a case, the following theorem plays an important role.

Theorem 2.5.2 (Hájek, 1968). *If*

(i) *the assumptions of Theorem* 2.5.1 *are satisfied, and for every* $\varepsilon > 0$, *there exists a* δ_ε *such that*

(ii) $\max_{i,j,x}|F_i(x) - F_j(x)| < \delta_\varepsilon$

and

(iii) $\sum_{i=1}^{N}(c_i - \bar{c}_N)^2 > \delta_\varepsilon^{-1}\max_{1\le i\le N}(c_i - \bar{c}_N)^2$,

then

$$\sup_x\left|P\left[\frac{L_N - EL_N}{d_N} < x\right] - \Phi(x)\right| < \varepsilon, \tag{2.5.19}$$

where

$$d_N^2 = \sum_{i=1}^{N}(c_i - \bar{c})^2\int_0^1[\phi(t) - \bar{\phi}]^2\,dt, \bar{\phi} = \int_0^1\phi(t)\,dt. \tag{2.5.20}$$

Proof. We rewrite $\int_{x_0}^{y}\phi'[H(x)]\,dF_j(x)$ as

$$\begin{aligned}\int_{x_0}^{y}\phi'[H(x)]\,dF_j(x) &= \int_{x_0}^{y}\phi'[F_i(x)]\,dF_i(x)\\ &\quad+\int_{x_0}^{y}\phi'[F_i(x)]\,d\big[F_j(x)-F_i(x)\big]\\ &\quad+\int_{x_0}^{y}\{\phi'[H(x)]-\phi'[F_i(x)]\}\,dF_j(x)\\ &=\phi[F_i(y)]+\text{const}+R_{ij}(y),\end{aligned}\tag{2.5.21}$$

where

$$\begin{aligned}|R_{ij}(y)| &= \left|\int_{x_0}^{y}\phi'[F_i(x)]\,d\big[F_j(x)-F_i(x)\big]\right.\\ &\quad\left.+\int_{x_0}^{y}[\phi'[H(x)]-\phi'[F_i(x)]]\,dF_j(x)\right|\\ &\le\left|\sup_t|\phi'(t)|\max_x|F_j(x)-F_i(x)|\right.\\ &\quad\left.+\int_{x_0}^{y}[H(x)-F_i(x)]\phi''(\zeta)\,dF_j(x)\right|\\ &\le\left|\sup_t|\phi'(t)|\max_{i,j,x}|F_j(x)-F_i(x)|\right.\\ &\quad\left.+\sup_t|\phi''(t)|\max_{i,j,x}|F_j(x)-F_i(x)|\right|\\ &=L\Big\{\max_{i,j,x}|F_j(x)-F_i(x)|\Big\},\qquad 1\le i,j\le N,\end{aligned}\tag{2.5.22}$$

and

$$L=L_\phi=\sup|\phi'(t)|+\sup|\phi''(t)|.\tag{2.5.23}$$

If we denote

$$V_i=(c_i-\bar{c}_N)\phi(F_i(X_i)),\qquad 1\le i\le N,\tag{2.5.24}$$

then on noting that $F_i(X_i)$ has the rectangular (0, 1) d.f., we have

$$EV_i = (c_i - \bar{c}_N)\bar{\phi}, \qquad \bar{\phi} = \int_0^1 \phi(u)\,du, \tag{2.5.25}$$

$$\operatorname{Var} V_i = (c_i - \bar{c}_N)^2 \int_0^1 [\phi(u) - \bar{\phi}]^2\,du, \qquad 1 \le i \le N. \tag{2.5.26}$$

Thus,

$$d_N^2 = \sum_{i=1}^{N} \operatorname{Var} V_i. \tag{2.5.27}$$

Also, from (2.4.49),

$$\begin{aligned} Z_i &= \frac{1}{N+1} \sum_{j=1}^{N} (c_j - c_i) \int [u(x - X_i) - F_i(x)]\,\phi'(H(x))\,dF_j(x) \\ &= \frac{1}{N+1} \sum_{j=1}^{N} (c_i - c_j) \int_{x_0}^{X_i} \phi'[H(y)]\,F_j(x) + \frac{c}{N+1} \sum_{j=1}^{N} (c_j - c_i) \\ &= V_i + \text{const} - \frac{1}{N+1} \sum_{j=1}^{N} (c_j - c_i) R_{ij}(X_i) + \frac{CN}{N+1}(\bar{c}_N - c_i) \\ &\quad + V_i + \text{const} + R_i^*(X_i)\left[N^{-1} \sum_{j=1}^{N} |c_j - c_i|\right], \end{aligned} \tag{2.5.28}$$

where c is a constant, and

$$|R_i^*(X_i)| = \left| \frac{\dfrac{-1}{N+1} \displaystyle\sum_{j=1}^{N} (c_j - c_i) R_{ij}(X_i)}{\dfrac{1}{N} \displaystyle\sum_{j=1}^{N} |c_j - c_i|} \right| \le L \max_{i,j,x} |F_j(x) - F_i(x)|.$$

(2.5.29)

Now

$$|\sigma_N - d_N| = \left|\sqrt{\operatorname{Var}\sum_{i=1}^{N} Z_i} - \sqrt{\operatorname{Var}\sum_{i=1}^{N} V_i}\right|$$

$$\leq \left[\sum_{i=1}^{N} \operatorname{Var}(Z_i - V_i)\right]^{1/2}$$

$$= \left[\sum_{i=1}^{N} \operatorname{Var}\left\{R_i^*(X_i)\frac{1}{N}\sum_{j=1}^{N}|c_j - c_i|\right\}\right]^{1/2}$$

$$\leq L \max_{i,j,x}|F_j(x) - F_i(x)| \cdot \left[2\sum_{i=1}^{N}(c_i - \bar{c}_N)^2\right]^{1/2}. \quad (2.5.30)$$

Now, we choose $K_{\varepsilon/2}$ such that $\sigma_N^2 > K_{\varepsilon/2}\max_i(c_i - \bar{c}_N)^2$. Then using Theorem 2.5.1,

$$\sup_x \left|P\left(\frac{L_N - EL_N}{\sigma_N} < x\right) - \phi(x)\right| < \frac{\varepsilon}{2}. \quad (2.5.31)$$

Also choose $\alpha > 0$ such that

$$|\Phi(x \pm \alpha) - \Phi(x)| < \frac{\varepsilon}{2} \quad (2.5.32)$$

and

$$\sup_x \left|P\left(\frac{L_N - EL_N}{\sigma_N} < x(1 \mp \alpha) - \Phi(x(1 \mp \alpha))\right|\right. < \frac{\varepsilon}{2}. \quad (2.5.33)$$

Then we get

$$\sup_x \left|P\left(\frac{L_N - EL_N}{\sigma_N} < x(1 \mp \alpha)\right) - \Phi(x)\right| < \varepsilon. \quad (2.5.34)$$

Now choose δ_ε such that $K_{\varepsilon/2} < \delta_\varepsilon^{-1}$. Then, since $\max_{i,j,x}|F_j(x) - F_i(x)| < \delta_\varepsilon$, we get, on using (2.5.30),

$$|d^{-1}\sigma_N - 1| < \alpha, \qquad \text{where} \quad \alpha = \frac{\sqrt{2}L}{K_{\varepsilon/2}}\int_0^1 [\phi(u) - \bar{\phi}]^2\, du. \quad (2.5.35)$$

Then, (2.5.34) and (2.5.35) imply (2.5.19), Q.E.D.

In the preceding two theorems, we have considered the asymptotic normality of rank statistics for score functions with bounded second derivatives. We shall now consider the general case of square-integrable score functions in our main theorem of this section. The basic trick is a fundamental polynomial approximation of monotonic and absolutely continuous score functions (or of the difference of two such monotonic score functions), which is considered in Lemma 2.5.4. Some refinements are studied in the next section.

Theorem 2.5.3 (Hájek, 1968). *Under* (2.2.2) *with* $\phi(t) = \phi_1(t) - \phi_2(t)$, $\phi_i(t)\uparrow$ *in* $t \in (0,1)$, $i = 1,2$, *and* $\phi(u)$ *square-integrable and absolutely continuous inside* $(0,1)$, *if for every* $\eta > 0$ *there exists an* N_η $(< \infty)$ *such that*

$$\operatorname{Var} L_N > \eta N \max_i (c_i - \bar{c})^2 \qquad \text{for} \quad N \geq N_\eta, \tag{2.5.36}$$

then

$$\sup_x \left| P\left\{ L_N - EL_N \leq x\sqrt{\operatorname{Var} L_N} \right\} - \Phi(x) \right| < \varepsilon, \tag{2.5.37}$$

$$\sup_x \left| P\{ L_N - EL_N \leq x\sigma_N \} - \Phi(x) \right| < \varepsilon, \tag{2.5.38}$$

where σ_N^2 *is given by* (2.5.3).

The proof of the theorem is based on the following lemma.

Lemma 2.5.4. If $\phi(t) = \phi_1(t) - \phi_2(t), 0 < t < 1$, where the $\phi_j(t)$, $j = 1,2$, are monotonic, absolutely continuous, and square-integrable inside $(0,1)$, then for every $\alpha > 0$ there exists a decomposition

$$\phi(t) = \psi(t) + \tilde{\phi}_1(t) - \tilde{\phi}_2(t), \qquad 0 < t < 1, \tag{2.5.39}$$

such that $\psi(t)$ is a polynomial, $\tilde{\phi}_1$ and $\tilde{\phi}_2$ are nondecreasing, and

$$\int_0^1 \tilde{\phi}_1^2(u)\,du + \int_0^1 \tilde{\phi}_2^2(u)\,du < \alpha. \tag{2.5.40}$$

Proof of Lemma 2.5.4. Without loss of generality, we may assume that $\phi(t)$ itself is nondecreasing. Take $\varepsilon > 0$, and define $\phi_0(t)$, $\phi_3(t)$, and $\phi_4(t)$

as follows:

	Interval $[0, \varepsilon]$	Interval $[\varepsilon, 1-\varepsilon]$	Interval $[1-\varepsilon, 1]$	
$\phi_0(t) =$	$\phi(\varepsilon)$	$\phi(t)$	$\phi(1-\varepsilon)$,	(2.5.41)
$\phi_3(t) =$	$\phi(t) - \phi(\varepsilon)$	0	0,	(2.5.42)
$\phi_4(t) =$	0	0	$\phi(t) - \phi(1-\varepsilon)$.	(2.5.43)

We observe that

$$\phi(t) = \phi_0(t) + \phi_3(t) + \phi_4(t); \tag{2.5.44}$$

$\phi_3(t)$ and $\phi_4(t)$ are both nondecreasing in t. Also, since $\phi(t)$ is absolutely continuous on $(\varepsilon, 1-\varepsilon)$, $\phi_0(t)$ is absolutely continuous in $(0,1)$. Thus, there exists a derivative $\phi_0'(t)$, such that

$$\phi_0(t) = \phi(\varepsilon) + \int_0^t \phi'(s)\, ds, \qquad 0 \le t \le 1. \tag{2.5.45}$$

Furthermore, since the set of polynomials is a dense subset of L_1, the space of integrable functions (Weierstrass theorem), it follows that for every $\beta > 0$ there exists a polynomial $q(s)$ such that

$$\int_0^1 |\phi_0'(s) - q(s)|\, ds < \beta. \tag{2.5.46}$$

Now denote

$$\psi(t) = \phi(\varepsilon) + \int_0^t q(s)\, ds, \tag{2.5.47}$$

$$\phi_1(t) = \phi_3(t) + \phi_4(t) + \int_0^t \max[0, \phi_0'(s) - q(s)]\, ds, \tag{2.5.48}$$

$$\phi_2(t) = \int_0^t \max[0, q(s) - \phi_0'(s)]\, ds. \tag{2.5.49}$$

Then $\psi(t) + \phi_1(t) - \phi_2(t) = \phi_0(t) + \phi_3(t) + \phi_4(t) = \phi(t)$. Here $\psi(t)$ is a polynomial, $\phi_1(t)$ and $\phi_2(t)$ are nondecreasing, and $\int_0^1 \phi_1^2(t)\, dt + \int_0^1 \phi_2^2(t)\, dt < \alpha$. Q.E.D.

Proof of Theorem 2.5.3. Take $\varepsilon > 0$ and $\eta > 0$. Then choose $\beta > 0$ and $\gamma > 0$ such that

$$\left| \Phi(x) - \Phi\left(\frac{x \pm \beta}{1 \pm \gamma}\right) \right| < \frac{\varepsilon}{4}. \tag{2.5.50}$$

Also choose $\alpha > 0$ such that

$$\alpha < \frac{\eta \min(\gamma^2, \beta^2\varepsilon/4)}{84}. \tag{2.5.51}$$

Now decompose ϕ according to Lemma 2.5.4 with α satisfying (2.5.40) and (2.5.51). Set

$$L_\psi = \sum_{j=1}^{N} c_j \psi\left(\frac{R_{Nj}}{N+1}\right), \qquad L_i = \sum_{j=1}^{N} c_j \phi_i\left(\frac{R_{Nj}}{N+1}\right), \qquad i = 1, 2. \tag{2.5.52}$$

Then

$$L_\psi + L_1 - L_2 = \sum_{j=1}^{N} c_j \phi\left(\frac{R_{Nj}}{N+1}\right) = L_N. \tag{2.5.53}$$

Using Theorem 2.3.1, we have

$$\begin{aligned}
&\left|\sqrt{\operatorname{Var} L_N} - \sqrt{\operatorname{Var} L_\psi}\right| \\
&\quad \le \sqrt{\operatorname{Var}(L - L_\psi)} \\
&\quad \le \sqrt{42}\, \max_i |c_i - \bar{c}| \left[\sum_{i=1}^{N} \left\{\phi_1^2\left(\frac{i}{N+1}\right) + \phi_2^2\left(\frac{i}{N+1}\right)\right\}\right]^{1/2} \\
&\quad \le \sqrt{\frac{42 \operatorname{Var} L_N}{N\eta}} \left[\sum_{i=1}^{N} \left\{\phi_1^2\left(\frac{i}{N+1}\right) + \phi_2^2\left(\frac{i}{N+1}\right)\right\}\right]^{1/2} \\
&\quad \le \sqrt{\frac{84 \operatorname{Var} L_N}{\eta}} \left[\int_0^1 \phi_1^2(t)\, dt + \int_0^1 \phi_2^2(t)\, dt\right]^{1/2} \\
&\quad \le \sqrt{\operatorname{Var} L_N} \sqrt{\min\left(\gamma^2, \frac{\beta^2\varepsilon}{4}\right)}.
\end{aligned} \tag{2.5.54}$$

Thus

$$\left|\sqrt{\operatorname{Var} L_N} - \sqrt{\operatorname{Var} L_\psi}\right| \le \sqrt{\operatorname{Var}(L_N - L_\psi)} \le \sqrt{\operatorname{Var} L_N \min\left(\gamma, \frac{\beta\varepsilon^{1/2}}{2}\right)}. \tag{2.5.55}$$

Now let $K_{\varepsilon/2} = K_{\varepsilon/2}(\psi)$ be the constant. (The existence of this constant was established in Theorem 2.5.1. Since ψ is a polynomial, it has a bounded second derivative.) Put $N_{\varepsilon\eta} = (1-\gamma)^{-2}\eta^{-1}K_{\varepsilon/2}$. Then

$$\begin{aligned} N > N_{\varepsilon\eta} \quad &\Rightarrow \quad N > (1-\gamma)^{-2}\eta^{-1}K_{\varepsilon/2} \quad \Rightarrow \quad N\eta > (1-\gamma)^{-2}K_{\varepsilon/2} \\ &\Rightarrow \quad \operatorname{Var} L_N > (1-\gamma)^{-2}K_{\varepsilon/2}\max_i (c_i - \bar{c}_N)^2 \\ &\Rightarrow \quad \operatorname{Var} L_\psi > K_{\varepsilon/2}(\psi)\max(c_i - \bar{c}_N). \qquad (2.5.56) \end{aligned}$$

Hence by Theorem 2.5.1 (since ψ possesses a second derivative),

$$\sup\left| P\left((L_\psi - EL_\psi)\Big/\sqrt{\operatorname{Var} L_\psi} < x\right) - \Phi(x)\right| < \frac{\varepsilon}{2}. \qquad (2.5.57)$$

Now let

$$\begin{aligned} A &= \left[L_N - EL_N < x\sqrt{\operatorname{Var} L_N}\right], \\ B &= \left[L_\psi - EL_\psi < x\sqrt{\operatorname{Var} L_N} + \beta\sqrt{\operatorname{Var} L_N}\right], \\ C &= \left[|L_N - EL_N - L_\psi + EL_\psi| > \beta\sqrt{\operatorname{Var} L_N}\right]. \end{aligned}$$

Then, since $A \subset (B \cup C)$, it follows that

$$\begin{aligned} P\left[\frac{L_N - EL_N}{\sqrt{\operatorname{Var} L_N}} < x\right] &\le P\left[\frac{L_\psi - EL_\psi}{\sqrt{\operatorname{Var} L_N}} < x + \beta\right] \\ &\quad + P\left[\frac{|L_N - EL_N - L_\psi + EL_\psi|}{\sqrt{\operatorname{Var} L_N}} > \beta\right] \\ &\le P\left\{L_\psi - EL_\psi \le (x+\beta)\sqrt{\operatorname{Var} L_N}\right\} \\ &\quad + \frac{\operatorname{Var}(L_N - L_\psi)}{\beta^2 \operatorname{Var} L_N} \qquad \text{(by the Chebyshev inequality)} \\ &\le P\left\{L_\psi - EL_\psi \le (x+\beta)\sqrt{\operatorname{Var} L_N}\right\} \\ &\quad + \frac{\min(\gamma^2, \beta^2\varepsilon)}{\beta^2} \\ &\le \Phi\left(\frac{x+\beta}{1-\gamma}\right) + \frac{\varepsilon}{2} + \frac{\varepsilon}{4} \le \Phi(x) + \varepsilon, \qquad (2.5.58) \end{aligned}$$

by proper choice of γ (> 0). Proceeding analogously, we obtain (see problem 2.5.2)

$$p\left[\frac{L_N - EL_N}{\sqrt{\operatorname{Var} L_N}} < x\right] \geq \Phi(x) - \varepsilon. \tag{2.5.59}$$

Using (2.5.58) and (2.5.59), we get (2.5.37).

Now to prove (2.5.38), we proceed as follows. Denote

$$Z_i^{(k)} = (N+1)^{-1} \sum_{j=1}^{N} (c_j - c_i) \int [u(y - X_i) - F_i(y)] \phi_k'(H(y))\, dF_j(y),$$
$$k = 1, 2, \tag{2.5.60}$$

and

$$W_i = (N+1)^{-1} \sum_{j=1}^{N} (c_j - c_i) \int [u(y - X_i) - F_i(y)] \psi(H(y))\, dF_j(y), \tag{2.5.61}$$

$$\sigma_\psi^2 = \sum_{i=1}^{N} \operatorname{Var} W_i; \qquad \sigma_k^2 = \sum_{i=1}^{N} \operatorname{Var} Z_i^{(k)}, \quad k = 1, 2. \tag{2.5.62}$$

Then corresponding to the decomposition $\phi = \psi + \phi_1 - \phi_2$, we decompose σ_N^2 into σ_ψ^2, σ_1^2, and σ_2^2 given by (2.5.62). Then we show that

$$|\sigma_N - \sigma_\psi| \leq \sigma_1 + \sigma_2. \tag{2.5.63}$$

Next, since ϕ_k in (2.5.60) are nondecreasing, the r.v.'s

$$\int [u(x - X_i) - F_i(x)] \phi_k'[H(x)]\, dF_j(x), \qquad j = 1, \ldots, N,$$

are nonnegatively correlated. Hence,

$$\sigma_k^2 \leq 4 \max_i (c_i - \bar{c}_N)^2 (N+1)^{-2}.$$
$$\times \sum_{i=1}^{N} \operatorname{Var}\left[\sum_{j=1}^{N} \int [u(x - X_i) - F_i(x)] \phi_k'(H(x))\, dF_j(x)\right]$$
$$\leq 4N \max_i (c_i - \bar{c}_N)^2 \int_0^1 \phi_k^2(t)\, dt, \qquad k = 1, 2. \tag{2.5.64}$$

Furthermore, using (2.5.40), (2.5.63), and (2.5.64), we obtain

$$|\sigma_N - \sigma_\psi|^2 \le 8N \max_i (c_i - \bar{c}_N)^2 \alpha. \tag{2.5.65}$$

Thus for α sufficiently small, $|\sigma_N - \sigma_\psi|$ can be made negligible in comparison with $N\eta \max(c_i - \bar{c}_N)^2$. Further, by virtue of (2.5.9) and (2.5.55), it is easy to show that $\left|\sigma_N - \sqrt{\operatorname{Var} L_N}\right|$ can be made very small with respect to σ_N if $\sigma_N^2 > N\eta \max_i (c_i - \bar{c}_N)^2$ and N is sufficiently large. The proof now follows.

Corollary 2.5.3. *If*

(i) *the conditions of Theorem* 2.5.3 *are satisfied, and*

(ii) *for every* $\varepsilon > 0$ *and* $\eta > 0$, *there exist* $N_{\varepsilon\eta}$ *and* $\delta_{\varepsilon\eta}$ *such that* $\sum_{i=1}^N (c_i - \bar{c}_N)^2 > \eta N \max_i (c_i - \bar{c}_N)^2$ *for* $N \ge N_{\varepsilon\eta}$, *and*

(iii) $\max_{i,j,x} |F_i(x) - F_j(x)| < \delta_{\varepsilon\eta}$,

then

$$\sup_x \left| P\left[\frac{L_N - EL_N}{d_N} < x\right] - \frac{1}{\sqrt{2\pi}} \int_{-\infty}^x e^{-t^2/2}\, dt \right| < \varepsilon \tag{2.5.66}$$

where d_N *is given by* (2.5.20).

The proof of this corollary is given as an exercise (Exercise 2.5.3). For contiguous alternatives, some parallel results are presented in Section 2.7.

2.6. CENTERING OF LINEAR RANK STATISTICS

In the setup of Hájek (1968), we have proved in Theorem 2.5.3 that $(L_N - EL_N)/\sigma_N$ [or $(L_N - EL_N)/d_N$], where σ_N and d_N are defined by (2.5.3) and (2.5.20), is asymptotically normally distributed with zero mean and unit variance. Looking at (2.2.3) and noting that

$$(N+1)^{-1} R_{Ni} = \frac{N}{N+1} H_N(X_i), \qquad 1 \le i \le N, \tag{2.6.1}$$

where $H_N(x)$ is the empirical d.f. of $X_1, \dots, X_N$, we have

$$L_N = \sum_{i=1}^N c_i a_N\left(\frac{N H_N(X_i)}{N+1}\right). \tag{2.6.2}$$

Also, by (2.2.2), for every u $(0 < u < 1)$,

$$\lim_{N\to\infty} a_N([uN]+1) = \phi(u), \qquad 0 < u < 1. \tag{2.6.3}$$

Also, from (2.2.2), (2.2.4), we have

$$EL_N = \sum_{i=1}^{N} c_i \left[\sum_{j=1}^{N} a_N(j) P\{R_{Ni} = j\} \right]. \tag{2.6.4}$$

As has been noted in Section 2.2, for arbitrary $F_1, \ldots, F_N$, (2.6.4) is a complicated function of these d.f.'s. On the other hand, a parallel version of the classical Glivenko–Cantelli theorem for independent but nonidentically distributed random variables (Wolfowitz, 1953), asserts that as $N \to \infty$

$$\sup_x \left| H_N(x) - \frac{1}{N} \sum_{i=1}^{N} F_i(x) \right| \to 0 \qquad \text{a.s.} \tag{2.6.5}$$

Thus, defining $H(x) = N^{-1}\sum_{i=1}^{N} F_i(x)$ $[= H_{(N)}(x)]$, we expect from (2.6.2), (2.6.3), and (2.6.5) that as $N \to \infty$,

$$|L_N - \mu_N| \to 0 \qquad \text{a.s.,} \tag{2.6.6}$$

where

$$\mu_N = \sum_{i=1}^{N} c_i \int_{-\infty}^{\infty} \phi(H(x))\, dF_i(x). \tag{2.6.7}$$

Indeed, (2.6.6) holds under fairly general regularity conditions (viz., Sen and Ghosh, 1972); see also Exercise 2.6.1. The centering constant μ_N is quite simple, and more easy to manipulate in asymptotic situations. In the context of two- or several-sample rank statistics, where the c_i are 0 or 1, we have seen in Chapter 3 of Puri and Sen (1971) that in the asymptotic normality of $(L_N - EL_N)/\sigma_N$ [or $(L_N - EL_N)/d_N$], EL_N can be replaced by the correspondingly simplified version of μ_N. The problem remains open for the general case when on $\phi(u)$ we only impose the regularity conditions of Theorem 2.5.3.

Hoeffding (1973) attacked the problem under slightly more restrictive regularity conditions. His treatment relies on the Bernstein polynomial approximation to absolutely continuous, square-integrable, and monotonic (or the difference of two monotonic) functions. Explicitly, Hoeffding proved the following theorem.

Theorem 2.6.1 (Hoeffding). *Let $\phi(t) = \phi_1(t) - \phi_2(t)$ satisfy the conditions of Theorem 2.5.3 with the square-integrability condition on ϕ_1 and ϕ_2 replaced by*

$$\int_0^1 t^{1/2}(1-t)^{1/2}\, d\phi_k(t) < \infty, \qquad k = 1, 2. \tag{2.6.8}$$

Then (2.5.37) and (2.5.38) hold with EL_N replaced by μ_N in the case of the scores $a_N(i) = E\phi(U_N^{(i)})$, and by

$$\mu'_N = \mu_N + \bar{c}\left\{\sum_{i=1}^{N} \phi\left(\frac{i}{N+1}\right) - N\int_0^1 \phi(t)\, dt\right\}, \tag{2.6.9}$$

in the case of scores $a_N(i) = \phi(i/(N+1))$. If $|\bar{c}_N|/\max_i|c_i - \bar{c}_N|$ is bounded, then EL_N may be replaced by μ_N in the case of scores $a_N(i) = \phi(i/(N+1))$.

Remark. If ϕ is nondecreasing, then integrating by parts in

$$J(\phi) = \int_0^1 t^{1/2}(1-t)^{1/2}\, d\phi(t), \tag{2.6.10}$$

we obtain

$$J(\phi) = \int_0^1 \phi(t)\left(t - \tfrac{1}{2}\right)t^{-1/2}(1-t)^{-1/2}\, dt, \tag{2.6.11}$$

which implies that the condition (2.6.8) is equivalent to

$$\int_0^1 |\phi_k(t)|\, t^{-1/2}(1-t)^{-1/2}\, dt < \infty, \qquad k = 1, 2. \tag{2.6.12}$$

Also note (Exercise 2.6.3) that if ϕ is nondecreasing, then the condition $J(\phi) < \infty$ implies the square-integrability of ϕ, and the condition

$$\int_0^1 \phi^2(t)\{\log(1 + |\phi(t)|)\}^{1+\delta}\, dt < \infty, \qquad \text{for some} \quad \delta > 0$$

implies $J(\phi) < \infty$. In this sense the condition (2.6.8) is not much stronger than square-integrability.

The proof of this theorem is based on the following three lemmas:

Lemma 2.6.2. There is a numerical constant C_1 such that if ϕ is nondecreasing, then

$$\sum_{i=1}^{N} \left| E\phi(U_N^{(i)}) - \phi\left(\frac{i}{N+1}\right)\right| \le C_1 N^{1/2} J(\phi). \tag{2.6.13}$$

Lemma 2.6.3. There is a numerical constant C_2 such that if ϕ is nondecreasing and $F_1, \ldots, F_N$ are any continuous distribution functions, then

$$\sum_{i=1}^{N} \left| E\phi\left(\frac{R_{Ni}}{N+1}\right) - \int_{-\infty}^{\infty} \phi(H(x))\, dF_i(x) \right| \leq C_2 N^{1/2} J(\phi). \quad (2.6.14)$$

Lemma 2.6.4. If ϕ satisfies the conditions of Theorem 2.6.1, then for every $\alpha > 0$ there exists a decomposition

$$\phi(t) = \psi(t) + \phi^{(1)}(t) - \phi^{(2)}(t), \qquad 0 < t < 1, \quad (2.6.15)$$

where ψ is a polynomial, $\phi^{(1)}$ and $\phi^{(2)}$ are nondecreasing, and

$$J(\phi^{(1)}) + J(\phi^{(2)}) < \alpha. \quad (2.6.16)$$

The proofs of Lemmas 2.6.2 and 2.6.3 are given in the Appendix. The proof of Lemma 2.6.4 is analogous to that of Lemma 2.5.4 and is therefore omitted.

Proof of Theorem 2.6.1. First let L_N be defined with $a_N(i) = \phi(i/(N+1))$. Then to prove the theorem with centering constant μ'_N, it suffices to show that for every $\beta > 0$ and $\eta > 0$, there exists a number $N' = N'_{\beta\eta}$ such that whenever

$$N > N' \quad \text{and} \quad \operatorname{Var} L_N > \eta N \max_i (c_i - \bar{c})^2, \quad (2.6.17)$$

we have

$$\frac{|EL_N - \mu'_N|}{(\operatorname{Var} L_N)^{1/2}} < \beta. \quad (2.6.18)$$

To see this select $\varepsilon > 0$ and $\eta > 0$. Then choose $\beta = \beta(\varepsilon)$ such that $\max_x |\Phi(x \pm \beta) - \Phi(x)| < \varepsilon/2$.

Let $N''_{\varepsilon\eta} = \max\{N'_{\beta(\varepsilon)\eta}, N_{\varepsilon/2\eta}\}$ with N defined in the proof of Theorem 2.5.3. Then (2.5.36) with N_η replaced by $N''_{\varepsilon\eta}$ implies (2.5.37) with EL_N replaced by μ'_N.

We write $L_N(\phi)$, $\mu'_N(\phi)$ for L_N and μ'_N to indicate the dependence on ϕ. Since

$$\sum_{i=1}^{N} \phi\left(\frac{R_{Ni}}{N+1}\right) = \sum_{i=1}^{N} \phi\left(\frac{i}{N+1}\right)$$

and

$$\sum_{i=1}^{N} \int \phi(H(x))\, dF_i(x) = N \int_0^1 \phi(t)\, dt,$$

we have from (2.6.9)

$$L_N(\phi) - \mu'_N(\phi) = \sum_{i=1}^{N} (c_i - \bar{c}_N)\left\{\phi\left(\frac{R_{Ni}}{N+1}\right) - \int \phi(H(x))\, dF_i(x)\right\}.$$

Hence

$$\left|EL_N(\phi) - \mu'_N(\phi)\right|$$

$$\leq \max_{1 \leq i \leq N} |c_i - \bar{c}_N| \sum_{i=1}^{N} \left| E\phi\left(\frac{R_{Ni}}{N+1}\right) - \int \phi(H(x))\, dF_i(x)\right|. \quad (2.6.19)$$

Now we apply Lemma 2.6.4 with α to be specified later. Note that

$$\left|EL_N(\phi) - \mu'_N(\phi)\right| \leq \left|EL_N(\psi) - \mu'_N(\psi)\right| + \left|\sum_{k=1}^{2} EL_N(\phi^{(k)}) - \mu'_N(\phi^{(k)})\right|. \quad (2.6.20)$$

Since ψ has a bounded second derivative, it follows by the Taylor expansion that there is a constant $k(\psi)$ such that for every $1 \leq i \leq N$,

$$\left| E\psi\left(\frac{R_{Ni}}{N+1}\right) - \int \psi(H(x))\, dF_i(x)\right| < k(\psi) N^{-1}. \quad (2.6.21)$$

Hence from (2.6.19), with $\phi = \psi$,

$$\left|EL_N(\psi) - \mu'_N(\phi)\right| \leq k(\psi) \max_{1 \leq i \leq N} |c_i - \bar{c}_N|. \quad (2.6.22)$$

From (2.6.19) with $\phi = \phi^{(k)}$, Lemma 2.6.3, and (2.6.16),

$$\sum_{k=1}^{2} \left|EL_N(\phi^{(k)}) - \mu'_N(\phi^{(k)})\right| \leq C_2 N^{1/2} \alpha \max_{1 \leq i \leq N} |c_i - \bar{c}_N|. \quad (2.6.23)$$

If $\operatorname{Var} L_N > N\eta \max_i (c_i - \bar{c}_N)^2$, it follows from (2.6.20), (2.6.22), and

(2.6.23) that

$$\frac{|EL_N(\phi) - \mu'_N(\phi)|}{(\operatorname{Var} L_N)^{1/2}} \le \eta^{-1/2} k(\psi) N^{-1/2} + C_2 \eta^{-1/2} \alpha. \quad (2.6.24)$$

Now, given $\beta > 0$ and $\eta > 0$, choose α in Lemma 2.6.4 so that $C_2\eta^{-1/2}\alpha = \beta/2$. This choice fixes $k(\psi) = k_1(\beta, \eta)$.

Define $N' = N'(\beta, \eta)$ by $\eta^{-1/2} k(\psi)(N')^{-1/2} = \beta/2$. Then (2.6.17) implies (2.6.18), as was to be proved.

We now prove (2.5.66) for the case $a_N(i) = \phi(i/(N+1))$. From (2.6.9)

$$|\mu'_N - \mu_N| \le |\bar{c}_N| \left| \sum_{i=1}^{N} \phi\left(\frac{i}{N+1}\right) - N\int_0^1 \phi(t)\,dt \right|$$

$$= |\bar{c}_N| N \left| \int_0^1 \left\{ \phi\left(\frac{[Nt]+1}{N+1}\right) - \phi(t) \right\} dt \right|. \quad (2.6.25)$$

Assume for the moment that ϕ is nondecreasing. Then

$$N\int_0^1 \left| \phi\left(\frac{[Nt]+1}{N+1}\right) - \phi(t) \right| dt$$

$$\le (N+1)\left\{ \int_{N/(N+1)}^1 \phi(t)\,dt - \int_0^{1/(N+1)} \phi(t)\,dt \right\}$$

$$\le (N+1)^{1/2} \left\{ \int_{N/(N+1)}^1 |\phi(t)| \{t(1-t)\}^{-1/2}\,dt \right.$$

$$\left. + \int_0^{1/(N+1)} |\phi(t)| \{t(1-t)\}^{-1/2}\,dt. \right. \quad (2.6.26)$$

Since $\phi = \phi_1 - \phi_2$ is the difference of two nondecreasing functions which satisfy the condition (2.6.8), it follows using (2.6.25) and (2.6.26) that

$$|\mu'_N - \mu_N| \le |\bar{c}_N| N k_N, \quad (2.6.27)$$

where $k_N = k_N(\phi) \to 0$ as $N \to \infty$. Hence if $\operatorname{Var} L_N > \eta N \max_i (c_i - \bar{c}_N)^2$, then

$$\frac{|\mu'_N - \mu_N|}{(\operatorname{Var} L_N)^{1/2}} < \frac{\eta^{-1/2} \cdot k_N \cdot |\bar{c}_N|}{\max_i |c_i - \bar{c}_N|}, \quad (2.6.28)$$

which is arbitrarily small for N large enough if $|\bar{c}_N|/\max_i|c_i - \bar{c}_N|$ is bounded. This implies (2.5.66).

Finally, consider L_N with $a_N(i) = E\phi(U^{(i)})$. In this case $\sum_{i=1}^N a_N(i) = N\int_0^1 \phi(t)\,dt$. Hence

$$L_N(\phi) - \mu_N(\phi) = \sum_{i=1}^{N} (c_i - \bar{c}_N)\left\{a_N(R_{Ni}) - \int \phi(H(x))\,dF_i(x)\right\} \tag{2.6.29}$$

and

$$\begin{aligned} &|EL_N(\phi) - \mu_N(\phi)| \\ &\qquad \le \max_i |c_i - \bar{c}_N| \sum_{i=1}^{N} \left| Ea_N(R_{Ni}) - \int \phi(H(x))\,dF_i(x)\right|. \end{aligned} \tag{2.6.30}$$

Now

$$\begin{aligned} &\sum_{i=1}^{N} \left| Ea_N(R_i) - \int \phi(H(x))\,dF_i(x)\right| \\ &\qquad \le \sum_{i=1}^{N} \left| E\phi\left(\frac{R_{Ni}}{N+1}\right) - \int \phi(H(x))\,dF_i(x)\right| \\ &\qquad\quad + \sum_{i=1}^{N} \left| E\phi(U_N^{(i)}) - \phi\left(\frac{i}{N+1}\right)\right|. \end{aligned} \tag{2.6.31}$$

For $\phi = \psi$, we apply Taylor's formula to the last term. Since $EU_N^{(i)} = i/(N+1)$ and $\operatorname{Var} U_N^{(i)} < N^{-1}$ for all i, we find there is constant $k'(\psi)$ such that

$$\left| E\psi(U_N^{(i)}) - \psi\left(\frac{i}{N+1}\right)\right| < k'(\psi)N^{-1}, \qquad i = 1, \dots, N.$$

Using (2.6.21), we obtain an inequality analogous to (2.6.22). Applying Lemmas 2.6.2 and 2.6.3 to (2.6.20) with $\phi = \phi^{(k)}$, $k = 1, 2$, and using Lemma 2.6.3, we obtain an inequality analogous to (2.6.23). The theorem follows as in the first part of the proof.

2.7. ASYMPTOTIC NORMALITY OF L_N UNDER CONTIGUOUS ALTERNATIVES

In the study of the asymptotic power and level of significance of statistical tests, the limiting distributions of allied test statistics play a vital role. It may be remarked that if a test is consistent against any (fixed) alternative, then the power of the test approaches its upper asymptotic value 1 as the sample sizes increase. Hence, for the study of the asymptotic power properties of a (consistent) test, often, one confines oneself to some local alternatives for which the asymptotic power is different from 1. Naturally, such local alternatives are specified in such a manner that as the sample sizes increase they converge to the null hypothesis at a certain rate so as to insure a nondegenerate limit for the power function of the allied tests. In the specification of such a sequence of local alternatives, the concept of contiguity of probability measures (originally developed by LeCam, 1960) has been very successfully employed by Håjek (1962). He has studied the asymptotic optimality and normality of linear rank statistics under such contiguous alternatives, and a very nice account of this is given in Chapter VI of Håjek and Šidåk (1967). Basically, contiguity of probability measures rests on somewhat more restrictive regularity conditions on the underlying d.f.'s (than the ones treated in Section 2.5), but allows the regularity conditions on the score functions to be weaker than the ones treated there, and furthermore, the ultimate results are simpler and easier to interpret. For the sake of completeness, we provide here a brief treatment of the notion of contiguity of probability measures and stress its role in the asymptotic theory of rank statistics. For a somewhat comprehensive study of this notation along with a variety of its statistical applications, the reader is referred to Roussas (1972).

Consider a sequence $\{p_N, q_N\}$ of simple hypotheses p_N and simple alternatives q_N defined on measure spaces $(\mathscr{X}_N, \mathscr{A}_N, \mu_N)$, $N \geq 1$, respectively. If for any sequence of events $\{A_N\} : A_N \in \mathscr{A}_N$,

$$P_N(A_N) \to 0 \quad \Rightarrow \quad Q_N(A_N) \to 0, \tag{2.7.1}$$

we say that the densities q_N are contiguous to p_N; here, $dP_N/d\mu_N = p_N$ and $dQ_N/d\mu_N = q_N$, $N \geq 1$. The contiguity of $\{q_N\}$ to $\{p_N\}$ (or $\{Q_N\}$ to $\{P_N\}$) implies that for any $\mathscr{A}_N$-measurable (r.v.) T_N, $T_N \to 0$ in P_N-probability insures that $T_N \to 0$ in Q_N-probability as well. Note that in (2.7.1) we have given a one-sided version: According to this version, contiguity of $\{Q_N\}$ to $\{P_N\}$ does not necessarily imply the contiguity of $\{P_N\}$ to $\{Q_N\}$. Further, if we consider the L_1-norm of $P_N - Q_N$, i.e., $\|P_N - Q_N\| = \sup\{|P_N(A_N) - Q_N(A_N)| : A_N \in \mathscr{A}_N\}$, then $\|P_N - Q_N\| \to 0$ implies

(2.7.1), but, the converse is not necessarily true. Thus, contiguity is weaker than the L_1-norm equivalence. Finally, the contiguity in (2.7.1) does not imply that the Q_N are absolutely continuous with respect to the P_N, but the singular part of Q_N, i.e., $Q_N(p_N = 0)$—must tend to 0 as $N \to \infty$ [as $P_N(p_N = 0) = 0$].

The LeCam (1960) characterization of the contiguity of $\{Q_N\}$ to $\{P_N\}$ rests on the following construction of the likelihood-ratio statistics L_N. Let L_N be equal to q_N/p_N, 1, or ∞, according as $p_N > 0$, $p_N = q_N = 0$, or $p_N = 0 < q_N$. Also, let $G_N(t) = P_N\{L_N \le t\}$, $t \in R^+$, be the d.f. of L_N under the P_N-probability. Further, let $\mathcal{N}(a, b)$ stand for the normal distribution with mean a and variance b. Then we have the following

Lemma 2.7.1. (LeCam's first lemma). If $\{G_N\}$ converges weakly to a distribution function G such that $\int_0^\infty t\,dG(t) = 1$, then $\{Q_N\}$ is contiguous to $\{P_N\}$. In particular, if $\log L_N$ has asymptotically $\mathcal{N}(-\frac{1}{2}\sigma^2, \sigma^2)$ for some $\sigma \in (0, \infty)$, then $\{Q_N\}$ is contiguous to $\{P_N\}$.

(Note that if $\log Y$ has the normal distribution with mean a and variance b, and if $a = -b/2$, then $EY = 1$.) For a formal proof of the lemma, we may refer to Hâjek and Šidak (1967, pp. 203–204). Verification of the second part of the lemma is greatly facilitated with the aid of the following result, where the p_N and q_N are conceived as product densities. Let $\mathbf{X}_N = (X_1, \ldots, X_N)$, $N \ge 1$, and assume that

$$p_N(\mathbf{X}_N) = \prod_{i=1}^{N} f_{Ni}(X_i) \quad \text{and} \quad q_N(\mathbf{X}_N) = \prod_{i=1}^{N} g_{Ni}(X_i). \tag{2.7.2}$$

Then

$$\log L_N = \sum_{i=1}^{N} \log \frac{g_{Ni}(X_i)}{f_{Ni}(X_i)}. \tag{2.7.3}$$

We assume that the summands in (2.7.3) are uniformly asymptotically negligible (UAN), i.e., for every $\varepsilon > 0$.

$$\lim_{N\to\infty} \max_{1 \le i \le N} P_N\left\{\left|\frac{g_{Ni}(X_i)}{f_{Ni}(X_i)} - 1\right| > \varepsilon\right\} = 0. \tag{2.7.4}$$

Let us then denote

$$W_N = 2\sum_{i=1}^{N}\left\{\left[\frac{g_{Ni}(X_i)}{f_{Ni}(X_i)}\right]^{1/2} - 1\right\}. \tag{2.7.5}$$

Note that the UAN condition in (2.7.4) underlies the asymptotic normality of W_N. The following lemma depicts the asymptotic relationship between $\log L_N$ and W_N.

Lemma 2.7.2. (LeCam's second lemma). Assume that (2.7.4) holds, and under P_N, W_N is asymptotically $\mathcal{N}(-\frac{1}{4}\sigma^2, \sigma^2)$. Then,

$$\lim_{N\to\infty} P_N\big(|\log L_N - W_N + \tfrac{1}{4}\sigma^2| > \varepsilon\big) = 0 \qquad \forall\varepsilon > 0, \qquad (2.7.6)$$

and

$$\log L_N \text{ is asymptotically } \mathcal{N}\big(-\tfrac{1}{2}\sigma^2, \sigma^2\big) \qquad \text{under } P_N. \qquad (2.7.7)$$

We again refer to Hájek and Šidák (1967, pp. 205–208) for a detailed proof of the lemma. The asymptotic normality of the statistic $\log L_N$ not only provides the means to verify the contiguity of $\{Q_N\}$ to $\{P_N\}$, but it can also be employed in providing the asymptotic distribution of other statistics under Q_N-probability as well. In this context, we have the following result due to LeCam (1960); for its proof we may again refer to Hájek and Šidák (1967, pp. 208–210).

Lemma 2.7.3 (LeCam's third lemma). Let S_N be an $\mathscr{A}_N$-measurable r.v., and $\log L_N$ be defined as above. Assume that under P_N, $(S_N, \log L_N)$ has asymptotically a bivariate normal distribution with mean vector $\boldsymbol{\mu} = (\mu_1, \mu_2)'$ and dispersion matrix $\boldsymbol{\Sigma} = ((\sigma_{ij}))$, where $\mu_2 = -\frac{1}{2}\sigma_{22}$. Then, under Q_N, S_N is asymptotically normal with mean $\mu_1 + \sigma_{12}$ and variance σ_{11}.

The ingenuity of the Hájek (1962) approach to the study of the asymptotic normality (and optimality) of linear rank statistics (under contiguous alternatives) lies in combining a convenient approximation of L_N by a sum of independent r.v.'s with Lemma 2.7.3 for deriving the desired results. With this in mind, we define

$$S_N = \sum_{i=1}^{N} (c_i - \bar{c}_N)\phi(F(X_i)), \qquad N \geq 1, \qquad (2.7.8)$$

$$\phi_N(u) = a_N(i) \quad \text{for } \frac{i-1}{N} < u \leq \frac{i}{N}, \qquad i = 1, \dots, N \quad N \geq 1, \qquad (2.7.9)$$

where the scores $a_N(i)$ and the score function ϕ are defined as in Section

2.2, and F is the d.f. under the null hypothesis $H_0: F_1 = \cdots = F_N = F$. We assume that

$$\lim_{N\to\infty}\left\{\int_0^1[\phi_N(u)-\phi(u)]^2\,du\right\} = 0. \tag{2.7.10}$$

Note that for (2.7.10), though we need the square-integrability of the score function, in contrast with Sections 2.4 and 2.5 we do not require their absolute continuity. Against the null hypothesis H_0, we consider an arbitrary sequence $\{K_N\}$ of alternatives, such that the two probability measures P_N and Q_N are contiguous in the sense of (2.7.1): thus we term $\{K_N\}$ a sequence of contiguous alternatives.

Theorem 2.7.4. *Under* (2.7.10) *and any contiguous* $\{K_N\}$,

$$C_N^{-1}|L_N - S_N| \to 0 \quad \textit{in probability} \qquad \textit{as} \quad N\to\infty, \tag{2.7.11}$$

where $C_N^2 = \sum_{i=1}^N (c_i - \bar{c}_N)^2$.

Proof. First, we consider the special case of the scores defined by (2.2.2). Let

$$a_N^0(i) = E\phi\big(U_n^{(i)}\big), \quad \phi_N^0(u) = a_N^0(i) \text{ for } \frac{i-1}{N} < u \le \frac{i}{N},$$
$$i = 1,\dots,N, \tag{2.7.12}$$

where the $U_n^{(i)}$ are defined after (2.2.2). Then

$$N^{-1}\sum_{i=1}^N a_N^0(i) = \bar{\phi} = \int_0^1 \phi(u)\,du \qquad \forall N \ge 1, \tag{2.7.13}$$

$$N^{-1}\sum_{i=1}^N \{a_N^0(i)\}^2 \le \int_0^1 \phi^2(u)\,du < \infty \qquad \forall N \ge 1, \tag{2.7.14}$$

while

$$\phi_N(u) \to \phi(u) \quad \text{as } N\to\infty \qquad \text{for every} \quad u \in (0,1). \tag{2.7.15}$$

Hence, by the Fatou Lemma, we obtain that

$$\int_0^1 \{\phi_N^0(u)\}^2\,du \to \int_0^1 \phi^2(u)\,du \qquad \text{as} \quad N\to\infty. \tag{2.7.16}$$

Thus, writing $L_N^0 = L_N$ in (2.2.3) with the scores in (2.7.12) and letting $\mathbf{R}_N = (R_{N1}, \ldots, R_{NN})$, we note that $L_N^0 = E_0\{S_N | \mathbf{R}_N\}$, so that

$$E_0\left(S_N - L_N^0\right)^2 = E_0 S_N^2 - E_0 L_N^{02} = C_N^2\left(\int_0^1 \phi^2(u)\,du - \int_0^1 \{\phi_N^0(u)\}^2\,du\right), \tag{2.7.17}$$

we obtain from (2.7.16), (2.7.17), and the Chebyshev inequality that under H_0,

$$C_N^{-1}\left(S_N - L_N^0\right) \to 0 \quad \text{in probability} \quad \text{as} \quad N \to \infty. \tag{2.7.18}$$

Let us next consider the general case of L_N for which (2.7.10) holds. Note that $L_N - L_N^0$ is a linear rank statistic with the scores $b_N(i) = a_N(i) - a_N^0(i)$, $i = 1, \ldots, N$, and hence,

$$\begin{aligned} E_0\left(L_N - L_N^0\right)^2 &\le (N-1)^{-1} N C_N^2 \left\{ N^{-1} \sum_{i=1}^{N} \left[a_N(i) - a_N^0(i)\right]^2 \right\} \\ &= (N-1)^{-1} N C_N^2 \left\{ \int_0^1 \left[\phi_N(u) - \phi_N^0(u)\right]^2 du \right\} \\ &\le 2(N-1)^{-1} N C_N^2 \left\{ \int_0^1 \left[\phi_N(u) - \phi(u) - \phi(u)\right]^2 du \right. \\ &\qquad \left. + \int_0^1 \left[\phi_N^0(u) - \phi(u)\right]^2 du \right\} \\ &= 2(N-1)^{-1} N C_N^2 \left\{ \int_0^1 \left[\phi_N(u) - \phi(u)\right]^2 du \right. \\ &\qquad \left. + \int_0^1 \phi^2(u)\,du - \int_0^1 \left[\phi_N^0(u)\right]^2 du \right\}, \end{aligned} \tag{2.7.19}$$

so that by (2.7.10), (2.7.16), and the Chebyshev inequality, under H_0,

$$C_N^{-1}\left(L_N - L_N^0\right) \to 0 \quad \text{in probability} \quad \text{as} \quad N \to \infty. \tag{2.7.20}$$

From (2.7.18) and (2.7.20), we obtain that under H_0 and (2.7.10),

$$C_N^{-1}\left(L_N - S_N\right) \to 0 \quad \text{in probability} \quad \text{as} \quad N \to \infty. \tag{2.7.21}$$

Finally, the contiguity of the alternative hypothesis $\{K_N\}$ (to H_0) insures that (2.7.21) also holds under $\{K_N\}$. Q.E.D.

By virtue of Theorem 2.7.4, in order to study the asymptotic normality of $C_N^{-1}L_N$ (under the null hypothesis or any contiguous alternative), it suffices to study the asymptotic normality of $C_N^{-1}S_N$. Towards this, we assume that the Noether condition, viz.

$$\lim_{N\to\infty} C_N^{-2}\left\{\max_{1\le i\le N}(c_i-\bar{c}_N)^2\right\}=0, \tag{2.7.22}$$

holds, and the score function ϕ satisfies the following:

$$0<A^2=\int_0^1\phi^2(u)\,du-\bar{\phi}^2<\infty. \tag{2.7.23}$$

Note that if we let $c_{Ni}=(c_i-\bar{c}_N)/C_N$ and $Z_{Ni}=c_{Ni}\phi(F(X_i))$, $i=1,\dots,N$, then $C_N^{-1}S_N=\sum_{i=1}^N Z_{Ni}$, where the Z_{Ni} are UAN, and further, by (2.7.22) and (2.7.23), letting $c_N^*=\max_{1\le i\le N}|c_{Ni}^*|$, under H_0, for every $\varepsilon>0$,

$$\begin{aligned}\sum_{i=1}^N P\{|Z_{Ni}|>\varepsilon\} &\le \varepsilon^{-2}\sum_{i=1}^N c_{Ni}^2 E\left\{\phi^2(F(X_i))I\left(|\phi(F(X_i))|>\frac{\varepsilon}{c_N^*}\right)\right\}\\ &\le \varepsilon^{-2}E\left\{\phi^2(F(X_1))I\left(|\phi(F(X_1))|>\frac{\varepsilon}{c_N^*}\right)\right\}\to 0\end{aligned} \tag{2.7.24}$$

as $N\to\infty$, and, similarly,

$$A^{-2}\sum_{i=1}^N\left(E\left[Z_{Ni}^2 I(|Z_{Ni}|<\varepsilon)\right]-\left\{E\left[Z_{Ni}I(|Z_{Ni}|<\varepsilon)\right]\right\}^2\right)\to 1 \tag{2.7.25}$$

as $N\to\infty$. Therefore, by the classical central limit theorem, under (2.7.22), (2.7.23), and H_0,

$$C_N^{-1}S_N \text{ is asymptotically } \mathcal{N}(0,A^2). \tag{2.7.26}$$

By (2.7.11) and (2.7.26), the asymptotic normality of $C_N^{-1}L_N$ (under H_0)

follows. Thus, it remains to introduce suitable families of contiguous alternatives and to establish the asymptotic normality of $C_N^{-1}S_N$ under such alternatives. For this purpose, we essentially follow Hájek (1962) and incorporate Lemmas 2.7.1, 2.7.2, and 2.7.3 in the formulation of these alternatives and the desired asymptotic results.

Consider a (double) sequence $\{X_{N1}, \ldots, X_{NN}, N \geq 1\}$ of (row-wise) independent r.v.'s having d.f.'s $\{F_{N1}, \ldots, F_{NN}, N \geq 1\}$, where

$$F_{Ni}(x) = F(x - d_{Ni}), \quad 1 \leq i \leq N, \qquad N \geq 1, \quad x \in R. \tag{2.7.27}$$

Suppose F possesses an absolutely continuous probability density function (pdf) f with a finite Fisher information

$$I(f) = \int_{-\infty}^{\infty} \left\{ \frac{f'(x)}{f(x)} \right\}^2 dF(x) \qquad (< \infty), \tag{2.7.28}$$

where $f'(x) = (d/dx)f(x) = (d^2/dx^2)F(x)$ exists a.e., and suppose the constants $\{d_{Ni}\}$ satisfy the following condition. Let

$$\bar{d}_N = N^{-1} \sum_{i=1}^{N} d_{Ni} \text{ and } D_N^2 = \sum_{i=1}^{N} (d_{Ni} - \bar{d}_N)^2, \qquad N \geq 1. \tag{2.7.29}$$

We assume that

$$\sup_N D_N^2 < \infty \quad \text{and} \quad \lim_{N \to \infty} \left\{ \max_{1 \leq i \leq N} (d_{Ni} - \bar{d}_N)^2 \right\} = 0. \tag{2.7.30}$$

Let $\{q_N\}$ be the sequence of (joint) probability densities under (2.7.27), and let $\{p_N\}$ be the corresponding sequence when in (2.7.27) we let $d_{Ni} = \bar{d}_N$ for $i = 1, \ldots, N$, $N \geq 1$. Then, as a first step, to establish the contiguity of $\{q_N\}$ to $\{p_N\}$, we appeal to Lemmas 2.7.1 and 2.7.2. [Note that $\{p_N\}$ corresponds to the null-hypothesis case where the $X_{N1}, \ldots, X_{NN}$ are i.i.d.r.v.'s with the d.f. $F(x - \bar{d}_N)$, $x \in R$.] To establish the contiguity of $\{q_N\}$ to $\{p_N\}$, it suffices to show that (2.7.7) holds, and for that, it suffices to show that (2.7.4) holds and for $g(X_{Ni})/f(X_{Ni}) = f(X_{Ni} - \bar{d}_N - (d_{Ni} - \bar{d}_N))/f(X_{Ni} - \bar{d}_N)$, $i = 1, \ldots, N$, W_N defined by (2.7.5) is asymptotically normal as in Lemma 2.7.2. Since f is absolutely continuous, for (2.7.4) to hold, (2.7.30) suffices. Hence, we proceed to verify the asymptotic normality of W_N. Towards this, let

$$S_N^* = \sum_{i=1}^{N} (d_{Ni} - \bar{d}_N) \frac{-f'(X_{Ni} - \bar{d}_N)}{f(X_{Ni} - \bar{d}_N)}, \qquad N \geq 1, \tag{2.7.31}$$

and

$$s(x) = [f(x)]^{1/2}, \qquad x \in R, \tag{2.7.32}$$

so that $s'(x) = \frac{1}{2}f'(x)/[f(x)]^{1/2}$ and by (2.7.28), $I(f) = 4\int_{-\infty}^{\infty}[s'(x)]^2\,dx$. Note that under (2.7.28), (2.7.30), and $\{P_N\}$,

$$\frac{S_N^*}{D_N} \text{ is asymptotically } \mathcal{N}(0, I(f)). \tag{2.7.33}$$

Hence, it suffices to show that under (2.7.28), (2.7.30), and $\{P_N\}$, as $N \to \infty$,

$$\left|EW_N + \tfrac{1}{4}D_N^2 I(f)\right| \to 0 \text{ and } \operatorname{Var}\left(W_N - S_N^*\right) \to 0. \tag{2.7.34}$$

Note that by (2.7.5), (2.7.27), and (2.7.32), under $\{P_N\}$,

$$\begin{aligned} EW_N &= 2\sum_{i=1}^{N}\int_{-\infty}^{\infty}[s(x - \bar{d}_N - (d_{Ni} - \bar{d}_N)) - s(x - \bar{d}_N)]s(x - \bar{d}_N)\,dx \\ &= -\sum_{i=1}^{N}(d_{Ni} - \bar{d}_N)^2\int_{-\infty}^{\infty}\left[\frac{s(y - (d_{Ni} - \bar{d}_N)) - s(y)}{d_{Ni} - \bar{d}_N}\right]^2 dy, \end{aligned} \tag{2.7.35}$$

and hence, for the first part of (2.7.34), it suffices to show that

$$\lim_{h \to 0}\int_{-\infty}^{\infty}\left[\frac{s(y+h) - s(y)}{h}\right]^2 dy = \int_{-\infty}^{\infty}[s'(y)]^2\,dy = \tfrac{1}{4}I(f) < \infty. \tag{2.7.36}$$

Towards this, note that by (2.7.28) and (2.7.32), $s(x)$ is absolutely continuous, so that $\lim_{h\to 0}\{s(x+h) - s(x)\}/h = s'(x)$ a.e., and further, for every $h > 0$,

$$\begin{aligned} \int_{-\infty}^{\infty}\left[\frac{s(x+h) - s(x)}{h}\right]^2 dx &= \int_{-\infty}^{\infty}\left[h^{-1}\int_0^h s'(x+t)\,dt\right]^2 dx \\ &\le h^{-1}\int_0^h\int_{-\infty}^{\infty}[s'(x+t)]^2\,dx\,dt \\ &= \int_{-\infty}^{\infty}[s'(y)]^2\,dy = \tfrac{1}{4}I(f) < \infty. \end{aligned} \tag{2.7.37}$$

A similar case holds for $h < 0$. Hence, (2.7.36) follows from the above and Fatou's lemma. To prove the second part of (2.7.34), note that both W_N and S_N^* involve independent summands, and hence,

$$\begin{aligned}
\operatorname{Var}\left(W_N - S_N^*\right) &= 4\sum_{i=1}^{N} \operatorname{Var}\left\{\frac{s(X_{Ni} - d_{Ni})}{s(X_{Ni} - \bar{d}_N)} - 1 \right. \\
&\qquad \left. + \tfrac{1}{2}(d_{Ni} - \bar{d}_N)\frac{f'(X_{Ni} - \bar{d}_N)}{f(X_{Ni} - \bar{d}_N)}\right\} \\
&\le 4\sum_{i=1}^{N} E\left\{\frac{s(X_{Ni} - d_{Ni})}{s(X_{Ni} - \bar{d}_N)} - 1 \right. \\
&\qquad \left. + (d_{Ni} - \bar{d}_N)\frac{s'(X_{Ni} - \bar{d}_N)}{s(X_{Ni} - \bar{d}_N)}\right\}^2 \\
&= 4\sum_{i=1}^{N} (d_{Ni} - \bar{d}_N)^2 \int_{-\infty}^{\infty} \left\{\frac{s(x - (d_{Ni} - \bar{d}_N))}{(d_{Ni} - \bar{d}_N)s(x)} - 1 \right. \\
&\qquad \left. + \frac{s'(x)}{s(x)}\right\}^2 f(x)\,dx \\
&= 4\sum_{i=1}^{N} (d_{Ni} - \bar{d}_N)^2 \\
&\quad \times \int_{-\infty}^{\infty} \left[\frac{s(x - (d_{Ni} - \bar{d}_N)) - s(x)}{d_{Ni} - \bar{d}_N} - s'(x)\right]^2 dx.
\end{aligned} \tag{2.7.38}$$

Now, as in (2.7.36)–(2.7.37), the integral in (2.7.38) converges to 0 (uniformly in $i: 1 \le i \le N$), as $N \to \infty$, while, by (2.7.30), $\sup_N D_N^2 < \infty$. Hence, $\operatorname{Var}(W_N - S_N^*)$ converges to 0 as $N \to \infty$ (when H_0 holds). Thus, under (2.7.27), (2.7.28), (2.7.30), and $H_0\colon F_{N1} = \cdots = F_{NN}$, $\forall N \ge 1$,

$$W_N \text{ is asymptotically } \mathcal{N}\left(-\tfrac{1}{4}D_N^2 I(f),\, D_N^2 I(f)\right), \tag{2.7.39}$$

so that by Lemma 2.7.2, the contiguity of $\{q_N\}$ to $\{p_N\}$ is established.

We now proceed to verify Lemma 2.7.3 for S_N. We may note that by (2.7.6) and (2.7.34), the joint asymptotic normality of $(C_N^{-1}S_N, \log L_N)$, under $\{P_N\}$, would follow from that of $(C_N^{-1}S_N, D_N^{-1}S_N^*)$, where, for the vector $(C_N^{-1}S_N, D_N^{-1}S_N^*)$, expressible as the sum of independent vectors, the (bivariate) central limit theorem directly holds under (2.7.22), (2.7.23), (2.7.28), and (2.7.30) (where we assume that H_0 holds), and hence,

$$\left(C_N^{-1}S_N, D_N^{-1}S_N^*\right) \text{ is asymptotically } \mathcal{N}_2(\mathbf{0}, \Gamma_N), \tag{2.7.40}$$

where

$$\Gamma_N = \begin{pmatrix} A^2 & \gamma\rho_N \\ \gamma\rho_N & I(f) \end{pmatrix}, \qquad \gamma = \int_{-\infty}^{\infty} \phi(F(x))\left[-\frac{f'(x)}{f(x)}\right] dF(x), \tag{2.7.41}$$

and

$$\rho_N = C_N^{-1}D_N^{-1}\sum_{i=1}^{N}(c_i - \bar{c}_N)(d_{Ni} - \bar{d}_N). \tag{2.7.42}$$

Let us now denote $\{K_N\}$ the sequence of alternative hypotheses in (2.7.27). Then, from Lemma 2.7.3 and the above discussion, we arrive at the following.

Theorem 2.7.5. [Hájek, 1962]. *Under* (2.7.22), (2.7.23), (2.7.28), (2.7.30), *and* $\{K_N\}$ *in* (2.7.27), $\{C_N^{-1}S_N - D_N\rho_N\gamma\}/A$ *has asymptotically the standard normal distribution, and hence additionally, by* (2.7.11), *under* (2.7.10),

$$C_N^{-1}L_N - D_N\rho_N\gamma \text{ is asymptotically } \mathcal{N}(0, A^2). \tag{2.7.43}$$

In (2.7.27), the constants d_{Ni} have been chosen rather arbitrarily [satisfying (2.7.30)]. In many practical problems, there are natural choices of these constants. For example, in the two-sample location problem, $N = n_1 + n_2$, $F_{N1} = F_{Nn_1} = F$, $F_{Nn_1+1} = \cdots = F_{NN} = G$, and $G(x) = F(x - \Delta)$. Here, the null hypothesis H_0 reduces to $\Delta = 0$. For local alternatives of the type $K_N: \Delta = N^{-1/2}\lambda$, $\lambda \neq 0$ (fixed), we have $d_{N1} = \cdots = d_{Nn_1} = 0$ and $d_{Nn_1+1} = \cdots = d_{NN} = N^{-1/2}\lambda$, for which (2.7.30) holds. Similarly, for the simple regression model, $F_i(x) = F(x - \beta_0 - \beta c_i)$, $i = 1, \ldots, N$, the null hypothesis H_0 reduces to $\beta = 0$. Again, for local alternatives of the type $K_N: \beta = C_N^{-1}\lambda$, $\lambda \neq 0$ (fixed) [where $C_N^2 = \sum_{i=1}^N (c_i - \bar{c}_N)^2$], we have $d_{Ni} = (\beta_0 + \lambda\bar{c}_N/C_N) + \lambda(c_i - \bar{c}_N)/C_N$, $i = 1, \ldots, N$, for which (2.7.30) holds under (2.7.22) and (2.7.23). Further, here ρ_N defined by (2.7.42) is equal to 1. In general, we may also consider a multiple-regression model where βc_i is to be replaced by $\boldsymbol{\beta}'\mathbf{c}_i$ (both $\boldsymbol{\beta}$ and $\mathbf{c}_i$ being q-vectors, for some $q \geq 1$) and define the d_{Ni} in an analogous way. Some of these will be treated in detail in Chapters 5, 6, and 7. For some other problems, such as the two-sample scale problem, a somewhat different formulation of (2.7.27) is needed, and we pose this in the form of an exercise at the end of this chapter.

We conclude this section with the remark that for (2.7.43), the score function ϕ is assumed to satisfy (2.7.10) and (2.7.23). These are weaker than the regularity conditions on ϕ, imposed in the earlier sections; in particular, the absolute continuity of the score function is not needed here. On the other hand, the regularity conditions on the underlying d.f.'s are more stringent here than in the earlier sections. Moreover, we confine ourselves here to local alternatives which are not needed in the earlier sections.

EXERCISES

2.3.1. Use the definitions of $\bar{\alpha}(X)$ and $\bar{\beta}(X)$ to verify (2.3.11).

2.3.2. A pair (X, Y) of r.v.'s or its (joint) d.f. F is said to be *positively quadrant-dependent* if $P(X \leq x,\ Y \leq y) \geq P(X \leq x)P(Y \leq y)$ for all x, y. The dependence is strict if inequality holds for at least some pair (x, y). Show that if (X, Y) is positively quadrant-dependent and the expectations in the following formula exist, then $E(XY) \geq E(X)E(Y)$, with the strict equality holding iff X and Y are independent. (Lehmann, 1966.)

2.3.3. Use the inequality in the preceding problem to provide an alternative proof of (2.3.4).

2.3.4. Use the decomposition that $\mathrm{Cov}(X, Y) = E(XY) - E(X)E(Y)$ for each of the two terms on the rhs of (2.3.6) and verify the equality of the two sides.

2.4.1. Write down the expressions for $\mathrm{Var}\, S_{Ni}$ and $\mathrm{Var}\, E(S_{Ni}|X_i)$, and hence (or otherwise) verify (2.4.11).

2.5.1. Verify the inequalities in (2.5.9), (2.5.10), (2.5.11), and (2.5.12). (Hájek, 1968.)

2.5.2. Provide a proof of (2.5.59). (Hájek, 1968.)

2.5.3. Provide a proof of Corollary 2.5.3. (Hájek, 1968.)

2.6.1. Suppose that the c_i satisfy the (Hájek) condition in Corollary 2.5.3 and for some $r > 2$, $\int_0^1 |\phi(u)|^r\, du < \infty$. Then show that (2.6.6) holds. (Sen and Ghosh, 1972.)

2.6.2. Show that the condition that $\phi \in L_r$ for some $r > 2$ in the preceding problem can be replaced by the following: ϕ is of bounded variation on any closed interval in [0, 1] with $\int_0^1|\phi(u)|\,du < \infty$. [For the special case of the two-sample problem, this result is due to Hájek (1974); the general result, due to M. Ghosh and Sen, is reported in Theorem 4.6.1 of Sen (1981a).]

2.6.3. Define $J(\phi)$ as in (2.6.11), and show that the condition $J(\phi) < \infty$ implies the square-integrability of ϕ, and the condition that $\int_0^1\phi^2(t)\{\log(1+|\phi(t)|)\}^{1+\delta}\,dt < \infty$ for some $\delta > 0$ implies that $J(\phi) < \infty$. (Hoeffding, 1973.)

2.6.4. Provide a proof of Lemma 2.6.4.

2.7.1. Provide a proof of Lemma 2.7.1. (Hájek, 1962.)

2.7.2. Provide a proof of Lemma 2.7.2. (Hájek, 1962.)

2.7.3. Provide a proof of Lemma 2.7.3. (Hájek, 1962.)

2.7.4. Under (2.7.24) through (2.7.30), with $g(X_{Ni}) = f(X_{Ni} - \bar{d}_N - (d_{Ni} - \bar{d}_N))$ and $f(X_{Ni}) = f(X_{Ni} - \bar{d}_N)$, $i \geq 1$, $N \geq 1$, verify (2.7.4), and hence (2.7.6) and (2.7.7). (Hájek, 1962.)

2.7.5. Consider the general multiple-regression model $F_i(x) = F(x - \beta_0 - \boldsymbol{\beta}'\mathbf{c}_i)$, $i \geq 1$, where the $\mathbf{c}_i$ are specified q-vectors ($q \geq 1$) and β_0 and $\boldsymbol{\beta}$ are unknown parameters. Let $\bar{\mathbf{c}}_N = N^{-1}\sum_{i=1}^N \mathbf{c}_i$ and $\mathbf{C}_N = \sum_{i=1}^N(\mathbf{c}_i - \bar{\mathbf{c}}_N)(\mathbf{c}_i - \bar{\mathbf{c}}_N)'$. For testing the null hypothesis $H_0: \boldsymbol{\beta} = \mathbf{0}$ against $H_N: \boldsymbol{\beta} = \mathbf{C}_N^{-1/2}\boldsymbol{\lambda}$, for some fixed $\boldsymbol{\lambda}$ ($\in R^q$), let $d_{Ni} = \boldsymbol{\lambda}'\mathbf{C}_N^{-1/2}(\mathbf{c}_i - \bar{\mathbf{c}}_N)$, $i \geq 1$. Verify the contiguity of the sequence of probability measures under $\{H_N\}$ to that under H_0 by imposing suitable conditions on the $\mathbf{c}_i$ and the pdf f. Show that the finite Fisher information on f suffices, and the Noether condition, extended to the vector case, for the $\mathbf{c}_i$ yields the desired result.

2.7.6. Consider the two-sample scale problem. Extend the notion of finite Fisher information to this problem, and verify the contiguity under this extended condition. (Hájek and Šidák, 1967.)

CHAPTER 3

Distribution Theory of Signed Rank-Order Statistics

3.1. INTRODUCTION

The linear rank statistics, considered in Chapter 2, depend on the observations only through their ranks. In dealing with the problem of testing the symmetry of a distribution or estimating its location—and in many other problems, as we shall see later on—in addition to the ranks, the signs of the observations play a vital role. In this chapter we consider statistics which depend on the observations through their ranks as well as their signs, and are therefore termed the signed rank statistics. Section 3.2 deals with the preliminary notions. The basic variance inequality and the projection approximation (for bounded scores) are studied in Section 3.3. The last section deals with the asymptotic normality results. Though the general case of arbitrary continuous distributions is considered in detail, a brief account for local alternatives is furnished for some uses in subsequent chapters. A substantial portion of the material in this chapter is due to Hušková (1970).

3.2. PRELIMINARY NOTIONS

Let $X_1, \ldots, X_N$ be independent random variables with continuous d.f.'s $F_1(x), \ldots, F_N(x)$, respectively, all defined on the real line $(-\infty, \infty)$. Define the sign function

$$\operatorname{sgn} \nu = s(\nu) = \begin{cases} 1, & \nu > 0, \\ 0, & \nu = 0, \\ -1, & \nu < 0, \end{cases} \tag{3.2.1}$$

and let $u(t)$ be equal to 1 or 0 according as $t \geq 0$ or < 0. Then

$$R_{Ni}^{+} = \sum_{j=1}^{N} u(|X_i| - |X_j|) \tag{3.2.2}$$

is the rank of $|X_i|$ among $|X_1|, \ldots, |X_N|$ for $i = 1, \ldots, N$. Here also, by virtue of the assumed continuity of $F_1, \ldots, F_N$, ties among $|X_1|, \ldots, |X_N|$ can be neglected with probability 1, so that $(R_{N1}^+, \ldots, R_{NN}^+)$ represent some permutation of $(1, \ldots, N)$. We shall be concerned with the statistics

$$S_N^+ = \sum_{i=1}^{N} c_i \operatorname{sgn}(X_i)\, a_N(R_{Ni}^+), \qquad N \geq 1, \tag{3.2.3}$$

where $\{c_1, \ldots, c_N\}$ are arbitrary (known) regression constants, and the scores $a_N(1), \ldots, a_N(N)$ are defined in the same manner as in (2.2.2). Here also, the underlying score function is denoted by $\phi(u)$, $0 < u < 1$.

In the particular case of $c_1 = \cdots = c_N = 1$, S_N^+ reduces to the one-sample rank-order statistic, which, in the general multivariate case, has been studied in detail in Chapter 4 of Puri and Sen (1971). Well-known cases of S_N^+ are the Wilcoxon signed rank statistic and the normal scores statistic, which correspond to $\phi(u) = u$ and $\phi(u) = \chi_1^{-1}(u)$, $0 < u < 1$, respectively, where $\chi_1(x)$ is the chi distribution with 1 degree of freedom. In this chapter, we consider the general case where $c_1, \ldots, c_N$ need not be identical.

We start with the following lemma, whose proof is analogous to that of Lemma 2.3.3 and is therefore left as an exercise.

Lemma 3.2.1. Let $F_i^*(x) = P\{|X_i| \leq x\}$, $0 \leq x < \infty$, $1 \leq i \leq N$, and let $X_1, \ldots, X_N$ be independent with d.f.'s $F_1, \ldots, F_N$. Then, for arbitrary scores $a_N(1), \ldots, a_N(N)$,

$$\begin{aligned} &E\big[a_N(R_{Ni}^+)s(X_i)|X_i = x,\ X_j = y\big] - E\big[a_N(R_{Ni}^+)s(X_i)|X_i = x\big] \\ &\quad = s(x)\big[u(|x| - |y|) - F_j^*(|x|)\big] \\ &\qquad \times \sum_{k=2}^{N} \{a_N(k) - a_N(k-1)\} P\big(R_{Ni}^+ = k|X_i = x,\ |X_j| < |x|\big). \end{aligned} \tag{3.2.4}$$

The following two lemmas will be useful in the sequel.

Lemma 3.2.2 (Hušková, 1970). If F is a continuous d.f., then for every $\varepsilon > 0$ there exists an $\alpha > 0$ such that

$$\sup_x |F(\xi x) - F(x)| < \varepsilon \qquad \text{whenever} \quad |\xi - 1| < \alpha. \tag{3.2.5}$$

Proof. Let K be such that $\int_{|x|>K} dF(x) < \varepsilon/2$, and let $\delta > 0$ be such that $|x - y| < \delta$ implies $|F(x) - F(y)| < \varepsilon/2$. Choose $\alpha < \min(\delta/K, 1)$, and let $|\xi - 1| \le \alpha$. Then for any real x,

$$|x| \le K \quad \Rightarrow \quad |\xi x - x| \le K|\xi - 1| < \delta$$

$$\Rightarrow \quad |F(\xi x) - F(x)| < \frac{\varepsilon}{2}, \tag{a}$$

$$|x| > K \text{ and } |\xi x| > K \quad \Rightarrow \quad |F(\xi x) - F(x)|$$

$$= \left| \int_x^{\xi x} dF(x) \right| \le \int_{|y|>K} dF(y) < \frac{\varepsilon}{2}, \tag{b}$$

$$|x| > K \text{ and } |\xi x| \le K \quad \Rightarrow \quad K \ge \xi x \operatorname{sgn} x > \xi K$$

$$\Rightarrow \quad |\xi x - K \operatorname{sgn} x| = |\xi x \operatorname{sgn} x - K| < K(1 - \xi) < \delta$$

$$\Rightarrow \quad |F(\xi x) - F(x)| \le |F(\xi x) - F(K \operatorname{sgn} x)|$$

$$+ |F(K \operatorname{sgn} x) - F(x)| < \varepsilon. \tag{c}$$

The proof is complete.

Lemma 3.2.3 (Hušková, 1970). Let W and V be random variables, F a continuous d.f., and $\varepsilon > 0$. Then there exist $\beta > 0$ and $\gamma > 0$ such that $|\beta'| \le \beta$ and $|\gamma' - 1| \le \gamma$ imply

$$\sup_x \left| F\left(\frac{x + \beta'}{\gamma'} \right) - F(x) \right| < \frac{\varepsilon}{4}. \tag{3.2.6}$$

If $|\gamma_1/\gamma_2 - 1| \le \gamma$, $E(W - V + \beta_1 - \beta_2)^2/\beta^2\gamma_2^2 \le \varepsilon/4$, and

$$\sup_x \left| P\left(\frac{V - \beta_1}{\gamma_1} \le x \right) - F(x) \right| < \frac{\varepsilon}{2}, \tag{3.2.7}$$

then

$$\sup_x \left| P\left(\frac{W - \beta_2}{\gamma_2} \le x \right) - F(x) \right| < \varepsilon. \tag{3.2.8}$$

The proof is left as an exercise.

3.3. PROJECTION AND VARIANCE INEQUALITY FOR BOUNDED SCORES

As in Section 2.4, we approximate S_N^+ by

$$L_N = \sum_{i=1}^{N} l_i(X_i) \quad \text{where} \quad El_i^2(X_i) < \infty, \quad 1 \le i \le N. \tag{3.3.1}$$

To this end, we have the following.

Lemma 3.3.1. If $E(S_N^+)^2 < \infty$, then, on letting

$$\hat{S}_N = \sum_{i=1}^{N} E(S_N^+ | X_i) - (N-1) ES_N^+, \tag{3.3.2}$$

we have

$$E\hat{S}_N = ES_N^+, \qquad E(S_N^+ - \hat{S}_N)^2 = \operatorname{Var} S_N^+ - \operatorname{Var} \hat{S}_N, \tag{3.3.3}$$

and

$$E(\hat{S}_N - L_N)^2 \le E(S_N^+ - L_N)^2. \tag{3.3.4}$$

The proof is analogous to that of Lemma 2.4.1 and is therefore left as an exercise.

Theorem 3.3.2. *If $\phi(u)$, $0 < u < 1$, possesses a bounded second derivative and $S_N^+ = \sum_{i=1}^N c_i s(X_i)\phi(R_{Ni}^+/(N+1))$, then there exists a constant $K = K(\phi)$ such that*

$$E(S_N^+ - \hat{S}_N)^2 \le KN^{-1} \sum_{i=1}^{N} c_i^2, \tag{3.3.5}$$

where $\hat{S}_N$ is defined by (3.3.2).

Proof. We proceed as in the proof of Theorem 2.4.3. Denote

$$\rho_i = \frac{R_{Ni}^+}{N+1}, \qquad 1 \le i \le N. \tag{3.3.6}$$

In the sequel, the subscript N in R_{Ni}^+, S_N^+, $\hat{S}_N$, etc. will be suppressed. Then

$$S^+ = \sum_{i=1}^{N} c_i \phi(\rho_i) s(X_i). \tag{3.3.7}$$

By the Taylor expansion,

$$\phi(\rho_i) = \phi(E[\rho_i|X_i]) + [\rho_i - E(\rho_i|X_i)]\phi'(E[\rho_i|X_i]) + (\rho_i - E[\rho_i|X_i])^2 \alpha_i(X_i, \rho_i), \tag{3.3.8}$$

where on letting $\phi^{(r)}(u) = (d^r/du^r)\phi(u)$, $r = 0, 1, 2,$

$$\alpha_i^2(X_i, \rho) \leq K_2 = \left[\sup_{0<u<1} |\phi^{(2)}(u)|\right]^2, \tag{3.3.9}$$

Thus, by (3.3.6) and (3.3.8), we have

$$S^+ = S_1 + S_2 + S_3, \tag{3.3.10}$$

where

$$S_1 = \sum_{i=1}^{N} c_i \phi(E[\rho_i|X_i])\rho_i s(X_i), \tag{3.3.11}$$

$$S_2 = \sum_{i=1}^{N} c_i \{[\rho_i - E(\rho_i|X_i)]\phi'[E(\rho_i|X_i)]\} s(X_i), \tag{3.3.12}$$

and

$$S_3 = \sum_{i=1}^{N} c_i \{\rho_i - E(\rho_i|X_i)\}^2 \alpha_i(X_i, \rho_i) s(X_i). \tag{3.3.13}$$

Now denote the projection of S_i by $\hat{S}_i$, $i = 1, 2, 3$. Thus

$$\hat{S}_i = \sum_{k=1}^{N} E(S_i|X_k) - (N-1)ES_i, \qquad i = 1, 2, 3. \tag{3.3.14}$$

Then, $\hat{S} = \hat{S}_1 + \hat{S}_2 + \hat{S}_3$. Since $\hat{S}_2 = S_2$, we have

$$E(S^+ - \hat{S})^2 = E(S_1 - \hat{S}_1 + S_3 - \hat{S}_3)^2 \leq 2E(S_1 - \hat{S}_1)^2 + 2E(S_3^2). \tag{3.3.15}$$

Using (3.3.9) and proceeding as in the proof of Theorem 2.4.3, we have

(Exercise 3.3.2)

$$E(S_3^2) \leq \frac{K_2}{4(N+1)} \sum_{i=1}^{N} c_i^2. \tag{3.3.16}$$

To bound $E(S_1 - \hat{S}_1)^2$, we let

$$S_{1i} = c_i \phi(E[\rho_i | X_i]) \rho_i s(X_i), \qquad 1 \leq i \leq N. \tag{3.3.17}$$

Then

$$S_1 = \sum_{i=1}^{N} S_{1i}. \tag{3.3.18}$$

Using Lemma 2.4.2, we obtain

$$\begin{aligned} E(S_1 - \hat{S}_1)^2 \leq & \sum_{i=1}^{N} E[S_{1i} - E(S_{1i}|X_i)]^2 \\ & + \sum\sum_{i \neq j} \Big\{ E\big([S_{1i} - E(S_{1i}|X_i)][S_{1j} - E(S_{1j}|X_j)]\big) \\ & - \sum_{k \neq i, j} \operatorname{Cov}\big[E(S_{1i}|X_k), E(S_{1j}|X_k)\big] \Big\}. \end{aligned} \tag{3.3.19}$$

We find the upper bound of each term on the right side of (3.3.19). Proceeding as in Theorem 2.4.3 [see (2.4.34)], we obtain

$$\sum_{i=1}^{N} E[S_{1i} - E(S_{1i}|X_i)]^2 \leq \frac{K_1}{4(N+1)} \sum_{i=1}^{N} c_i^2, \tag{3.3.20}$$

where K_1 is the least upper bound for $|\phi'(t)|^2$, $0 < t < 1$. Next we write [cf. (2.4.35)]

$$\begin{aligned} E\big([S_{1i} - E(S_{1i}|X_i)][S_{1j} - E(S_{1j}|X_j)]\big) = & \, E \operatorname{Cov}\big[(S_{1i}, S_{1j})|X_i, X_j\big] \\ & + E\big[\{E(S_{1i}|X_i, X_j) - E(S_{1i}|X_i)\}\{E(S_{1j}|X_i, X_j) - E(S_{1j}|X_j)\}\big], \end{aligned} \tag{3.3.21}$$

and proceeding as in (2.4.35) through (2.4.39), we have

$$E\,\mathrm{Cov}\Big[\,(S_{1i}, S_{1j})\big|X_i, X_j\Big] = c_i c_j (N+1)^{-2} \sum_{k \neq i,\,j} \mathrm{Cov}\big[l_{ik}(X_k), l_{jk}(X_k)\big], \tag{3.3.22}$$

where

$$l_{ik}(s) = \int [u(x-s) - F_k(x)]\,\phi'(E[\rho_i|X_i = x])\,dF_i(x),$$

$$\Big|E\big[\{E(S_{1i}|X_i, X_j) - E(S_{1i}|X_i)\}\{E(S_{1j}|X_i, X_j) + E(S_{1j}|X_j)\}\big]\Big| \le \frac{K_1}{4}|c_i c_j|(N+1)^{-2}, \tag{3.3.23}$$

$$\mathrm{Cov}\big[E(S_{1i}|X_k), E(S_{1j}|X_k)\big] = c_i c_j (N+1)^{-2}\mathrm{Cov}\big[l_{ik}(X_k), l_{jk}(X_k)\big], \quad i \neq j, \quad k \neq i, j. \tag{3.3.24}$$

Using (3.3.20) through (3.3.24) in (3.3.15), we obtain that

$$E(S_1 - \hat{S}_1)^2 \le \frac{K_1}{4}(N+1)^{-1}\sum_{i=1}^{N} c_i^2 + \frac{K_1}{4}(N+1)^{-2}\sum_{i \neq j = 1}^{N} |c_i c_j| \le \frac{K_1}{2}(N+1)^{-1}\sum_{i=1}^{N} c_i^2. \tag{3.3.25}$$

The proof follows then by using (3.3.16) and (3.3.25) in (3.3.15) with $K = K_1 + \frac{1}{2}K_2$.

Next, we show that $\hat{S}$ can be replaced by $\Sigma_{i=1}^N Z_i$, which is simpler in form.

Theorem 3.3.3. *Under the hypothesis of Theorem* 3.3.2, *there exists a positive constant* $M = M(\phi)$ *such that for every* $N (\ge 1)$, $c_1, \ldots, c_N$, *and* $F_1, \ldots, F_N$,

$$E\left(S^+ - ES^+ - \sum_{i=1}^{N} Z_i\right)^2 \le MN^{-1}\sum_{i=1}^{N} c_i^2, \tag{3.3.26}$$

$$(ES^+ - \mu)^2 \le MN^{-1}\sum_{i=1}^{N} c_i^2, \tag{3.3.27}$$

where for every i $(= 1, \ldots, N)$,

$$Z_i = \frac{1}{N+1} \sum_{j=1}^{N} c_j \int_{-\infty}^{\infty} s(x)\left[u(|x| - |X_i|) - F_i^*(|x|)\right]$$

$$\times \phi'(H^*(|x|))\, dF_j(x)$$

$$+ c_i \left\{ \left[s(X_i)\phi(H^*(|X_i|))\right] - E\left[s(X_i)\phi(H^*(|X_i|)\right]\right\}, \qquad (3.3.28)$$

$$\mu = \sum_{i=1}^{N} c_i \int_{-\infty}^{\infty} s(x)\phi(H^*(|x|))\, dF_i(x), \qquad (3.3.29)$$

and $H^*(x) = N^{-1}\sum_{i=1}^{N} F_i^*(x)$, $0 \le x < \infty$.

Proof. We only give the outline of the proof of this theorem, because of its similarity with the proof of Theorem 2.4.5. Let $\rho_i = R_{Ni}^+/(N+1)$, $1 \le i \le N$. Then, by the Taylor expansion,

$$\phi(\rho_i) - \phi(H^*(|X_i|)) = \left[\rho_i - H^*(|X_i|)\right]\phi'(H^*(|X_i|))$$

$$+ \left[\rho_i - H^*(|X_i|)\right]^2 \alpha_i(\rho_i, X_i), \qquad (3.3.30)$$

where

$$\alpha_i^2(\rho, X_i) \le \frac{K_2}{4} < K_2. \qquad (3.3.31)$$

Since

$$E\left[\rho_i - H^*\,(|X_i|)|X_i\right] = \frac{1}{N+1}(1 - F_i^*(|X_i|)) - \frac{1}{N+1} H^*(|X_i|),$$

it follows that

$$\left|E\left[\rho_i - H^*\,(|X_i|)|X_i\right]\right| \le (N+1)^{-1}. \qquad (3.3.32)$$

Furthermore, after routine computations, we obtain (Exercise 3.3.3) that

$$E\left\{\left[\rho_i - H^*(|X_i|)\right]^2 \middle| X_i\right\} \le (N+1)^{-1}. \qquad (3.3.33)$$

Using (3.3.30) through (3.3.33) together with the definition of S^+ and μ, we obtain after some computations that

$$E(S^+ - \mu)^2 \le \left(\frac{\sqrt{K_1} + \sqrt{K_2}}{N+1} \right)^2 N \sum_{i=1}^{N} c_i^2 \le \left(\sqrt{K_1} + \sqrt{K_2} \right)^2 N^{-1} \sum_{i=1}^{N} c_i^2, \tag{3.3.34}$$

where K_1 and K_2 are given by (3.3.20) and (3.3.9) respectively. Equation (3.3.34) yields (3.3.27) with $M = (\sqrt{K_1} + \sqrt{K_2})^2$.

We now prove (3.3.26). By (3.3.5),

$$E(S^+ - \hat{S})^2 \le KN^{-1} \sum_{i=1}^{N} c_i^2, \tag{3.3.35}$$

where $\hat{S}$ is given by (3.3.2). Now

$$\begin{aligned} E\left(S^+ - ES^+ - \sum_{i=1}^{N} Z_i \right)^2 &= E\left(S^+ - \hat{S} + \hat{S} - E\hat{S} - \sum_{i=1}^{N} Z_i \right)^2 \\ &\le 2E(S^+ - \hat{S})^2 + 2E\left(\hat{S} - E\hat{S} - \sum_{i=1}^{N} Z_i \right)^2. \end{aligned} \tag{3.3.36}$$

Thus by virtue of (3.3.35), (3.3.26) will follow if we show that

$$E\left(\hat{S} - E\hat{S} - \sum_{i=1}^{N} Z_i \right)^2 \le CN^{-1} \sum_{i=1}^{N} c_i^2, \tag{3.3.37}$$

where $C = C(\phi)$ is a constant depending on ϕ. Now

$$\hat{S} - E\hat{S} = \sum_{i=1}^{N} \sum_{j=1}^{N} c_j \left\{ E\left[\phi(\rho_j) s(X_j) | X_i \right] - E\left[\phi(\rho_j) s(X_j) \right] \right\}. \tag{3.3.38}$$

Using Lemma 3.2.1, we obtain (Exercise 3.3.4) that

$$
\begin{aligned}
&E\left[\phi(\rho_j)s(X_j)\middle|X_i\right] - E\left[\phi(\rho_j)s(X_j)\right] \\
&\quad = E\left\{E\left[\phi(\rho_j)s(X_j)\middle|X_i, X_j\right] - E\left[\phi(\rho_j)s(X_j)\middle|X_j\right]\right\} \\
&\quad = (N+1)^{-1}\int s(x)\left[u(|x| - |X_i|) - F_i^*(|x|)\right] \\
&\quad \times E\left[\phi'(\rho_j)\middle|X_j = x, |X_i| < |X_j|\right] dF_j(x) \\
&\quad -(N+1)^{-2}\int s(x)\left[u(|x| - |X_i|) - F_i^*(|x|)\right] \\
&\quad \times E\left[\phi''(\xi_j)\middle|X_j = x, |X_i| < |X_j|\right] dF_j(x),
\end{aligned}
\tag{3.3.39}
$$

where $\rho_j - 1/(N+1) < \xi_j < \rho_j$. Now since the absolute value of the second term on the right $\leq (N+1)^{-2}K_2^{1/2} \leq N^{-3/2}K_2^{1/2}$, it follows that

$$
\begin{aligned}
&(N+1)^{-2}\int_{-\infty}^{\infty} s(x)\left[u(|x| - |X_i|) - F_i^*(|x|)\right] \\
&\quad \times E\left[\phi''(\xi_j)\middle|X_j = x, |X_i| < |X_j|\right] dF_j(x) \\
&= N^{-3/2}K_2^{1/2}\gamma_{ij}(X_i),
\end{aligned}
\tag{3.3.40}
$$

where $|\gamma_{ij}(X_i)| \leq 1$. Also, $\phi'(\rho_j) = \phi'(H^*(|X_j|)) + [\rho_j - H^*(|X_j|)]\phi''(\theta_j\rho_j + (1-\theta_j)H^*(|X_j|))$, $0 < \theta_j < 1$. Hence, using Lemma 3.2.1, we have

$$
E\left[\left\{\rho_j - H^*(|X_j|)\right\}^2\middle|X_j = x, |X_i| < |X_j|\right] \leq \frac{N-2+8}{2(N+1)^2} \leq N^{-1},
\tag{3.3.41}
$$

so that

$$
\begin{aligned}
&\Big|E\Big[\left\{\rho_j - H^*(|X_j|)\right\}\phi''\big(H^*(|X_j|) \\
&\qquad +\theta_j\left(\rho_j - H^*(|X_j|)\right)\Big|X_j = x, |X_i| < |X_j|\Big]\Big| \\
&\leq \left\{E\left[\left\{\rho_j - H^*(|X_j|)\right\}\middle|X_i = x, |X_j| < |X_i|\right]\right\}^{1/2}\sqrt{K_2} \\
&\leq N^{-1/2}\sqrt{K_2}\,.
\end{aligned}
$$

Thus

$$E\left[\phi'(\rho_j) \middle| X_j = x, |X_i| < |X_j|\right] = \phi'(H^*(|x|)) + \sqrt{K_2}\, N^{-1/2}\delta_{ij}(x),$$

$$|\delta_{ij}(x)| \le 1. \tag{3.3.42}$$

Hence,

$$(N+1)^{-1}\int s(x)\left[u(|x|) - |X_i|) - F_i^*(|x|)\right]\sqrt{K_2}\, N^{-1/2}\delta_{ij}(x)\, dF_i(x)$$

$$= \sqrt{K_2}\, N^{-3/2}\gamma'_{ij}(X_i), \qquad |\gamma'_{ij}(X_i)| \le 1. \tag{3.3.43}$$

Using (3.3.40) through (3.3.43), we find that (3.3.39) equals

$$(N+1)^{-1}\int s(x)\left[u(|x| - |X_i|) - F_i^*(|x|)\right]\phi'(H^*(|x|))\, dF_j(x)$$

$$+ 2\sqrt{K_2}\, N^{-3/2}\delta'_{ij}(X_i), \tag{3.3.44}$$

where $\delta'_{ij}(x) = \frac{1}{2}[\gamma_{ij}(x) + \gamma'_{ij}(x)]$, so that $|\delta'_{ij}(x)| \le 1$ for all x. For convenience, set

$$a_{ij}(X_i) = \int_{-\infty}^{\infty} s(x)\left[u(|x| - |X_i|) - F_i^*(|x|)\right]\phi'(H^*(|x|))\, dF_j(x),$$

$$i, j = 1, \ldots, N. \tag{3.3.45}$$

Then

$$\hat{S} - E\hat{S} = \sum_{\substack{i=1 \\ i \ne j}}^{N} \sum_{j=1}^{N} c_j \left[\frac{a_{ij}(X_i)}{N+1} + 2\sqrt{K_2}\, N^{-3/2}\delta'_{ij}(X_i)\right]$$

$$+ \sum_{i=1}^{N} c_i \left\{E\left[\phi(\rho_i)s(X_i) \middle| X_i\right] - E\left[\phi(\rho_i)s(X_i)\right]\right\}. \tag{3.3.46}$$

Now

$$\sum_{i=1}^{N} Z_i = \sum_{\substack{i=1 \\ 1 \ne j}}^{N} \sum_{j=1}^{N} c_j \left[\frac{a_{ij}(X_i)}{N+1}\right] + \sum_{i=1}^{N} \frac{c_i a_{ii}(X_i)}{N+1}$$

$$+ \sum_{i=1}^{N} c_i \left\{E\left[s(X_i)\phi(H^*(|X_i|))\middle| X_i\right) - E\left[s(X_i)\phi(H^*(|X_i|))\right]\right\}.$$

$$\tag{3.3.47}$$

Furthermore, omitting routine computations, we obtain

$$E\left\{\sum_{i=1}^{N} c_i\Big\{E\left[\phi(\rho_i)s(X_i)|X_i\right] - E\left[\phi(\rho_i)s(X_i)\right]\right.$$

$$\left. - E\left[s(X_i)\phi\left(H^*(|X_i|)\right)|X_i\right] + E\left[s(X_i)\phi\left(H^*(|X_i|)\right)\right]\Big\}\right\}^2$$

$$\leq \left(\sqrt{K_1} + \sqrt{K_2}\right)^2 (N+1)^{-2} \sum_{i=1}^{N} c_i^2. \tag{3.3.48}$$

The desired result follows by using (3.3.46)–(3.3.48) and performing some routine computations.

3.4. ASYMPTOTIC NORMALITY OF S_N

As in Chapter 2, we consider first the case of score functions with bounded second derivatives, and then the general case.

Theorem 3.4.1. *If $\phi(u)$, $0 < u < 1$, has a bounded second derivative, and if for every $\varepsilon > 0$ there exists a constant K_ε ($< \infty$) such that*

$$\operatorname{Var} S_N^+ > K_\varepsilon \left[\max_{1 \leq i \leq N} c_i^2\right], \tag{3.4.1}$$

then for every real x,

$$\lim_{N \to \infty} P\left\{\frac{S_N^+ - ES_N^+}{\sqrt{\operatorname{Var} S_N^+}} \leq x\right\} = \Phi(x), \tag{3.4.2}$$

where $\Phi(x)$ is defined by (2.5.2). The assertion remains true if in (3.4.2) we replace ES_N^+ by μ defined in (3.3.29), or $\operatorname{Var} S_N^+$ by

$$\sigma_N^2 = \sum_{i=1}^{N} \operatorname{Var} Z_i, \tag{3.4.3}$$

where the Z_i are defined by (3.3.28).

Proof. Let $\varepsilon > 0$, and let $G_i(x)$ be the c.d.f. of Z_i. Then by the Lindeberg central limit theorem, there exists a $\delta > 0$ such that

$$\sigma_N^{-2} \sum_{i=1}^{N} \int_{|x|>\delta\sigma} x^2 \, dG_i(x) < \delta \tag{3.4.4}$$

implies

$$\sup_x \left| P\left(\sum_{i=1}^{N} \frac{Z_i}{\sigma_N} \leq x \right) - \Phi(x) \right| < \frac{\varepsilon}{2}. \tag{3.4.5}$$

Note that, by definition,

$$|Z_i| \leq (N+1)^{-1} \sum_{j=1}^{N} |c_j| K_1^{1/2} + 2|c_i| K_0^{1/2}$$

$$\leq \left(2K_0^{1/2} + K_1^{1/2}\right)\left[\max_{1 \leq i \leq N} |c_i|\right], \tag{3.4.6}$$

$$K_r = \sup_{u \in (0,1)} \left|\phi^{(r)}(u)\right|^2, \qquad r = 0, 1, 2.$$

Thus, to prove (3.4.4), it suffices to show that

$$\delta\sigma_N > \left(2K_0^{1/2} + K_1^{1/2}\right)\left[\max_{1 \leq i \leq N} |c_i|\right], \tag{3.4.7}$$

and the proof of this follows directly by using (3.4.3), Theorem 3.3.3, and (3.4.1), where we choose K_ε adequately large, depending on δ (> 0) and K_0, K_1. The proof of the theorem follows then from (3.4.5) and Theorem 3.3.3.

We now consider the case of general score functions.

Theorem 3.4.2. *Let* $\phi(u) = \phi_1(u) - \phi_2(u)$, $0 < u < 1$, *where* $\phi_i(u)$, $i = 1, 2$, *are both nondecreasing, square-integrable and absolutely continuous inside* $(0, 1)$. *If for every* $\varepsilon > 0$ *and* $\eta > 0$ *there exists an integer* $N_{\varepsilon\eta}$ *such that*

$$\operatorname{Var} S_N^+ > N\eta\left[\max_{1 \leq i \leq N} c_i^2\right] \qquad \textit{for every} \quad N \geq N_{\varepsilon\eta}, \tag{3.4.8}$$

then

$$\sup_x \left| P\left\{ \frac{S_N^+ - ES_N^+)}{\sqrt{\operatorname{Var} S_N^+}} \le x \right\} - \Phi(x) \right| < \varepsilon \qquad \forall N \ge N_{\varepsilon\eta}, \tag{3.4.9}$$

The assertion remains true if we replace $\operatorname{Var} S_N^+$ *by* σ_N^2, *defined by* (3.4.3).

We closely follow the treatment of Section 2.5 and first consider the following two lemmas, which will be subsequently used in the proof of the theorem.

Lemma 3.4.3 (Variance inequality). For arbitrary $c_1, \ldots, c_N$ and $a_N(1) \le \cdots \le a_N(N)$,

$$\operatorname{Var} S_N^+ \le 40\left[\max_{1 \le i \le N} c_i^2\right]\left[\sum_{i=1}^{N} a_N^2(i)\right]. \tag{3.4.10}$$

Lemma 3.4.4. Let the Z_i be defined by (3.3.28) and $\phi(u)$ be nondecreasing, square-integrable, and absolutely continuous inside $(0, 1)$. Then there exists a constant $C = C(\phi)$ such that

$$\sum_{i=1}^{N} \operatorname{Var} Z_i \le CN\left[\max_{1 \le i \le N} c_i^2\right] \int_0^1 \phi^2(u)\, du. \tag{3.4.11}$$

Proof of Lemma 3.4.3. Clearly,

$$\begin{aligned}
\operatorname{Var} S_N^+ &= E\Bigg\{ \sum_{i=1}^{N} c_i \{ s(X_i) a_N(R_{Ni}^+) - s(X_i) E[a_N(R_{Ni}^+)|X_i] \} \\
&\quad + \sum_{i=1}^{N} c_i \{ s(X_i) E[a_N(R_{Ni}^+)|X_i] - E[s(X_i) a_N(R_{Ni}^+)] \} \Bigg\}^2 \\
&\le 2E\left\{ \sum_{i=1}^{N} c_i [s(X_i) \{ a_N(R_{Ni}^+) - E[a_N(R_{Ni}^+)|X_i] \}] \right\}^2 \\
&\quad + 2E\left\{ \sum_{i=1}^{N} c_i [s(X_i) E[a_N(R_{Ni}^+)|X_i] - E[s(X_i) a_N(R_{Ni}^+)]] \right\}^2 .
\end{aligned} \tag{3.4.12}$$

In the sequel, as in Section 3.3, the subscript N in R_{Ni}^+, S_N^+, etc. will be suppressed. The first term on the rhs of (3.4.12) is bounded from above by

$$\begin{aligned}
&2\sum_{i=1}^{N} c_i^2 E\left\{a_N(R_i^+) - E\left[a_N(R_i^+)|X_i\right]\right\}^2 \\
&\qquad + 2\sum_{\substack{i=1\\ i\neq j}}^{N}\sum_{j=1}^{N} c_i c_j E\left[s(X_i)s(X_j)\right] \\
&\qquad \times\left\{a_N(R_i^+) - E\left[a_N(R_i^+)|X_i\right]\right\}\left\{a_N(R_j^+) - E\left[a_N(R_j^+)|X_j\right]\right\} \\
&\le \left(\max_{1\le i\le N} c_i^2\right)\cdot 2\sum_{i=1}^{N} E\left\{E\left[a_N^2(R_i^+)|X_i\right]\right\} \\
&\qquad + 2\left|\sum_{\substack{i=1\\ i\neq j}}^{N}\sum_{j=1}^{N} c_i c_j E\left[s(X_i)s(X_j)\right]\right. \\
&\qquad \left.\times\left\{a_N(R_i^+) - E\left[a_N(R_i^+)|X_i\right]\right]\left[a_N(R_j^+) - E\left[a_N(R_j^+)|X_j\right]\right\}\right| \\
&\le 2\left(\max_{1\le i\le N} c_i^2\right)\sum_{i=1}^{N} a_N^2(i) + 2|A_{12}|, \qquad (3.4.13)
\end{aligned}$$

where

$$\begin{aligned}
A_{12} = &\sum_{i\neq j=1} c_i c_j E\Big(s(X_i)s(X_j)\left\{a_N(R_i^+) - E\left[a_N(R_i^+)|X_i\right]\right\} \\
&\times\left\{a_N(R_j^+) - E\left[a_N(R_j^+)|X_j\right]\right\}\Big). \qquad (3.4.14)
\end{aligned}$$

The second term on the rhs of (3.4.12) is bounded by

$$\begin{aligned}
&2E\left\{\sum_{i=1}^{N} c_i\left\{s(X_i)E\left[a_N(R_i^+)|X_i\right] - E\left[s(X_i)a_N(R_i^+)\right]\right\}\right\}^2 \\
&= 2\sum_{i=1}^{N} c_i^2 E\left\{s(X_i)E\left[a_N(R_i^+)|X_i\right] - E\left[s(X_i)a_N(R_i^+)\right]\right\}^2 \\
&\le 2\sum_{i=1}^{N} c_i^2 E\left\{E\left[s(X_i)a_N(R_i^+)|X_i\right]\right\}^2 \\
&\le 2\sum_{i=1}^{N} c_i^2 E\left\{E\left[a_N(R_i^+)|X_i\right]\right\}^2 \le 2\left(\max_{1\le i\le N} c_i^2\right)\sum_{i=1}^{N} a_N^2(i). \qquad (3.4.15)
\end{aligned}$$

Thus, to prove the lemma, it suffices to show that

$$|A_{12}| \leq 18\Big\{\max_{1\leq i\leq N} c_i^2\Big\}\Big\{\sum_{i=1}^{N} a_N^2(i)\Big\}. \tag{3.4.16}$$

For this note that

$$A_{12} = B_{12} + C_{12}, \tag{3.4.17}$$

$$B_{12} = \sum_{i\neq j=1} c_i c_j E\Big\{s(X_i)s(X_j)\mathrm{Cov}\big[a_N(R_i^+), a_N(R_j^+)|X_i X_j\big]\Big\}, \tag{3.4.18}$$

$$C_{12} = \sum_{i\neq j=1} c_i c_j E\Big(s(X_i)s(X_j)\Big\{E\big[a_N(R_i^+)|X_i, X_j\big] - E\big[a_N(R_i^+)|X_i\big]\Big\}$$

$$\times\Big\{E\big[a_N(R_j^+)|X_i, X_j\big] - E\big[a_N(R_j^+)|X_j\big]\Big\}\Big). \tag{3.4.19}$$

Now,

$$|B_{12}| \leq \Big(\max_{1\leq i\leq N} c_i^2\Big)\sum_{i\neq j=1}^{N} E\Big|\mathrm{Cov}\big[\big(a_N(R_i^+), a_N(R_j^+)\big)|X_i, X_j\big]\Big|$$

$$\leq \Big(\max_{1\leq i\leq N} c_i^2\Big)\sum_{i\neq j=1}\mathrm{Cov}\Big\{E\big[a_N(R_i^+)|X_i, X_j\big], E\big[a_N(R_j^+)|X_i, X_j\big]\Big\}$$

$$\leq 10\Big(\max_{1\leq i\leq N} c_i^2\Big)\sum_{i=1}^{N} a_N^2(i), \tag{3.4.20}$$

proceeding as in (2.3.25) ff. Thus, it remains only to show that

$$|C_{12}| \leq 8\Big(\max_{1\leq i\leq N} c_i^2\Big)\sum_{i=1}^{N} a_N^2(i). \tag{3.4.21}$$

To prove this, first note that using Lemma 3.2.1 and the fact that $P(R_i^+ = k|X_i = x,\ |X_j| < |X_i|) \leq P(R_i^+ = k|X_i = x) + P(R_i^+ = k-1|X_i = x)$, we have

$$\begin{aligned}
&\left|s(X_i)s(X_j)\left\{E\left[a_N(R_i^+)|X_i, X_j\right] - E\left[a_N(R_i^+)|X_i\right]\right\}\right.\\
&\quad\left.\times\left\{E\left[a_N(R_j^+)|X_i, X_j\right] - E\left[a_N(R_j^+)|X_j\right]\right\}\right|\\
&\quad\le \sum_{k=2}^{N}\sum_{h=2}^{N}[a_N(k) - a_N(k-1)][a_N(h) - a_N(h-1)]\\
&\quad\times\left[P(R_i^+ = k|X_i) + P(R_i^+ = k-1|X_i)\right]\\
&\quad\times\left[P(R_j^+ = h|X_j) + P(R_j^+ = h-1|X_j)\right]. \qquad (3.4.22)
\end{aligned}$$

Now taking expectation on both sides of (3.4.22) and substituting in (3.4.19), we get that

$$\begin{aligned}
|C_{12}| &\le \left(\max_{1\le i\le N} c_i^2\right)\sum_{k=2}^{N}\sum_{h=2}^{N}[a_N(k) - a_N(k-1)][a_N(h) - a_N(h-1)]\\
&\quad\times\sum_{i\ne j=1}^{N}[P(R_i^+ = k) + P(R_i^+ = k-1)]\\
&\quad\times\left[P(R_j^+ = h) + P(R_j^+ = h-1)\right]\\
&\le 4\left(\max_{1\le i\le N} c_i^2\right)\sum_{k=2}^{N}\sum_{h=2}^{N}[a_N(k) - a_N(k-1)][a_N(h) - a_N(h-1)]\\
&= 4\left(\max_{1\le i\le N} c_i^2\right)[a_N(N) - a_N(1)]^2\\
&\le 8\left(\max_{1\le i\le N} c_i^2\right)\sum_{i=1}^{N} a_N^2(i). \qquad (3.4.23)
\end{aligned}$$

Proof of Lemma 3.4.4. From (3.3.28), we get that

$$\begin{aligned}
\sum_{i=1}^{N}\operatorname{Var}(Z_i) &\le 2\sum_{i=1}^{N} V\left[\sum_{j=1}^{N} c_j\frac{1}{N+1}\int_{-\infty}^{\infty} s(x)\left[u(|x| - |X_i|)\right.\right.\\
&\qquad\left.\left. - F_i^*(|x|)\right]\phi'\left(H^*(|x|)\right) dF_j(x)\right]\\
&\quad + 2\sum_{i=1}^{N}\operatorname{Var}\left(c_i s(X_i)\phi\left(H^*(|X_i|)\right)\right) \qquad (3.4.24)
\end{aligned}$$

and

$$\sum_{i=1}^{N} \operatorname{Var}\bigl(c_i s(X_i)\phi(H^*(|X_i|))\bigr) \le \sum_{i=1}^{N} c_i^2 E\phi^2(H^*(|X_i|))$$
$$\le \Bigl(\max_{1\le i\le N} c_i^2\Bigr) N \int_0^1 \phi^2(u)\,du. \tag{3.4.25}$$

Furthermore,

$$\operatorname{Var}\Biggl(\sum_{j=1}^{N} c_j \frac{1}{N+1}\int_{-\infty}^{\infty} s(x)\bigl[u(|x|-|X_i|) - F_i^*(|x|)\bigr]\phi'(H^*(|x|))\,dF_j(x)\Biggr)$$
$$= \sum_{j=1}^{N}\sum_{k=1}^{N} \frac{c_j c_k}{(N+1)^2}\int_{-\infty}^{\infty}\int_{-\infty}^{\infty} s(x)s(y)$$
$$\times E\bigl\{\bigl[u(|x|-|X_i|) - F_i^*(|x|)\bigr]\bigl[u(|y|-|X_i|) - F_i^*(|y|)\bigr]\bigr\}$$
$$\times \phi'(H^*(|x|))\phi'(H^*(|y|))\,dF_j(x)\,dF_k(y) \tag{3.4.26}$$

where by Lemma 2.3.2, $E\{[u(|x| - |X_i|) - F_i^*(|x|)][u(|y| - |X_i|) - F_i^*(|y|)]\} = \operatorname{Cov}[u(|x| - |X_i|), u(|y| - |X_i|)] \ge 0$, and as ϕ is nondecreasing, $\phi' \ge 0$ a.e. Thus, the rhs of (3.4.26) is

$$\le \Bigl(\max_{1\le i\le N} c_i^2\Bigr)(N+1)^{-2}\int_{-\infty}^{\infty}\int_{-\infty}^{\infty} E\bigl\{\bigl[u(|x|-|X_i|) - F_i^*(|x|)\bigr]$$
$$\times\bigl[u(|y|-|X_i|) - F_i^*(|y|)\bigr]\bigr\}$$
$$\times \phi'(H^*(|x|))\phi'(H^*(|y|)) \sum_{j=1}^{N}\sum_{k=1}^{N} dF_j(x)\,dF_k(y).$$
$$= \Bigl(\max_{1\le i\le N} c_i^2\Bigr)\int_{-\infty}^{\infty}\int_{-\infty}^{\infty} E\bigl\{\bigl[u(|x|-|X_i|) - F_i^*(|x|)\bigr]$$
$$\times\bigl[u(|y|-|X_i|) - F_i^*(|y|)\bigr]$$
$$\times \phi'(H^*(|x|))\phi'(H^*(|y|))\,dH(x)\,dH(y)\bigr\}$$
$$\le \Bigl(\max_{1\le i\le N} c_i^2\Bigr)\operatorname{Var}\Bigl\{\int_{-\infty}^{\infty} u(|x|-|X_i|)\phi'(H^*(|x|))\,dH(x)\Bigr\}$$
$$\le \Bigl(\max_{1\le i\le N} c_i^2\Bigr) E\Bigl\{\int_{|X_i|\le x} \phi'(H^*(|x|))\,dH(x)\Bigr\}^2$$
$$= \Bigl(\max_{1\le i\le N} c_i^2\Bigr) E\phi^2(H^*(|X_i|)). \tag{3.4.27}$$

Thus, the first term on the rhs of (3.4.24) is bounded by

$$2N\Big(\max_{1\le i\le N} c_i^2\Big)\frac{1}{N}\sum_{i=1}^{N} E\phi^2\big(H^*(|X_i|)\big) = 2N\Big(\max_{1\le i\le N} c_i^2\Big)\int_0^1 \phi^2(u)\,du. \tag{3.4.28}$$

The proof is completed by (3.4.24), (3.4.25), and (3.4.28).

Proof of Theorem 3.4.2. For convenience, for the function ϕ absolutely continuous inside (0, 1), we denote

$$S_\phi^+ = \sum_{i=1}^{N} c_i \phi\left(\frac{R_i^+}{N+1}\right)s(X_i), \tag{3.4.29}$$

$$T_\phi^+ = \sum_{i=1}^{N} c_i E\left[\phi\left(U_N^{(R_i^+)}\right)\right]s(X_i), \tag{3.4.30}$$

$$Z_{\phi i} = \frac{1}{N+1}\sum_{j=1}^{N} c_j \int s(x)\big[u(|x| - |X_i|) - F_i^*(|x|)\big]\phi'\big[H^*(|x|)\big]\,dF_j(x)$$

$$+ c_i\big[s(X_i)\phi(H^*(|X_i|))\big] - E\big[s(X_i)\phi(H^*(|X_i|))\big], \tag{3.4.31}$$

and

$$\sigma_\phi^2 = \operatorname{Var}\sum_{i=1}^{N} Z_{\phi i}. \tag{3.4.32}$$

We prove the asymptotic normality of S_ϕ^+. The proof for the asymptotic normality of T_ϕ^+ is given as an exercise.

Let ϕ satisfy the assumptions of Theorem 3.4.2. Let $\varepsilon > 0$ and $\eta > 0$ be given. Let $\beta > 0$ and $\gamma > 0$ be such that (cf. Lemma 3.2.3) $|\beta'| \le \beta$ and $|\gamma' - 1| \le \gamma$ imply

$$\sup_x \left|\Phi\left(\frac{x+\beta'}{\gamma'}\right) - \Phi(x)\right| < \frac{\varepsilon}{4},$$

and let $\alpha > 0$ satisfy

$$\alpha < \frac{\eta}{160}\min\left(\gamma^2, \frac{\beta^2\varepsilon}{4}\right). \tag{3.4.33}$$

By Lemma 2.5.4, there exists a decomposition

$$\phi(t) = \psi(t) + \phi_{(1)}(t) - \phi_{(2)}(t), \qquad 0 < t < 1, \tag{3.4.34}$$

where $\psi(t)$ is a polynomial [thus ψ has a bounded second derivative in $(0,1)$], $\phi_{(1)}$ and $\phi_{(2)}$ are nondecreasing, and

$$\int_0^1 \phi_{(1)}^2(t)\,dt + \int_0^1 \phi_{(2)}^2(t)\,dt < \alpha. \tag{3.4.35}$$

From the proof of Theorem 3.4.1, it follows that there is a constant $K_{\varepsilon/2}$ such that

$$\operatorname{Var} S_\psi^+ \geq K_{\varepsilon/2} \max_i c_i^2 \quad \left[\text{or} \quad \sigma_\psi^2 \geq K_{\varepsilon/2} \max_i c_i^2\right] \tag{3.4.36}$$

implies

$$\sup_x \left| P\left(\frac{S_\psi^+ - ES_\psi^+}{\sqrt{\operatorname{Var} S_\psi^+}} < x \right) - \Phi(x) \right| < \frac{\varepsilon}{2}$$

$$\left[\text{or} \quad \sup_x \left| P\left(\frac{S_\psi^+ - ES_\psi^+}{\sigma_\psi} < x \right) - \Phi(x) \right| < \frac{\varepsilon}{2}\right]. \tag{3.4.37}$$

Let

$$N_{\varepsilon\eta} = (1-\gamma)^{-2} \eta^{-1} K_{\varepsilon/2}(\psi). \tag{3.4.38}$$

Then we shall show that

$$\operatorname{Var} S_\phi^+ > N\eta \max_i c_i^2 \qquad \text{for} \quad N > N_{\varepsilon\eta} \tag{3.4.39}$$

implies

$$\left| \frac{\sqrt{\operatorname{Var} S_\psi^+}}{\sqrt{\operatorname{Var} S_\phi^+}} - 1 \right| \leq \gamma, \tag{3.4.40}$$

$$\frac{E\left(S_\phi^+ - S_\psi^+ + ES_\psi^+ - ES_\phi^+\right)^2}{B^2 \operatorname{Var} S_\phi^+} \leq \frac{\varepsilon}{4}, \tag{3.4.41}$$

and

$$\sup_x \left| P\left(\frac{S_\psi^+ - ES_\psi^+}{\sqrt{\operatorname{Var} S_\psi^+}} < x \right) - \Phi(x) \right| < \frac{\varepsilon}{2}. \tag{3.4.42}$$

The theorem will then follow by virtue of Lemma 3.2.3.

Let λ be a nondecreasing, square-integrable function in $(0, 1)$. Then it is easy to check that

$$\sum_{i=1}^{N} \lambda^2\left(\frac{i}{N+1}\right) \leq 2N\int_0^1 \lambda^2(t)\,dt \quad \text{and} \quad \sum_{i=1}^{N} \left[E\lambda\left(U_N^{(i)}\right)\right]^2 \leq N\int_0^1 \lambda^2(t)\,dt. \tag{3.4.43}$$

Also it is clear that

$$S_\phi^+ = S_\psi^+ + S_{\phi_{(1)}}^+ - S_{\phi_{(2)}}^+, \qquad Z_{\phi i} = Z_{\psi i} + Z_{\phi_{(1)i}} - Z_{\phi_{(2)i}}. \tag{3.4.44}$$

Thus

$$\begin{aligned}
\left|\sqrt{\operatorname{Var} S_\phi^+} - \sqrt{\operatorname{Var} S_\psi^+}\right| &\leq \sqrt{\operatorname{Var}\left(S_\phi^+ - S_\psi^+\right)} \\
&= \sqrt{E\left(S_{\phi_{(1)}}^+ - S_{\phi_{(2)}}^+ + ES_{\phi_{(2)}}^+ - ES_{\phi_{(1)}}^+\right)^2} \\
&\leq \sqrt{\operatorname{Var} S_{\phi_{(1)}}^+} + \sqrt{\operatorname{Var} S_{\phi_{(2)}}^+} \\
&\leq \sqrt{160}\, N\alpha \max_i |c_i| \leq \left\{ N\eta \max\left(\gamma^2, \frac{\beta^2\varepsilon}{4}\right)\right\}^{1/2} \max_i |c_i|,
\end{aligned} \tag{3.4.45}$$

by using Lemma 3.4.3, (3.4.43), and (3.4.33).

Now let (3.4.38) hold. Then using (3.4.43), we obtain (3.4.40), which implies

$$\operatorname{Var} S_\psi^+ \geq (1-\gamma)^2 \operatorname{Var} S_\phi^+ \geq (1-\gamma)^2 N_{\varepsilon\eta}\eta \max_i c_i^2 \geq K_{\varepsilon/2}(\psi) \max c_i^2$$

by (3.4.38), and this implies (3.4.37), i.e.,

$$\sup_x \left| P\left(\frac{S_\psi^+ - ES_\psi^+}{\sqrt{\operatorname{Var} S_\phi^+}} < x \right) - \Phi(x) \right| < \frac{\varepsilon}{2}.$$

Furthermore,

$$\frac{E\left(S_\phi^+ - S_\psi^+ + ES_\psi^+ - ES_\phi^+\right)^2}{\beta^2 \operatorname{Var} S_\phi^+} \le \frac{N\eta \max\left(\gamma^2, \beta^2\varepsilon/4\right) \max_i c_i^2}{\beta^2 N\eta \max c_i^2} \le \frac{\varepsilon}{4},$$

using (3.4.38) and (3.4.45). Thus (3.4.40), (3.4.41), and (3.4.42) hold. Hence,

$$\sup_x \left| P\left(\frac{S_\phi^+ - ES_\phi^+}{\sqrt{\operatorname{Var} S_\phi^+}} < x \right) - \Phi(x) \right| < \varepsilon. \qquad (3.4.46)$$

The proof is complete.

We now consider the specializations of Theorems 3.4.1 and 3.4.2 for the cases when the distribution functions $F_1, \ldots, F_N$ are "nearly" identical or "nearly" symmetric or identical. Such are most of the cases encountered in applications (see e.g. Hájek and Šidák, 1967; Puri and Sen, 1971).

First we prove the following lemmas.

Lemma 3.4.5. Assume that for every $\varepsilon > 0$ and $\eta > 0$, there exists a $\delta_{\varepsilon\eta} > 0$ such that

$$\sum_{i=1}^N c_i^2 > \left(\max_{1 \le i \le N} c_i^2 \right) \delta_{\varepsilon\eta}^{-1}, \qquad (3.4.47)$$

$$\max_{i,j,x} \left| F_i(x) - F_j(x) \right| < \delta_{\varepsilon\eta}, \qquad (3.4.48)$$

and

$$\operatorname{Var} U_1 \ge \eta, \qquad (3.4.49)$$

where

$$U_1 = s(X_i)\phi\left(F_1^*(|X_1|)\right) + \int s(x)(u|x| - |X_1|)\phi'\left(F_1^*(|x|)\right) dF_1(x).$$

Then

$$d^2 > \varepsilon^{-1} \max_{1 \le i \le N} c_i^2, \qquad (3.4.50)$$

where

$$d^2 = \sum_{i=1}^N \operatorname{Var} V_i \qquad (3.4.51)$$

and

$$V_i = c_i s(X_i)\phi(F_i^*(|X_i|)) + \bar{c}\int s(x)u(|x| - |X_i|)\phi'(F_i^*(|x|))\,dF_i(x). \tag{3.4.52}$$

Proof. Set

$$Y_i = s(X_i)\phi(F_i^*(|X_i|)),$$

$$Q_i = \int_{x \geq |X_i|} \phi'(F_i^*(|x|))\,d[F_i(x) + F_i(-x)] \tag{3.4.53}$$

Then we note that

$$\max_{i,j}|\operatorname{Var} Y_i - \operatorname{Var} Y_j| < C_1 \max_{i,j,x}|F_i(x) - F_j(x)|, \tag{3.4.54}$$

$$\max_{i,j}|\operatorname{Var} Q_i - \operatorname{Var} Q_j| < C_2 \max_{i,j,x}|F_i(x) - F_j(x)|, \tag{3.4.55}$$

$$\max_{i,j}\left|\operatorname{Cov}(Y_i, Q_i) - \operatorname{Cov}(Y_j, Q_j)\right| < C_3 \max_{i,j,x}|F_i(x) - F_j(x)|, \tag{3.4.56}$$

and for every i $(= 1, \ldots, N)$,

$$\operatorname{Var} Y_i \geq \int_0^1 \phi^2(u)\,du - \left(\int_0^1 |\phi(u)|\,du\right)^2, \tag{3.4.57}$$

where C_1, C_2, C_3 are constants depending on ϕ only. Now

$$\begin{aligned} d^2 &= \sum_{i=1}^{N} \operatorname{Var}(c_i Y_i + \bar{c} Q_i) \\ &= \sum_{i=1}^{N} c_i^2 \operatorname{Var} Y_1 + N\bar{c}^2 \operatorname{Var} Q_1 + 2N\bar{c}^2 \operatorname{Cov}(Q_1, Y_1) \\ &\quad + \sum_{i=1}^{N} c_i^2(\operatorname{Var} Y_i - \operatorname{Var} Y_1) + \bar{c}^2 \sum_{i=1}^{N} (\operatorname{Var} Q_i - \operatorname{Var} Q_1) \\ &\quad + 2\bar{c} \sum_{i=1}^{N} c_i \{\operatorname{Cov}(Q_i, Y_i) - \operatorname{Cov}(Q_1, Y_1)\}. \end{aligned} \tag{3.4.58}$$

Using (3.4.54) to (3.4.56), we find that the right side of (3.4.58) is not less than

$$\sum_{i=1}^{N} c_i^2 \operatorname{Var} Y_1 + Nc^{-2}[\operatorname{Var} Q_1 + 2\operatorname{Cov}(Q_1, Y_1)] - \sum_{i=1}^{N} c_i^2 (C_1 + C_2 + 2C_3) M, \tag{3.4.59}$$

where

$$M = \max_{i,j,x} |F_i(x) - F_j(x)|. \tag{3.4.60}$$

Now either $\operatorname{Var} Q_1 + 2\operatorname{Cov}(Q_1, Y_1) \geq 0$ or < 0. In the former case,

$$d^2 \geq \sum_{i=1}^{N} c_i^2 [\operatorname{Var} Y_1 - M(C_1 + C_2 + 2C_3)], \tag{3.4.61}$$

and in the latter case,

$$d^2 \sum_{i=1}^{N} c_i^2 [\operatorname{Var}(Y_1 + Q_1) - M(C_1 + C_2 + 2C_3)]. \tag{3.4.62}$$

Let now $C^* = \min\{\operatorname{Var} Y_1, \operatorname{Var}(Y_1 + Q_1)\}$. Then from (3.4.60) and (3.4.61), we infer on using (3.4.47) and (3.4.48) that

$$\begin{aligned} d^2 &\geq \sum_{i=1}^{N} c_i^2 \{C^* - M(C_1 + C_2 + 2C_3)\} \\ &\geq \max_i c_i^2 \delta_{\varepsilon\eta}^{-1} \{C^* - \delta_{\varepsilon\eta}(C_1 + C_2 + 2C_3)\} \\ &\geq \varepsilon^{-1} \max_i c_i^2 \quad \text{if} \quad \delta_{\varepsilon\eta} < C^*(C_1 + C_2 + 2C_3 + \varepsilon^{-1})^{-1}. \end{aligned} \tag{3.4.63}$$

This proves (3.4.50).

Lemma 3.4.6. There is a constant $C_4 = C_4(\phi)$ such that

$$\left| \frac{\sigma}{d} - 1 \right| \leq C_4 M (C^* - M)^{-1/2}, \tag{3.4.64}$$

where σ and M are given by (3.4.3) and (3.4.60) respectively.

Proof. First note that (Exercise 3.4.2)

$$\operatorname{Var}(Z_i - Y_i) \le 8M^2\left(c_i^2 K_1 + N^{-1}\left(2\sqrt{K_2} + \sqrt{K_1}\right)^2 \sum_{j=1}^{N} c_j^2\right), \tag{3.4.65}$$

where K_1 and K_2 are given by (3.3.20) and (3.3.9) respectively. Using this relation, we obtain

$$|\sigma - d| \le \left\{\operatorname{Var}\left(\sum_{i=1}^{N}(Z_i - V_i)\right)\right\}^{1/2} \le \left\{\sum_{i=1}^{N}\operatorname{Var}(Z_i - V_i)\right\}^{1/2}$$

$$\le M\left\{8\left[K_1 + \left(2\sqrt{K_2} + \sqrt{K_1}\right)^2\right]\right\}^{1/2} \sum_{j=1}^{N} c_j^2. \tag{3.4.66}$$

The proof follows by using (3.4.63), and (3.4.64) with $C_4 = \{8[K_1 + (2\sqrt{K_2} + \sqrt{K_1})^2]\}^{1/2}$.

Theorem 3.4.7. *If the assumptions of Theorem* 3.4.1 *and of Lemma* 3.4.5 *are satisfied and if ϕ is not a constant, then*

$$\lim_{N\to\infty} P\left[\frac{S_N^+ - ES_N^+}{d} < x\right] = \Phi(x), \tag{3.4.67}$$

where d is given by (3.4.51) *and* (3.4.52). *The assertion* (3.4.67) *remains true if we replace ES_N^+ by*

$$\mu = \sum_{i=1}^{N} c_i \int s(x)\phi(H^*(|x|))\, dF_i(x). \tag{3.4.68}$$

Proof. By Theorem 3.4.1, if there exists a $K_{\varepsilon/2}$ such that

$$\sigma_N^2 > K_{\varepsilon/2} \max_i c_i^2 \tag{3.4.69}$$

then

$$\sup_x \left| P\left(\frac{S_N^+ - ES_N^+}{\sigma_N} < x\right) - \Phi(x)\right| < \frac{\varepsilon}{2}, \tag{3.4.70}$$

and

$$\sup_x \left| P\left(\frac{S_N^+ - \mu}{\sigma} < x \right) - \Phi(x) \right| < \frac{\varepsilon}{2}. \tag{3.4.71}$$

Now choose $\beta > 0$, $\gamma > 0$ as in Lemma 3.2.3. Then by Lemma 3.4.5, it follows that

$$d^2 > (1-\gamma)^{-2} K_{\varepsilon/2} \left\{ \max_{1 \le i \le N} c_i^2 \right\}, \tag{3.4.72}$$

by taking $\delta_{\varepsilon\eta}$ sufficiently small. Let

$$\delta_{\varepsilon\eta} < \min\left\{ \frac{\gamma^2 C^*}{C_4^2 + (C_1 + C_2 + 2C_3)\gamma^2}, 1 \right\};$$

then

$$\gamma \ge C_4^2 \delta_{\varepsilon\eta} \left[C^* - (C_1 + C_2 + 2C_3)\delta_{\varepsilon\eta} \right]^{-1},$$

and hence, using (3.4.47), (3.4.48), and Lemma 3.4.6,

$$\left| \frac{\sigma}{d} - 1 \right| \le \gamma, \tag{3.4.73}$$

which implies

$$\sigma^2 \ge (1-\gamma)^2 d^2 \ge K_{\varepsilon/2} \left\{ \max_{1 \le i \le N} c_i^2 \right\}, \tag{3.4.74}$$

so that (3.4.69) holds. Hence (3.4.69) and (3.4.70) hold. The result now follows as a simple application of Lemma 3.2.3.

In Theorem 3.4.7 we considered the "nearly identical" case under the assumption that the score function ϕ has a bounded second derivative. We now consider the same case under the assumption that ϕ is absolutely continuous inside (0, 1). Precisely, we have the following analogue of Theorem 3.4.3.

Theorem 3.4.8. *Under the assumptions of Theorem* 3.4.2, (3.4.47), (3.4.48), *and*

$$\sum_{i=1}^N c_i^2 > N\eta \max_{1 \le i \le N} c_i^2 \quad \textit{for} \quad N > N_{\varepsilon\eta}, \tag{3.4.75}$$

then

$$\lim_{N\to\infty} P\left[\frac{S_N^+ - ES_N^+}{d} < x\right] = \Phi(x). \tag{3.4.76}$$

Since the proof of this theorem follows from Theorem 3.4.7 as the proof of Theorem 3.4.2 did from Theorem 3.4.1, it is left as an exercise.

We now consider the case when $F_1, \ldots, F_N$ are identical.

Theorem 3.4.9 (Hušková, 1970). *Let $X_1, \ldots, X_N$ be independent r.v.'s each having a continuous c.d.f. F. Under the assumptions of Theorem 3.4.2, if, for every $\varepsilon > 0$, there exists a $\delta_\varepsilon > 0$ such that*

$$\sum_{i=1}^{N} c_i^2 > \delta_\varepsilon^{-1} \max_{1\le i\le N} c_i^2, \tag{3.4.77}$$

then (3.4.76) holds.

The proof of this theorem is given as an exercise.

We now state the theorem for the case when $F_1, \ldots, F_N$ are "nearly symmetrical" as well as "nearly identical".

Theorem 3.4.10 (Hušková, 1970). *Let the assumptions of Theorem 3.4.1 be satisfied. If, for every $\varepsilon > 0$, there exists an $\eta_\varepsilon > 0$ such that*

$$\sum_{i=1}^{N} c_i^2 > \eta_\varepsilon^{-1} \max_{1\le i\le N} c_i^2 \tag{3.4.78}$$

and

$$\max_{i,j,x} \left| F_i(x) + F_j(-x) - 1 \right| < \eta_\varepsilon, \tag{3.4.79}$$

then

$$\lim_{N\to\infty} P\left(\frac{S_N^+ - ES_N^+}{d^*} < x\right) = \Phi(x), \tag{3.4.80}$$

where

$$d^{*2} = \left(\int_0^1 \phi^2(u)\, du\right) \sum_{i=1}^{N} c_i^2. \tag{3.4.81}$$

Proof. To avoid inessential trivialities, we assume ϕ is not a constant. Denote

$$C_\phi = \int_0^1 \phi^2(u)\,du - \left(\int_0^1 |\phi(u)|\,du\right)^2 > 0, \qquad \eta_1 = C_{\phi/2}, \tag{3.4.81a}$$

and

$$V_i = c_i Y_i + \bar{c} Q_i, \tag{3.4.81b}$$

where Y_i and Q_i are given by (3.4.53).

To prove the theorem, it suffices to show that the assumptions of Theorem 3.4.7 are satisfied and that we can replace d by d^* in (3.4.67). The assumption (3.4.47) holds because of (3.4.78), and the assumption (3.4.48) holds because of (3.4.79), since $\max_{i,j,x}|F_i(x) - F_j(x)| \le 2\max_{i,j,x}|F_i(x) + F_j(-x) - 1|$. To establish (3.4.64), note that

$$|Q_1| \le \left(\sqrt{K_1} + \sqrt{K_1}\right)\eta_\varepsilon, \tag{3.4.82}$$

and thus

$$\sqrt{\mathrm{Var}(Y_i + Q_i)} \ge \sqrt{\mathrm{Var}\,Y_1} - \sqrt{\mathrm{Var}\,Q_1} \ge \sqrt{C_\phi} - \left(\sqrt{K_1} + \sqrt{K_2}\right)\eta_\varepsilon \ge \eta, \tag{3.4.83}$$

which implies (3.4.62). Thus the assumptions of Theorem 3.4.7 are satisfied. Hence (3.4.67) holds. The theorem now follows by showing that $|d/d^* - 1| \le \gamma$.

Remark. The assertion (3.4.80) also holds under the assumptions of Theorem 3.4.2, (3.4.75), and (3.4.79). The reader is invited to verify this.

In Theorem 3.4.7, we have considered the possibility of replacing ES_N^+ by μ [defined by (3.4.68)] when the score function has a bounded second derivative. As in Theorem 2.6.1, this result holds for a more general case of unbounded score functions. Towards this, we present the following theorem (cf. Puri and Ralescu, 1984a).

Theorem 3.4.11. *Let $\phi(t) = \phi_1(t) - \phi_2(t)$, $t \in (0,1)$, satisfy the conditions of Theorem 3.4.2 with the square-integrability condition on ϕ_1 and ϕ_2 replaced* *by*

$$\int_0^1 \{t(1-t)\}^{1/2}\, d\phi_j(t) < \infty, \qquad j = 1,2. \tag{3.4.84}$$

Then the assertion (3.4.67) *remains true if we replace* ES_N^+ *by* μ, *defined in* (3.4.68).

The proof of this theorem runs parallel to that of Theorem 2.6.1 (considered in detail in Section 2.6), and hence is left as an exercise. The results in the last two theorems simplify to some extent for local alternatives satisfying the contiguity criterion discussed in Section 2.7. These are presented briefly in the concluding section of this chapter.

3.5. ASYMPTOTIC NORMALITY OF S_N^+ UNDER CONTIGUOUS ALTERNATIVES

As in Section 2.7, the basic idea in this section is to approximate S_N^+ by a statistic (T_N) expressible as a sum of independent r.v.'s (when the null hypothesis or some contiguous alternative holds) and to incorporate the asymptotic-normality results on T_N to yield parallel results on S_N^+. With this in mind, we define F_i^* as in Lemma 3.2.1 and assume that under $H_0: F_1 = \cdots = F_N = F$, F^* is the common form of the F_i^*. Let then

$$T_N = \sum_{i=1}^{N} c_i s(X_i)\phi(F^*(|X_i|)), \qquad N \geq 1, \tag{3.5.1}$$

$$\phi_N(u) = a_N(i) \qquad \text{for} \quad \frac{i-1}{N} < u \leq \frac{i}{N}, \quad i = 1,\dots,N, \tag{3.5.2}$$

where the scores $a_N(i)$ and the score generating function ϕ are defined as in Section 3.2. We assume further that under H_0, F is symmetric about 0 and

$$\int_0^1 \{\phi_N(u) - \phi(u)\}^2\, du \to 0 \qquad \text{as} \quad N \to \infty. \tag{3.5.3}$$

Note that in (3.5.3), the square-integrability of ϕ is tacitly assumed, though the absolute continuity of ϕ may not be needed. Against H_0, we consider an arbitrary sequence $\{K_N\}$ of alternative hypotheses, such that Q_N, the probability measure under K_N, is contiguous to P_N, that under H_0. We term $\{K_N\}$ a sequence of contiguous alternatives.

Theorem 3.5.1. *Under* (3.5.3) *and any contiguous alternative* $\{K_N\}$,

$$\frac{|S_N^+ - T_N|}{C_N^*} \to 0 \quad \textit{in probability} \qquad \textit{as} \quad N \to \infty, \tag{3.5.4}$$

where $C_N^{*2} = \sum_{i=1}^N c_i^2$.

Outline of the Proof. Note that under H_0 (with F symmetric about 0), sgn X_i and X_i are independent, while the ranks R_{Ni}^+ do not depend on the signs of the X_i. Hence, under H_0,

$$C_N^{*-2} E_0 (S_N^+ - T_N)^2 = E_0 \{ a_N(R_{N1}^+) - \phi(F^*(|X_1|)) \}^2. \tag{3.5.5}$$

Therefore, proceeding as in (2.7.12) through (2.7.19), we conclude that under (3.5.3), the right-hand side of (3.5.5) converges to 0 as $N \to \infty$, so that, by (3.5.5) and the Chebyshev inequality, under H_0 and (3.5.3),

$$C_N^{*-1} |S_N^+ - T_N| \to 0 \quad \text{in probability} \qquad \text{as} \quad N \to \infty. \tag{3.5.6}$$

Finally, the contiguity of Q_N to P_N and (3.5.6) insure that (3.5.4) holds under $\{K_N\}$ as well, Q.E.D.

Let us now assume that

$$\lim_{N \to \infty} \left\{ \max_{1 \le i \le N} (c_i^2 / C_N^{*2}) \right\} = 0 \tag{3.5.7}$$

and

$$0 < A^2 = \int_0^1 \phi^2(u)\, du < \infty. \tag{3.5.8}$$

Then, proceeding as in (2.7.23)–(2.7.26), we note that T_N / C_N^* involves independent summands which satisfy the UAN (uniformly asymptotically negligible) condition, and under H_0, (3.5.7), and (3.5.8),

$$(T_N / C_N^*) \text{ is asymptotically } \mathcal{N}(0, A^2). \tag{3.5.9}$$

Note that (3.5.6) and (3.5.9) insure the asymptotic normality of S_N^+ / C_N^* under H_0. We proceed next to define suitable alternatives for which contiguity holds and T_N / C_N^* is asymptotically normal, and then use Theorem 3.5.1 to derive the asymptotic normality of S_N^+ / C_N^* under the same alternatives.

Consider a (double) sequence $\{X_{N1}, \ldots, X_{NN}\}$ of (row-wise) independent r.v.'s having d.f.'s $\{F_{N1}, \ldots, F_{NN}\}$, $N \geq 1$, where

$$F_{Ni}(x) = F(x - d_{Ni}), \qquad i = 1, \ldots, N, \quad x \in R; \tag{3.5.10}$$

F is an absolutely continuous d.f. symmetric about 0, and its pdf f is also absolutely continuous with a finite Fisher information $I(f)$, defined by (2.7.28); and the constants d_{Ni} satisfy the following:

$$\sup_N D_N^* < \infty \quad \text{and} \quad \lim_{N \to \infty} \left\{ \max_{1 \leq i \leq N} \left(d_{Ni}^2 / D_N^{*2} \right) \right\} = 0, \tag{3.5.11}$$

where

$$D_N^{*2} = \sum_{i=1}^{N} d_{Ni}^2. \tag{3.5.12}$$

Note that the null hypothesis H_0 relates to the case of $d_{N1} = \cdots = d_{NN} = 0$. Since (3.5.10) is a special case of (2.7.27), with the additional assumption of symmetry of F around 0, the contiguity of probability measures under (3.5.10)–(3.5.11) (to that under H_0) follows precisely as in Section 2.7, and hence the details are omitted. We may note that as under H_0 we have $\bar{d}_N = 0$, in (2.7.31) we need to replace $\bar{d}_N$ by 0, and in subsequent formulae D_N by D_N^*. Let us denote the resulting statistics in (2.7.31) with $\bar{d}_N = 0$ as T_N^*. Then, (2.7.34) holds with S_N^* replaced by T_N^*, and in (2.7.40)–(2.7.42) we need to replace $C_N^{-1} S_N$, $D_N^{-1} S_N^*$, γ, and ρ_N by T_N / C_N^*, T_N^* / D_N^*, γ^*, and ρ_N^*, respectively, where

$$\gamma^* = \int_{-\infty}^{\infty} \operatorname{sgn}(x) \phi(F^*(|x|)) \left\{ -\frac{f'(x)}{f(x)} \right\} dF(x), \tag{3.5.13}$$

$$\rho_N^* = \left(\sum_{i=1}^{N} c_i d_{Ni} \right) \Big/ (C_N^* D_N^*). \tag{3.5.14}$$

Therefore, by Lemma 2.7.3 and the above discussion, we arrive at the following.

Theorem 3.5.2. *Under* (3.5.10) *and* (3.5.11), $\{T_N / C_N^* - D_N^* \rho_N^* \gamma^*\}$ *has asymptotically the standard normal distribution, and hence, additionally, under* (3.5.3),

$$S_N^+ / C_N^* - D_N^* \rho_N^* \gamma^* \textit{ is asymptotically } \mathcal{N}(0, A^2). \tag{3.5.15}$$

In (3.5.10), the constants d_{Ni} have been chosen rather arbitrarily. In many practical problems (such as we shall encounter in Chapters 5, 6, and 7), there are some natural choices for these constants. For example, in the one-sample location problem, we have $d_{Ni} = N^{-1/2}\theta$, $i = 1, \dots, N$, where θ is some real constant and the c_i are all equal to 1. Thus, here ρ_N^* reduces to 1, and D_N^* to θ. Further, if we let $\phi(u) = \phi^*((1+u)/2)$, $0 < u < 1$, where ϕ^* is skew-symmetric [i.e., $\phi^*(u) + \phi^*(1-u) = 0$, $0 < u < 1$], then γ^* reduces to

$$\int_{-\infty}^{\infty} \phi^*(F(x))\left\{-\frac{f'(x)}{f(x)}\right\} dF(x). \tag{3.5.16}$$

These simplifications apply to most practical problems. Finally, as has been noted earlier, the absolute continuity of ϕ is not needed in Theorem 3.5.2.

EXERCISES

3.2.1. Provide a formal proof of Lemma 3.2.1.

3.2.2. Provide a proof of Lemma 3.2.2.

3.3.1. Prove the inequality in (3.3.14).

3.3.2. Prove the inequality in (3.3.16).

3.3.3. Prove the inequality in (3.3.33).

3.3.4. Prove the identity in (3.3.39).

3.4.1. Use the decomposition in (3.4.34) to write $T_\phi^+ = T_\psi^+ + T_{\phi_{(1)}}^+ - T_{\phi_{(2)}}^+$. Show that for the second and third components, the variance inequality and (3.4.8) lead to their asymptotic negligibility in the normality of the standardized form of T_ϕ^+. For T_ψ^+, use the fact that ψ is a continuous function (with continuous derivatives) inside (0, 1) and the results in Theorem 3.4.1 to establish its asymptotic normality. (Hušková, 1970.)

3.4.2. Prove the variance inequality in (3.4.65). (Hušková, 1970.)

3.4.3. Provide a proof of Theorem 3.4.8. (Hušková, 1970.)

3.4.4. Provide a proof of Theorem 3.4.9. (Hušková, 1970.)

3.4.5. Prove that (3.4.80) holds under the assumptions of Theorem 3.4.2, (3.4.75), and (3.4.79). (Hušková, 1970.)

3.4.6. Provide a formal proof of Theorem 3.4.11.

Hints. Analogous to Lemma 2.6.3, show that under the hypothesis of Theorem 3.4.11,

$$\sum_{i=1}^{N}\left|E\left[s(X_i)\phi\left(\frac{R_i^+}{N+1}\right)-\int_{-\infty}^{\infty}s(x)\phi(F^*(|x|))\,dF_i(x)\right]\right|\le K_1N^{1/2}J(\phi),$$

where K_1 is some positive constant. Use then Lemma 2.6.4, and proceed precisely as in the proof of Theorem 2.6.1.

3.5.1. Establish the contiguity of the sequence of probability measures under (3.5.10)–(3.5.12) to that under $H_0: d_{N1} = \cdots = d_{NN} = 0$. (Hâjek, 1962.)

3.5.2. Provide a formal proof of Theorem 3.5.2.

3.5.3. Let $\phi(u) = \phi_1(u) - \phi_2(u)$, $0 < u < 1$, satisfy the conditions of Theorem 3.4.2 with the square-integrability condition on ϕ_1 and ϕ_2 replaced by (2.6.8). Assume further that $a_N(i) = E\phi(U_N^{(i)})$ or $\Sigma_{i=1}^N|a_N(i) - \phi(i/(N+1))| = o(N^{1/2})$, where $U_N^{(i)}$ is the ith order statistic in a random sample of size N from the rectangular (0, 1) distribution. Then the assertions of Theorems 3.4.2 hold with $E(S_N^+)$ replaced by $\mu_N^+ = \Sigma_{i=1}^N c_i \int_{-\infty}^{\infty} \operatorname{sgn}(x)\,\phi(H^*(|x|))\,dF_i(x)$. (Puri and Ralescu, 1984a.)

CHAPTER 4

Distribution Theory of Multivariate Rank-Order Statistics

4.1. INTRODUCTION

The results concerning the distribution of simple linear and signed rank statistics studied in the preceding two chapters are extended here to vector-valued random variables. In Section 4.2, for independent p-dimensional random vectors ($p \geq 1$) with continuous distributions, the joint distribution of the coordinatewise linear rank statistics is studied, and this extends the results of Chapter 2 to the multivariate case. The theory is mainly discussed in Puri and Sen (1969a), and is mostly adopted from there. Section 4.3 is devoted to the distribution theory of coordinatewise signed rank statistics in the general multivariate case, and the results are mainly due to Hušková (1971). The last section deals with the characterization of the multivariate multisample problem and the multivariate paired-comparison problem in terms of the general linear model of the preceding two sections. Unlike the univariate case, the null-hypothesis distributions of rank statistics in the multivariate case depend on the underlying distributions, and this calls for the development of the corresponding permutational (conditional) distribution theory because of its vital role in hypotheses testing. This theory will be developed in the next chapter; in the present one, we confine ourselves exclusively to unconditional asymptotic distribution-theory problems.

4.2. MULTIVARIATE LINEAR RANK STATISTICS

Let $\mathbf{X}_1, \ldots, \mathbf{X}_N$ be independent stochastic vectors with continuous $p(\geq 1)$-variate d.f.'s $F_1(\mathbf{x}), \ldots, F_N(\mathbf{x})$ respectively, all defined on the p-dimensional Euclidean space R^p. We denote

$$\mathbf{X}_\alpha = (X_{\alpha 1}, \ldots, X_{\alpha p})', \qquad \alpha = 1, \ldots, N \tag{4.2.1}$$

and the jth marginal d.f. of $\mathbf{X}_\alpha$ by $F_{\alpha[j]}(x) = P\{X_{\alpha j} \le x\}$, $i \le j \le p$, $1 \le \alpha \le N$. By assumption, all the $F_{\alpha[j]}$ are continuous, so that for every j $(= 1, \dots, p)$, the observations $X_{1j}, \dots, X_{Nj}$ are distinct with probability one. As in (2.2.1), we define the coordinatewise ranks by

$$R_{Ni}^{(j)} = \sum_{\alpha=1}^{N} u(X_{ij} - X_{\alpha j}), \qquad 1 \le i \le N, \tag{4.2.2}$$

so that $(R_{N1}^{(j)}, \dots, R_{NN}^{(j)})$ is some permutation of the first N integers $(1, \dots, N)$, for each $j = 1, \dots, p$. The corresponding matrix

$$\underset{(p\times N)}{\mathbf{R}_N} = \begin{pmatrix} R_{N1}^{(1)} & \cdots & R_{NN}^{(1)} \\ \vdots & & \vdots \\ R_{N1}^{(p)} & \cdots & R_{NN}^{(p)} \end{pmatrix} = \begin{pmatrix} \mathbf{R}_N^{(1)} \\ \vdots \\ \mathbf{R}_N^{(p)} \end{pmatrix} \tag{4.2.3}$$

is called a *rank-collection matrix*. The statistics to be considered here depend on $\mathbf{X}_1, \dots, \mathbf{X}_N$ only through $\mathbf{R}_N$. To define them, we consider for each j $(= 1, \dots, p)$ a set of N *rank scores*

$$a_{Nj}(i) = E\phi_j(U_N^{(i)}) \text{ or } \phi_j\left(\frac{i}{N+1}\right), \qquad 1 \le i \le N, \tag{4.2.4}$$

where the order statistics $U_N^{(1)} < \cdots < U_N^{(N)}$ and the score functions $\phi_j(u)$, $0 < u < 1$, for $j = 1, \dots, p$, are defined as in Section 2.2. Note that the rank-collection matrix $\mathbf{R}_N$ generates the *score-collection matrix*

$$\mathbf{A}_N = \begin{pmatrix} a_{N1}(R_{N1}^{(1)}) & \cdots & a_{N1}(R_{NN}^{(1)}) \\ \vdots & & \vdots \\ a_{Np}(R_{N1}^{(p)}) & \cdots & a_{Np}(R_{NN}^{(p)}) \end{pmatrix} = \begin{pmatrix} \mathbf{a}_N^{(1)} \\ \vdots \\ \mathbf{a}_N^{(p)} \end{pmatrix}. \tag{4.2.5}$$

A multivariate linear rank statistic is a vector of simple linear rank statistics constructed from $\mathbf{a}_N^{(1)}, \dots, \mathbf{a}_N^{(p)}$ (the rows of $\mathbf{A}_N$) is the same manner as in (2.2.3). For this reason we introduce the matrix of (known) regression constants

$$\begin{pmatrix} c_{11} & \cdots & c_{1q} \\ \vdots & \ddots & \vdots \\ c_{N1} & \cdots & c_{Nq} \end{pmatrix} = (\mathbf{c}_1, \dots, \mathbf{c}_q), \qquad q \ge 1. \tag{4.2.6}$$

Then, we define the desired rank statistics by

$$\mathbf{L}_N = ((L_{N,jl})), \qquad j = 1,\dots,p, \quad l = 1,\dots,q, \tag{4.2.7}$$

where

$$L_{N,jl} = \mathbf{a}_N^{(j)\prime}\mathbf{c}_l = \sum_{i=1}^{N} c_{il} a_{Nj}\left(R_{Ni}^{(j)}\right), \qquad 1 \le j \le p, \quad 1 \le l \le q. \tag{4.2.8}$$

Note that if for some l $(1 \le l \le q)$, $\mathbf{c}_l$ can be expressed as a linear combination of $\mathbf{c}_{l'}$, $l'(\neq l) = 1,\dots,q$, then the corresponding $L_{N,jl}$ can be expressed as a linear combination of $L_{N,jl'}$, $l'(\neq l) = 1,\dots,q$, for every $1 \le j \le p$. Consequently, the q columns of $\mathbf{L}_N$ will not be linearly independent. Thus, without any loss of generality, we assume that $(\mathbf{c}_1,\dots,\mathbf{c}_q)$ are all linearly independent, as otherwise we can always work with some q' $(< q)$ for which $(c_1,\dots,c_{q'})$ are linearly independent. This requires that $N \ge q$, since otherwise $\mathbf{c}_1,\dots,\mathbf{c}_q$ will be linearly dependent. Now let us define

$$C_{N,ll'} = \frac{1}{N}\sum_{i=1}^{N}(c_{il} - \bar{c}_{N,l})(c_{il'} - \bar{c}_{N,l'}), \qquad l, l' = 1,\dots,q, \tag{4.2.9}$$

where

$$\bar{c}_{N,l} = \frac{1}{N}\sum_{i=1}^{N} c_{il}, \qquad l = 1,\dots,q. \tag{4.2.10}$$

Then, our first assumption is that for every $N \ge N_0$,

$$\mathbf{C}_N = ((C_{N,ll'})) \text{ is positive definite (p.d.).} \tag{4.2.11}$$

Also note that though each row of $\mathbf{R}_N$ is some permutation of $(1,\dots,N)$ [so that $\mathbf{R}_N$ has $(N!)^p$ possible realizations], the distribution of $\mathbf{R}_N$ depends on the joint distribution of the $(R_{Ni}^{(1)},\dots,R_{Ni}^{(p)})$, which in turn depends on the d.f.'s $F_1,\dots,F_N$. An exceptional case occurs when $F_1 = \cdots = F_N \equiv F$ and the p variates are independent—in which case $\mathbf{R}_N$ has all $(N!)^p$ realizations equally likely. However, this special case is of limited interest, and in general we have the distribution of $\mathbf{R}_N$ dependent on $F_1,\dots,F_N$. Consequently, $\mathbf{L}_N$ is not generally a distribution-free statistic.

In the spirit of Chapter 2, we shall develop here the asymptotic distribution theory of $\mathbf{L}_N$ when $F_1,\dots,F_N$ are arbitrary. Our main concern will be to establish the asymptotic multinormality of $\mathbf{L}_N$. To this end, we first introduce the following

Definition 4.2.1. Let $\{X_\nu\}$ be a sequence of stochastic p-vectors such that there exist a sequence $\{\mu_\nu\}$ of p-vectors and a sequence $\{D_\nu\}$ of p.d. symmetric matrices for which for every $\mathbf{l} \neq \mathbf{0}$,

$$\mathscr{L}\left(\frac{\mathbf{l}'[X_\nu - \mu_\nu]}{(\mathbf{l}'\mathbf{D}_\nu\mathbf{l})^{1/2}}\right) \to \mathscr{N}(0,1) \quad \text{as} \quad \nu \to \infty. \tag{4.2.12}$$

Then we say that $\mathbf{X}_\nu$ is *asymptotically p-variate normal* with mean $\boldsymbol{\mu}_\nu$ and dispersion matrix $\mathbf{D}_\nu$; in symbols, we write

$$\mathbf{X}_\nu \sim \mathscr{N}_p(\boldsymbol{\mu}_\nu, \mathbf{D}_\nu) \quad \text{as} \quad \nu \to \infty. \tag{4.2.13}$$

If $\mathbf{X}_\nu$ is a $p \times q$ matrix and $\boldsymbol{\mu}_\nu$ is a $p \times q$ matrix, we can convert $\mathbf{X}_\nu$ as well as $\boldsymbol{\mu}_\nu$ into a $pq \times 1$ matrix (i.e. a pq-vector), and then define a corresponding $\mathbf{D}_\nu$ of order $pq \times pq$ in which (4.2.13) holds with p replaced by pq. In that case, we write

$$\mathbf{X}_\nu \sim \mathscr{N}_{pq}(\boldsymbol{\mu}_\nu, \mathbf{D}_\nu) \quad \text{as} \quad \nu \to \infty. \tag{4.2.14}$$

Definition 4.2.2. For a $p \times q$ stochastic matrix $\mathbf{X} = ((X_{ij}))_{i=1,\dots,p;\, j=1,\dots,q}$, we denote by

$$\underset{pq \times pq}{V(\mathbf{X})} = E\big\{[X_{11} - EX_{11}, \dots, X_{pq} - EX_{pq}]' \\ \times [X_{11} - EX_{11}, \dots, X_{pq} - EX_{pq}]\big\} \tag{4.2.15}$$

the dispersion matrix of $\mathbf{X}$.

Definition 4.2.3. For a $p \times p$ matrix $\mathbf{A}$, the roots of the equation

$$|\mathbf{A} - \lambda \mathbf{I}| = 0, \tag{4.2.16}$$

arranged in descending order of magnitudes and denoted by $\lambda_1 \geq \lambda_2 \geq \cdots \geq \lambda_p$, are termed the *characteristic roots* of $\mathbf{A}$. We denote them by

$$\mathrm{Ch}_j(\mathbf{A}) = \lambda_j \quad \text{for} \quad j = 1, \dots, p. \tag{4.2.17}$$

If A is positive semidefinite (p.s.d.), then $\lambda_j \geq 0$ for $1 \leq j \leq p$, while if A is p.d., then $\lambda_j > 0$ for $1 \leq j \leq p$.

We are now in a position to state our first theorem, which extends Theorem 2.5.1 to the multivariate multiparameter case.

Theorem 4.2.1. *For the scores* $\{a_{Nj}(i),\ 1 \le i \le N,\ 1 \le j \le p\}$ *defined by* (4.2.4), *whenever* $\phi_j(u)$ *has a bounded second derivative inside* $(0,1)$ (*for* $1 \le j \le p$), *and the* c_{il}, $1 \le i \le N$, $1 \le l \le q$ *are such that for every* $\varepsilon > 0$ *there exists a positive* K_ε $(< \infty)$ *and an integer* $N_0(\varepsilon)$ *such that for* $N \ge N_0(\varepsilon)$

$$\nu_N = \mathrm{Ch}_{pq}(V(\mathbf{L}_N)) \ge K_\varepsilon \left\{ \max_{1 \le l \le q} \max_{1 \le i \le N} [c_{il} - \bar{c}_{N,l}]^2 \right\}, \tag{4.2.18}$$

then, as $N \to \infty$,

$$\mathbf{L}_N \sim \mathcal{N}_{pq}(E\mathbf{L}_N, V(\mathbf{L}_N)). \tag{4.2.19}$$

The assertion remains true if, in (4.2.19), $V(\mathbf{L}_N)$ *is replaced by* $V(\Sigma_{i=1}^N \mathbf{Z}_i)$, *where*

$$\underset{p \times q}{\mathbf{Z}_i} = ((Z_{i,jl}))$$

with

$$Z_{i,jl} = (N+1)^{-1} \sum_{\alpha=1}^{N} (c_{\alpha l} - c_{il})$$
$$\times \int_{-\infty}^{\infty} \left[u(x - X_{ij}) - F_{i[j]}(x)\right] \phi_j'(H_{[j]}(x))\, dF_{\alpha[j]}(x), \tag{4.2.20}$$

$$H_{[j]}(x) = N^{-1} \sum_{i=1}^{N} F_{i[j]}(x), \qquad j = 1, \dots, p, \tag{4.2.21}$$

and $F_{i[j]}(x)$ *is the d.f. of* X_{ij} *for* $1 \le j \le p$, $1 \le i \le N$.

Proof. By (4.2.18), we have for every $1 \le j \le p$, $1 \le l \le q$,

$$\mathrm{Var}(L_{N,jl}) > K_\varepsilon \max_{1 \le i \le N} (c_{il} - \bar{c}_{N,l})^2, \tag{4.2.22}$$

Now, by using for each j $(= 1, \dots, p)$ and l $(= 1, \dots, q)$ the results of Section 2.4 [namely, (2.4.62)], we conclude that under the hypothesis of the theorem, there exists a positive constant M ($< \infty$) such that

$$E\left(L_{N,jl} - EL_{N,jl} - \sum_{i=1}^{N} Z_{i,jl}\right)^2 \leq MN^{-1}\sum_{i=1}^{N}(c_{il} - \bar{c}_{N,l})^2, \tag{4.2.23}$$

for every $1 \leq j \leq p$, $1 \leq l \leq q$. Therefore, by (4.2.22) and (4.2.23),

$$\frac{E\left(L_{N,jl} - EL_{N,jl} - \sum_{i=1}^{N} Z_{i,jl}\right)^2}{\text{Var}(L_{N,jl})} \leq \frac{M\left\{\sum_{i=1}^{N}(c_{il} - \bar{c}_{N,l})^2\right\}}{K_\varepsilon N\left\{\max_{1\leq i\leq N}[c_{il} - \bar{c}_{N,l}]^2\right\}} < \varepsilon,$$

$$\text{for every} \quad 1 \leq j \leq p, \quad 1 \leq l \leq q. \tag{4.2.24}$$

We now consider the following lemmas:

Lemma 4.2.2. Let $\mathbf{X}_\nu$ and $\mathbf{Y}_\nu$ be two stochastic p-vectors such that $\text{Var}(X_{\nu j} - Y_{\nu j})/\text{Var}(X_{\nu j}) \to 0$ (for every $j = 1, \ldots, p$) as $\nu \to \infty$. Then

$$\frac{[\text{Cov}(X_{\nu j}, X_{\nu j'}) - \text{Cov}(Y_{\nu j}, Y_{\nu j'})]}{\{\text{Var}(X_{\nu j})\text{Var}(X_{\nu j'})\}^{1/2}} \to 0 \quad \text{as} \quad \nu \to \infty \tag{4.2.25}$$

for every $j, j' = 1, \ldots, p$.

The proof follows by expressing

$$\begin{aligned}\text{Cov}(Y_{\nu j}, Y_{\nu j'}) &= \text{Cov}(X_{\nu j}, X_{\nu j'}) + \text{Cov}(Y_{\nu j} - X_{\nu j}, X_{\nu j'}) \\ &\quad + \text{Cov}(X_{\nu j}, Y_{\nu j'} - X_{\nu j'}) \\ &\quad + \text{Cov}(Y_{\nu j} - X_{\nu j}, Y_{\nu j'} - X_{\nu j'}),\end{aligned}$$

applying the Schwarz inequality to the second, third, and fourth terms on the rhs, and making use of the condition stated in the lemma.

Lemma 4.2.3. If $\mathbf{X}_\nu \sim \mathcal{N}_p(E\mathbf{X}_\nu, V(\mathbf{X}_\nu))$ and if $\text{Var}(X_{\nu j} - Y_{\nu j})/\text{Var}(X_{\nu j}) \to 0$ as $\nu \to \infty$ for every $1 \leq j \leq p$, then $Y_\nu \sim \mathcal{N}_p(E\mathbf{Y}_\nu, V(\mathbf{X}_\nu))$ as $\nu \to \infty$.

The proof follows by standard arguments and is left as an exercise (see Exercise 4.2.9).

We now return to the proof of Theorem 4.2.1 and note that by virtue of (4.2.22), (4.2.24), and Lemmas 4.2.2 and 4.2.3, it suffices to show that as $N \to \infty$,

$$\sum_{i=1}^{N} Z_i \sim \mathcal{N}_{pq}(\mathbf{0}, V(\mathbf{L}_N)), \tag{4.2.26}$$

where by (4.2.20) and Fubini's theorem,

$$E\mathbf{Z}_i = \mathbf{0} \qquad \text{for every} \quad 1 \leq i \leq N. \tag{4.2.27}$$

Also, by (4.2.22) and (4.2.24), in (4.2.26) one can replace $V(\mathbf{L}_N)$ by $V(\Sigma_{i=1}^{N}\mathbf{Z}_i)$. To prove (4.2.26), we use the classical Cramér–Wold tool of an arbitrary linear combination of the components of $\Sigma_{i=1}^{N}\mathbf{Z}_i$ and establish its asymptotic normality. With this end in view, let

$$Z_N^*(\mathbf{d}) = \sum_{j=1}^{p}\sum_{l=1}^{q} d_{jl}\sum_{i=1}^{N} Z_{i,jl}, \qquad \mathbf{d} \neq \mathbf{0}, \tag{4.2.28}$$

where the d_{jl} are real constants. Then, by (4.2.20) and (4.2.28),

$$Z_N^*(\mathbf{d}) = \sum_{i=1}^{N}\left\{\sum_{j=1}^{p}\sum_{l=1}^{q} d_{jl}Z_{i,jl}\right\} = \sum_{i=1}^{N} g(\mathbf{X}_i;\mathbf{d}), \quad \text{say}, \tag{4.2.29}$$

where

$$g(\mathbf{X}_i,\mathbf{d}) = (N+1)^{-1}\sum_{\alpha=1}^{N}\sum_{j=1}^{p}\sum_{l=1}^{q} d_{jl}(c_{\alpha l} - c_{il})B_{j(i,\alpha)}(X_{ij}), \tag{4.2.30}$$

$$B_{j(i,\alpha)}(X_{ij}) = \int_{-\infty}^{\infty}\left[u(x - X_{ij}) - F_{i[j]}(x)\right] \times \phi_j'(H_{[j]}(x))\, dF_{\alpha[j]}(x), \tag{4.2.31}$$

for $j = 1, \ldots, p$; $i, \alpha = 1, \ldots, N$. Note that $EB_{j(i,\alpha)}(X_{ij}) = 0$ for all j, i, and α;

$$\begin{aligned} &E\left[B_{j(i,\alpha)}(X_{ij}) \cdot B_{j(i,\beta)}(X_{ij})\right] \\ &= \iint_{-\infty<x<y<\infty} F_{i[j]}(x)\left[1 - F_{i[j]}(y)\right]\phi_j'(H_{[j]}(x))\phi_j'(H_{[j]}(y)) \\ &\qquad \times\left[dF_{\alpha[j]}(x)\, dF_{\beta[j]}(y) + dF_{\beta[j]}(x)\, dF_{\alpha[j]}(y)\right] \end{aligned} \tag{4.2.32}$$

for $i, \alpha, \beta = 1, \ldots, N$, $j = 1, \ldots, p$; and

$$E\left[B_{j(i,\alpha)}(X_{ij})B_{j'(i',\beta)}(X_{i'j'})\right]$$

$$= \begin{cases} \int_{-\infty}^{\infty}\int_{-\infty}^{\infty}\left[F_{i[jj']}(x,y) - F_{i[j]}(x)F_{i[j']}(y)\right]\phi_j'(H_{[j]}(x)) \\ \times\phi_{j'}'(H_{[j']}(y))\,dF_{\alpha[j]}(x)\,dF_{\beta[j']}(y), & \text{if } i = i', \quad j \neq j', \\ 0 & \text{if } i \neq i', \end{cases}$$

(4.2.33)

where $F_{i[jj']}(x, y) = P[X_{ij} \le x,\ X_{ij'} \le y]$ for $j \neq j' = 1, \ldots, p$, $1 \le i \le N$. These expressions will be used subsequently for certain asymptotic simplifications of $V(\mathbf{L}_N)$.

Now, since by assumption ϕ_j has bounded first and second derivatives inside (0, 1), it follows that $B_{j(i,\alpha)}(X_{ij})$, $1 \le j \le p$; $i, \alpha = 1, \ldots, N$, are all bounded random variables. Thus, it suffices to show that the random variables $g(\mathbf{X}_i, \mathbf{d})$, $1 \le i \le N$, are independent and bounded, and

$$\sum_{i=1}^{N} \operatorname{Var}[g(\mathbf{X}_i, \mathbf{d})]$$

can be made arbitrarily large as $N \to \infty$. By (2.5.12) and (4.2.30), it follows that

$$\max_{1 \le i \le N} |g(\mathbf{X}_i, \mathbf{d})| \le \left\{\max_{j,l}|d_{jl}|\right\}\left[\sum_{j=1}^{p}\sum_{l=1}^{q}\left\{\max_{1\le i\le N}|Z_{ij,l}|\right\}\right]$$

is bounded, while by (4.2.24), (4.2.29), and Lemma 4.2.2, for the second assertion we have only to show that for every $\mathbf{d} \neq \mathbf{0}$, as $N \to \infty$,

$$\sum_{j=1}^{p}\sum_{j'=1}^{p}\sum_{l=1}^{q}\sum_{l'=1}^{q}\sum_{i=1}^{N}\operatorname{Cov}(L_{N,jl}, L_{N,j'l'}) \to \infty. \tag{4.2.34}$$

Now the lhs of (4.2.34) is bounded from below by

$$\left(\sum_{j=1}^{p}\sum_{l=1}^{q} d_{jl}^2\right)\operatorname{Ch}_{pq}[V(\mathbf{L}_N)] = \left(\sum_{j=1}^{p}\sum_{l=1}^{q} d_{jl}^2\right)\gamma_N, \tag{4.2.35}$$

which by (4.2.18) can be made arbitrarily large by choosing ε (> 0) arbitrarily small. The theorem follows.

As in Section 2.5, we now consider the case where the marginal and bivariate distributions corresponding to $F_1, \ldots, F_N$ are very close to each other. Thus, we assume that for every $\varepsilon > 0$, there exists a $\delta_\varepsilon > 0$ such that

$$\sup_x \left\{ \max_{i \neq i'} \left| F_{i[j]}(x) - F_{i'[j]}(x) \right| \right\} < \delta_\varepsilon, \tag{4.2.36}$$

$$\sup_{x,y} \left\{ \max_{i \neq i'} \max_{j \neq j'} \left| F_{i[j,j']}(x,y) - F_{i'[j,j']}(x,y) \right| \right\} < \delta_\varepsilon. \tag{4.2.37}$$

Then, under (4.2.36), (4.2.32) reduces to

$$2 \iint_{-\infty < x < y < \infty} H_{[j]}(x)\left[1 - H_{[j]}(y)\right] \phi_j'\left(H_{[j]}(x)\right) \phi_j'\left(H_{[j]}(y)\right) dH_{[j]}(y)$$

$$= \int_{-\infty}^{\infty} \phi_j^2\left(H_{[j]}(x)\right) dH_{[j]}(x) - \left(\int_{-\infty}^{\infty} \phi_j\left(H_{[j]}(x)\right) dH_{[j]}(x) \right)^2$$

$$= \int_0^1 \phi_j^2(u)\, du - \left(\int_0^1 \phi_j(u)\, du \right)^2 = \nu_{jj}, \quad \text{say.} \tag{4.2.38}$$

Also, under (4.2.37) and (4.2.38), (4.2.33) reduces to

$$\int_{-\infty}^{\infty} \int_{-\infty}^{\infty} \left[H_{[j,j']}(x,y) - H_{[j]}(x) H_{[j']}(y) \right]$$

$$\times \phi_j'\left(H_{[j]}(x)\right) \phi_j'\left(H_{[j']}(y)\right) dH_{[j]}(x)\, dH_{[j]}(y)$$

$$= \int_{-\infty}^{\infty} \int_{-\infty}^{\infty} \phi_j\left(H_{[j]}(x)\right) \phi_{j'}\left(H_{[j']}(y)\right) dH_{[j,j']}(x,y)$$

$$- \left(\int_{-\infty}^{\infty} \phi_j\left(H_{[j]}(x)\right) dH_{[j]}(x) \right) \left(\int_{-\infty}^{\infty} \phi_{j'}\left(H_{[j']}(y)\right) dH_{[j']}(y) \right)$$

$$= \int_{-\infty}^{\infty} \int_{-\infty}^{\infty} \phi_j\left(H_{[j]}(x)\right) \phi_{j'}\left(H_{[j']}(y)\right) dH_{[j,j']}(x,y)$$

$$- \left(\int_0^1 \phi_j(u)\, du \right) \left(\int_0^1 \phi_{j'}(u)\, du \right)$$

$$= \nu_{jj'}(H), \quad \text{say.} \tag{4.2.39}$$

Let then

$$\boldsymbol{\nu}(H) = \left(\left(\nu_{jj'}(H)\right)\right)_{j,j'=1,\ldots,p}. \tag{4.2.40}$$

Then, by (4.2.29), (4.2.30), (4.2.31), (4.2.32), (4.2.33), (4.2.38), and (4.2.39), it follows that as $N \to \infty$

$$V(\mathbf{L}_N) - \boldsymbol{\nu}(H) \otimes \mathbf{C}_N \to \mathbf{0}, \tag{4.2.41}$$

where $\otimes$ is the Kronecker product and $\mathbf{C}_N$ is defined in (4.2.9)–(4.2.11). Since $\mathbf{C}_N$ is assumed to be p.d., $V(\mathbf{L}_N)$ will be asymptotically p.d. whenever $\boldsymbol{\nu}(H)$ is p.d. Finally, $\boldsymbol{\nu}(H)$ does not depend on $\mathbf{C}_N$, and hence, for (4.2.18) to hold, we may formulate a parallel condition in terms of $\mathrm{Ch}_q[\mathbf{C}_N]$, i.e., for every $\varepsilon > 0$, there exists a $0 < K_\varepsilon < \infty$ and an integer $N_0(\varepsilon)$ such that for $N \geq N_0(\varepsilon)$,

$$\mathrm{Ch}_q[\mathbf{C}_N] \geq K_\varepsilon \left\{ \max_{1 \leq l \leq q} \max_{1 \leq i \leq N} \left(c_{il} - \bar{c}_{N,l}\right)^2 \right\}. \tag{4.2.42}$$

This leads to the following theorem.

Theorem 4.2.4. *If the* $a_{Nj}(i)$, $1 \leq i \leq N$, $1 \leq j \leq p$, *are defined by* (4.2.4) *and the score functions* ϕ_j, $1 \leq j \leq p$, *all have bounded second derivative inside* $(0,1)$, *then under* (4.2.36), (4.2.37), *and* (4.2.42), *as* $N \to \infty$,

$$\mathbf{L}_N \sim \mathcal{N}_{pq}\left(E\mathbf{S}_N, \boldsymbol{\nu}(H) \otimes \mathbf{C}_N\right), \tag{4.2.43}$$

where H *is the limiting d.f.*

So far we considered the case of score functions with bounded second derivative inside $(0,1)$. In the following, we consider the general case of square-integrable absolutely continuous score functions, and provide a multivariate multiparameter generalization of Theorem 2.5.3.

Theorem 4.2.5. *If the* $a_{Nj}(i)$, $1 \leq i \leq N$, $1 \leq j \leq p$, *are defined by* (4.2.4) *with* $\phi_j(u) = \phi_{j,k}(u)$, $0 < u < 1$, $1 \leq j \leq p$, *where the* $\phi_{j,k}(u)$, $k = 1, 2$, *are all nondecreasing, absolutely continuous, and square-integrable inside* $[0,1]$, *and if for every* $\varepsilon > 0$ *and* $\eta > 0$, *there exists an* $N_{\varepsilon\eta}(< \infty)$ *such that for* $N \geq N_{\varepsilon\eta}$,

$$\mathrm{Ch}_{pq}\left[V(\mathbf{L}_N)\right] > \eta N \left\{ \max_{1 \leq l \leq q} \max_{1 \leq i \leq N} \left(c_{il} - \bar{c}_{N,l}\right)^2 \right\}, \tag{4.2.44}$$

then, as $N \to \infty$,

$$\mathbf{L}_N \sim \mathcal{N}_{pq}(E\mathbf{L}_N, V(\mathbf{L}_N)), \tag{4.2.45}$$

where $V(\mathbf{L}_N)$ *can also be replaced by* $V(\Sigma_{i=1}^N \mathbf{Z}_i)$ *with the* $\mathbf{Z}_i$, $i = 1, \dots, N$, *defined by* (4.2.20).

Proof. We closely follow the line of proof of Theorem 2.5.3. For each j $(= 1, \dots, p)$, by Lemma 2.5.4, we have for every $\varepsilon > 0$

$$\phi_j(u) = \psi_j(u) + \phi_{j(1)}(u) - \phi_{j(2)}(u), \qquad 0 < u < 1, \tag{4.2.46}$$

where ψ_j is a polynomial, $\phi_{j(1)}$ and $\phi_{j(2)}$ are nondecreasing, and

$$\int_0^1 \{\phi_{j(1)}^2(u) + \phi_{j(2)}^2(u)\}\, du < p^{-1}\varepsilon, \qquad 1 \le j \le p. \tag{4.2.47}$$

Now, replacing ϕ_j in (4.2.4) by $\psi_j + \phi_{j,1} - \phi_{j,2}$, we have

$$a_{Nj}(i) = a_{Nj}(i, \psi_j) + a_{Nj}(i, \phi_{j(1)}) - a_{Nj}(i, \phi_{j(2)}), \qquad 1 \le j \le p, \tag{4.2.48}$$

for $1 \le i \le N$. Then $L_{N,jl}$ in (4.2.8) can be written as

$$L_{N,j} = \sum_{i=1}^{N} \left[c_{il} a_{Nj}(R_{Ni}^{(j)}, \psi_j) + a_{Nj}(R_{Ni}^{(j)}, \phi_{j(1)}) - a_{Nj}(R_{Ni}^{(j)}, \phi_{j(2)}) \right]$$

$$= L_{N,jl}(\psi_j) + L_{N,jl}(\phi_{j(1)}) - L_{N,j}(\phi_{j(2)}). \tag{4.2.49}$$

Proceeding then as in the proof of Theorem 2.5.3, namely, as in (2.5.54) and (2.5.55), it can be easily seen that under (4.2.44),

$$\frac{V(L_{N,jl} - L_{N,jl}(\psi_j))}{V(L_{N,jl})} < \eta, \qquad \eta > 0, \quad N \ge N_{\varepsilon\eta}, \tag{4.2.50}$$

where η can be made arbitrarily small and $N_{\varepsilon\eta}$ depends on ε and η. As a result, by Lemma 4.2.3, it suffices to show that as $N \to \infty$,

$$\mathbf{L}_N(\psi) = ((L_{N,jl}(\psi))) \sim \mathcal{N}_{pq}(E\mathbf{L}_N(\psi), V(\mathbf{L}_N)). \tag{4.2.51}$$

Now, ψ_j, $1 \le j \le p$, are all polynomials, and hence they have bounded first and second derivatives. As a result, Theorem 4.2.1 holds for $\mathbf{L}_N(\psi)$, and the desired asymptotic multinormality of $\mathbf{L}_N(\psi)$ follows from there. Finally, by (4.2.50) and Lemma 4.2.2, it follows that in the asymptotic distribution of $\mathbf{L}_N(\psi)$, $V(\mathbf{L}_N(\psi))$ can be replaced by $V(\mathbf{L}_N)$. Hence, (4.2.51) holds. Finally, by (4.2.44), (4.2.46), and (4.2.50), it follows that $V(\mathbf{L}_N)$ can be replaced by $V(\Sigma_{i=1}^N \mathbf{Z}_i)$, as by (4.2.24) and Lemma 4.2.3, $V(\mathbf{L}_N(\psi))$ can be replaced by $V(\Sigma_{i=1}^N \mathbf{Z}_i)$. The proof is complete.

In passing we may remark that as in Theorem 2.6.1, when $\phi_j(u)$ is the difference of two monotone $\phi_{j,1}(u)$ and $\phi_{j,2}(u)$ which are absolutely continuous inside $(0,1)$, such that for each k $(= 1, 2)$,

$$\int_0^1 \{|\phi_{j,k}(u)|\}\{u(1-u)\}^{-1/2}\,du < \infty, \qquad 1 \le j \le p, \tag{4.2.52}$$

then in Theorem 4.2.5 we may replace $E\mathbf{L}_N$ by $\boldsymbol{\mu}_N^* = ((\mu_{N,jl}^*))$, where

$$\mu_{N,jl}^* = \mu_{N,jl} + \bar{c}_{N,l}\left[\sum_{i=1}^N a_{Nj}(u) - N\int_0^1 \phi_j(u)\,du\right] \tag{4.2.53}$$

and

$$\mu_{N,jl} = \sum_{i=1}^N c_{il}\int_{-\infty}^{\infty} \phi_j\big(H_{[j]}(x)\big)\,dF_{i[j]}(x), \tag{4.2.54}$$

for $1 \le j \le p$, $1 \le l \le q$. In practical problems the commonly used score functions all satisfy (4.2.52).

4.3. MULTIVARIATE SIGNED RANK STATISTICS

Let $\mathbf{X}_i = (X_{i1}, \ldots, X_{ip})'$, $i = 1, \ldots, N$, be independent stochastic vectors with continuous d.f.'s $F_1(x), \ldots, F_N(x)$ respectively, all defined on the p-dimensional Euclidean space R^p. As in Chapter 3, we define the coordinatewise rank of the absolute values by

$$\tilde{R}_{ni}^{(j)} = \sum_{\alpha=1}^N u\big(|X_{ij}| - |X_{\alpha j}|\big), \qquad 1 \le j \le P, \quad 1 \le i \le N. \tag{4.3.1}$$

By virtue of the assumed continuity of the F_i, ties among the $|X_{ij}|$, $1 \le i \le N$ (for each $j = 1, \ldots, p$) may be neglected in probability, so that $(\tilde{R}_{Ni}^{(j)}, \ldots, \tilde{R}_{NN}^{(j)}$ is some permutation of the first N integers $(1, \ldots, N)$ for each $j = 1, \dots, p$. The corresponding matrix (of order $p \times N$)

$$\tilde{\mathbf{R}}_N = \begin{pmatrix} \tilde{R}_{N1}^{(1)} & \cdots & \tilde{R}_{NN}^{(1)} \\ \vdots & \ddots & \vdots \\ \tilde{R}_{N1}^{(p)} & \cdots & \tilde{R}_{NN}^{(p)} \end{pmatrix} = \begin{pmatrix} \tilde{\mathbf{R}}_N^{(1)} \\ \vdots \\ \tilde{\mathbf{R}}_N^{(p)} \end{pmatrix} \tag{4.3.2}$$

is called a (*modulus*) *rank-collection matrix*. Also, let

$$s(u) = \begin{cases} 1, & u > 0, \\ 0, & u = 0, \\ -1, & u < 0, \end{cases} \tag{4.3.3}$$

and with each $\mathbf{X}_i$ associate the vector

$$\mathbf{s}_i = \big(s(X_{i1}), \dots, s(X_{ip})\big)' = (S_{i1}, \dots, S_{ip})', \quad \text{say}, \tag{4.3.4}$$

so that we have a sign matrix

$$\mathbf{S}_N = (\mathbf{s}_1, \dots, \mathbf{s}_N) = \begin{pmatrix} S_{11} & \cdots & S_{N1} \\ \vdots & \ddots & \vdots \\ S_{1p} & \cdots & S_{Np} \end{pmatrix} \tag{4.3.5}$$

The statistics to be considered in this section depend on the observations $\mathbf{X}_1, \dots, \mathbf{X}_N$ only through $\tilde{\mathbf{R}}_N$ and $\mathbf{S}_N$. To define them, we consider for each j $(= 1, \dots, p)$ a set of N rank scores

$$a_{Nj}(i) = E\phi_j\big(U_N^{(i)}\big) \text{ or } \phi_j\left(\frac{i}{N+1}\right), \qquad 1 \le i \le N, \tag{4.3.6}$$

where the order statistics $U_N^{(1)} < \cdots < U_N^{(N)}$ and the score functions $\phi_j(u)$, $0 < u < 1$, $j = 1, \dots, p$, are all defined as in Section 2.2. Let then

$$\mathbf{A}_N = \begin{pmatrix} a_{N1}\big(\tilde{R}_{N1}^{(1)}\big) & \cdots & a_{N1}\big(\tilde{R}_{NN}^{(1)}\big) \\ \vdots & & \vdots \\ a_{Np}\big(\tilde{R}_{N1}^{(p)}\big) & \cdots & a_{Np}\big(\tilde{R}_{NN}^{(p)}\big) \end{pmatrix} \tag{4.3.7}$$

be the corresponding score-collection matrix, which has a one-to-one correspondence with $\tilde{\mathbf{R}}_N$. Finally, consider a matrix of known (regression) constants

$$\begin{pmatrix} c_{11} & \cdots & c_{1q} \\ \vdots & \ddots & \vdots \\ c_{N1} & \cdots & c_{Nq} \end{pmatrix} = (\mathbf{c}_1, \ldots, \mathbf{c}_q)', \qquad q \geq 1. \tag{4.3.8}$$

Then the proposed signed statistics are

$$\mathbf{S}_N^+ = \left(\left(S_{N,jl}^+ \right) \right)_{j=1,\ldots,p,\, l=1,\ldots,q}, \tag{4.3.9}$$

where

$$S_{N,jl}^+ = \sum_{i=1}^{N} s_{ij} c_{il} a_{Nj}\left(\tilde{R}_{Ni}^{(j)} \right), \qquad 1 \leq j \leq p, \quad 1 \leq l \leq q. \tag{4.3.10}$$

Now, as in the discussion following (4.2.8), we assume without any loss of generality that

$$\mathbf{c}_1, \ldots, \mathbf{c}_q \text{ are linearly independent,} \tag{4.3.11}$$

so that for $N \geq q$,

$$\mathbf{C}_N^* = \left(\left(c_{Nll'}^* \right) \right) = \left(\left(\sum_{i=1}^{N} c_{il} c_{il'} \right) \right)_{l,l'=1,\ldots,q} \quad \text{is p.d.} \tag{4.3.12}$$

Note that, as in the case of multivariate linear rank statistics, the distribution of $\mathbf{S}_N^+$ depends on the underlying $F_1, \ldots, F_N$, unless for each i, $X_{i1}, \ldots, X_{ip}$ are mutually independent, and all the N d.f.'s $F_1, \ldots, F_N$ are symmetric about the origin. In the latter case, $\mathbf{S}_N$ and $\tilde{\mathbf{R}}_N$ are mutually independent, and $\mathbf{S}_N$ has 2^{Np} equally likely realizations (on all possible sign inversions of its elements) which generate the distribution of $\mathbf{S}_N^+$. However, this special case is of limited interest. In practical problems of multivariate observations, the p-variates are not generally mutually independent, so that the task of studying the distribution of $\mathbf{S}_N^+$ remains of statistical importance.

We shall see in Chapter 5 that for testing suitable hypotheses of diagonal symmetry of the d.f.'s $F_1, \ldots, F_N$ under a certain permutational invariance structure, $\mathbf{S}_N^+$ is a conditional distribution-free statistic. On the other hand, the unconditional distribution of $\mathbf{S}_N^+$ depends on $F_1, \ldots, F_N$. In this section, we study the distribution theory of $\mathbf{S}_N^+$ in the general (unconditional) setup, including the cases where the d.f.'s $F_1, \ldots, F_N$ are "close" to each other or are "nearly symmetric". Of course, as in Section 4.2, our main emphasis is on the asymptotic distribution theory, because of its fundamental role in the subsequent chapters. Here also we retain Definitions 4.2.1, 4.2.2, and 4.2.3.

As in the preceding chapters, we first consider the score functions with bounded second derivatives, and then pass on to the general case of square-integrable, absolutely continuous score functions. Our first theorem in this direction is the following:

Theorem 4.3.1. *For the scores defined by* (4.3.6), *suppose* $\phi_j(u)$ *has a bounded second derivative inside* $(0, 1)$ (*for each* $j = 1, \ldots, p$), *and the constants* c_{il}, $1 \le l \le q$, $1 \le i \le N$, *satisfy the condition that for every* $\varepsilon > 0$ *there exists a* K_ε ($< \infty$) *and an integer* $N_0(\varepsilon)$ *such that for* $N \ge N_0(\varepsilon)$,

$$\gamma_N = \mathrm{Ch}_{pq}\left[V(\mathbf{S}_N^+)\right] \ge K_\varepsilon \left\{ \max_{1 \le l \le q} \max_{1 \le i \le N} c_{il}^2 \right\}. \tag{4.3.13}$$

Then as $N \to \infty$,

$$\mathbf{S}_N^+ \sim \mathscr{N}_{pq}\left(E\mathbf{S}_N^+, V(\mathbf{S}_N^+)\right). \tag{4.3.14}$$

The assertion remains true if, in (4.3.14) *we replace* $V(\mathbf{S}_N^+)$ *by* $V(\sum_{i=1}^N \mathbf{Z}_i^+)$, *where* $\mathbf{Z}_i^+ = ((Z_{i,jl}^+)_{j=1,\ldots,p,\, l=1,\ldots,q}$,

$$\begin{aligned} Z_{i,jl}^+ = c_{il}\Big\{ &\phi_j\left(H_j^*(|X_{ij}|)\right)S_{ij} - E\left[\phi_j\left(H_j^*(|X_{ij}|)\right)S_{ij}\right]\Big\} \\ &+ \frac{1}{N}\sum_{k=1}^{N} c_{kl}\int_{-\infty}^{\infty} s(x)\left[u(|x| - |X_{ij}|) - F_{[ij]}^*(|x|)\right] \\ &\times \phi_j'\left(H_j^*(|x|)\right) dF_{k[j]}(x), \end{aligned} \tag{4.3.15}$$

$F_{i[j]}(x) = P\{X_{ij} \le x\}$ *is the marginal d.f. of* X_{ij}, *and for* $x \ge 0$,

$$H_j^*(x) = N^{-1}\sum_{i=1}^{N} F_{i[j]}^*(x), \quad \textit{for} \quad 1 \le j \le p, \tag{4.3.16}$$

$$F_{i[j]}^*(x) = F_{i[j]}(x) = F_{i[j]}(-x), \quad 1 \le i \le N, \quad 1 \le j \le p. \tag{4.3.17}$$

Proof. It follows from Theorem 3.3.3 that under the hypothesis of the theorem, there exists a positive constant $M = M(\phi_1, \ldots, \phi_p)$ such that for every $\mathbf{c}_1, \ldots, \mathbf{c}_q$, $F_1, \ldots, F_N$, and N (≥ 1),

$$E\left[S_{N,jl}^{+} - ES_{N,jl}^{+} - \sum_{i=1}^{N} Z_{i,jl}^{+}\right]^2 \leq MN^{-1}\sum_{i=1}^{N} c_{il}^2, \tag{4.3.18}$$

$$\left[ES_{N,jl}^{+} - \mu_{jl}\right]^2 \leq MN^{-1}\sum_{i=1}^{N} c_{il}^2,\ j = 1, \ldots, p, l = 1, \ldots, q, \tag{4.3.19}$$

where for every $1 \leq j \leq p$, $1 \leq l \leq q$,

$$\mu_{jl} = \sum_{i=1}^{N} c_{il}\int_{-\infty}^{\infty} s(x)\phi_j\left(H_j^*(|x|)\right) dF_{i[j]}(x), \tag{4.3.20}$$

On the other hand, (4.3.13) implies that for every $N \geq N_0(\varepsilon)$,

$$\begin{aligned}\operatorname{Var}\left(S_{N,jl}\right) &\geq \gamma_N \geq k_\varepsilon\left\{\max_{1\leq l\leq q}\max_{1\leq i\leq N} c_{il}^2\right\}\\ &\geq k_\varepsilon N^{-1}\sum_{i=1}^{N} c_{il}^2, \qquad j = 1, \ldots, p, \quad l = 1, \ldots, q,\end{aligned} \tag{4.3.21}$$

Consequently, by (4.3.18) through (4.3.21), we have for $N \geq N_0(\varepsilon)$

$$\frac{E\left[S_{N,jl}^{+} - ES_{N,jl}^{+} - \sum_{i=1}^{N} Z_{i,jl}\right]^2}{\operatorname{Var}\left(S_{N,jl}^{+}\right)} \leq MK_\varepsilon^{-1} < \varepsilon, \tag{4.3.22}$$

$$\frac{\left[ES_{N,jl}^{+} - \mu_{jl}\right]^2}{\operatorname{Var}\left(S_{N,jl}^{+}\right)} < \varepsilon, \qquad j = 1, \ldots, p, \quad l = 1, \ldots, q \tag{4.3.23}$$

(by suitable choice of K_ε). Using Lemmas 4.2.2 and 4.2.3, it follows by (4.3.22) and (4.3.23) that we have only to show that as $N \to \infty$,

$$\sum_{i=1}^{N} \mathbf{Z}_i \sim \mathcal{N}_p\left(\mathbf{0}, V(\mathbf{S}_N^{+})\right), \tag{4.3.24}$$

where again, by virtue of (4.3.22), one may replace $V(\mathbf{S}_N^{+})$ by $V(\Sigma_{i=1}^{N}\mathbf{Z}_i^{+})$.

To prove (4.3.24), we consider a linear combination

$$Z_N^+(\mathbf{d}) = \sum_{j=1}^{p} \sum_{l=1}^{q} d_{jl} \left(\sum_{i=1}^{N} Z_{i,jl}^+ \right), \qquad \text{where} \quad \mathbf{d} \neq \mathbf{0}, \tag{4.3.25}$$

and show that $Z_N^+(\mathbf{d})$ is asymptotically

$$\mathscr{N}\left(0, \sum_{j=1}^{p} \sum_{l=1}^{q} \sum_{j'=1}^{p} \sum_{l'=1}^{q} d_{jl} d_{j'l'} \operatorname{Cov}(Z_{i,jl}, Z_{i,j'l'})\right).$$

By (4.3.15) and (4.3.25)

$$Z_N^+(\mathbf{d}) = \sum_{i=1}^{N} \left\{ \sum_{j=1}^{p} \sum_{l=1}^{q} d_{jl} Z_{i,jl} \right\} = \sum_{i=1}^{N} g(\mathbf{X}_i, \mathbf{d}), \tag{4.3.26}$$

where by the same technique as in (3.4.6),

$$|g(X_i, d)| \leq \left(2K_0^{1/2} + K_1^{1/2}\right) \left(\sum_{j=1}^{p} \sum_{l=1}^{q} |d_{jl}| \right) \left[\max_{1 \leq l \leq q} \max_{1 \leq i \leq N} |c_{il}| \right], \tag{4.3.27}$$

and

$$K_r = \max_{1 \leq j \leq p} \left\{ \sup_{0 < u < 1} \left| \phi_j^{(r)}(u) \right|^2 \right\}, \qquad r = 0, 1, 2. \tag{4.3.28}$$

On the other hand, by (4.3.13) and (4.3.18), for every $\mathbf{d} \neq \mathbf{0}$, and for $N \geq N_0(\varepsilon)$,

$$\begin{aligned} &\sum_{j=1}^{p} \sum_{j'=1}^{p} \sum_{l=1}^{q} \sum_{l'=1}^{q} d_{jl} d_{j'l'} \operatorname{Cov}\left[\sum_{i=1}^{N} Z_{i,jl}, \sum_{i=1}^{N} Z_{i,j'l'} \right] \\ &\geq \left(\sum_{j=1}^{p} \sum_{l=1}^{q} d_{jl}^2 \right) \operatorname{Ch}_{pq}\left[V\left(\sum_{i=1}^{N} \mathbf{Z}_i \right) \right] \\ &\geq \left(\sum_{j=1}^{p} \sum_{l=1}^{q} d_{jl}^2 \right) K_\varepsilon \left\{ \max_{1 \leq l \leq q} \max_{1 \leq i \leq N} c_{il}^2 \right\}. \end{aligned} \tag{4.3.29}$$

Comparing (4.3.27) and (4.3.29) and choosing K_ε sufficiently large, it follows that for every $\mathbf{d} \neq \mathbf{0}$ $\{g(\mathbf{X}_i, \mathbf{d}),\ 1 \leq i \leq N\}$ satisfy the Lindeberg condition of the central limit theorem, and hence $Z_N^+(\mathbf{d})$ has asymptotically a normal distribution. This proves the asymptotic multinormality of $\sum_{i=1}^N \mathbf{Z}_i$ in (4.3.24), and hence the theorem follows.

Theorem 4.3.2. *For the scores defined by* (4.3.6), *if* $\phi_j(u) = \phi_{j,1}(u) - \phi_{j,2}(u)$, $0 < u < 1$, $1 \le j \le p$, *where the* $\phi_{j,k}(u)$, $k = 1, 2$, *are all nondecreasing, absolutely continuous, and square-integrable inside* $(0,1)$, *and if for every* $\varepsilon > 0$, $\eta > 0$, *there exists an* $N_{\varepsilon\eta}$ $(< \infty)$ *such that for* $N \ge N_{\varepsilon\eta}$

$$\mathrm{Ch}_{pq}\left[V(\mathbf{S}_N^+)\right] > \eta N\left\{\max_{1\le l\le q}\max_{1\le i\le N} c_{il}^2\right\}, \tag{4.3.30}$$

then as $N \to \infty$,

$$\mathbf{S}_N^+ \sim \mathcal{N}_{pq}\left(E\mathbf{S}_N^+, V(\mathbf{S}_N^+)\right). \tag{4.3.31}$$

The assertion remains true, if, in (4.3.31), $V(\mathbf{S}_N^+)$ *is replaced by* $V(\sum_{i=1}^N \mathbf{Z}_i)$, *where the* $\mathbf{Z}_i$ *are defined by* (4.3.16).

Proof. For each j $(= 1, \dots, p)$, we use for $\phi_j(u)$ the decomposition in (3.4.34) and (3.4.35), so that

$$\mathbf{S}_N^+ = \mathbf{S}_N^+(\psi) + \mathbf{S}_N^+(\phi_{(1)}) - \mathbf{S}_N^+(\phi_{(2)}), \tag{4.3.32}$$

where ψ represents the polynomial component and $\phi_{(1)}$ and $\phi_{(2)}$ represent (nondecreasing) residual components such that

$$\sum_{j=1}^{p}\left\{\int_0^1\left[\phi_{j(1)}^2(u) + \phi_{j(2)}^2(u)\right]du\right\} < \alpha \tag{4.3.33}$$

for some preassigned small α (> 0). Then, as in (3.4.41), for each j $(= 1, \dots, p)$ and l $(= 1, \dots, q)$, for $N \ge N_{\varepsilon\eta}$,

$$\frac{E\left[S_{N,jl}^+ - S_{N,jl}^+(\psi) - ES_{N,jl}^+ + ES_{N,jl}^+(\psi)\right]^2}{\beta^2 \mathrm{Var}\left(S_{N,jl}^+\right)} < \frac{\varepsilon}{4}, \tag{4.3.34}$$

where $\beta > 0$ and $\varepsilon > 0$ are arbitrary small. As a result, by Lemma 4.2.3, it suffices to show that as $N \to \infty$,

$$\mathbf{S}_N^+(\psi) \sim \mathcal{N}_{pq}\left(E\mathbf{S}_N^+(\psi), V(\mathbf{S}_N^+)\right). \tag{4.3.35}$$

Since ψ represents the polynomial component (having bounded second derivative), the proof of Theorem 3.4.1 holds, and hence we have for

$N \to \infty$

$$\mathbf{S}_N(\psi) \sim \mathcal{N}_{pq}(E\mathbf{S}_N^+(\psi), V(\mathbf{S}_N^+(\psi))). \tag{4.3.36}$$

So to prove the theorem, it remains only to show that (4.3.36) implies (4.3.35). For this we require to show that for every $\mathbf{d} \neq \mathbf{0}$,

$$\frac{\left|\operatorname{Var}\left(\sum_{j=1}^{p}\sum_{l=1}^{q} d_{jl}S_{N,jl}^+\right) - \operatorname{Var}\left(\sum_{j=1}^{p}\sum_{l=1}^{q} d_{jl}S_{N,jl}^+(\psi)\right)\right|}{\operatorname{Var}\left(\sum_{j=1}^{p}\sum_{l=1}^{q} d_{jl}S_{N,jl}^+\right)} < \varepsilon. \tag{4.3.37}$$

Since, by (4.3.30), for $N \geq N_{\varepsilon\eta}$,

$$\operatorname{Var}\left(\sum_{j=1}^{p}\sum_{l=1}^{q} d_{jl}S_{N,jl}^+\right) \geq \left(\sum_{j=1}^{p}\sum_{l=1}^{q} d_{jl}^2\right)\operatorname{Ch}_{pq}[V(\mathbf{S}_N^+)]$$

$$\geq \left(\sum_{j=1}^{p}\sum_{l=1}^{q} d_{jl}^2\right)\eta N\left(\max_{1\leq l\leq q}\max_{1\leq i\leq N} c_{il}^2\right), \tag{4.3.38}$$

where $\eta > 0$, and by (3.4.10), for each j, l,

$$\operatorname{Var}\left(S_{N,jl}^+\right) \leq 40\left[\max_{1\leq i\leq N} c_{il}^2\right]\left[\sum_{i=1}^{N} a_{N,j}^2(i)\right]$$

$$\leq C\left[\max_{1\leq i\leq N}\max_{1\leq l\leq q} c_{il}^2\right], \tag{4.3.39}$$

(4.3.37) follows from (4.3.34), (4.3.38), (4.3.39), and Lemma 4.2.2. The proof is complete.

In passing we may remark that under the conditions of Theorem 4.3.1, in (4.3.14) we may replace $E\mathbf{S}_N^+$ by $\boldsymbol{\mu}_N^+ = ((\mu_{N,jl}^+))$, where for every $1 \leq j \leq p$, $1 \leq l \leq q$,

$$\mu_{N,jl}^+ = \sum_{i=1}^{N} c_{il}\int_{-\infty}^{\infty} s(x)\phi_j\left(H_j^*(|x|)\right) dF_{i[j]}(x), \tag{4.3.40}$$

Under the conditions of Theorem 4.3.2, in general, it is not possible to replace $E\mathbf{S}_N^+$ by $\boldsymbol{\mu}_N^+$. However, as in Theorem 3.4.11, if we assume that for each j (= $1,\ldots,p$) we have $\phi_j(u) = \phi_{j,1}(u) - \phi_{j,2}(u)$, where the $\phi_{j,k}(u)$, $k = 1, 2$, are all nondecreasing, and for each k (= 1, 2),

$$\int_0^1 |\phi_{j,k}(u)|\{u(1-u)\}^{-1/2}\,du < \infty, \qquad j = 1,\ldots,p, \tag{4.3.41}$$

then this replacement is possible. Since the $ES_{N,jl}^{+}$ depends only on the marginal d.f.'s $F_{i[j]}$, $1 \le i \le N$, the proof is the same as in Theorem 3.4.11, and hence is omitted.

In subsequent chapters dealing with the asymptotic behavior of rank-order tests and estimates, we require to simplify the results in Theorem 4.3.1 for "nearby alternatives", i.e. when the d.f.'s $F_1,\ldots,F_N$ are "close" to each other and are "nearly symmetric". Consider a d.f. $F(\mathbf{x})$, $\mathbf{x} \in R^p$, with univariate marginals $F_{[j]}$, $1 \le j \le p$, and bivariate marginals $F_{[jj']}$, $1 \le j \le j' \le p$, and let us denote by $F_{i[j]}$ and $F_{i[jj']}$ the corresponding marginals for the d.f. F_i, $i = 1,\ldots,N$. Then we assume that for every $\varepsilon > 0$ and $\eta > 0$, there exists a $\delta_{\varepsilon\eta} > 0$ such that

$$\sup_{x,y}\left\{\max_{1\le i\le N}|F_{I[jj']}(x,y) - F_{[jj']}(x,y)|\right\} < \delta_{\varepsilon\eta} \tag{4.3.42}$$

for every $1 \le j \le j' \le p$, which implies that

$$\sup_{x}\left\{\max_{1\le i\le N}|F_{i[j]}(x) - F_{[j]}(x)|\right\} < \delta_{\varepsilon\eta} \tag{4.3.43}$$

for every $1 \le j \le p$, and, on letting

$$\mathbf{C}_N^* = \left(\left(\sum_{i=1}^{N} c_{il}c_{il'}\right)\right)_{l,l'=1,\ldots,q},$$

that

$$\mathrm{Ch}_q[\mathbf{C}_N^*] \ge \delta_{\varepsilon\eta}^{-1}\left[\max_{1\le l\le q}\max_{1\le i\le N} c_{il}^2\right]. \tag{4.3.44}$$

Finally, we assume that

$$\mathrm{Ch}_p[V(\mathbf{U})] \ge \eta > 0, \tag{4.3.45}$$

where $\mathbf{U} = (U_1, \ldots, U_p)$ with

$$U_j = s(X_{1j})\phi_j\left(F^*_{[j]}(X^0_{1j})\right) + \int_{-\infty}^{\infty} s(x)\left[u(|x| - |X^0_{1j}|) - F^*_{[j]}(|x|)\right]\phi'_j\left(F^*_{[j]}(|x|)\right) dF_{[j]}(x), \tag{4.3.46}$$

where $F^*_{[j]}(x) = F_{[j]}(x) - F_{[j]}(-x)$, $x \geq 0$, $j = 1, \ldots, p$, and X^0_{ij} has the d.f. $F_{[j]}(x)$, $j = 1, \ldots, p$. Note that for each j $(= 1, \ldots, p)$,

$$\begin{aligned} EU_j &= \int_{-\infty}^{\infty} s(x)\phi_j\left(F^*_{[j]}(|x|)\right) dF_{[j]}(x) \\ &= \int_0^{\infty} \phi_j\left(F^*_{[j]}(x)\right) d\left[F_{[j]}(x) + F_{[j]}(-x)\right], \end{aligned} \tag{4.3.47}$$

so that if the $F_{[j]}$ are symmetric about 0 [i.e., $F_{[j]}(x) + F_{[j]}(-x) = 1$ for every $x \geq 0$, $1 \leq j \leq p$], then $EU_j = 0$, $1 \leq j \leq p$. Now let

$$A_{jj'}(x, y) = \begin{cases} F^*_{[j]}(\min[x, y]) - F^*_{[j]}(x)F^*_{[j]}(y), & j = j', \\ F^*_{[jj']}(x, y) - F^*_{[j]}(x)F^*_{[j']}(y), & j \neq j', \end{cases} \tag{4.3.48}$$

where $x \geq 0$, $y \geq 0$, and $F^*_{[jj']}(x, y) = P\{|X^0_{ij}| \leq x, |X^0_{1j'}| \leq y\}$. Then, by (4.3.36),

$$\begin{aligned} \operatorname{Cov}[U_j, U_{j'}] &= \int_{-\infty}^{\infty}\int_{-\infty}^{\infty} s(x)s(y)\phi_j\left(F^*_{[j]}(|x|)\right)\phi_{j'}\left(F^*_{[j']}(|y|)\right) dF_{[jj']}(x, y) \\ &\quad + \int_{-\infty}^{\infty}\int_{-\infty}^{\infty} s(x)s(y)A_{jj'}(x, y)\phi'_j\left(F^*_{[j]}(|x|)\right) \\ &\qquad \times \phi'\left(F^*_{[j']}(|y|)\right) dF_{[jj']}(x, y) \\ &\quad + \int_{-\infty}^{\infty}\int_{-\infty}^{\infty} s(y)\phi_j\left(F^*_{[j]}(|x|)\right)\phi'_{j'}\left(F^*_{[j']}(|y|)\right) \\ &\quad \times E\left\{s(X^0_{1j})\left[u(|y| - |X^0_{1j'}| - F^*_{[j']}(|y|)\right]\right\} dF_{[j']}(y) \\ &\quad + \int_{-\infty}^{\infty}\int_{-\infty}^{\infty} s(y)\phi_{j'}\left(F^*_{[j']}(|x|)\right)\phi'_j\left(F^*_{[j]}(|y|)\right) \\ &\quad \times E\left\{s(X^0_{1j'})\left[u(|y| - |X^0_{1j}|) - F^*_{[j]}(|y|)\right]\right\} dF_{[j]}(y). \end{aligned} \tag{4.3.49}$$

If $F_{[jj']}(x, y)$ is diagonally symmetric about (0, 0), i.e.

$$\left(X_{1j}^0, X_{1j'}^0\right) \quad \text{and} \quad \left(-X_{1j}^0, -X_{1j'}^0\right) \tag{4.3.50}$$

have the same d.f. $F_{[jj']}$ and the score functions $\phi_j(u)$ satisfy

$$\phi_j(u) = \phi_j^*\left(\frac{1+u}{2}\right), \tag{4.3.51}$$

where $\phi_j^*(u) + \phi_j^*(1-u) = 0$, $0 < u < 1$, for $1 \le j \le p$, then (4.3.49) simplifies to

$$\int_{-\infty}^{\infty}\int_{-\infty}^{\infty} \phi_j^*\left(F_{[j]}(x)\right)\phi_{j'}^*\left(F_{[j']}(y)\right) dF_{[jj']}(x, y), \qquad j \neq j', \tag{4.3.52}$$

and

$$\int_{-\infty}^{\infty} \left[\phi_j^*\left(F_{[j]}(x)\right)\right]^2 dF_{[j]}(x) = \int_0^1 \left[\phi_j^*(u)\right]^2 du, \qquad j = j'. \tag{4.3.53}$$

Thus, in this case, (4.3.45) amounts to saying that $\phi_j^*(F_{[j]}(X_{ij}^0))$, $1 \le j \le p$ are all linearly independent on a set of positive probability, so that the variance matrix is p.d. Thus, if we define the dispersion matrix in (4.3.49) by $\boldsymbol{\nu}(F)$, and assume that $\boldsymbol{\nu}(F)$ is p.s.d., then for $F_1, \ldots, F_N$ "nearly equal to F" [in the sense of (4.3.42)], using (4.3.15), (4.3.42), and some standard computations, it follows that as $N \to \infty$,

$$\text{Ch}_1\left[V\left(\sum_{i=1}^{N} \mathbf{Z}_i\right) - \mathbf{C}_N^* \otimes \boldsymbol{\nu}(F)\right] \Big/ \text{Ch}_q\left[\mathbf{C}_N^*\right] \to 0, \tag{4.3.54}$$

The details are left as an exercise (Exercise 4.3.9). As a result, we arrive at the following theorem.

Theorem 4.3.3. *For the scores defined by* (4.3.6), *whenever the score functions* $\phi_j(u)$, $0 < u < 1$, $j = 1, \ldots, p$, *satisfy the condition of Theorem* 4.3.1 *or Theorem* 4.3.2, *then under* (4.3.42), (4.3.44), *and* (4.3.45), *as* $N \to \infty$,

$$\mathbf{S}_N^+ \sim \mathcal{N}_{pq}\left(E\mathbf{S}_N^+, \mathbf{C}_N^* \otimes \boldsymbol{\nu}(F)\right). \tag{4.3.55}$$

If, in addition, (4.3.41) *holds, then* $E\mathbf{S}_N^+$ *may be replaced by* $\boldsymbol{\mu}_N^+$ *defined by* (4.3.40).

4.4. CHARACTERIZATION OF CERTAIN MULTIVARIATE PROBLEMS

In Sections 4.2 and 4.3, we have considered the case of arbitrary continuous $F_1, \ldots, F_N$. In many practical problems, we have a finite set of subsamples where within each set the distributions are identical, and the rank statistics in these situations are of fundamental importance. In what follows, we briefly introduce these problems and show how the corresponding distribution theory follows from our theorems in Sections 4.2 and 4.3. This will lead us to a natural transition to general linear models to be studied in the subsequent chapters.

4.4.1. Multivariate Two-Sample Problem

Here we assume that $N = n_1 + n_2$, where n_1 and n_2 are positive integers, and that

$$F_1(\mathbf{x}) = \cdots = F_{n_1}(\mathbf{x}) = F_1(\mathbf{x}), \qquad F_{n_1+1}(\mathbf{x}) = \cdots = F_N(\mathbf{x}) = F_0(\mathbf{x}). \tag{4.4.1}$$

The problems of interest are (1) to test the hypothesis $F_1 = F_0$, (2) to estimate the difference of locations (assuming of course that F_0 and F_1 differ by locations only) and (3) to find the confidence regions for the vector of location differences. For these problems, rank-order procedures are based on the statistics

$$\mathbf{L}_N = \left(L_{N,1}, \ldots, L_{N,p}\right)', \tag{4.4.2}$$

where

$$L_{N,j} = \sum_{i=1}^{n_1} a_{Nj}\left(R_{Ni}^{(j)}\right), \qquad j = 1, \ldots, p, \tag{4.4.3}$$

and the scores $a_{Nj}(i)$, $1 \le i \le N$, and where the ranks $R_{Ni}^{(j)}$, $1 \le i \le N$, $1 \le j \le p$, are defined by (4.2.4) and (4.2.2) respectively. Note that (4.4.3) is a special case of (4.2.8) when $q = 1$ and

$$c_{1,1} = \cdots = c_{n_1,1} = 1, \qquad c_{n_1+1,1} = \cdots = c_{N,1} = 0. \tag{4.4.4}$$

Here $c_{N,11}^* = n_1 n_2 / N^2$, and $\max_{1 \le i \le N} |c_i - \bar{c}_N| = N^{-1}[\max(n_1, n_2)] < 1$, so that the results of Theorem 4.2.1–4.2.4 simplify considerably. For the detailed study of these problems refer to Puri and Sen (1971, Chapters 5, 6).

4.4.2. Multivariate Multisample Problem

Here we assume that for some $c\ (\geq 2)$, $N = n_1 + \cdots + n_c$, where the n_k are all positive integers and

$$F_i(\mathbf{x}) = G_j(\mathbf{x}) \qquad \text{for all} \quad i = \sum_{k=0}^{j-1} n_k + \alpha, \quad \alpha = 1, \ldots, n_j, \quad 1 \leq j \leq c, \tag{4.4.5}$$

$n_0 = 0$ and $G_1(\mathbf{x}), \ldots, G_c(\mathbf{x})$ are c unknown distributions. The problem is to test the identity of $G_1, \ldots, G_c$ against specified alternatives—e.g. that these distributions differ in location vectors or dispersion matrices. Alternatively, assuming that

$$G_j(\mathbf{x}) = G(\mathbf{x} - \boldsymbol{\theta}_j), \qquad 1 \leq j \leq c, \quad G \text{ unknown}, \tag{4.4.6}$$

one may be interested in estimating contrasts among $\boldsymbol{\theta}_1, \ldots, \boldsymbol{\theta}_c$. Rank-order procedures for these problems are discussed in detail in Chapters 5, 6, and 9 of Puri and Sen (1971). They are based on the statistics

$$\mathbf{T}_{Nk} = \left(T_{Nk}^{(1)}, \ldots, T_{Nk}^{(p)}\right)', \qquad k = 1, \ldots, c, \tag{4.4.7}$$

where

$$T_{Nk}^{(j)} = \sum_{\alpha=1}^{n_k} a_{Nj}\left(R_{Nk,\alpha}^{(j)}\right), \qquad 1 \leq j \leq p, \quad 1 \leq k \leq c, \tag{4.4.8}$$

and $R_{Nk,\alpha}^{(j)}$ is the rank of $X_{k\alpha}^{(j)}$ among the N jth variate observations $X_{q\alpha}^{(j)}$, $1 \leq \alpha \leq n_q$, $1 \leq q \leq c$ in the combined sample. Note that the $\mathbf{T}_{Nk}$ correspond to the special case of (4.2.8) where $q = c$ and

$$c_{il} = \begin{cases} 1 & \text{if} \quad \sum_{j=0}^{l-1} n_j + 1 \leq i \leq \sum_{j=0}^{l} n_j, \quad n_0 = 0, \\ 0 & \text{otherwise} \end{cases} \tag{4.4.9}$$

for $l = 1, \ldots, c$. Note that $\sum_{k=1}^{c} \mathbf{T}_{Nk} = \sum_{i=1}^{N} [a_{N,1}(i), \ldots, a_{N,p}(i)]'$ is nonstochastic, and hence only $c - 1$ of the vectors in (4.4.7) are linearly independent. Thus, for (4.2.11) to hold, we need to work only with the set

$$\{\mathbf{T}_{Nk}, \ k = 1, \ldots, c - 1\}, \tag{4.4.10}$$

derive the joint asymptotic distribution theory of these $p(c-1)$ variables, and then, using the fact that

$$\mathbf{T}_{Nc} = \sum_{i=1}^{N} \left[a_{N,1}(i), \ldots, a_{N,p}(i)\right]' - \sum_{k=1}^{c-1} \mathbf{T}_{Nk},$$

derive the joint asymptotic distribution of all the pc variables in (4.4.7). Thus Theorems 4.2.1–4.2.4 provide the necessary results. Compared to the parallel theorems in Chapter 5 of Puri and Sen (1971), these theorems require less stringent regularity conditions.

4.4.3. Multivariate One-Sample Problem

Assume that the $\mathbf{X}_\alpha$, $\alpha = 1, \ldots, N$, are independent stochastic vectors with the distributions $F_\alpha(\mathbf{x})$, $\alpha = 1, \ldots, N$. We assume further that all the F_α's are diagonally symmetric about their respective medians. Our first problem is to test the null hypothesis that all these median vectors are equal to some specified vector, which without loss of generality can be taken to be Q, by suitable translation of the vectors. Conditional (permutationally) distribution-free tests for this hypothesis based on signed rank statistics are discussed in detail in Chapter 4 of Puri and Sen (1971). Alternatively, one may be interested in estimating the median vector, assuming that all these vectors have a common (unknown) median. Rank-order estimates for this problem are discussed in detail in Chapter 6 of Puri and Sen (1971). The testing and estimation procedures involve the use of signed rank statistics which are particular cases of $\mathbf{S}_N^+$ in (4.3.9) and (4.3.10), where $q = 1$, $\mathbf{c}_1 = (1, \ldots, 1)$. Thus the necessary distribution theory of these one-sample rank-order statistics follows from our Theorems 4.3.1 and 4.3.2, which require less stringent regularity conditions than the parallel theorems studied in Chapter 4 of Puri and Sen (1971).

4.4.4. Multivariate Paired-Comparison Procedures Based on Rank Statistics

Consider a paired-comparison design involving t (≥ 2) treatments such that for $1 \leq i < j \leq t$, the comparisons made on the (i, j)th pair (in n_{ij} encounters) yield the random vectors (difference of observations) $\mathbf{X}_{ij,l}, \ldots, \mathbf{X}_{ij,n_{ij}}$, $1 \leq i < j \leq t$, and let $N = \sum_{1 \leq i < j \leq t} n_{ij}$. We assume that the $\mathbf{X}_{ij,k}$ are independently and identically distributed with a distribution function $G_{ij}(\mathbf{x})$, $\mathbf{x} \in R^p$, for $1 \leq i < j \leq t$. The hypothesis

$$H_0: G_{12} = \cdots = G_{t-1,t} = G \text{ (unknown)} \qquad (4.4.11)$$

of no treatment effects is to be tested by using suitable rank tests. In the general multivariate setup, such tests have been studied by Sen and David (1968), Davidson and Bradley (1969, 1970), Shane and Puri (1969), Puri and Shane (1970), and Russel and Puri (1974). Basically, these procedures involve the use of all the 2^t one-sample rank-order statistics for the (i, j)th pair, $1 \leq i < j \leq t$, where (a) either separate ranking is made for each pair, or (b) joint ranking is made. In case (a), all these 2^t one-sample rank-order statistics are stochastically independent, and their distribution theory follows readily from our Theorems 4.3.1 and 4.3.2. In case (b), because of the joint ranking, the 2^t one-sample rank-order statistics will not be stochastically independent. However, this corresponds to the special case of (4.3.9) and (4.3.10) where $q = 2^t$ and the c_{il} are more elementary. Thus, here also, the distribution theory readily follows from Theorems 4.3.1 and 4.3.2.

Conceptually, the paired-comparison model leads to more general incomplete block designs for which the paired comparisons extend to comparisons of aligned observations. Some aligned rank procedures for such (multivariate) incomplete block designs are discussed in Sen (1968, 1971a) and Mehra and Sen (1969). However, these procedures rest on different distribution theory, discussed in detail in Puri and Sen (1971, Chapter 7), and are outside the scope of the present chapter.

EXERCISES

4.2.1. Prove that if $\mathbf{X}_1, \ldots, \mathbf{X}_N$ are independent and identically distributed, then all the $(N!)^p$ realizations of $\mathbf{R}_N$ are equally likely.

4.4.2. Provide a formal proof of the assertion (4.2.23).

4.2.3. Provide a formal proof of Lemma 4.2.2.

4.2.4. Provide a formal proof of Lemma 4.2.3.

4.2.5. Prove that

$$\max_{1 \leq i \leq N} |g(\mathbf{X}_i, \mathbf{d})| \leq \left\{ \max_{j,l} |d_{j,l}| \right\} \sum_{j=1}^{p} \sum_{l=1}^{q} \left\{ \max_{1 \leq i \leq N} |Z_{ij,l}| \right\}.$$

4.2.6. Provide a formal proof of the assertion (4.2.41).

4.2.7. Provide a formal proof of the assertion (4.2.50).

4.2.8. Prove that if $\phi_j(u) = \phi_{j,1}(u) - \phi_{j,2}(u)$, where $\phi_{j,k}(u)$, $k = 1, 2$, are (1) monotontically nondecreasing, (2) absolutely continuous in $(0, 1)$, and (3) such that $\int_0^1 \{|\phi_{j,k}(u)|\}\{u(1-u)\}^{-1/2}\,du < \infty$, $1 \leq j \leq p$, then $E\mathbf{L}_N$ in

Theorem 4.2.5 can be replaced by $\boldsymbol{\mu}_N^* = ((\mu_{N,jl}^*))$ given by (4.2.53) and (4.2.54).

4.3.1. Prove that if for each $1 \le i \le N$, $X_{i1}, \ldots, X_{ip}$ are independent random variables, and $F_1, \ldots, F_N$ are diagonally symmetric about $\mathbf{0}$, then all the 2^{Np} realizations of $\mathbf{S}_N$ defined in (4.3.5) are equally likely.

4.3.2. Provide formal proofs of the assertions (4.3.18), (4.3.19), (4.3.21), and (4.3.23).

4.3.3. Provide formal proofs of the assertions (4.3.27) and (4.3.29).

4.3.4. Prove that for every $\mathbf{d} \neq \mathbf{0}$, $\{g(\mathbf{X}_i, \mathbf{d}), 1 \le i \le N\}$ defined in (4.3.26) satisfy the Lindeberg condition of the central limit theorem.

4.3.5. Provide a formal proof of the assertion (4.3.34).

4.3.6. Provide a formal proof of the assertion (4.3.36).

4.3.7. Provide that under the assumptions of Exercise 4.2.8, $E\mathbf{S}_N^+$ in (4.3.31) can be replaced by $\boldsymbol{\mu}_N^+ = ((\mu_{N,jl}^+))$ given by (4.3.40).

4.3.8. Prove that if $F_{[jj']}(x, y)$ is diagonally symmetric about $(0, 0)$, and the score functions $\phi_j(u)$, $1 \le j \le p$, satisfy (4.3.51), then $\text{Cov}[\mathbf{U}_j, U_j']$ is given by (4.3.52) and (4.3.53).

4.3.9. Provide a formal proof of the assertion (4.3.54).

4.4.1. Let $\mathbf{X}_i = (X_{i1}, \ldots, X_{ip})'$, $1 \le i \le N$, be independent p-vectors with continuous d.f.'s $F_i(\mathbf{x})$, $1 \le i \le N$. Let $\mathbf{R}_{Ni} = (\tilde{R}_{N1}^{(i)}, \ldots, \tilde{R}_{Np}^{(i)})'$, $1 \le i \le N$. Let $\mathbf{a}^{(N)}(\cdot)$ be a multivariate score generating function taking values in R^p, and given by $\mathbf{a}_N(\mathbf{R}) = (a_{N1}(\tilde{R}_1), \ldots, a_{Np}(\tilde{R}_p))'$ for $\mathbf{R}$ in R^p. For each $\mathbf{x}$ in R^p, define $\mathbf{s}(\mathbf{X}) = (s(X_1), \ldots, s(X_p))$, where $s(\cdot)$ is defined in (4.3.3); and for $\boldsymbol{\alpha}, \boldsymbol{\beta}$ in R^p, let $\boldsymbol{\alpha} * \boldsymbol{\beta} = (\alpha_1\beta_1, \ldots, \alpha_p\beta_p)'$ be the Hadamard product of $\boldsymbol{\alpha}$ and $\boldsymbol{\beta}$. Let

$$\mathbf{S}^+ = S^+(X_1, \ldots, X_N; \tilde{R}_{N1}, \ldots, \tilde{R}_{NN}) = \sum_{j=1}^{N} c_j \mathbf{s}(\mathbf{X}_j) * \mathbf{a}_N(\tilde{R}_j)$$

be a $p \times 1$ statistic such that $E(\mathbf{S}^{+\prime}\mathbf{S}^+)$ is finite. Let

$$\hat{\mathbf{S}}^+ = \sum_{i=1}^{N} E(\mathbf{S}^+ | \mathbf{X}_i) - (N-1) E\mathbf{S}^+.$$

Then:

(a) $E\mathbf{S}^+ = E\hat{\mathbf{S}}^+$.
(b) $E[(\mathbf{S} - \hat{\mathbf{S}}^+)(\mathbf{S}^+ - \hat{\mathbf{S}}^+)'] = \text{Cov}\,\mathbf{S}^+ - \text{Cov}\,\hat{\mathbf{S}}^+$.

(c) Moreover, if $\mathbf{L} = \sum_{i=1}^{N} l_i(\mathbf{X}_i)$ with $E[l_i(\mathbf{X}_i)'l_i(\mathbf{X}_i)]$ finite, $1 \le i \le N$, then

$$E\left[(\mathbf{S}^+ - \mathbf{L})(\mathbf{S}^+ - \mathbf{L})'\right] = E\left[(\mathbf{S}^+ - \hat{\mathbf{S}}^+)(\mathbf{S}^+ - \hat{\mathbf{S}}^+)'\right] + E\left[(\hat{\mathbf{S}}^+ - \mathbf{L})(\hat{\mathbf{S}}^+ - \mathbf{L})'\right].$$

Note: This is a multivariate version of the Hájek (1968) projection lemma. (Koziol, 1978.)

4.4.2. Consider the statistic $\mathbf{S}^+$ in Exercise 4.4.1, and assume that $\boldsymbol{\phi} = (\phi_1, \ldots, \phi_p)'$, where ϕ_i, $1 \le i \le p$, are defined by (4.3.6). Assume that each ϕ_i, $1 \le i \le p$, has bounded second derivative. Then there exists a constant M depending only on $\boldsymbol{\phi}$ such that for any N, $(c_1, \ldots, c_N)$, and continuous $F_1, F_2, \ldots, F_N$,

$$E\left[\left(\mathbf{S}^+ - E\mathbf{S}^+ - \sum_{k=1}^{N} \mathbf{Z}_k^+\right)\left(\mathbf{S}^+ - E\mathbf{S}^+ - \sum_{k=1}^{N} \mathbf{Z}_k^+\right)'\right] \le MN^{-1} \sum_{i=1}^{N} c_j^2 \mathbf{I}_{p\times p},$$

where

$$\mathbf{Z}_k^+ = \mathbf{Z}_k^+(\mathbf{X}_k)$$

$$= (N+1)^{-1} \sum_{j=1}^{N} c_j \int \mathbf{s}(\mathbf{x}) * \begin{bmatrix} u(|x_1| - |X_{k1}|) - F_{[k1]}(|x_1|) \\ \vdots \\ u(|x_p| - |X_{kp}|) - F_{[kp]}(|x_p|) \end{bmatrix}$$

$$* \boldsymbol{\Phi}_N'(\mathbf{H}^+(|\mathbf{x}|))\, dF_j$$

$$+ c_k\left[\mathbf{s}(\mathbf{X}_k) * \boldsymbol{\Phi}(H^+(|\mathbf{X}_k|)) - E\mathbf{S}(\mathbf{X}_k) * \boldsymbol{\Phi}(\mathbf{H}^+(|\mathbf{X}_k|))\right],$$

$$k = 1, \ldots, N.$$

Here $*$ denotes the Hadamard product, $\mathbf{H}^+(x) = (H_1^+(x_1), \ldots, H_p^+(x_p))'$, and

$$H_i^+(x) = N^{-1} \sum_{j=1}^{N} F_{[ij]}(x).$$

(Koziol, 1978.)

PART 2

Nonparametric Inference in Linear Models

CHAPTER 5

Distribution-Free Rank-Order Tests for Some Linear Hypotheses

5.1. INTRODUCTION

In the preceding three chapters, for univariate as well as multivariate models, simple linear and signed rank statistics were introduced and their distribution theory studied. In the present chapter, these statistics are incorporated in the formulation of distribution-free tests for some general linear models. For general linear hypotheses, strictly distribution-free rank tests may not always exist. But it is possible to construct asymptotically distribution-free tests. These will be studied in Chapters 7 and 8.

In Section 5.2, we start with the simple linear regression model with nonstochastic predictors. In this case, under suitable null hypotheses, the joint distribution of the allied random variables remain invariant under appropriate groups of transformations, and this generates distribution-free rank tests. The results are extended in Section 5.3 to some linear hypotheses in the multiple linear regression model (univariate case). In the multivariate case, the basic invariance structure of the joint distribution of the sample stochastic vectors holds, but in general, due to mutual (stochastic) dependence of the different variates, rank-order tests are not genuinely distribution-free. However, using the derived permutational invariance structure, permutationally (conditionally) distribution-free rank-order tests can be constructed, and these are treated in Section 5.4. The asymptotic power and relative efficiency of rank-order tests for univariate as well as multivariate linear models are studied in Sections 5.5, 5.6, and 5.7. For univariate linear models, under fairly general regularity conditions, asymptotic optimality of rank-order tests (for suitable families of local alternatives) can be established. However, for general multivariate models this requires more stringent regularity conditions, which are considered in Section 5.8. The last section deals with the rank-order tests (and their properties) for the specific problems treated in Section 4.4. The case of stochastic predictors will be taken up in Chapter 8. Rank tests for subhypotheses in general linear models will be considered in Chapter 7.

5.2. RANK TESTS FOR THE SIMPLE REGRESSION MODEL

Let $X_1, \ldots, X_N$ be independent random variables with continuous c.d.f.'s $F_1, \ldots, F_N$, respectively, all defined on the real line R; it is assumed that

$$F_i(x) = F(x - \beta_0 - \beta c_i), \qquad i = 1, \ldots, N, \quad x \in R, \tag{5.2.1}$$

F is continuous, $\mathbf{c}_N = (c_1, \ldots, c_N)$ is a vector of (known) regression constants (not all equal), and (β_0, β) are the unknown parameters; β_0 is termed the *intercept* of the regression line (of X_i on c_i), and β, the *regression coefficient*.

Our first hypothesis of interest is that of no regression, viz.,

$$H_0^{(1)}: \beta = 0 \quad \text{vs.} \quad H^{(1)}: \beta \neq 0 \quad \left(\text{or} \quad H_+^{(1)}: \beta > 0\right), \tag{5.2.2}$$

where β_0 is treated as a nuisance parameter. We may also be interested in the joint hypothesis

$$H_0^{(2)}: (\beta_0, \beta) = \mathbf{0} \quad \text{vs.} \quad H^{(2)}: (\beta_0, \beta) \neq \mathbf{0}. \tag{5.2.3}$$

In either case we shall see that distribution-free rank-order tests exist. In passing, we may also note that in (5.2.2) or (5.2.3), one could have taken $\beta_0 = \beta_0^*$ and $\beta = \beta^*$ for some specified (β_0^*, β^*). In that case, on working with $X_i^* = X_i - \beta^* c_i$ or $X_i - \beta_0^* - \beta^* c_i$, $i = 1, \ldots, N$, one can reduce the model to (5.2.2) or (5.2.3). Hence, so long as the parameters are specified under the null hypothesis, there is no loss of generality in the formulation in (5.2.2) or (5.2.3). A third hypothesis of interest may be

$$H_0^{(3)}: \beta_0 = 0 \quad \text{vs.} \quad H^{(3)}: \beta_0 \neq 0 \tag{5.2.4}$$

(or $H_+^{(3)}: \beta_0 > 0$, or $H_-^{(3)}: \beta_0 < 0$), where β is treated as a nuisance parameter. For β unknown and different from 0, signed rank statistics are not distribution-free (even under $H_0^{(3)}$), and as a result, neither are rank-order tests. However, in Chapter 7 we shall develop some aligned rank-order tests which are asymptotically distribution-free.

5.2.1. Rank-Order Tests for $H_0^{(1)}$

Let us define

$$L_N = \sum_{i=1}^{N} (c_i - \bar{c}_N) a_N(R_{Ni}) \quad \text{and} \quad \bar{c}_N = N^{-1} \sum_{i=1}^{N} c_i, \tag{5.2.5}$$

where the ranks $R_{N1}, \ldots, R_{NN}$ and the rank scores $a_N(1), \ldots, a_N(N)$ are defined by (2.2.1) and (2.2.2) respectively. Under $H_0^{(1)}$ in (5.2.2), $F_1(x) = \cdots = F_N(x) = F(x - \beta_0) = F_0(x)$, say, where F_0 is continuous. Hence, $X_1, \ldots, X_N$ are i.i.d.r.v.'s and since F_0 is continuous, ties among $X_1, \ldots, X_N$ can be neglected, in probability. Thus, $\mathbf{R}_N = (R_{N1}, \ldots, R_{NN})$ assumes all possible $N!$ permutations of $(1, \ldots, N)$ with the common probability $(N!)^{-1}$. Let $\mathscr{R}_N = \{(i_1, \ldots, i_N) : 1 \leq i \neq \cdots \neq i_N \leq N\}$, and let $\mathbf{r}_N$ denote a typical element of $\mathscr{R}_N$. Then, under $H_0^{(1)}$, we have

$$P\{\mathbf{R}_N = \mathbf{r}_N | H_0^{(1)}\} = \frac{1}{N!} \qquad \forall \mathbf{r}_N \in \mathscr{R}_N. \tag{5.2.6}$$

This implies that

$$P\{R_{Ni} = k | H_0^{(1)}\} = N^{-1} \qquad \forall 1 \leq i, k \leq N, \tag{5.2.7}$$

$$P\{R_{Ni} = k,\ R_{Nj} = q | H_0^{(1)}\} = \frac{1}{N(N-1)} \qquad \forall 1 \leq i \neq j, k \neq q \leq N. \tag{5.2.8}$$

Consequently, by (5.2.5), (5.2.7), and (5.2.8),

$$\begin{aligned} E\{L_N | H_0^{(1)}\} &= \sum_{i=1}^{N} (c_i - \bar{c}_N) E\{a_N(R_{Ni}) | H_0^{(1)}\} \\ &= \sum_{i=1}^{N} (c_i - \bar{c}_N) \left\{ N^{-1} \sum_{i=1}^{N} a_N(i) \right\} = 0, \end{aligned} \tag{5.2.9}$$

$$\begin{aligned} V\{L_N | H_0^{(1)}\} &= \sum_{i=1}^{N} \sum_{j=1}^{N} (c_i - \bar{c}_N)(c_j - \bar{c}_N) \\ &\quad \times E\{[a_N(R_{Ni}) - \bar{a}_N]\,[a_N(R_{Nj}) - \bar{a}_N] | H_0^{(1)}\} \\ &= A_N^2 C_N^2, \end{aligned} \tag{5.2.10}$$

where

$$A_N^2 = (N-1)^{-1} \sum_{i=1}^{N} \{a_N(i) - \bar{a}_N\}^2, \qquad \bar{a}_N = N^{-1} \sum_{i=1}^{N} a_N(i), \quad (5.2.11)$$

$$C_N^2 = \sum_{i=1}^{N} (c_i - \bar{c}_N)^2. \quad (5.2.12)$$

Also, L_N is a completely specified function of $\mathbf{c}_N$, $a_N(1), \ldots, a_N(N)$, and $\mathbf{R}_N$. Since $\mathbf{c}_N$ and the $a_N(i)$ are given, and by (5.2.6), $\mathbf{R}_N$ has a specified (discrete) probability distribution under $H_0^{(1)}$, we conclude that the distribution of L_N, under $H_0^{(1)}$, does not depend on the underlying d.f. F_0. Hence, L_N is distribution-free under $H_0^{(1)}$. This characterizes the existence of genuinely distribution-free rank-order tests for $H_0^{(1)}$. For small N and given $\mathbf{c}_N$, the null-hypothesis distribution of L_N can be evaluated by direct computations, using (5.2.5) and (5.2.6). This process becomes laborious as N increases. For this reason, we shall use the asymptotic distribution of L_N when $H_0^{(1)}$ holds. But first we remark on the behavior of L_N when $H_0^{(1)}$ does not hold, so that an appropriate critical region can be constructed. Suppose that $H_+^{(1)}$ holds, i.e., $\beta > 0$. Then, for some $\beta > 0$, $X_i - \beta c_i$ [or equivalently, $X_i - \beta(c_i - \bar{c}_N)$], $i = 1, \ldots, N$, are i.i.d.r.v.'s Thus, if for $i \neq j$, $c_i - \bar{c}_N$ is greater than $c_j - \bar{c}_N$, then X_i is stochastically larger than X_j, so that $R_{Ni} > R_{Nj}$ is more probable than $R_{Ni} < R_{Nj}$. An opposite conclusion holds when $\beta < 0$. Thus, rewriting (5.2.5) as

$$L_N = \frac{1}{2N} \sum_{1 \le i \neq j \le N} (c_i - c_j)\left[a_N(R_{Ni}) - a_N(R_{Nj})\right], \quad (5.2.13)$$

it follows that if $a_N(1) \le \cdots \le a_N(N)$ (with at least one strict inequality), then, for $\beta > 0$ (< 0), L_N will be stochastically positive (negative). In fact, $A_N^{-1} C_N^{-2} L_N$ may even converge in probability to a positive (negative) quantity when $\beta > 0$ (< 0) and the c_i satisfy certain regularity conditions (see Exercise 5.2.1). Also, if F admits an absolutely continuous density function f having a finite Fisher information $I(f) = \int_{-\infty}^{\infty} \{f'(x)/f(x)\}^2 dF(x)$ ($< \infty$), where f' is the derivative of f, then, for testing $H_0^{(1)}$ against $H_+^{(1)}$, the test with the critical region

$$\sum_{i=1}^{N} (c_i - \bar{c}_N) a_N^*(R_{Ni}) \ge k_\alpha \quad (5.2.14)$$

[where $a_N^*(i) = E\phi^*(U_N^{(i)})$, $i = 1, \ldots,$; the $U_N^{(i)}$ are defined as in (2.2.2); and $\phi^*(u) = -f'(F^{-1}(u))/f(F^{-1}(u))$, $0 < u < 1$] is the locally most powerful rank test (see Exercise 5.2.2) at the significance level α $(0 < \alpha < 1)$. Thus (5.2.14) motivates a rank test; however, since F and hence ϕ^* are not known, we may not be in a position to use this particular test. On the other hand, motivated by this, we may employ L_N as a test statistic and by suitable choice of $a_N(i)$, $i = 1, \dots, N$, we may even claim its local optimality for some specific density; in any case, L_N remains valid over a broad class of d.f.'s. We therefore propose the following.

For testing $H_0^{(1)}$ vs. $H_+^{(1)}$, consider the *critical region* specified by

$$A_N^{-1} C_N^{-1} L_N \geq k_{\alpha,N}^+, \tag{5.2.15}$$

where $k_{\alpha,N}^+$ is so chosen that $P\{L_N \geq k_{\alpha,N}^+ A_N C_N | H_0^{(1)}\} \leq \alpha$, the desired level of significance. (A randomized test procedure may be used if it is desired to have the type I error exactly equal to α.) For testing $H_0^{(1)}$ vs. $H_-^{(1)}$ and $H_0^{(1)}$ vs. $H^{(1)}$, the respective critical regions are

$$A_N^{-1} C_N^{-1} L_N \leq k_{\alpha,N}^- \quad \text{and} \quad |A_N^{-1} C_N^{-1} L_N| \geq k_{\alpha,N}^*, \tag{5.2.16}$$

where $k_{\alpha,N}^-$ and $k_{\alpha,N}^*$ are so chosen that the type I error is bounded by α. For small values of n, given $\mathbf{c}_n$, (5.2.6) can be used to enumerate the exact d.f. of L_N (under $H_0^{(1)}$), and hence to evaluate the constants $k_{\alpha,N}^+$, $k_{\alpha,N}^-$, and $k_{\alpha,N}^*$. The process becomes exceedingly laborious as N increases. For this reason, we provide suitable asymptotic approximations to these constants; these are quite useful when N is at least moderately large. These rest on the classical permutational central limit theorem, in its most general form due to Hájek (1961); earlier works are due to Wald and Wolfowitz (1944), Madow (1948), Noether (1949), Hoeffding (1951a), and Motoo (1957), among others.

Theorem 5.2.1. *Let $\mathbf{R}_N = (R_{N1}, \dots, R_{NN})$ be a random vector which takes on the $N!$ permutations of $(1, \dots, N)$ with equal probability. Let $\{a_N(i), 1 \leq i \leq N\}$ and $\{b_N(i), 1 \leq i \leq N\}$ be two sets of real numbers, and let $L_N = \sum_{i=1}^N b_N(i) a_N(R_{Ni})$. Suppose that as $N \to \infty$,*

$$\max_{1 \leq i \leq N} \frac{\{a_N(i) - \bar{a}_N\}^2}{\sum_{i=1}^N \{a_N(i) - \bar{a}_N\}^2} \to 0, \qquad \max_{1 \leq i \leq N} \frac{\{b_N(i) - \bar{b}_N\}^2}{\sum_{i=1}^N \{b_N(i) - \bar{b}_N\}^2} \to 0. \tag{5.2.17}$$

Then $(L_N - EL_N)/\sqrt{V(L_N)}$ *is asymptotically normally distributed iff, for every* $\varepsilon > 0$,

$$\lim_{N\to\infty}\left\{N^{-1}\sum_{i=1}^{N}\sum_{j=1}^{N}\delta_{Nij}^2 I(|\delta_{Nij}| > \varepsilon)\right\} = 0, \tag{5.2.18}$$

where $I(A)$ *stands for the indicator function of the set* A *and*

$$\delta_{Nij} = \frac{\{a_N(i) - \bar{a}_N\}\{b_N(j) - \bar{b}_N\}}{\left\{N^{-1}\sum_{i=1}^{N}\{a_N(i) - \bar{a}_N\}^2\sum_{j=1}^{N}\{b_N(j) - \bar{b}_N\}^2\right\}^{1/2}} \tag{5.2.19}$$

for $i, j = 1, \ldots, N$.

A second (projection) theorem due to Hájek (1961) is also stated below.

Theorem 5.2.2. *If* $a_N(1) \leq \cdots \leq a_N(N)$ *and* $\max_{1\leq i\leq N}\{a_N(i) - \bar{a}_N\}^2/\Sigma_{i=1}^N\{a_N(i) - \bar{a}_N\}^2 \to 0$ *as* $N \to \infty$, *then* L_N, *defined in Theorem* 5.2.1, *is asymptotically equivalent in the quadratic mean to the statistic*

$$T_N = \sum_{i=1}^{N}\{b_N(i) - \bar{b}_N\}a_N^0(U_i) + N\bar{b}_N\bar{a}_N, \tag{5.2.20}$$

where the U_i *are i.i.d.r.v.'s with the uniform* $(0,1)$ *d.f. and*

$$a_N^0(u) = a_N(i) \quad \text{for } \frac{i-1}{N} < u \leq \frac{i}{N}, \qquad i = 1,\ldots,N. \tag{5.2.21}$$

The proofs of these theorems are sketched in the Appendix (Section A.3). The condition (5.2.17) is known as the *Noether condition*, while (5.2.18) is the classical *Lindeberg condition*. For a martingale formulation of these results, we may refer to Sen (1981a, Chapter 4). The following is a corollary to the preceding two theorems.

Theorem 5.2.3. *If the scores* $a_N(1),\ldots,a_N(N)$ *and* c_N *in* (5.2.5) *all satisfy the Noether condition and* (5.2.18) *holds for* $\delta_{Nij} = [a_N(i) - \bar{a}_N](c_i - \bar{c}_N)/(A_N C_N)$, $i, j = 1,\ldots,N$, *then* $L_N/(A_N C_N)$ *is asymptotically (under* $H_0^{(1)}$*) normally distributed with* 0 *mean and unit variance. Hence, for every*

$\alpha: 0 < \alpha < 1$,

$$\lim_{N\to\infty} k_{\alpha,N}^{+} = \lim_{N\to\infty} k_{\alpha,N}^{-} = \tau_\alpha \quad \textit{and} \quad \lim_{N\to\infty} k_{\alpha,N}^{*} = \tau_{\alpha/2}, \tag{5.2.22}$$

where τ_ε *is the upper* $100\varepsilon\%$ *point of the standard normal d.f.*

In practice, usually we specify conditions on $\mathbf{c}_N$ and $\{a_N(1), \dots, a_N(N)\}$ which are easily verifiable and imply the ones in Theorem 2.5.3. For example, if the underlying score function $\phi(u)$, $0 < u < 1$, either has a bounded second derivative inside $(0, 1)$ or is the difference of two nondecreasing, square-integrable, and absolutely continuous score functions, then by the theorems of Section 2.5, $A_N^{-1} C_N^{-1} L_N$ is asymptotically normal. Alternatively, for arbitrary scores $\{a_N(1), \dots, a_N(N)\}$, if we impose the condition that

$$\sup_N \left\{ N^{-1} \sum_{i=1}^{N} |a_N(i)|^{2+\delta} \right\} < \infty \qquad \text{for some} \quad \delta > 0, \tag{5.2.23}$$

then, without imposing the continuity of $\phi(u)$ or its differentiability, under the Noether condition on $\mathbf{c}_N$, the conditions of Theorem 2.5.3 can be verified easily (see Exercise 5.2.3). Similarly, if the Noether condition on $\mathbf{c}_N$ is replaced by the Hájek condition, viz.,

$$NC_N^{-2}\left\{ \max_{1\le i\le N} (c_i - \bar{c}_N)^2 \right\} = O(1), \tag{5.2.24}$$

then the Noether condition on the $a_N(i)$ alone insures the conditions of Theorem 5.2.3. (See Exercise 5.2.4). Equation (5.2.24) holds for a broad class of designs.

5.2.2. Rank-Order Tests for $H_0^{(2)}$

We define

$$S_{N,1}^{+} = \sum_{i=1}^{N} a_N^{*}(R_{Ni}^{+}) \operatorname{sgn} X_i, \qquad S_{N,2}^{+} = \sum_{i=1}^{N} c_i a_N^{*}(R_{Ni}^{+}) \operatorname{sgn} X_i, \tag{5.2.25}$$

where R_{Ni}^{+} is the rank of $|X_i|$ among $|X_1|, \dots, |X_N|$ for $i = 1, \dots, N$, and $a_N^{*}(1), \dots, a_N^{*}(N)$ are the rank scores generated by a score function $\phi^{*}(u)$, $0 < u < 1$, in the same manner as in (2.2.2). Here, for the model (5.2.1), we assume further that the d.f. F is symmetric about 0, i.e.,

$$F(x) + F(-x) = 1 \qquad \text{for every real } x. \tag{5.2.26}$$

Let us also denote

$$\lambda_{11}^{(N)} = N, \qquad \lambda_{12}^{(N)} = \lambda_{21}^{(N)} = \sum_{i=1}^{N} c_i, \quad \lambda_{22}^{(N)} = \sum_{i=1}^{N} c_i^2, \tag{5.2.27}$$

and let

$$\mathbf{S}_N^+ = (S_{N,1}^+, S_{N,2}^+)' \quad \text{and} \quad \Lambda^{(N)} = \left(\left(\lambda_{ij}^{(N)}\right)\right)_{i,j=1,2}. \tag{5.2.28}$$

Since the c_i are not all equal, the rank of $\Lambda^{(N)}$ is 2, and we denote its inverse by

$$\Lambda_N^{-1} = (\Lambda^{(N)})^{-1}. \tag{5.2.29}$$

Under $H_0^{(2)}$ in (5.2.3) and (5.2.26), the X_i are i.i.d.r.v.'s with a continuous d.f. F, symmetric about 0. Consider now a finite group $\mathscr{G}_N$ of transformations $\{g_N\}$ which map the sample space onto itself. If we let $\mathbf{X}_N = (X_1, \dots, X_N)$, then typically a transformation g_N is characterized by

$$g_N \mathbf{X}_N = \left((-1)^{i_1} X_{j_1}, \dots, (-1)^{i_n} X_{j_n}\right), \tag{5.2.30}$$

where

$$i_k = 0, 1 \qquad \text{for} \quad k = 1, \dots, N \tag{5.2.31}$$

and $(j_1, \dots, j_N)$ is any permutation of $(1, \dots, N)$. Thus, there are $N!\,2^N$ possible elements of $\mathscr{G}_N$. Under $H_0^{(2)}$, $g_N \mathbf{X}_N$ has the same distribution as $\mathbf{X}_N$ has, for every $g_N \in \mathscr{G}_N$, and hence the conditional distribution over the set of $N!\,2^N$ realizations (i.e. $\{g_N \mathbf{X}_N : g_N \in \mathscr{G}_N\}$) is uniform, each having the common probability $(N!\,2^N)^{-1}$. Let us denote this permutational probability measure by $\mathscr{P}_N$. Note that under $\mathscr{P}_N$, sgn X_i, $i = 1, \dots, N$, are independent r.v.'s, each sgn X_i assumes the values ± 1 with the probability $\frac{1}{2}$, and (sgn $X_1, \dots,$ sgn X_N) and $(R_{N1}^+, \dots, R_{NN}^+)$ are stochastically independent, where $(R_{N1}^+, \dots, R_{NN}^+)$ assumes each permutation of $(1, \dots, N)$ with the common probability $(N!)^{-1}$. Hence,

$$E(\mathbf{S}_N^+ \mid \mathscr{P}_N) = \mathbf{0}, \tag{5.2.32}$$

$$E(\mathbf{S}_N^+ \mathbf{S}_N^{+\prime} \mid \mathscr{P}_N) = \Lambda^{(N)} \left\{ N^{-1} \sum_{i=1}^{N} \left[a_N^*(i)\right]^2 \right\} = A_N^{*2} \cdot \Lambda^{(N)}, \tag{5.2.33}$$

where

$$A_N^{*2} = N^{-1} \sum_{i=1}^{N} \left[a_N^*(i)\right]^2. \tag{5.2.34}$$

Therefore, the use of the Mahalanobis distance of $\mathbf{S}_N^+$ from its center of gravity (under $H_0^{(2)}$) leads us to the following test statistic:

$$\mathscr{L}_N^+ = \mathbf{S}_N^{+\prime}\left(E_{\mathscr{P}_N}\mathbf{S}_N^+\mathbf{S}_N^{+\prime}\right)^{-1}\mathbf{S}_N^+ = \left(\mathbf{S}_N^{+\prime}\mathbf{\Lambda}_N^{-1}\mathbf{S}_N^+\right)/A_N^{*2}. \tag{5.2.35}$$

Note that under $\mathscr{P}_N$, A_N^* and $\mathbf{\Lambda}_N^{-1}$ remain invariant, while the distribution of $\mathbf{S}_N^+$ (and hence, of $\mathscr{L}_N^+$) is generated by the $N!\,2^N$ possible (equally likely) realizations of $(R_{N1}^+, \ldots, R_{NN}^+)$ and $(\text{sgn } X_1, \ldots, \text{sgn } X_N)$. Thus, under $H_0^{(2)}$, the distribution of $\mathscr{L}_N^+$ does not depend on the unknown F when (5.2.26) holds. Hence, $\mathscr{L}_N^+$ is distribution-free under $H_0^{(2)}$ and (5.2.26). Here, also, for small N and specific $\mathbf{c}_N$, one can evaluate the exact c.d.f. (under $H_0^{(2)}$) of $\mathscr{L}_N^+$. Since for large N this becomes prohibitively laborious, we shall provide, as in the case of $H_0^{(1)}$, asymptotic simplifications to the distribution of $\mathscr{L}_N^+$ when $H_0^{(2)}$ holds. These developments rest on the following two theorems, which are extensions of Theorems 5.2.1 and 5.2.2, the proofs of which are given in the appendix.

Theorem 5.2.4. *Let $X_1, \ldots, X_N$ be N i.i.d.r.v.'s with a continuous d.f. $F(x)$ satisfying (5.2.26), and let $R_{N1}^+, \ldots, R_{NN}^+$ be the ranks of $|X_1|, \ldots, |X_N|$ among themselves. Let $\{a_N(i), 1 \le i \le N\}$ and $\{b_N(i), 1 \le i \le N\}$ be two sequences of real numbers, and let*

$$S_N^+ = \sum_{i=1}^{N} b_N(i)\, a_N(R_{Ni}^+) \text{sgn } X_i. \tag{5.2.36}$$

Suppose that as $N \to \infty$,

$$\frac{\max_{1 \le i \le N} b_N^2(i)}{\sum_{i=1}^{N} b_N^2(i)} \to 0, \qquad \frac{\max_{1 \le i \le N} a_N^2(i)}{\sum_{i=1}^{N} a_N^2(i)} \to 0. \tag{5.2.37}$$

Then $(S_N^+ - ES_N^+)/\sqrt{V(S_N^+)}$ is asymptotically normally distributed iff, for every $\varepsilon > 0$,

$$\lim_{N\to\infty} \left\{ N^{-1} \sum_{i=1}^{N} \sum_{j=1}^{N} \delta_{Nij}^2 I\left(|\delta_{Nij}| > \varepsilon\right) \right\} = 0, \tag{5.2.38}$$

where for every $1 \le i, j \le N$,

$$\delta_{Nij} = \frac{a_N(i) b_N(j)}{\left\{ N^{-1} \sum_{i=1}^{N} a_N^2(i) \sum_{j=1}^{N} b_N^2(j) \right\}^{1/2}}, \tag{5.2.39}$$

Theorem 5.2.5. *If* $a_N(1) \le \cdots \le a_N(N)$ *and* $\max_{1 \le i \le N} a_N^2(i) / \sum_{i=1}^N a_N^2(i) \to 0$ *as* $N \to \infty$, *then under* $H_0^{(2)}$ *and* (5.2.26), S_N^+, *defined by* (5.2.36), *is asymptotically equivalent in quadratic mean to the statistic*

$$T_N = \sum_{i=1}^{N} b_N(i) a_N^0(U_i) \operatorname{sgn} X_i, \tag{5.2.40}$$

where U_i $(= F^*(|X_i|)$, F^* *being the d.f. of* $|X_i|)$, $i = 1, \ldots, N$, *are i.i.d.r.v.'s having the uniform* $(0, 1)$ *d.f., and*

$$a_N^0(u) = a_N(i) \qquad \text{for} \quad \frac{i-1}{N} < u \le \frac{i}{N}, \quad i = 1, \ldots, N. \tag{5.2.41}$$

By virtue of Theorem 5.2.4, it follows that if the c_i satisfy (5.2.37), and in conjunction with $\{a_N(1), \ldots, a_N(N)\}$, the condition (5.2.38) holds, then under $H_0^{(2)}$, the vector $\mathbf{S}_N^+$ has asymptotically a bivariate normal distribution with null mean vector and dispersion matrix $A_N^{*2} \cdot \Lambda^{(N)}$, defined by (5.2.33). Hence, under $H_0^{(2)}$, $\mathscr{L}_N^+$ has asymptotically a chi-square distribution with 2 degrees of freedom (DF). Thus if $\chi_{t,\alpha}^2$ stands for the upper $100\alpha\%$ point of the chi-square d.f. with t DF, then we have the following test procedure for large N:

$$\textit{Reject or accept } H_0^{(2)}, \textit{ according as } \mathscr{L}_N^+ \ge \textit{ or } < \chi_{2,\alpha}^2 \tag{5.2.42}$$

We illustrate the test procedures considered in this section by the following example. Two groups of ninth-grade students were chosen at random from a school, and a quantitative test was made on their language skill. These two groups were subsequently treated with two different methods of instruction (say, A and B) for a period of 6 weeks in the summer. A second test was made at the termination of this program, and the differences in the scores for each individual (for the two tests) were recorded. These are as follows:

Group A: 5.2, −0.7, −2.3, 3.2, −1.5, 4.7, 1.8, −0.4, 0.6, 6.6.
Group B: −0.9, 1.7, −0.3, 2.4, 4.2, −1.6, −4.3, 0.8, −0.5, −0.2.

It is desired to test for the effectiveness of the 6-week instruction courses as well as for the relative difference of the two methods. This model can be fitted to (5.2.1) where $c_i = +1$ for $1 \le i \le 10$ and -1 for $11 \le i \le 20$; thus β_0 stands for the shift of location over the pre-instruction level and 2β stands for the difference of locations of methods A and B. For simplicity, we take the Wilcoxon scores, where $a_N(i)$ or $a_N^*(i)$ is equal to $i/(N+1)$, $i = 1, \dots, N$. Then we have the table shown opposite. Thus, we have $L_N = +\frac{26}{21}$, $C_N^2 = 20$, and $A_N^2 = 20/(12 \times 21)$, so that

$$\frac{L_N}{A_N C_N} = 0.9826.$$

Observation No.	Response	c_i	R_{Ni}	R_{Ni}^+	$c_i R_{Ni}$	$\operatorname{sgn} X_i R_{Ni}^+$
1	5.2	+1	19	19	19	19
2	−0.7	+1	6	6	6	−6
3	−2.3	+1	2	13	2	−13
4	3.2	+1	16	15	16	15
5	−1.5	+1	4	9	4	−9
6	4.7	+1	18	18	18	18
7	1.8	+1	14	12	14	12
8	−0.4	+1	8	3	8	−3
9	0.6	+1	11	5	11	5
10	6.6	+1	20	20	20	20
11	−0.9	−1	5	8	−5	−8
12	1.7	−1	13	11	−13	11
13	−0.3	−1	9	2	−9	−2
14	2.4	−1	15	14	−15	14
15	4.2	−1	17	16	−17	16
16	−1.6	−1	3	10	−3	−10
17	−4.3	−1	1	17	−1	−17
18	0.8	−1	12	7	−12	7
19	−0.5	−1	7	4	−7	−4
20	−0.2	−1	10	1	−10	−1

For this particular (equal-sample-size, two-sample) problem, the normal approximation in Theorem 5.2.3 works out well for $N = 20$, and hence the 5% critical value may well be approximated as 1.95. Thus, we accept the hypothesis $H_0^{(1)}$ that there is no difference in the two methods of instructions with respect to the responses observed. Similarly, here

$$S_{N,1}^+ = \tfrac{64}{21}, \quad S_{N,2}^+ = \tfrac{52}{21}, \quad \Lambda^{(N)} = 20\begin{pmatrix} 1 & 0 \\ 0 & 1 \end{pmatrix}, \text{ and } A_N^{*2} = \tfrac{41}{126}.$$

Thus,

$$\mathscr{L}_N^+ = \tfrac{680}{287} = 2.368.$$

Also, $\chi^2_{2,.05} = 5.99$, so that we also accept $H_0^{(2)}$ and conclude that the 6-week training has not been effective in improving the language skill of the students.

5.3. RANK TESTS FOR SOME MULTIPLE LINEAR REGRESSION MODELS

Let $X_1, \dots, X_N$ be independent (real-valued) r.v.'s with continuous d.f.'s $F_1, \dots, F_N$, respectively, where it is assumed that for some $q \geq 1$,

$$F_i(x) = F(x - \beta_0 - \boldsymbol{\beta}'\mathbf{c}_i), \qquad i = 1, \dots, N, \quad x \in R, \tag{5.3.1}$$

with β_0 and $\boldsymbol{\beta}' = (\beta_1, \dots, \beta_q)$ all unknown parameters and the $\mathbf{c}_i$ specified q-vectors. We let $\boldsymbol{\beta}^* = (\beta_0, \boldsymbol{\beta}')'$. Our first hypothesis of interest is

$$H_0^{(1)}: \boldsymbol{\beta} = \mathbf{0} \quad \text{vs.} \quad H^{(1)}: \boldsymbol{\beta} \neq \mathbf{0}. \tag{5.3.2}$$

Secondly, we may also be interested in

$$H_0^{(2)}: \boldsymbol{\beta}^* = \mathbf{0} \quad \text{vs.} \quad H^{(2)}: \boldsymbol{\beta}^* \neq \mathbf{0}. \tag{5.3.3}$$

These hypotheses are analogous to the ones in Section 5.2, and we shall derive here parallel rank-order tests.

5.3.1. Rank Tests for $H_0^{(1)}$

For each j $(= 1, \dots, q)$, let us define $L_{Nj} = \sum_{i=1}^N (c_{ij} - \bar{c}_{Nj}) a_N(R_{Ni})$, $\bar{c}_{Nj} = N^{-1}\sum_{i=1}^N c_{ij}$, where the scores $a_N(1), \dots, a_N(N)$ and the ranks $R_{N1}, \dots, R_{NN}$ are defined as in (2.2.1) and (2.2.2). Let then

$$\mathbf{L}_N = \sum_{i=1}^N (\mathbf{c}_i - \bar{\mathbf{c}}_N) a_N(R_{Ni}) = (L_{N1}, \dots, L_{Nq})', \tag{5.3.4}$$

Now, under $H_0^{(1)}$, $F_1(x) = \cdots = F_N(x) = F(x - \beta_0) = F_0(x)$, say, $-\infty < x < \infty$, so that the X_i are i.i.d.r.v.'s. Also, by virtue of the assumed continuity of the d.f. F, ties among the X_i can be neglected in probability. Thus, under $H_0^{(1)}$, $(R_{N1}, \dots, R_{NN})$ assumes all possible permutations of

$(1, \ldots, N)$ with equal probability $(N!)^{-1}$. Hence, as in Section 5.2,

$$E\left(\mathbf{L}_N | H_0^{(1)}\right) = \mathbf{0} \quad \text{and} \quad E\left(\mathbf{L}_N \mathbf{L}_N' | H_0^{(1)}\right) = A_N^2 \mathbf{C}_N, \tag{5.3.5}$$

where A_N^2 is defined by (5.2.11) and $\mathbf{C}_N = ((C_{N,jj'}))_{j,j'=1,\ldots,q}$ is defined by

$$\begin{aligned} \mathbf{C}_N &= \sum_{i=1}^{N} (\mathbf{c}_i - \bar{\mathbf{c}}_N)(\mathbf{c}_i - \bar{\mathbf{c}}_N)' \\ &= \left(\left(\sum_{i=1}^{N} (c_{ij} - \bar{c}_{Nj})(c_{ij'} - \bar{c}_{Nj'}) \right)\right). \end{aligned} \tag{5.3.6}$$

The Mahalanobis distance of $\mathbf{L}_N$ from its center of gravity under $H_0^{(1)}$ (i.e. $\mathbf{0}$) is given by

$$\mathscr{L}_N = A_N^{-2} \{ \mathbf{L}_N' \mathbf{C}_N^- \mathbf{L}_N \}, \tag{5.3.7}$$

and $\mathbf{C}_N^-$ is a generalized inverse of $\mathbf{C}_N$. Our test for $H_0^{(1)}$ is based on the statistic $\mathscr{L}_N$, and we reject (or accept) $H_0^{(1)}$ for numerically large (or small) values of $\mathscr{L}_N$.

Now, $\mathbf{L}_N$ is a function of $\mathbf{c}_1, \ldots, \mathbf{c}_N$, $a_N(1), \ldots, a_N(N)$, and the stochastic vector $(R_{N1}, \ldots, R_{NN})$. Hence, by the same arguments as in Section 5.2, under $H_0^{(1)}$, the distribution of $\mathbf{L}_N$ does not depend on the unknown F. Also, A_N is a constant and $\mathbf{C}_N$ is a known matrix. Hence, the null-hypothesis distribution of $\mathscr{L}_N$ does not depend on the underlying d.f. F. Thus, the test based on $\mathscr{L}_N$ is genuinely distribution-free. Here also, for small N and given $\mathbf{c}_1, \ldots, \mathbf{c}_N$, the exact null distribution of $\mathscr{L}_N$ can be obtained by direct enumeration, but the task becomes prohibitively laborious as N increases. For this reason, we consider the following large-sample approach.

We assume that as $N \to \infty$,

$$\xi_N = \max_{1 \le i \le N} (\mathbf{c}_i - \bar{\mathbf{c}}_N)' \mathbf{C}_N^- (\mathbf{c}_i - \bar{\mathbf{c}}_N) \to 0. \tag{5.3.8}$$

We also write

$$\mathbf{C}_N = \mathbf{D}_N \mathbf{Q}_N \mathbf{D}_N, \qquad \text{where} \quad \mathbf{D}_N = \operatorname{diag}\left(C_{N,11}^{1/2}, \ldots, C_{N,qq}^{1/2}\right), \tag{5.3.9}$$

so that the diagonal elements of $\mathbf{Q}_N$ are all equal to unity and $0 < |Q_N| \le 1$. We further assume that there exists a $Q^* > 0$ such that

$$\operatorname{Ch}_q(\mathbf{Q}_N) \ge Q^* > 0 \qquad \text{for every} \quad N \ge N^* \tag{5.3.10}$$

i.e., $\mathbf{Q}_N$ is strictly positive definite for almost all N. In fact, if $\mathbf{Q}_N$ is not of full rank, then by (5.3.3), the L_{Nj} are not all linearly independent, and hence there exists a subset of q' ($< q$) elements of $\mathbf{L}_N$ which are linearly independent. In that case, one may consider the corresponding minor of $\mathbf{C}_N$ (of order $q' \times q'$), which will be positive definite, and replace q by q' everywhere. Hence, without any essential loss of generality, we may assume that $q = q'$ and (5.3.10) holds.

Theorem 5.3.1. *Suppose* (1) (5.3.8) *and* (5.3.10) *hold*, (2) $a_N(1), \dots, a_N(N)$ *satisfy the Noether condition, and* (3) *for each* s ($= 1, \dots, q$), $\delta_{Nij,s}$, *defined by* (5.2.19) *with the* $b_N(j)$ *and* $\bar{b}_N$ *replaced by* c_{js} *and* $\bar{c}_{Ns}$, *respectively, satisfy the Lindeberg condition in* (5.2.18). *Then, under* $H_0^{(1)}$ *in* (5.3.2), $\mathscr{L}_N$ *has asymptotically chi-square distribution with* q *DF.*

Proof. By virtue of (5.3.5), (5.3.7), and the well-known results on asymptotic distributions of quadratic forms in asymptotically normally distributed random vectors, it suffices to show that under $H_0^{(1)}$ in (5.3.2), $\mathbf{L}_N$ has asymptotically the q-variate normal distribution with null mean vector and dispersion matrix $A_N^2 \mathbf{C}_N$. For this, consider an arbitrary $\boldsymbol{\lambda} = (\lambda_1, \dots, \lambda_q)' \neq \mathbf{0}$, and let

$$Z_N(\boldsymbol{\lambda}) = \boldsymbol{\lambda}'\mathbf{L}_N = \sum_{i=1}^{N} \left[\boldsymbol{\lambda}'(\mathbf{c}_i - \bar{\mathbf{c}}_N)\right] a_N(R_{Ni})$$

$$= \sum_{i=1}^{N} d_i(\boldsymbol{\lambda}) a_N(R_{Ni}), \quad \text{say.} \tag{5.3.11}$$

Then, by an appeal to Theorem 5.2.1, it suffices to show that the $d_i(\boldsymbol{\lambda})$ satisfy (5.2.17) and (5.2.18), in conjunction with the $a_N(i)$. Note that

$$\frac{d_i^2(\boldsymbol{\lambda})}{\sum_{i=1}^{N} d_i^2(\boldsymbol{\lambda})} = \frac{\boldsymbol{\lambda}'(\mathbf{c}_i - \bar{\mathbf{c}}_N)(\mathbf{c}_i - \bar{\mathbf{c}}_N)'\boldsymbol{\lambda}}{\boldsymbol{\lambda}'\mathbf{C}_N\boldsymbol{\lambda}}$$

$$\leq \mathrm{Ch}_1\left((\mathbf{c}_i - \bar{\mathbf{c}}_N)(\mathbf{c}_i - \bar{\mathbf{c}}_N)'\right)\mathbf{C}_N^{-1}$$

$$= (\mathbf{c}_i - \bar{\mathbf{c}}_N)'\mathbf{C}_N^{-1}(\mathbf{c}_i - \bar{\mathbf{c}}_N) \qquad \forall \boldsymbol{\lambda} \neq \mathbf{0}, \tag{5.3.12}$$

and hence, (5.3.8) insures that the $d_i(\boldsymbol{\lambda})$ satisfy the Noether condition. Also, if we let [after noting that $\Sigma_{i=1}^{N} d_i(\boldsymbol{\lambda}) = 0$] for every $1 \leq i, j \leq N$,

$$\delta_{Nij}(\boldsymbol{\lambda}) = \frac{d_i(\boldsymbol{\lambda})(a_N(j) - \bar{a}_N)}{N^{-1} \sum_{i=1}^{N} d_i^2(\boldsymbol{\lambda}) \sum_{j=1}^{N} (a_N(j) - \bar{a}_N)^2}, \tag{5.3.13}$$

then, by (5.3.10),

$$\begin{aligned} \delta_{Nij}^2(\boldsymbol{\lambda}) &= \frac{\boldsymbol{\lambda}'(\mathbf{c}_i - \bar{\mathbf{c}}_N)(\mathbf{c}_i - \bar{\mathbf{c}}_N)\boldsymbol{\lambda}[a_N(j) - \bar{a}_N]^2}{N^{-1} \sum_{i=1}^{N} [a_N(i) - \bar{a}_N]^2 \boldsymbol{\lambda}'\mathbf{C}_N\boldsymbol{\lambda}} \\ &\leq \frac{(\mathbf{c}_i - \bar{\mathbf{c}}_N)'\mathbf{C}_N^{-1}(\mathbf{c}_i - \bar{\mathbf{c}}_N)[a_N(j) - \bar{a}_N]^2}{N^{-1} \sum_{i=1}^{N} (a_N(i) - \bar{a}_N)^2} \\ &\leq \frac{q}{Q^*}\left\{ \max_{1 \leq s \leq q} \delta_{Nij,s}^2 \right\}. \end{aligned} \tag{5.3.14}$$

Consequently, if (5.2.18) holds for each s $(= 1, \ldots, q)$, it will also hold for the $\delta_{Nij}(\boldsymbol{\lambda})$, for every $\boldsymbol{\lambda} \neq \mathbf{0}$. Hence, the conditions of Theorem 5.2.1 hold and the result follows.

The remarks made after Theorem 5.2.3 also apply to Theorem 5.3.1. Thus, under fairly general regularity conditions, for large N, the critical values of $\mathscr{L}_N$ can be approximated by the corresponding $\chi_{q,\alpha}^2$.

5.3.2. Rank Tests for $H_0^{(2)}$

For each j $(= 0, 1, \ldots, q)$, we define

$$S_{N,j}^+ = \sum_{i=1}^{N} c_{ij} \operatorname{sgn} X_i a_N^*(R_{Ni}^+), \tag{5.3.15}$$

where the scores $a_N^*(1), \ldots, a_N^*(N)$ and the ranks $R_{N1}^+, \ldots, R_{NN}^+$ are defined as in Section 5.2 and $c_{i0} = 1$ for every $1 \leq i \leq N$. Let then

$$\mathbf{S}_N^+ = (S_{N,0}^+, S_{N,1}^+, \ldots, S_{N,q}^+)'. \tag{5.3.16}$$

As in Section 5.2.2, here we assume that (5.2.26) holds. Then under $H_0^{(2)}$, the invariance structure of the joint d.f. of $X_1, \dots, X_N$, discussed in (5.2.30)–(5.2.31), holds, and hence, repeating the same arguments, we obtain that

$$E(\mathbf{S}_N^+ | \mathscr{P}_N) = \mathbf{0} \quad \text{and} \quad E(\mathbf{S}_N^+ \mathbf{S}_N^{+\prime} | \mathscr{P}_N) = A_N^{*2} \mathbf{C}_N^*, \tag{5.3.17}$$

where A_N^* is defined by (5.2.34) and $\mathbf{C}_N^* = ((C_{N,jj'}^*))_{j,j'=0,1,\dots,q}$ is defined by

$$\mathbf{C}_N^* = \sum_{i=1}^N \mathbf{c}_i^* \mathbf{c}_i^{*\prime} = \left(\left(\sum_{i=1}^N c_{ij} c_{ij'} \right) \right)_{j,j'=0,1,\dots,q}, \tag{5.3.18}$$

with $\mathbf{c}_i^* = (c_{i0}, c_{i1}, \dots, c_{iq})'$, $i = 1, \dots, N$. Proceeding as in Section 5.2.2, we consider the test statistic

$$\mathscr{L}_N^+ = A_N^{*-2} \left(\mathbf{S}_N^{+\prime} (\mathbf{C}_N^*)^{-1} \mathbf{S}_N^+ \right), \tag{5.3.19}$$

and the hypothesis $H_0^{(2)}$ is rejected (or accepted) for large (or small) values of $\mathscr{L}_N^+$. Here also, note that under $\mathscr{P}_N$, $\mathbf{C}_N^*$ and A_N^* are invariant, while the distribution of $\mathbf{S}_N^+$ (and hence of $\mathscr{L}_N^+$) is generated by the $N!\,2^N$ equally likely realizations of (sgn $X_1, \dots,$ sgn X_N) and $(R_{N1}^+, \dots, R_{NN}^+)$. Hence, $\mathscr{L}_N^+$ is a distribution-free statistic. As in the case of $\mathscr{L}_N^+$ in Section 5.2.2, the evaluation of the exact null distribution of $\mathscr{L}_N^+$ for large N becomes cumbersome, and hence we proceed to provide some asymptotic simplifications.

We assume that

$$\xi_N^* = \max_{1 \le i \le N} \mathbf{c}_i^{*\prime} (\mathbf{C}_N^*)^{-1} \mathbf{c}_i^* \to 0 \quad \text{as} \quad N \to \infty, \tag{5.3.20}$$

where we express the matrix $\mathbf{C}_N^*$ in (5.3.18) as

$$\mathbf{C}_N^* = \mathbf{D}_N^* \mathbf{Q}_N^* \mathbf{D}_N^*; \quad \mathbf{D}_N^* = \operatorname{diag}\left(N^{1/2}, C_{N,11}^{*1/2}, \dots, C_{N,qq}^{*1/2} \right), \tag{5.3.21}$$

so that the diagonal elements of $\mathbf{Q}_N^*$ are all equal to unity and $0 \le |Q_N^*| \le 1$. Parallel to (5.3.10), we assume here that there exists a positive Q_0^* such that

$$\operatorname{Ch}_{q+1}(\mathbf{Q}_N^*) \ge Q_0^* > 0 \quad \text{for every} \quad N \ge N^*, \tag{5.3.22}$$

i.e., $\mathbf{Q}_N^*$ is strictly p.d. for large N. Then we have the following.

Theorem 5.3.2. *If (i) (5.3.20) and (5.3.22) hold, (ii) $a_N^*(1), \ldots, a_N^*(N)$ satisfy the Noether condition, and (iii) for each s $(= 1, \ldots, q)$, $\delta_{Nij,s}^* = c_{js} a_N^*(i)/A_N^* C_{N,ss}^{*1/2}$, $i, j = 1, \ldots, N$, satisfy the Lindeberg condition in (5.2.18), then under $H_0^{(2)}$ in (5.3.3), $\mathscr{L}_N^+$ has asymptotically chi-square distribution with $q + 1$ DF.*

The proof of this theorem runs parallel to that of Theorem 5.3.1 and hence is left as an exercise (Exercise 5.3.1). Thus, for large N, the percentile points of the null-hypothesis distribution of $\mathscr{L}_N^+$ can be approximated by $\chi_{q+1,\alpha}^2$, and the test may be based on the critical region based on this approximation.

We conclude this section with an example. Suppose that in the language-testing problem of Section 5.2, there is an additional group of 10 students who were not given any extra instruction during the summer but were asked to appear in the second test, and for them, the differences (from the first test) in scores are also available. In this case, we can conceive of the model (5.3.1) with $\mathbf{c}_i^{*\prime} = (1,1,0)$ for $i = 1, \ldots, 10$, $(1,0,1)$ for $i = 1, \ldots, 20$, and $(1,0,0)$ for $i = 21, \ldots, 30$. Thus, here $\mathbf{C}_N$ and $\mathbf{C}_N^*$ are given by

$$\mathbf{C}_N = \begin{pmatrix} \frac{20}{3} & -\frac{10}{3} \\ -\frac{10}{3} & \frac{20}{3} \end{pmatrix}, \qquad \mathbf{C}_N^* = \begin{pmatrix} 30 & 10 & 10 \\ 10 & 10 & 0 \\ 10 & 0 & 10 \end{pmatrix},$$

and the conditions of Theorems 5.3.1 and 5.3.2 are easy to verify. The computation of the $L_{N,j}$ or $S_{N,j}^+$ is very similar to that in Section 5.2, and the test statistics $\mathscr{L}_N$ or $\mathscr{L}_N^+$ can be easily computed by using the formulae (5.3.7) and (5.3.19).

5.4. RANK-ORDER TESTS FOR MULTIVARIATE LINEAR MODELS

Consider a sequence $\{\mathbf{X}_i = (X_{i1}, \ldots, X_{ip})', \ i \geq 1\}$ of stochastic p-vectors, distributed independently according to continuous d.f.'s $\{F_i, \ i \geq 1\}$, where

$$F_i(\mathbf{x}) = F(\mathbf{x} - \boldsymbol{\alpha} - \boldsymbol{\beta}\mathbf{c}_i), \qquad i \geq 1, \quad x \in R^p \ (p \geq 1); \tag{5.4.1}$$

$\boldsymbol{\alpha} = (\alpha_1, \ldots, \alpha_p)'$ and $\boldsymbol{\beta} = ((\beta_{jk}))_{j=1,\ldots,p,\, k=1,\ldots,q}$ are unknown parameters; $\mathbf{c}_i = (c_{i1}, \ldots, c_{iq})'$, $i \geq 1$, are specified regression vectors; and $q \geq 1$. Having observed $\mathbf{E}_N = (\mathbf{X}_1, \ldots, \mathbf{X}_N)$ and assuming suitable conditions on F and the $\mathbf{c}_i$, we desire to test the following hypotheses:

$$H_0^{(1)}: \boldsymbol{\beta} = \mathbf{0} \quad \text{vs.} \quad H^{(1)}: \boldsymbol{\beta} \neq \mathbf{0}, \tag{5.4.2}$$

$$H_0^{(2)}: \boldsymbol{\alpha} = \mathbf{0},\ \boldsymbol{\beta} = \mathbf{0} \quad \text{vs.} \quad H^{(2)}: (\boldsymbol{\alpha}, \boldsymbol{\beta}) \neq (\mathbf{0}, \mathbf{0}). \tag{5.4.3}$$

Since the case of $p = 1$ has been dealt with in the preceding two sections, in the sequel it will be assumed that $p > 1$.

5.4.1. Rank Tests for $H_0^{(1)}$

As in (4.2.2), let $R_{Ni}^{(j)}$ be the rank of X_{ij} among $X_{1j}, \ldots, X_{Nj}$ for $i = 1, \ldots, N$ and $j = 1, \ldots, p$. As in (4.2.8), define

$$L_{N,jl} = \sum_{i=1}^{N} c_{il} a_{Nj}\left(R_{Ni}^{(j)}\right), \qquad j = 1, \ldots, p, \quad l = 1, \ldots, q, \quad (5.4.4)$$

where for each j $(= 1, \ldots, p)$, the rank scores $a_{Nj}(1), \ldots, a_{Nj}(N)$ are defined by (4.2.4). The test to be considered here are based on

$$\mathbf{L}_N = \left(L_{N,jl}\right)_{j=1,\ldots,p,\, l=1,\ldots,q}. \quad (5.4.5)$$

Note that, in general, the p coordinates of $\mathbf{X}_i$ are not mutually stochastically independent, and consequently the joint distribution of $\mathbf{L}_N$ depends on the unknown F even when (5.4.2) holds. Thus, in general, $\mathbf{L}_N$ is not genuinely distribution-free. However, the Chatterjee–Sen (1964) rank permutation principle yields a class of permutationally (conditionally) distribution-free tests, studied in detail by Puri and Sen (1969a), which we present below.

Under $H_0^{(1)}$ in (5.4.2), $\mathbf{E}_N$ is composed of N i.i.d. stochastic vectors. Hence, the joint distribution of $\mathbf{E}_N$ (which is the product of the distributions of $\mathbf{X}_i, \ldots, \mathbf{X}_N$) remains invariant under the $N!$ permutations of the N vectors $\mathbf{X}_1, \ldots, \mathbf{X}_N$ among themselves. We define

$$\mathbf{R}_N = (\mathbf{R}_{N1}, \ldots, \mathbf{R}_{NN}) = \begin{pmatrix} R_{N1}^{(1)} & \cdots & R_{NN}^{(1)} \\ \vdots & & \vdots \\ R_{N1}^{(p)} & \cdots & R_{NN}^{(p)} \end{pmatrix}, \quad (5.4.6)$$

where $\mathbf{R}'_{Ni} = (R_{Ni}^{(1)}, \ldots, R_{Ni}^{(p)})$, $i = 1, \ldots, N$. Each row of $\mathbf{R}_N$ is a permutation of the numbers $1, \ldots, N$, there being in all $(N!)^p$ possible realizations of $\mathbf{R}_N$. (By virtue of the assumed continuity of F, ties among $X_{1j}, \ldots, X_{Nj}$ can be neglected in probability for each $j = 1, \ldots, p$.) Let us rearrange the columns of $\mathbf{R}_N$ in such a manner that the first row has the elements $1, \ldots, N$ in the natural order, and denote the corresponding matrix by $\mathbf{R}_N^*$. Then $\mathbf{R}_N$ is said to be *permutationally equivalent* to $\mathbf{R}_N^*$ if it is possible to obtain $\mathbf{R}_N^*$ from $\mathbf{R}_N$ only by permutations of the columns of the latter. Then, for each

$\mathbf{R}_N^*$, there will be a set $\Sigma(R_N^*)$ of $N!$ possible realizations of $\mathbf{R}_N$, such that any member of this set is permutationally equivalent to $\mathbf{R}_N^*$. Now, as was explained earlier, unless $F(\mathbf{x})$ is the product of its p univariate marginals, the probability distribution of $\mathbf{R}_N$ over the $(N!)^p$ possible realizations will depend on $F(\mathbf{x})$. However, given a particular set $\Sigma(\mathbf{R}_N^*)$, the conditional distribution of $\mathbf{R}_N$ over the $N!$ permutations of the columns of $\mathbf{R}_N^*$ would be uniform under $H_0^{(1)}$ in (5.4.2), that is,

$$\mathbf{P}\{\mathbf{R}_N = \mathbf{r} | \Sigma(\mathbf{R}_N^*), H_0^{(1)}\} = \frac{1}{N!} \qquad \forall \mathbf{r} \in \Sigma(\mathbf{R}_N^*), \tag{5.4.7}$$

whatever be $F(\mathbf{x})$. Let us denote by $\mathscr{P}_N$ the permutational (conditional) probability measure generated by the conditional law in (5.4.7). Then it follows that under $\mathscr{P}_N$, $R_{Ni}^{(j)}$ assumes each of the N values $1, \dots, N$ with the common probability $1/N$; $(R_{Ni}^{(j)}, R_{Ni}^{(l)})$ assumes the N possible values $(R_{N\alpha}^{(j)}, R_{N\alpha}^{(l)})$, $\alpha = 1, \dots, N$, with the common conditional probability $1/N$; and $(R_{Ni}^{(j)}, R_{Ni'}^{(l)})$, $i \neq i'$, assumes the $N(N-1)$ possible values $(R_{N\alpha}^{(j)}, R_{N\beta}^{(l)})$, $\alpha \neq \beta$ $(= 1, \dots, N)$, with the common conditional probability $1/N(N-1)$. Hence, it follows by (5.4.4), (5.4.5), and a few routine steps (Exercise 5.4.1) that

$$E[\mathbf{L}_N | \mathscr{P}_N] = ((N\bar{c}_l \bar{a}_{Nj})) \tag{5.4.8}$$

[where $\bar{c}_l = N^{-1}\Sigma_{i=1}^N c_{il}$ and $\bar{a}_{Nj} = N^{-1}\Sigma_{i=1}^N a_{Nj}(i)$], and

$$V(\mathbf{L}_N | \mathscr{P}_N) = \mathbf{V}_N \otimes \mathbf{C}_N, \tag{5.4.9}$$

where $\mathbf{C}_N = ((C_{N,ll'}))$ is defined by (4.2.9) and $\mathbf{V}_N = ((v_{Njj'}))_{j,j'=1,\dots,p}$ is defined by

$$v_{Njj'} = (N-1)^{-1} \sum_{i=1}^{N} \left[a_{Nj}(R_{Ni}^{(j)}) - \bar{a}_{Nj}\right]\left[a_{Nj'}\left(R_{Ni}^{(j')}\right) - \bar{a}_{Nj'}\right] \tag{5.4.10}$$

for $j, j' = 1, \dots, p$. Note that $\mathbf{C}_N$ is nonstochastic, while $\mathbf{V}_N$ is stochastic. However, $\mathbf{V}_N$ is invariant under $\mathscr{P}_N$. At this stage, we assume that both $\mathbf{C}_N$ and $\mathbf{V}_N$ are p.d., and denote by $\mathbf{C}_N^{-1}$ and $\mathbf{V}_N^{-1}$ the corresponding reciprocal matrices. Then we have

$$[\mathbf{V}_N \otimes \mathbf{C}_N]^{-1} = \mathbf{V}_N^{-1} \otimes \mathbf{C}_N^{-1} = \mathbf{D}_N^{-1}, \quad \text{say}, \tag{5.4.11}$$

where

$$\mathbf{D}_N^{-1} = \left(\left(d_N^{jj',kk'}\right)\right)_{j,j'=1,\dots,p,\ k,k'=1,\dots,q}. \tag{5.4.12}$$

Now, as in Section 5.3, as a measure of divergence of $\mathbf{L}_N$ from its permutational center of gravity, we consider the following test statistic:

$$\mathscr{L}_N = \sum_{j=1}^{p} \sum_{j'=1}^{p} \sum_{k=1}^{q} \sum_{k'=1}^{q} d_N^{jj',kk'}\left(L_{N,jk} - N\bar{a}_{Nj}\bar{c}_k\right)\left(L_{N,j'k'} - N\bar{a}_{Nj'}\bar{c}_{k'}\right). \tag{5.4.13}$$

Now, $\mathscr{L}_N$ will be stochastically large if $\mathbf{L}_N - N((\bar{a}_{Nj}\bar{c}_l))$ if stochastically different from $\mathbf{0}$, a case that is more probable when $H^{(1)}$ holds. Hence, our test procedure consists in rejecting $H_0^{(1)}$ when $\mathscr{L}_N$ is large. For small N, (5.4.7) may be used to find the permutation distribution of $\mathscr{L}_N$ [over the set $\Sigma(\mathbf{R}_N^*)$] and construct a conditionally distribution-free test based on $\mathscr{L}_N$. This, however, requires the $N!$ possible realizations of $\mathbf{L}_N$ (under $\mathscr{P}_N$), and the task becomes prohibitively laborious as N increases. For this reason, we proceed now to study the asymptotic distribution theory of $\mathbf{L}_N$ and $\mathscr{L}_N$ when $H_0^{(1)}$ in (5.4.2) holds.

Let us define

$$\bar{H}_N(\mathbf{x}) = N^{-1}\sum_{i=1}^{N} F_i(\mathbf{x}), \qquad \mathbf{x} \in R^p, \quad N \geq 1, \tag{5.4.14}$$

and let $\bar{H}_{N[j]}(x)$ and $\bar{H}_{N[j,j']}(x, y)$ be respectively the jth marginal and (j, j')th bivariate marginal d.f.'s for $j \neq j' = 1, \ldots, p$. Now set

$$\boldsymbol{\nu}(\bar{H}_N) = \left(\left(\nu_{jj'}(\bar{H}_N)\right)\right)_{j,j'=1,\ldots,p}, \tag{5.4.15}$$

where for every $1 \leq j, j' \leq p$,

$$\nu_{jj}(\bar{H}_N) = \nu_{jj} = \int_0^1 \phi_j^2(u)\,du - \left(\int_0^1 \phi_j(u)\,du\right)^2, \tag{5.4.16}$$

$$\nu_{jj'}(\bar{H}_N) = \int_{-\infty}^{\infty}\int_{-\infty}^{\infty} \phi_j\left(\bar{H}_{N[j]}(x)\right)\phi_{j'}\left(\bar{H}_{N[j']}(y)\right)\,d\bar{H}_{N[j,j']}(x, y)$$
$$- \left(\int_0^1 \phi_j(u)\,du\right)\left(\int_0^1 \phi_{j'}(u)\,du\right), \tag{5.4.17}$$

and the functions $\phi_j(u): 0 < u < 1$, $1 \leq j \leq p$ correspond to the scores a_{Nj} as defined in (4.2.4).

Theorem 5.4.1. *Under (4.2.4) with $\phi_j(u) = \phi_{j,1}(u) - \phi_{j,2}(u)$, where $\phi_{j,k}(u)$ is nondecreasing in u $(0 < u < 1)$, $k = 1, 2$, and $\phi_j(u)$ square-integrable and absolutely continuous inside $[0, 1]$ for $j = 1, \ldots, p$ as $N \to \infty$, we have*

$$[\mathbf{V}_N - \mathbf{v}(\overline{H}_N)] \to \mathbf{0} \qquad \textit{in probability}, \tag{5.4.18}$$

and hence, under $H_0^{(1)}$,

$$\mathbf{V}_N \xrightarrow{P} \mathbf{v}(F) \qquad \textit{as} \quad N \to \infty. \tag{5.4.19}$$

Proof. By virtue of (5.4.10) and (5.4.16), we obtain by a few routine steps (Exercise 5.4.2) that for each $j(= 1, \ldots, p)$

$$v_{Njj} = (N-1)^{-1}\left\{\sum_{i=1}^{N} a_{Nj}^2(i) - \frac{1}{N}\left(\sum_{i=1}^{N} a_{Nj}(i)\right)^2\right\} \to \nu_{jj} \qquad \text{as} \quad N \to \infty.$$

So, to prove the theorem, we need only to show that as $N \to \infty$,

$$\left|v_{N,jj'} - \nu_{jj'}(\overline{H}_N)\right| \xrightarrow{P} 0 \qquad \text{for} \quad j \neq j' = 1, \ldots, p. \tag{5.4.20}$$

Now, under the hypothesis of the theorem, by Lemma 2.5.4, for every $\varepsilon > 0$, we have for every j $(= 1, \ldots, p)$,

$$\phi_j(t) = \psi_{j,1}^*(t) - \phi_{j,2}^*(t), \qquad 0 < t < 1. \tag{5.4.21}$$

where $\psi_j(t)$ is a polynomial, $\phi_{j,k}^*(t)$ are nondecreasing in $t: 0 < t < 1$, and

$$\sum_{k=1}^{2} \int_0^1 \left[\phi_{j,k}^*(u)\right]^2 du < \varepsilon \qquad \text{for} \quad j = 1, \ldots, p. \tag{5.4.22}$$

We consider the proof for the scores $a_{Nj}(i) = \phi_j(i/(N+1))$, $1 \leq i \leq N$, $j = 1, \ldots, p$; the other case in (4.2.4) follows similarly and is left as an exercise (Exercise 5.4.3).

We have for every $j \neq j'$

$$v_{N,jj'} = \frac{1}{N-1}\left\{\sum_{i=1}^{N} \phi_j\left(\frac{R_{Ni}^{(j)}}{N+1}\right)\phi_{j'}\left(\frac{R_{Ni}^{(j')}}{N+1}\right) - N\bar{a}_{Nj}\bar{a}_{Nj'}\right\}, \tag{5.4.23}$$

where for every $j(= 1, \ldots, p)$,

$$\bar{a}_{Nj} = \frac{1}{N} \sum_{i=1}^{N} \phi_j\left(\frac{i}{N+1}\right) \to \int_0^1 \phi_j(u)\, du \quad \text{as } N \to \infty, \tag{5.4.24}$$

It suffices to show that for $j \neq j'$, as $N \to \infty$,

$$\left| \frac{1}{N} \sum_{i=1}^{N} \phi_j\left(\frac{R_{Ni}^{(j)}}{N+1}\right) \phi_{j'}\left(\frac{R_{Ni}^{(j')}}{N+1}\right) - \int_{-\infty}^{\infty} \int_{-\infty}^{\infty} \phi_j\left(\bar{H}_{N[j]}(x)\right) \phi_{j'}\left(\bar{H}_{N[j']}(y)\right) d\bar{H}_{N[j,j']}(x,y) \right| \xrightarrow{P} 0. \tag{5.4.25}$$

Now, by (5.4.21) and (5.4.22),

$$\begin{aligned} N^{-1} \sum_{i=1}^{N} \phi_j\left(\frac{R_{Ni}^{(j)}}{N+1}\right) \phi_{j'}\left(\frac{R_{Ni}^{(j')}}{N+1}\right) &= N^{-1} \sum_{i=1}^{N} \left\{ \psi_j\left(\frac{R_{Ni}^{(j)}}{N+1}\right) + \phi_{j,1}^{*}\left(\frac{R_{Ni}^{(j)}}{N+1}\right) - \phi_{j,2}^{*}\left(\frac{R_{Ni}^{(j)}}{N+1}\right) \right\} \\ &\quad \times \left\{ \psi_{j'}\left(\frac{R_{Ni}^{(j')}}{N+1}\right) + \phi_{j',1}^{*}\left(\frac{R_{Ni}^{(j')}}{N+1}\right) - \phi_{j',2}^{*}\left(\frac{R_{Ni}^{(j')}}{N+1}\right) \right\} \\ &= N^{-1} \sum_{i=1}^{N} \psi_j\left(\frac{R_{Ni}^{(j)}}{N+1}\right) \psi_{j'}\left(\frac{R_{Ni}^{(j')}}{N+1}\right) + \Omega_N, \end{aligned} \tag{5.4.26}$$

where Ω_N involves eight terms with at least one factor $\phi_{j,k}$ or $\phi_{j',k}$ $(k = 1, 2)$ in each of them. Applying the Cauchy–Schwarz inequality and (5.4.22), it can be shown (see Exercise 5.4.4) that

$$|\Omega_N| < \varepsilon' \quad \text{where} \quad \varepsilon'(> 0) \to 0 \quad \text{as } \varepsilon \to 0. \tag{5.4.27}$$

Let us now write for $j, j' = 1, \ldots, p, (x, y) \in R^2$,

$$H_{N[j]}(x) = N^{-1} \sum_{i=1}^{N} u(x - X_{ij}), \qquad x \in R, \tag{5.4.28}$$

$$H_{N[j,j']}(x, y) = N^{-1} \sum_{i=1}^{N} u(x - X_{ij}) u(y - X_{ij'}), \quad j \neq j'. \tag{5.4.29}$$

Then, the first term on the rhs of (5.4.26) can be written as

$$\int_{-\infty}^{\infty} \int_{-\infty}^{\infty} \psi_j\left(\frac{N}{N+1} H_{N[j]}(x)\right) \psi_{j'}\left(\frac{N}{N+1} H_{N[j']}(y)\right) dH_{N[j,j']}(x, y). \tag{5.4.30}$$

Note that by a direct and trivial extension of the Glivenko–Cantelli lemma to the non-identically-distributed (independent) random variables, as $N \to \infty$, we have [Exercise 5.4.5], for every $j \neq j' = 1, \ldots, p$,

$$\sup_{-\infty < x < y} \left| H_{N[j]}(x) - \overline{H}_{N[j]}(x) \right| \to 0 \qquad \text{a.s.,} \tag{5.4.31}$$

$$\sup_{-\infty < x, y < \infty} \left| H_{N[j,j']}(x, y) - \overline{H}_{N[j,j']}(x, y) \right| \to 0 \qquad \text{a.s.} \tag{5.4.32}$$

Let us rewrite (5.4.30) as the sum of the following two terms:

$$\int_{-\infty}^{\infty} \int_{-\infty}^{\infty} \left\{ \psi_j\left(\frac{N}{N+1} H_{N[j]}(x)\right) \psi_{j'}\left(\frac{N}{N+1} H_{N[j']}(y)\right) \right.$$
$$\left. - \psi_j\left(\overline{H}_{N[j]}(x)\right) \psi_{j'}\left(\overline{H}_{N[j]}(y)\right) \right\} dH_{N[j,j']}(x, y), \tag{5.4.33}$$

and

$$N^{-1} \sum_{i=1}^{N} \psi_j\left(\overline{H}_{N[j]}(X_{ij})\right) \psi_{j'}\left(\overline{H}_{N[j']}(X_{ij'})\right) \tag{5.4.34}$$

and where, by the weak law of large numbers, (5.4.34) is stochastically equivalent to $\int_{-\infty}^{\infty}\int_{-\infty}^{\infty}\psi_j(\overline{H}_{N[j]}(x))\psi_{j'}(\overline{H}_{N[j']}(y))\, d\overline{H}_{N[j,j']}(x, y)$ (as $N \to \infty$). Since the ψ_j are polynomials, they are absolutely continuous and bounded functions, and hence, by (5.4.31), (5.4.33) goes to 0, in probability, as $N \to \infty$. The proof of (5.4.25) follows from (5.4.26), (5.4.27), (5.4.33), (5.4.34), and the above arguments.

Corollary 5.4.1. *Suppose that* $\mathbf{v}(F)$ *is p.d. and the rank of* $\mathbf{C}_N$ *is* q *for every* $N \geq N_0$. *Then, under the conditions of Theorem* 5.4.1, *the rank of* $\mathbf{D}_N$ *is* pq, *in probability, as* $N \to \infty$.

For notational simplification, we may assume (without any loss of generality) that

$$\bar{a}_{Nj} = 0 \qquad \text{for} \quad j = 1, \dots, p; \tag{5.4.35}$$

then, by (5.4.8), $E(\mathbf{L}_N | \mathscr{P}_N) = \mathbf{0}$.

Theorem 5.4.2. *Suppose that* (1) (5.3.8)–(5.3.10) *hold*; (2) *for each* j $(= 1, \dots, p)$, *the scores* $a_{Nj}(1), \dots, a_{Nj}(N)$ *satisfy the conditions of Theorem* 5.4.1, *and* $\mathbf{v}(F)$ *is p.d. Then, under* $H_0^{(1)}$ *in* (5.4.2) *and the permutation model* $\mathscr{P}_N$, $\mathbf{L}_N$ *has asymptotically (in probability) a multinormal distribution with mean* $\mathbf{0}$ *and dispersion matrix* $\mathbf{D}_N$ *or rank* pq.

A proof of the theorem resting on some martingale arguments [Sen (1983a)] is given in the Appendix (Section A.3). By virtue of the preceding two theorems, we arrive at the following.

Theorem 5.4.3. *Under the conditions of Theorem* 5.4.2, $\mathscr{L}_N$, *defined by* (5.4.13), *has a permutation distribution asymptotically converging (in probability) to chi-square distribution with pq DF.*

The above theorem leads us to the following large-sample procedure:

> Reject or accept $H_0^{(1)}$ in (5.4.2) according as $\mathscr{L}_N$ is greater than or less than $\chi^2_{pq,\alpha}$.

(5.4.36)

Since the conditional level of significance is asymptotically equal to α in probability, the unconditional one is also asymptotically equal to α.

5.4.2. Rank Tests for $H_0^{(2)}$

As in (4.3.1), let $\tilde{R}_{Ni}^{(j)}$ be the rank of $|X_{ij}|$ among $|X_{1j}|, \dots, |X_{Nj}|$ for $i = 1, \dots, N$, $j = 1, \dots, p$. Also, as in (4.3.3)–(4.3.4), let $S_i = s(X_{ij})$ for

$i = 1, \ldots, N$, $j = 1, \ldots, p$. Define then $\mathbf{S}_N^+ = ((S_{N,jj'}^+))$ by letting

$$S_{N,jj'}^+ = \sum_{i=1}^{N} S_{ij} c_{ij'} a_{Nj}^*\left(\tilde{R}_{Ni}^j\right), \qquad j = 1, \ldots, p, \quad j' = 0$$

$$c_{i0} = 1 \ \forall i \geq 1, \quad (5.4.37)$$

and let the rank scores $a_{Nj}^*(1), \ldots, a_{Nj}^*(N)$ be generated by the score function $\phi_j^*(u)$, $0 < u < 1$, $j = 1, \ldots, p$, in the same manner as in (4.3.6). Here also, the interdependence (stochastic) of the p coordinates of the $\mathbf{X}_i$ makes the distribution of $\mathbf{S}_N^+$ dependent on F, even when $H_0^{(2)}$ holds. However, the following group of sign-invariant permutations leads to a class of permutationally (conditionally) distribution-free tests based on $\mathbf{S}_N^+$. Our additional assumption is that F is diagonally symmetric about $\mathbf{0}$, so that under $H_0^{(2)}$, both $\mathbf{X}_i$ and $(-1)\mathbf{X}_i$ have the same d.f. F. Denote by

$$\tilde{\mathbf{R}}_N = \left(\left(\tilde{R}_{Ni}^j\right)\right)_{j=1,\ldots,p,\, i=1,\ldots,N}, \qquad \mathbf{A}_N^* = \left(\left(a_{Nj}^*\left(\tilde{R}_{Ni}^j\right)\right)\right)_{j=1,\ldots,p,\, i=1,\ldots,N} \quad (5.4.38)$$

the *rank*- and the *score-collection matrices* respectively, and define $\Lambda^{(N)}$ as in (5.3.22). Let then $\mathscr{G}_N^*$ be the group of transformations $\{g_N^*\}$, where typically a g_N^* is such that

$$\begin{aligned} g_N^* \mathbf{E}_N &= g_N^*(\mathbf{X}_1, \ldots, \mathbf{X}_N) \\ &= \left((-1)^{j_1}\mathbf{X}_{i_1}, \ldots, (-1)^{j_N}\mathbf{X}_{i_N}\right), \end{aligned} \quad (5.4.39)$$

with $j_k = 0, 1$, $k = 1, \ldots, N$, and $(i_1, \ldots, i_N)$ any permutation of $(1, \ldots, N)$; here $\mathbf{E}_N = (X_1, \ldots, X_N)$ is the sample point. Also, let

$$\mathscr{E}_N = \left\{ g_N^* \mathbf{E}_N : g_N^* \in \mathscr{G}_N^* \right\}. \quad (5.4.40)$$

Note that under $H_0^{(2)}$ and for diagonally symmetric F, $g_N^* \mathbf{E}_N$ has the same distribution as $\mathbf{E}_N$ for all $g_N^* \in \mathscr{G}_N^*$. Consequently, the conditional distribution of $\mathbf{E}_N$ over the set $\mathscr{E}_N$ is uniform, all the $2^N N!$ elements being conditionally equally likely. Let us denote this probability measure by $\mathscr{P}_N$. Then, by routine steps, it follows that

$$E\left[S_{ij} a_{Nj}^*\left(\tilde{R}_{Ni}^{(j)}\right) \middle| \mathscr{P}_N \right] = 0 \qquad \forall 1 \leq j \leq p, 1 \leq i \leq N, \quad (5.4.41)$$

$$E\left[S_{ij} a_{Nj}^*\left(\tilde{R}_{Ni}^{(j)}\right) S_{i'j'} a_{Nj'}^*\left(\tilde{R}_{Ni'}^{(j')}\right) \middle| \mathscr{P}_N \right]$$

$$= \delta_{ii'} \left\{ N^{-1} \sum_{i=1}^{N} S_{ij} S_{ij'} a_{Nj}^*\left(\tilde{R}_{Ni}^{(j)}\right) a_{Nj'}^*\left(\tilde{R}_{Ni}^{(j')}\right) \right\} \quad (5.4.42)$$

for $j, j' = 1, \ldots, p$ and $i, i' = 1, \ldots, N$, where δ_{rs} is the usual Kronecker delta. Let then

$$\tilde{v}_{N,jj'} = \frac{1}{N} \sum_{i=1}^{N} S_{ij} S_{ij'} a^*_{Nj}\left(\tilde{R}^{(j)}_{Ni}\right) a^*_{Nj'}\left(\tilde{R}^{(j')}_{Ni}\right), \tag{5.4.43}$$

$$\tilde{\mathbf{V}}_N = \left(\left(\tilde{v}_{N,jj'}\right)\right)_{j,j'=1,\ldots,p}. \tag{5.4.44}$$

From (5.4.37) and (5.4.41)–(5.4.44), we obtain that

$$E\left[\mathbf{S}^+_N | \mathscr{P}_N\right] = \mathbf{0} \quad \text{and} \quad V(\mathbf{S}^+_N | \mathscr{P}_N) = \tilde{\mathbf{V}}_N \otimes \mathbf{\Lambda}^{(N)}. \tag{5.4.45}$$

Let us denote

$$\tilde{\mathbf{D}}_N = \tilde{\mathbf{V}}_N \otimes \mathbf{\Lambda}^{(N)}, \qquad \tilde{\mathbf{D}}_N^{-1} = \tilde{\mathbf{V}}_N^{-1} \otimes \left(\mathbf{\Lambda}^{(N)}\right)^{-1} = \left(\left(\tilde{d}_N^{jj',ll'}\right)\right), \tag{5.4.46}$$

and assume for the time being that both $\tilde{\mathbf{V}}_N$ and $\mathbf{\Lambda}^{(N)}$ are positive definite. As a direct multivariate extension of (5.3.23), we consider the test statistic

$$\mathscr{L}^+_N = \sum_{j=1}^{p} \sum_{j'=1}^{p} \sum_{l=0}^{q} \sum_{l'=0}^{q} \tilde{d}_N^{jj',ll'} S^+_{N,jl} S^+_{N,j'l'}. \tag{5.4.47}$$

Note that $X_{i1}, \ldots, X_{ip}$ are, in general, mutually stochastically dependent, and as a result, $S^+_{N,jl}$, $1 \le j \le p$, $0 \le l \le q$, have a joint distribution dependent on the unknown F. Hence, $\mathscr{L}^+_N$ is not genuinely distribution-free under $H_0^{(2)}$. However, under the conditional model (i.e. under $\mathscr{P}_N$), $\tilde{\mathbf{V}}_N$ and $\mathbf{\Lambda}^{(N)}$ (and hence $\tilde{\mathbf{D}}_N$) are invariant, while the conditional distribution of $\mathbf{S}^+_N$ is generated by the $2^N N!$ equally (conditionally) likely realizations in (5.4.40). Hence, under $\mathscr{P}_N$, $\mathscr{L}^+_N$ has a distribution which does not depend on F. This leads to a conditionally distribution-free test for $H_0^{(2)}$. For small N and specific $\mathbf{c}_1, \ldots, \mathbf{c}_N$, the conditional d.f. of $\mathscr{L}^+_N$ (given $\mathscr{P}_N$) can be enumerated to construct the exact critical points. In view of the prohibitive amount of labor involved (for large N) for enumerating this conditional d.f., we proceed now to provide the following asymptotic simplifications to the large-sample permutation distribution of $\mathbf{S}^+_N$ and $\mathscr{L}^+_N$.

Let $F^*_{i[j]}(x) = P\{|X_{ij}| \le x\}$ be the c.d.f. of $|X_{ij}|$, $1 \le j \le p$, $1 \le i \le N$, and for each j $(= 1, \ldots, p)$, let

$$H^*_{Nj}(x) = N^{-1} \sum_{i=1}^{N} F^*_{i[j]}(x), \qquad x \ge 0. \tag{5.4.48}$$

Let then $\tilde{\boldsymbol{\nu}}(H_N^*) = ((\tilde{\nu}_{jj'}(H_N^*)))_{j,j'=1,\ldots,p}$ be defined by

$$\tilde{\nu}_{jj}(H_N^*) = \tilde{\nu}_{jj} = \int_0^1 [\phi_j^*(u)]^2\,du, \qquad j = 1,\ldots,p, \tag{5.4.49}$$

$$\tilde{\nu}_{jj'}(H_N^*) = N^{-1}\sum_{i=1}^{N} E\{S_{ij}S_{ij'}\phi_j^*(H_{Nj}^*(|X_{ij}|))\phi_{j'}^*(H_{Nj'}^*(|X_{ij'}|))\} \tag{5.4.50}$$

for $j \neq j' = 1,\ldots,p$. Note that under $H_0^{(2)}$, $F_{i[j]}^* = F_{[j]}^*$ $\forall i \geq 1$, $j = 1,\ldots,p$, so that

$$\tilde{\boldsymbol{\nu}}(H_N^*) = \tilde{\boldsymbol{\nu}}(F^*), \qquad \text{where } F^* \text{ is the common d.f.} \tag{5.4.51}$$

Then, along the same lines as in the proof of Theorem 5.4.1, we may prove the following theorem (the proof is therefore left as an exercise).

Theorem 5.4.4. *If ϕ_j^* is the difference of two nondecreasing, absolutely continuous, and square-integrable score functions (inside [0, 1]) for each $j = 1,\ldots,p$, then*

$$\tilde{\mathbf{V}}_N - \tilde{\boldsymbol{\nu}}(H_N^*) \to \mathbf{0} \quad \textit{in probability} \quad \textit{as} \quad N \to \infty. \tag{5.4.52}$$

Also, we have the following theorem, whose proof follows along the lines of that of Theorem 5.4.2 (and hence is also left as an exercise).

Theorem 5.4.5. *Assume that* (1) (5.3.24), (5.3.26) *hold*, (2) *for each j (= 1,…, p), the scores $a_{Nj}^*(1),\ldots,a_{Nj}^*(N)$ satisfy the conditions of Theorem* 5.4.4, *and* (3) *$\tilde{\boldsymbol{\nu}}(F^*)$ is p.d. Then, under $H_0^{(2)}$, the permutation distribution of $\mathbf{S}_N^+$ is asymptotically (in probability) multinormal with null mean and dispersion matrix $\mathbf{D}_N$ of rank $p(q+1)$, in probability.*

From the preceding two theorems, it follows that under $\mathscr{P}_N$, $\mathscr{L}_N^+$ has asymptotically (in probability) a chi-square distribution with $p(q+1)$ DF. Hence, we have the following large-sample test (whose conditional as well as unconditional level of significance is equal to α asymptotically):

Reject or accept $H_0^{(2)}$ according as $\mathscr{L}_N^+$ is greater or less than $\chi^2_{p(q+1),\alpha}$.

(5.4.53)

In passing, we may remark that for both $H_0^{(1)}$ and $H_0^{(2)}$, the permutation principle discussed here provides a convenient way of constructing suitable tests, though for large samples, one need not carry out such conditional tests.

5.5. ASYMPTOTIC POWER PROPERTIES OF RANK-ORDER TESTS

We shall study now the power properties of the rank-order tests studied in the preceding three sections. Note that the study of the exact power function of any rank statistic demands the knowledge of the exact distribution of the ranks of the observations when the underlying distributions are specified by the alternative hypothesis and hence are not necessarily the same. Again, when the null hypothesis is not true, the invariance structure studied in the earlier sections does not hold, and hence the distribution of the ranks depends on the underlying d.f. F. In fact, in general, the distribution of the ranks when the null hypothesis does not hold is quite cumbersome, and its evaluation requires extensive numerical quadrature procedures. Except in some very specialized cases and for simple forms of F, very little has been done on the computation of the exact distribution of rank-order statistics. Indeed, the prospect does not seem to be very bright, even with the enormous computational facilities now available. On the other hand, for large sample sizes, we have studied in Chapters 2, 3, and 4 the non-null distribution theory of rank-order statistics. These results are utilized here to study the large-sample power properties of the rank tests considered in Sections 5.2, 5.3, and 5.4.

We consider here the power functions (in the asymptotic case) of the general multivariate rank-order statistics; the univariate case, following as a particular case, is briefly summarized at the end. For the asymptotic power properties of $\mathscr{L}_N$ and $\mathscr{L}_N^+$, we confine ourselves to local alternatives. For this, with regard to the model (5.4.1), while testing $H_0^{(1)}$ in (5.4.2), we either regard $\boldsymbol{\beta}$ to have elements all close to zero, or equivalently, we fix $\boldsymbol{\beta}$ and replace the $\mathbf{c}_i$ by a sequence $\{\mathbf{c}_i^{(N)}, 1 \le i \le N\}$ whose elements all go to zero as $N \to \infty$. From operational point of view we may express this as follows. For Pitman-type local alternatives, we let $H_N^{(1)}: \boldsymbol{\beta} = \boldsymbol{\beta}_N = N^{-1/2}\boldsymbol{\beta}^0$, where $\boldsymbol{\beta}^0$ is fixed. If then we let

$$\mathbf{c}_i^{(N)} = N^{-1/2}(\mathbf{c}_i - \bar{\mathbf{c}}_N), \qquad i = 1, \ldots, N, \tag{5.5.1}$$

then we have

$$F(\mathbf{x} - \boldsymbol{\alpha} - \boldsymbol{\beta}_N \mathbf{c}_i) = F(\mathbf{x} - \boldsymbol{\alpha}_N - \boldsymbol{\beta}^0 \mathbf{c}_i^{(N)}), \tag{5.5.2}$$

where $\boldsymbol{\alpha}_N = \boldsymbol{\alpha} + N^{-1/2}\boldsymbol{\beta}^0\bar{\mathbf{c}}_N$, and by (5.5.1), under mild restrictions on the $\mathbf{c}_i$, the elements of $\mathbf{c}_i^{(N)}$ all go to 0. For testing $H_0^{(2)}$ in (5.4.3), we further make $\boldsymbol{\alpha}$ close to $\mathbf{0}$, so that in (5.5.2), we let

$$\boldsymbol{\alpha}_N = N^{-1/2}\boldsymbol{\alpha}^0 \qquad \text{for some fixed } \boldsymbol{\alpha}^0. \tag{5.5.3}$$

With this introduction, we consider a (double) sequence $\{\mathbf{X}_{N1},\ldots,\mathbf{X}_{NN}\}$, $N \geq 1$, of independent stochastic vectors, where $\mathbf{X}_{Ni}$, $i = 1,\ldots,N$, are independently distributed according to continuous p-variate d.f.'s $\{F_{N_i}(\mathbf{x}), 1 \leq i \leq N\}$, $\mathbf{x} \in R^p$, $N \geq 1$, and

$$F_{N_i}(\mathbf{x}) = F\left(\mathbf{x} - \alpha_N - \boldsymbol{\beta}^0\mathbf{c}_i^{(N)}\right), \qquad 1 \leq i \leq N. \tag{5.5.4}$$

Here $\boldsymbol{\alpha}_N$ and $\boldsymbol{\beta}^0$ are unknown parameters, and $\mathbf{c}_i^{(N)}$, $1 \leq i \leq N$, are known vectors of constants which satisfy the following conditions:

$$\lim_{N\to\infty}\left\{\max_{1\leq j\leq q}\max_{1\leq i\leq N}\frac{\left(c_{ij}^{(N)} - \bar{c}_{jN}\right)^2}{\tilde{C}_{N,jj}}\right\} = 0, \tag{5.5.5}$$

$$\sup_N\left\{\tilde{C}_{N,jj}\right\} < \infty \qquad \text{for} \quad 1 \leq j \leq q, \tag{5.5.6}$$

$$\mathrm{Ch}_q\left[\tilde{\mathbf{C}}_N\right] > \eta > 0 \qquad \text{for every} \quad N \geq N_\eta, \tag{5.5.7}$$

where $\tilde{\mathbf{C}}_N = (\tilde{C}_{N,jj})$ is defined by (4.2.9), with $\mathbf{c}_i$ replaced by $\mathbf{c}_i^{(N)}$. For a given N, the hypothesis specified by (5.5.4)–(5.5.7), treating $\boldsymbol{\alpha}_N$ as a nuisance parameter, is denoted by $H_N^{(1)}$, so that we have the sequence $H_N^{(1)}$ of alternative hypotheses against the null hypothesis $H_0^{(1)}$ in (5.4.2). For testing $H_0^{(2)}$ in (5.4.3), we consider the sequence $\{H_N^{(2)}\}$ of alternative hypotheses, where under $H_N^{(2)}$, for every $i(= 1,\ldots,N)$,

$$F_{Ni}(\mathbf{x}) = F\left(\mathbf{x} - \boldsymbol{\alpha}_N - \boldsymbol{\beta}^0\mathbf{c}_i^{(N)}\right), \qquad \boldsymbol{\alpha}_N = N^{-1/2}\boldsymbol{\alpha}^0, \tag{5.5.8}$$

and (5.5.5)–(5.5.7) hold for the $\mathbf{c}_i^{(N)}$.

Consider $\mathscr{L}_N$ first the case of, and let us study the asymptotic behavior of $\mathbf{L}_N$ and $\mathbf{D}_N$ defined by (5.4.5) and (5.4.12) respectively; as in (4.2.20), we replace the $\mathbf{c}_i$ by $\mathbf{c}_{i(N)}$, $1 \leq i \leq N$, and because of the dependence of the $\mathbf{c}_i^{(N)}$ and the d.f. F_i $(= F_{Ni})$ on N, we denote these matrices by $\mathbf{Z}_{Ni}$, $1 \leq i \leq N$. Note that under (5.5.5) and (5.5.6), as $N \to \infty$,

$$\sup_x\left|N^{-1}\sum_{i=1}^N F_{Ni}(\mathbf{x}) - F(\mathbf{x} - \boldsymbol{\alpha}_N)\right| \to 0, \tag{5.5.9}$$

and that on defining $\boldsymbol{\nu}(F)$ as in (5.4.15)–(5.4.17), the elements of $\boldsymbol{\nu}(F)$ are invariant under translation, so that $\boldsymbol{\nu}(F)$ does not depend on $\boldsymbol{\alpha}_N$. Then we obtain from (5.5.9) and (4.2.20) that as $N \to \infty$,

$$V\left(\sum_{i=1}^{N} \mathbf{Z}_{Ni}\right) - \boldsymbol{\nu}(F) \otimes \tilde{\mathbf{C}}_N \to \mathbf{0}. \tag{5.5.10}$$

Also, by (4.2.50), Lemma 4.2.2, and (5.5.10), we have

$$V(\mathbf{L}_N) - \boldsymbol{\nu}(F) \otimes \tilde{\mathbf{C}}_N \to \mathbf{0} \quad \text{as} \quad N \to \infty. \tag{5.5.11}$$

Thus, if we assume that $\boldsymbol{\nu}(F)$ is p.d., then (4.2.44) holds whenever (5.5.7) and the following (classically known as the *Hájek condition*) hold: for some $\eta > 0$, for every $j(= 1, \dots, q)$,

$$\tilde{C}_{N,jj} > \eta N \max_{1 \le i \le N} \left(c_{ij}^{(N)} - \bar{c}_{jN}\right)^2 \qquad \forall\, N \ge N\eta, \tag{5.5.12}$$

which by virtue of (5.5.6) reduces to

$$\max_{1 \le i \le N} N\left(c_{ij}^{(N)} - \bar{c}_{jN}\right)^2 < K_\eta < \infty \tag{5.5.13}$$

for some $K_\eta < \infty$, $1 \le j \le q$, and $N \ge N_\eta$. On the other hand, by Theorem 4.2.4, for score functions with bounded second derivative, the asymptotic multinormality of $\mathbf{L}_N$ follows under (5.5.4)–(5.5.7).

Let us now simplify the expression for $E(\mathbf{L}_N | H_N^{(1)})$. For this purpose, we assume that the *Hoeffding condition* (4.2.52) holds, so that by (4.2.53)–(4.2.54), we have

$$\begin{aligned}
&E\left[L_{N,jl} - N\bar{a}_{Nj}\bar{c}_{lN} | H_N^{(1)}\right] \\
&= \sum_{i=1}^{N} c_{il}^{(N)}\left[\int_{-\infty}^{\infty} \phi_j\left(\frac{1}{N}\sum_{i=1}^{N} F_{Ni[j]}(x)\right) dF_{Ni[j]}(x) - \int_0^1 \phi_j(u)\, du\right] \\
&= \sum_{i=1}^{N} c_{il}^{(N)}\left[\int_{-\infty}^{\infty} \phi_j\left(\frac{1}{N}\sum_{i=1}^{N} F_{[j]}\left(x - \alpha_{Nj} - \boldsymbol{\beta}_j^0\mathbf{c}_i^{(N)}\right)\right)\right. \\
&\qquad \left. \times dF_{[j]}\left(x - \alpha_{Nj} - \boldsymbol{\beta}_j^0\mathbf{c}_i^{(N)}\right) - \int_0^1 \phi_j(u)\, du\right],
\end{aligned} \tag{5.5.14}$$

where $\boldsymbol{\beta}_j^0$ is the jth row of $\boldsymbol{\beta}^0$, for $j = 1, \ldots, p$, $l = 1, \ldots, q$. To simplify (5.5.14), we further assume that for each j $(= 1, \ldots, p)$,

(a) $F_{[j]}(x)$ is absolutely continuous with an absolutely continuous density function $f_{[j]}(x)$,

(b) $\phi_j(u)$ has a continuous derivative $\phi_j'(u)$ for all $0 < u < 1$, and

(c) $\lim_{x \to \pm\infty} \phi_j'(F_{[j]}(x)) f_{[j]}(x)$ is bounded.

Let then

$$B(F_{[j]}, \phi_j) = B_j = \int_{-\infty}^{\infty} \phi_j'(F_{[j]}(x)) f_{[j]}^2(x)\, dx \tag{5.5.15}$$

for $j = 1, \ldots, p$. Note that for monotonic increasing ϕ_j, we have $B_j > 0$, while (a), (b), and (c) insure that $B_j < \infty$. Then, the rhs of (5.5.14) may be written as

$$\begin{aligned}
&\sum_{i=1}^{N} c_{il}^{(N)} \Bigg\{ \int_{-\infty}^{\infty} \phi_j \left(\frac{1}{N} \sum_{i=1}^{N} F_{[j]}\left(x - \alpha_{Nj} - \boldsymbol{\beta}_j^0 \mathbf{c}_i^{(N)}\right) \right) \\
&\qquad \times d \Bigg[F_{[j]}\left(x - \alpha_{Nj} - \boldsymbol{\beta}_j^0 \mathbf{c}_i^{(N)}\right) \\
&\qquad\qquad - \frac{1}{N} \sum_{i=1}^{N} F_{[j]}\left(x - \alpha_{Nj} - \boldsymbol{\beta}_j^0 \mathbf{c}_i^{(N)}\right) \Bigg] \Bigg\} \\
&\quad = \sum_{l'=1}^{q} \beta_{jl'}^0 \left(\sum_{i=1}^{N} c_{il}^{(N)} \left(c_{il'}^{(N)} - \bar{c}_{l'N} \right) \right) \\
&\qquad \times \int_{-\infty}^{\infty} f_{[j]}^2(x) \phi_j'\left(F_{[j]}(x)\right) dx + o(1) \\
&\quad = \sum_{l'=1}^{q} \beta_{jl'}^0 \tilde{C}_{N,ll'} B_j + o(1),
\end{aligned} \tag{5.5.16}$$

for every $1 \le j \le p$, $1 \le l \le q$, so that

$$E\left(\mathbf{L}_N - N\left((\bar{a}_{Nj}\bar{c}_{lN})\right) \middle| H_N^{(1)}\right) - \mathbf{B}\boldsymbol{\beta}^0 \tilde{\mathbf{C}}_N \to \mathbf{0} \qquad \text{as} \quad N \to \infty, \tag{5.5.17}$$

where

$$\mathbf{B} = \operatorname{diag}(B_1, \ldots, B_p). \tag{5.5.18}$$

Finally, by Theorem 5.4.1 and (5.5.9), it follows that under $\{H_N\}$, as $N \to \infty$,

$$\mathbf{D}_N - \boldsymbol{\nu}(F) \otimes \tilde{C}_N \to \mathbf{0}, \tag{5.5.19}$$

where $\mathbf{D}_N$ is defined by (5.4.12). Consequently, by the definition of $\mathscr{L}_N$ in (5.4.13), the asymptotic normality of $\mathbf{L}_N$, (5.5.11), (5.5.12), and (5.5.17), we arrive at the following.

Lemma 5.5.1. Under $\{H_N^{(1)}\}$ and the conditions stated before, as $N \to \infty$,

$$\mathscr{L}_N \overset{P}{\sim} \mathscr{L}_N^* \tag{5.5.20}$$

where

$$\mathscr{L}_N^* = \sum_{j=1}^{p} \sum_{j'=1}^{p} \sum_{l=1}^{q} \sum_{l'=1}^{q} \nu^{jj'}(F) \tilde{C}_N^{ll'}$$

$$\times \left(L_{N,jl} - N\bar{a}_{Nj}\bar{c}_{lN}\right)\left(L_{N,j'l'} - N\bar{a}_{Nj'}\bar{c}_{l'N}\right), \tag{5.5.21}$$

and $((\tilde{C}_N^{ll'})) = \tilde{\mathbf{C}}_N^{-1}$.

We now strengthen (5.5.6)–(5.5.7) a little, and assume that

$$\lim_{N\to\infty} \tilde{\mathbf{C}}_N = \tilde{\mathbf{C}} = ((\tilde{C}_{ll'})) \tag{5.5.22}$$

exists and is positive definite. Let then

$$\tilde{\mathbf{C}}^{-1} = ((\tilde{C}^{ll'})). \tag{5.5.23}$$

Theorem 5.5.2. *Under* (5.5.4), (5.5.12), (5.5.22), *and the conditions* (a), (b) *and* (c) *on F and* $\boldsymbol{\phi}$, *when* $\{H_N^{(1)}\}$ *holds,* $\mathscr{L}_N$ *has asymptotically a noncentral chi-square distribution with pq DF and noncentrality parameter*

$$\Delta_{\mathscr{L}} = \sum_{j=1}^{p} \sum_{j'=1}^{p} \sum_{l=1}^{q} \sum_{l'=1}^{q} \beta_{jl}^0 \beta_{j'l'}^0 \tilde{C}_{ll'} \tau^{jj'}(F), \tag{5.5.24}$$

where

$$\mathbf{T}(F) = \left(\left(\tau_{jj'}(F)\right)\right) = \left(\left(\frac{\nu_{jj'}(F)}{B_j B_{j'}}\right)\right), \qquad \mathbf{T}^{-1}(F) = \left(\left(\tau^{jj'}(F)\right)\right). \quad (5.5.25)$$

Proof. By virtue of Lemma 5.5.1, it suffices to prove the theorem for $\mathscr{L}_N^*$ defined by (5.5.21). Again, by virtue of (5.5.11), the theorem follows from the asymptotic multinormality (under $\{H_N^{(1)}\}$) of $[\mathbf{L}_N - N((\bar{a}_{Nj}\bar{c}_{lN}))]$ and the fact that by (5.5.17) and (5.5.22),

$$\sum_{j=1}^{p}\sum_{j'=1}^{p}\sum_{l=1}^{q}\sum_{l'=1}^{q} \nu^{jj'}(F)\tilde{C}_N^{ll'}\left[E\left(L_{N,jl}|H_N^{(1)}\right) - N\bar{a}_{Nj}\bar{c}_{lN}\right]$$

$$\times\left[E\left(L_{N,j'l'}|H_N^{(1)}\right) - N\bar{a}_{Nj'}\bar{c}_{l'N}\right]$$

$$= \sum_{j=1}^{p}\sum_{j'=1}^{p}\sum_{l=1}^{q}\sum_{l'=1}^{q} \nu^{jj'}(F)C_N^{ll'}\left(\sum_{k=1}^{q}\beta_{jk}\tilde{C}_{N,jk}B_j + o(1)\right)$$

$$\times\left(\sum_{k'=1}^{q}\beta_{j'k'}B_{j'}\tilde{C}_{N,l'k'} + o(1)\right)$$

$$= \sum_{j=1}^{p}\sum_{j'=1}^{p}\sum_{k=1}^{q}\sum_{k'=1}^{q} B_j B_{j'}\nu^{jj'}(F)$$

$$\times \sum_{l'=1}^{q}\left\{\sum_{l=1}^{q}\tilde{C}_N^{ll'}\tilde{C}_{N,lk}\right\}\tilde{C}_{N,l'k'}\beta_{jk}^0\beta_{j'k'}^0 + o(1)$$

$$= \sum_{j=1}^{p}\sum_{j'=1}^{p}\sum_{k=1}^{q}\sum_{k'=1}^{q} B_j B_{j'}\beta_{jk}^0\beta_{j'k'}^0\nu^{jj'}(F)\tilde{C}_{N,kk'} + o(1)$$

$$= \sum_{j=1}^{p}\sum_{j'=1}^{p}\sum_{k=1}^{q}\sum_{k'=1}^{q} \beta_{jk}^0\beta_{j'k'}^0\tilde{C}_{N,kk'}\tau^{jj'}(F) + o(1)$$

$$= \Delta_{\mathscr{L}} + o(1) \quad \text{as} \quad N \to \infty, \qquad \text{Q.E.D.} \qquad (5.5.26)$$

Let us now consider the power of the test for $H_0^{(1)}$ based on the rank-order statistic $\mathscr{L}_N$. It follows from Theorem 5.5.2 that for large N and alternative hypotheses $\{H_N^{(1)}\}$ "close" to the null one, the power of the test based on $\mathscr{L}_N$ can be approximated by

$$P\left\{\chi_{pq}^2(\Delta_{\mathscr{L}}) \geq \chi_{pq,\alpha}^2\right\}, \qquad (5.5.27)$$

where $\chi_t^2(\Delta)$ has the noncentral chi-square distribution with t DF and noncentrality parameter Δ.

Let us now proceed to test for $H_0^{(2)}$ in (5.4.3) based on $\mathscr{L}_N^+$. Here, we confine ourselves to (5.5.8) with the side conditions that F is diagonally symmetric about $\mathbf{0}$,

$$\lim_{N\to\infty}\left\{\max_{1\le j\le q}\max_{1\le i\le N}|c_{ij}^{(N)}|\right\}=0, \tag{5.5.28}$$

$$0<Q^*<\inf_{N>q}\mathrm{Ch}_q(\tilde{\mathbf{C}}_N^*)\le\sup_{N>q}\mathrm{Ch}_1(\tilde{\mathbf{C}}_N^*)<\infty, \tag{5.5.29}$$

and

$$\tilde{\mathbf{C}}_N^*=\left(\left(\sum_{i=1}^N c_{ij}^{(N)}c_{ij'}^{(N)}\right)\right)_{j,j'=0,\dots,q}, \qquad c_{i0}^{(N)}=N^{-1/2},\quad i\ge 1. \tag{5.5.30}$$

Then, by using (4.3.15) and Theorem 4.3.2, and proceeding as in (5.5.3)–(5.5.11), it can be shown by some standard steps that under $\{H_N^{(2)}\}$, specified by (5.5.8), (5.5.28), and (5.5.29), as $N\to\infty$,

$$V(\mathbf{S}_N^+)-\tilde{\boldsymbol{\nu}}(F^*)\otimes\mathbf{C}_N^*\to\mathbf{0}, \tag{5.5.31}$$

where $\tilde{\nu}(F^*)$ is defined by (5.4.58)–(5.4.60). As a result, the asymptotic normality of $\mathbf{S}_N^+$ follows under the modified Hájek condition that

$$\max_{1\le i\le N}\max_{1\le j\le q}N^{1/2}|c_{ij}^{(N)}|<K_\eta<\infty \qquad \forall N\ge N_\eta. \tag{5.5.32}$$

On the other hand, for score functions with bounded second derivatives, the asymptotic normality of $\mathbf{S}_N^+$ follows from Theorem 4.3.1 and (5.5.31) under the conditions (5.5.28)–(5.5.29).

To simplify the expression for $E(\mathbf{S}_N^+|H_N^{(2)})$, we recal that the scores $a_{Nj}(i)$, $1\le i\le N$, are generated by $\phi_j^*(u)$, $0<u<1$ for $j=1,\dots,p$. We assume that for each j:

1. ϕ_j^* satisfies (4.3.41), i.e., the Hoeffding condition holds.
2. There exists a skew symmetric function $\phi_j(u)$, $0<u<1$ [i.e., $\phi_j(u)+\phi_j(1-u)=0$ $\forall 0<u<1$] such that

$$\phi_j^*(u)=\phi_j\left(\frac{1+u}{2}\right), \qquad 0<u<1. \tag{5.5.33}$$

3. $\phi_j(u)$, $0 < u < 1$, and $F_{[j]}(x)$, the jth marginal d.f. of $F(\mathbf{x})$, satisfy the conditions (a), (b), and (c) stated after (5.5.14).

We define the $B_j = B(F_{[j]}, \phi_j)$ as in (5.5.15). If we let $\overline{H}^*_{N[j]}(x) = N^{-1}\sum_{i=1}^{N} F^*_{Ni[j]}(x) = N^{-1}\sum_{i=1}^{N} P\{|X_{ij}| \le x | H_N^{(2)}\}$, $1 \le j \le p$, we have by (4.3.41)

$$E\left(S^+_{N,jl} | H_N^{(2)}\right) = \sum_{i=1}^{N} c_{il}^{(N)} \int_{-\infty}^{\infty} s(x)\phi_j^*\left(\overline{H}^*_{N[j]}(|x|)\right) dF_{Ni[j]}(x)$$

$$= \sum_{i=1}^{N} c_{il}^{(N)} \int_{-\infty}^{\infty} s(x)\phi_j\left(\tfrac{1}{2}\left[1 + \overline{H}^*_{N[j]}(|x|)\right]\right) dF_{Ni[j]}(x).$$

$$(5.5.34)$$

Let us rewrite (5.5.8) as $F(\mathbf{x} - \boldsymbol{\beta}^*\mathbf{c}^*_{Ni})$, $1 \le i \le N$, where $\boldsymbol{\beta}^* = (\boldsymbol{\alpha}^0, \boldsymbol{\beta}^0)$ and $\mathbf{c}^*_{Ni} = (N^{-1/2}, \mathbf{c}_i^{(N)\prime})'$, $1 \le i \le N$, so that $\sum_{i=1}^{N} \mathbf{c}^*_{Ni}\mathbf{c}^{*\prime}_{Ni} = \tilde{\mathbf{C}}^*_N$. Then, on using the fact that for $x \ge 0$,

$$\overline{H}^*_{N[j]}(x) = N^{-1} \sum_{i=1}^{N} \left[F_{Ni[j]}(x) - F_{Ni[j]}(-x)\right]$$

$$= N^{-1} \sum_{i=1}^{N} \left[F_{[j]}\left(x - \boldsymbol{\beta}_j^*\mathbf{c}^*_{Ni}\right) - F_{[j]}\left(-x - \boldsymbol{\beta}_j^*\mathbf{c}^*_{Ni}\right)\right]$$

$$= 2F_{[j]}(x) - 1 - 2\boldsymbol{\beta}_j^*\bar{\mathbf{c}}^*_N f_{[j]}(x + \eta_N), \qquad (5.5.35)$$

where $\bar{\mathbf{c}}^*_N = \dfrac{1}{N}\sum_{i=1}^{N} \mathbf{c}^*_{iN}$, $\eta_N \to 0$ as $N \to \infty$ and $\boldsymbol{\beta}_j^*$ is the jth row of $\boldsymbol{\beta}^*$, we obtain by using (5.5.28), (5.5.29), (5.5.34), and (5.5.35) that

$$\left| E\left(S^+_{N,jl} | H_N^{(2)}\right) - 2\sum_{l'=0}^{q} \beta^*_{jl'} B_j \tilde{C}^*_{N,ll'} \right| \to 0 \qquad (5.5.36)$$

for every $1 \le j \le p$ and $0 \le l \le q$, where $\boldsymbol{\beta}^* = ((\beta^*_{jl}))$. Thus,

$$E\left(\mathbf{S}^+_N | H_N^{(2)}\right) - 2\mathbf{B}\boldsymbol{\beta}^*\tilde{\mathbf{C}}^*_N \to 0 \quad \text{as} \quad N \to \infty, \qquad (5.5.37)$$

where $\mathbf{B}$ is defined by (5.5.18). Finally, by Theorem 5.4.4, (5.5.28), and

(5.5.35), it follows that under $\{H_N^{(2)}\}$, as $N \to \infty$,

$$\tilde{\mathbf{D}}_N - \tilde{\boldsymbol{\nu}}(F^*) \otimes \tilde{\mathbf{C}}_N^* \to 0, \tag{5.5.38}$$

where $\tilde{\mathbf{D}}_N$ is defined by (5.4.55). Further, on using (5.5.33) and the diagonal symmetry of F [so that $F_{[j]}^* = 2F_{[j]} - 1$, $1 \le j \le p$, and $dF_{[j,j']}(x, y) = dF_{[j,j']}(-x, -y)$ $\forall x, y$, $1 \le j \le j' \le p$], we obtain by (5.4.17) and (5.4.59) that

$$\tilde{\nu}_{jj'}(F^*) = \nu_{jj'}(F) \qquad \text{for every} \quad j \ne j' = 1, \ldots, p, \tag{5.5.39}$$

and by (5.4.16), (5.4.58), and (5.5.33),

$$\tilde{\nu}_{jj}(F^*) = \nu_{jj}(F) = \nu_{jj}, \qquad 1 \le j \le p. \tag{5.5.40}$$

Therefore from (5.5.38), (5.5.39), and (5.5.40), under $\{H_N^{(2)}\}$,

$$\tilde{\mathbf{D}}_N - \boldsymbol{\nu}(F) \otimes \mathbf{C}_N^* \to \mathbf{0} \qquad \text{as} \quad N \to \infty. \tag{5.5.41}$$

Then assuming that

$$\lim_{N\to\infty} \tilde{\mathbf{C}}_N^* = \tilde{\mathbf{C}}^* \text{ exists and is of rank } q+1, \tag{5.5.42}$$

and following the same line of proof as in the case of $\mathscr{L}_N$, we arrive at the following.

Theorem 5.5.3. *Under $\{H_N^{(2)}\}$ (i.e., (5.5.8), (5.5.28) and (5.5.29)) and the conditions (a), (b) and (c) on F and $\phi_1, \ldots, \phi_p$, when (4.3.41), (5.5.33), and (5.5.42) hold, $\mathscr{L}_N^+$ has asymptotically a noncentral chi-square distribution with $p(q+1)$ DF and noncentrality parameter*

$$\Delta_{\mathscr{L}^+} = \sum_{j=1}^{p} \sum_{j'=1}^{p} \sum_{l=0}^{q} \sum_{l'=0}^{q} \beta_{jl}^* \beta_{j'l'}^* C_{ll'}^* \tau^{jj'}(F), \tag{5.5.43}$$

where $((\tau^{jj'}(F)))$ is defined by (5.5.25).

From the above theorem, we conclude that for large N and local alternatives, the power of the test based on $\mathscr{L}_N^+$ can be approximated by

$$P\{\chi^2_{p(q+1)}(\Delta_{\mathscr{L}^+}) \ge \chi^2_{p(q+1),\alpha}\}. \tag{5.5.44}$$

Later on, (5.5.27) and (5.5.44) will be used to study the asymptotic power properties of the rank tests.

5.6. ASYMPTOTIC THEORY OF NORMAL-THEORY TESTS FOR GENERAL LINEAR MODELS

Consider first the case of $H_0^{(1)}$ in (5.4.2). Let us write

$$\mathbf{A}_N = N^{-1}\sum_{i=1}^{N}(\mathbf{X}_i - \overline{\mathbf{X}}_N)(\mathbf{X}_i - \overline{\mathbf{X}}_N)', \qquad \overline{\mathbf{X}}_N = N^{-1}\sum_{i=1}^{N}\mathbf{X}_i, \tag{5.6.1}$$

$$\mathbf{H}_N = N^{-1}\sum_{i=1}^{N}(\mathbf{X}_i - \overline{\mathbf{X}}_N)(\mathbf{c}_i - \bar{\mathbf{c}}_N)' = N^{-1}\sum_{i=1}^{N}\mathbf{X}_i(\mathbf{c}_i - \bar{\mathbf{c}}_N)', \tag{5.6.2}$$

and let $\mathbf{C}_N$ be defined as in (4.2.9). Then the least-squares estimator of $\boldsymbol{\beta}$ in (5.4.1) is given by

$$\hat{\boldsymbol{\beta}}_N = \mathbf{H}_N\mathbf{C}_N^{-1}. \tag{5.6.3}$$

Under the assumption that F in (5.4.1) is a multinormal d.f., the *likelihood-ratio criterion* for testing $H_0^{(1)}$ in (5.4.2) is

$$\lambda_N = \left\{\frac{\|N\mathbf{A}_N - \hat{\boldsymbol{\beta}}_N\mathbf{C}_N\hat{\boldsymbol{\beta}}_N'\|}{\|N\mathbf{A}_N\|}\right\}^{N/2} \tag{5.6.4}$$

We shall study the limiting behavior of λ_N when F is not necessarily normal. It is assumed that F has finite second-order moments and

$$\mathbf{V}(\mathbf{X}_i) = \boldsymbol{\Sigma} \text{ is positive definite.} \tag{5.6.5}$$

Lemma 5.6.1. Under (5.6.5), as $N \to \infty$,

$$\mathbf{A}_N \to \boldsymbol{\Sigma} \qquad \text{in probability.} \tag{5.6.6}$$

Proof. Let us write $\mathbf{X}_i = \boldsymbol{\alpha} + \boldsymbol{\beta}\mathbf{c}_i + \boldsymbol{\varepsilon}_i$, $1 \le i \le N$, so that the $\boldsymbol{\varepsilon}_i$ are i.i.d.r.v.'s with mean $\mathbf{0}$ and $V(\boldsymbol{\varepsilon}_i) = \boldsymbol{\Sigma}$ for all $i \ge 1$. Then, by (5.6.1), on writing $\bar{\boldsymbol{\varepsilon}}_N = N^{-1}\Sigma_{i=1}^N \boldsymbol{\varepsilon}_i$,

$$\mathbf{A}_N = N^{-1}\sum_{i=1}^N [\boldsymbol{\beta}(\mathbf{c}_i - \bar{\mathbf{c}}_N) + (\boldsymbol{\varepsilon}_i - \bar{\boldsymbol{\varepsilon}}_N)][\boldsymbol{\beta}(\mathbf{c}_i - \bar{\mathbf{c}}_N) + (\boldsymbol{\varepsilon}_i - \bar{\boldsymbol{\varepsilon}}_N)]',$$

$$= \boldsymbol{\beta}\left[\frac{1}{N}\sum_{i=1}^N (\mathbf{c}_i - \bar{\mathbf{c}}_N)(\mathbf{c}_i - \bar{\mathbf{c}}_N)'\right]\beta' + \frac{1}{N}\sum_{i=1}^N (\boldsymbol{\varepsilon}_i - \bar{\boldsymbol{\varepsilon}}_N)(\boldsymbol{\varepsilon}_i - \bar{\boldsymbol{\varepsilon}}_N)'$$

$$+\boldsymbol{\beta}\left[N^{-1}\sum_{i=1}^N (\mathbf{c}_i - \bar{\mathbf{c}}_N)(\boldsymbol{\varepsilon}_i - \bar{\boldsymbol{\varepsilon}}_N)'\right] + \left[N^{-1}\sum_{i=1}^N (\boldsymbol{\varepsilon}_i - \bar{\boldsymbol{\varepsilon}}_N)(\mathbf{c}_i - \bar{\mathbf{c}}_N)'\right]\boldsymbol{\beta}'$$

$$= \boldsymbol{\beta}[N^{-1}\mathbf{C}_N]\boldsymbol{\beta}' + \mathbf{A}_N^* + \mathbf{R}_N^{(1)} + \mathbf{R}_N^{(2)}, \text{ say.} \tag{5.6.7}$$

Note that

$$\mathbf{A}_N^* = N^{-1}\sum_{i=1}^N \boldsymbol{\varepsilon}_i\boldsymbol{\varepsilon}_i' - (\bar{\boldsymbol{\varepsilon}}_N\bar{\boldsymbol{\varepsilon}}_N'), \tag{5.6.8}$$

and by the Khinchine law of large numbers, as $N \to \infty$,

$$N^{-1}\sum_{i=1}^N \boldsymbol{\varepsilon}_i\boldsymbol{\varepsilon}_i' \xrightarrow{P} \boldsymbol{\Sigma}, \qquad \bar{\boldsymbol{\varepsilon}}_N \xrightarrow{P} \mathbf{0} \left(\Rightarrow \bar{\boldsymbol{\varepsilon}}_N\bar{\boldsymbol{\varepsilon}}_N' \xrightarrow{P} \mathbf{0}\right), \tag{5.6.9}$$

so that by (5.6.8) and (5.6.9), as $N \to \infty$,

$$\mathbf{A}_N^* \xrightarrow{P} \boldsymbol{\Sigma}, \qquad \text{whenever } \boldsymbol{\Sigma} \text{ exists.} \tag{5.6.10}$$

Also, by (5.4.2), under $H_0^{(1)}$, $\boldsymbol{\beta} = \mathbf{0}$, so that $\boldsymbol{\beta}(N^{-1}\mathbf{C}_N)\boldsymbol{\beta}' = \mathbf{0}$. Under $\{H_N^{(1)}\}$, note that by (5.5.6), $N^{-1}\tilde{\mathbf{C}}_N \to \mathbf{0}$ as $N \to \infty$, so that for every fixed $\boldsymbol{\beta}$, $\boldsymbol{\beta}[N^{-1}\tilde{\mathbf{C}}_N]\boldsymbol{\beta}' \to \mathbf{0}$ as $N \to \infty$. Similarly, on using the fact that under $H_0^{(1)}$, $\boldsymbol{\beta} = \mathbf{0}$ or under $\{H_N^{(1)}\}$, $\max_{1 \le i \le N}|\mathbf{c}_i^{(N)} - \bar{\mathbf{c}}_N| \to \mathbf{0}$ as $N \to \infty$, it follows that both $\mathbf{R}_N^{(1)}$ and $\mathbf{R}_N^{(2)} \xrightarrow{P} \mathbf{0}$ as $N \to \infty$. Consequently, (5.6.6) holds, Q.E.D.

For matrix $\mathbf{B}_N = \mathbf{O}_p(N^{-1})$ stand for the fact that each element of $\mathbf{B}_N$ is $O_p(N^{-1})$. Then we have the following.

Lemma 5.6.2. Under $\{H_N^{(1)}\}$, as $N \to \infty$,

$$N^{-1}\hat{\boldsymbol{\beta}}_N \mathbf{C}_N \hat{\boldsymbol{\beta}}_N' = \mathbf{O}_p(N^{-1}). \tag{5.6.11}$$

Proof. Note that by (5.6.3), $\hat{\boldsymbol{\beta}}_N$ is a linear combination of $\mathbf{X}_1, \ldots, \mathbf{X}_N$. Some routine computations yield that

$$E[\hat{\boldsymbol{\beta}}_N] = \boldsymbol{\beta} \quad \text{and} \quad NV(\hat{\boldsymbol{\beta}}_N) = \boldsymbol{\Sigma} \otimes \tilde{\mathbf{C}}_N^{-1}. \tag{5.6.12}$$

Since by (5.5.6) and (5.5.7) the elements of $\tilde{\mathbf{C}}_N^{-1}$ are finite, and $\boldsymbol{\Sigma}$ is p.d., N times the variance of each $\hat{\beta}_{Njl}$ is finite, so that by the Chebyshev inequality

$$|\hat{\beta}_{Njl} - \beta_{jl}| = O_p(N^{-1/2}) \qquad \forall 1 \le j \le p, 1 \le l \le q. \tag{5.6.13}$$

The proof of the lemma follows directly from (5.6.13) and (5.5.6).

Lemma 5.6.3. Let $\mathbf{A}$ and $\mathbf{B}$ be two p.s.d. matrices with $\|A\| > 0$. Then as $N \to \infty$,

$$\log\left\{\frac{\|\mathbf{A} - N^{-1}\mathbf{B}\|}{\|\mathbf{A}\|}\right\} = -N^{-1}\sum_{i=1}^{p}\sum_{j=1}^{p} b_{ij}a^{ij} + O(N^{-2}), \tag{5.6.14}$$

where $a^{ij} = A_{ij}/\|A\|$ is the inverse of a_{ij}, $1 \le i, j \le p$.

Proof. Note that by definition

$$\|\mathbf{A} - N^{-1}\mathbf{B}\| = \sum(-1)^{j_1 + \cdots + j_p}\left[\prod_{l=1}^{p}\left(a_{lj_l} - N^{-1}b_{lj_l}\right)\right], \tag{5.6.15}$$

where the summation extends over all possible $(j_1, \ldots, j_p)$ which are the permutations of $(1, \ldots, p)$. By expansion of (5.6.15) one obtains that

$$\|\mathbf{A} - N^{-1}\mathbf{B}\| = \|\mathbf{A}\| - N^{-1}\sum_{l=1}^{p}\sum_{k=1}^{p} b_{lk}\|A_{lk}\| + O(N^{-2}), \tag{5.6.16}$$

where

$$\|A_{lk}\| = \text{cofactor of } a_{lk} \text{ in } \|A\|. \tag{5.6.17}$$

The lemma follows from (5.6.16), (5.6.17), and the fact that $a^{jk} = \|A_{jk}\|/\|A\|$ for all $1 \le j, k \le p$.

Now from Lemmas 5.6.1, 5.6.2, and 5.6.3, we conclude that under $\{H_N^{(1)}\}$,

$$-2\log\lambda_N = -N\log\left\{\frac{\|\mathbf{A}_N - N^{-1}\hat{\boldsymbol\beta}_N\mathbf{C}_N^*\hat{\boldsymbol\beta}_N'\|}{\|A_N\|}\right\}$$

$$= \sum_{j=1}^{p}\sum_{j'=1}^{p} a_N^{jj'}\left(\sum_{l=1}^{q}\sum_{l'=1}^{q}\hat\beta_{Njl}C_{N,ll'}^*\hat\beta_{Nj'l'}\right) + O_p(N^{-1})$$

$$= \sum_{j=1}^{p}\sum_{j'=1}^{p}\sum_{l=1}^{q}\sum_{l'=1}^{q}\hat\beta_{Njl}\hat\beta_{Nj'l'}C_{N,ll'}^*a_N^{jj'} + O_p(N^{-1}). \qquad (5.6.18)$$

Now, by (5.6.3), we have

$$\hat{\boldsymbol\beta}_N = \sum_{i=1}^{N}\mathbf{X}_i\left\{N^{-1}(\mathbf{c}_i - \bar{\mathbf{c}}_N)'\tilde{\mathbf{C}}_N^{-1}\right\} = \sum_{i=1}^{N}\mathbf{X}_i\mathbf{W}_{Ni}, \quad \text{say}, \qquad (5.6.19)$$

where

$$\mathbf{W}_{Ni} = N^{-1}(\mathbf{c}_i - \bar{\mathbf{c}}_N)'\tilde{\mathbf{C}}_N^{-1}, \qquad 1 \le i \le N, \qquad (5.6.20)$$

are q-vectors. We may then write

$$\hat{\boldsymbol\beta}_N - \boldsymbol\beta = \sum_{i=1}^{N}\boldsymbol\varepsilon_i\mathbf{W}_{Ni} = \sum_{i=1}^{N}\boldsymbol\eta_i, \quad \text{say}, \qquad (5.6.21)$$

and hence, using the multivariate version of the central limit theorem on the independent matrices $\boldsymbol\eta_1, \ldots, \boldsymbol\eta_N$ (each of order $p \times q$), we conclude that as $N \to \infty$,

$$\hat{\boldsymbol\beta}_N - \boldsymbol\beta \text{ is asymptotically } \mathcal{N}_{pq}\left(0, \boldsymbol\Sigma \otimes \tilde{\mathbf{C}}_N^{-1}\right). \qquad (5.6.22)$$

Finally, by Lemma 5.6.1 and the assumed positive definiteness of $\boldsymbol\Sigma$, we have

$$\left(\left(a_N^{jj'}\right)\right) = \mathbf{A}_N^{-1} \xrightarrow{P} \boldsymbol\Sigma^{-1} \quad \text{as} \quad N \to \infty. \qquad (5.6.23)$$

Thus, as in $\mathcal{L}_N$, if we assume that (5.5.22) holds, then by the same method of proof as in Theorem 5.5.2, we arrive at the following theorem by using (5.6.18), (5.6.22), (5.5.23), and (5.5.22).

Theorem 5.6.4. *Under $\{H_N^{(1)}\}$ and for a d.f. F possessing a positive definite dispersion matrix* $\boldsymbol{\Sigma}$, *when* (5.5.22) *holds,* $-2\log\lambda_N$ *has asymptotically a noncentral chi-square distribution with pq DF and noncentrality parameter*

$$\Delta_\lambda = \sum_{j=1}^{p}\sum_{j'=1}^{p}\sum_{l=0}^{q}\sum_{l'=0}^{q}\beta_{jl}\beta_{j'l'}\sigma^{jj'}\tilde{C}_{ll'}. \tag{5.6.24}$$

(*Naturally, under* $H_0^{(1)}$, $-2\log\lambda_N$ *has asymptotically central chi-square distribution with pq DF.*)

This shows that for a broad class of d.f.'s, including the multivariate normal one, the chi-square approximation to the distribution of the normal-theory likelihood-ratio statistic works out asymptotically; and for local alternatives, asymptotically, the power of the test is given by

$$P\{\chi^2_{pq}(\Delta_\lambda) \geq \chi^2_{pq,\alpha}\}. \tag{5.6.25}$$

Let us now consider the case of $H_0^{(2)}$ in (5.4.3). Here the normal-theory likelihood-ratio criterion is

$$\lambda_N^+ = \left\{\frac{\left\|\sum_{i=1}^{N}\mathbf{X}_i\mathbf{X}_i' - N\overline{\mathbf{X}}_N\overline{\mathbf{X}}_N' - \hat{\boldsymbol{\beta}}_N\tilde{\mathbf{C}}^*_{N00}\hat{\boldsymbol{\beta}}_N'\right\|}{\left\|\sum_{i=1}^{N}\mathbf{X}_i\mathbf{X}_i'\right\|}\right\}^{N/2}, \tag{5.6.26}$$

where $\tilde{\mathbf{C}}^*_{N00}$ is the cofactor of the first diagonal element of $\tilde{\mathbf{C}}^*_N$. Proceeding then as in the case of λ_N, it can be shown that under $\{H_N^{(2)}\}$, $N^{-1}\sum_{i=1}^N\mathbf{X}_i\mathbf{X}_i' \xrightarrow{P} \boldsymbol{\Sigma}$ as $N\to\infty$, $\overline{\mathbf{X}}_N\overline{\mathbf{X}}_N' \xrightarrow{P} \mathbf{0}$, $\hat{\boldsymbol{\beta}}_N\tilde{\mathbf{C}}^*_{N00}\hat{\boldsymbol{\beta}}_N'$ has elements all bounded in probability, and hence, as in (5.6.18), $-2\log\lambda_N^+$ can be expressed as a quadratic form in $(\overline{\mathbf{X}}_N, \hat{\boldsymbol{\beta}}_N)$. Therefore the following holds.

Theorem 5.6.5. *Under $\{H_N^{(2)}\}$ and for F with p.d. (finite)* $\boldsymbol{\Sigma}$, *when* (5.5.42) *holds,* $-2\log\lambda_N^+$ *has asymptotically a noncentral chi-square distribution with* $p(q+1)$ *DF and noncentrality parameter*

$$\Delta_{\lambda^+} = \sum_{j=1}^{p}\sum_{j'=1}^{p}\sum_{l=0}^{q}\sum_{l'=0}^{q}\beta_{jl}\beta_{j'l'}\sigma^{jj'}\tilde{C}^*_{ll'}, \tag{5.6.27}$$

where $\beta_{j0} = \alpha_j^0$, $1 \leq j \leq p$, *and* $\boldsymbol{\alpha}^0$ *is defined by* (5.5.8).

The proof is left as an exercise (Exercise 5.6.2).

By Theorem 5.6.5, for large N and local alternatives, the power of the test based on λ_N^+ can be approximated by

$$P\left\{\chi^2_{p(q+1)}(\Delta_{\lambda^+}) \geq \chi^2_{p(q+1),\alpha}\right\}. \tag{5.6.28}$$

5.7. ASYMPTOTIC RELATIVE EFFICIENCY OF RANK TESTS

The family of rank-order tests and the normal-theory likelihood-ratio test enjoy the common property that under null hypothesis they have the common asymptotic distribution, namely, chi-square distribution with pq or $p(q+1)$ DF depending on $H_0^{(1)}$ or $H_0^{(2)}$. Also, under $\{H_N^{(k)}\}$, $k = 1, 2$, these statistics have all noncentral chi-square distribution with the same number of degrees of freedom but possibly different noncentrality parameters. Since the tail probability of a noncentral chi-square distribution is a monotonically increasing function of the noncentrality parameter (when the number of degrees of freedom is held fixed), for two sequences of tests statistics, say $\{Q_N\}$ and $\{Q_N^*\}$, having asymptotically (under a sequence $\{H_N\}$ of alternative hypotheses) noncentral chi-square distributions with t (≥ 1) DF and noncentrality parameters Δ and Δ^* respectively, the asymptotic relative efficiency (ARE) of $\{Q_N\}$ with respect to $\{Q_N^*\}$ is defined by

$$e(Q, Q^*) = \frac{\Delta}{\Delta^*}. \tag{5.7.1}$$

Indeed, such a definition can be justified from the viewpoint of Pitman efficiency. This approach has been treated in a fairly detailed manner in Section 3.8.3 of Puri and Sen (1971), to which the reader is referred for background.

From Theorems 5.5.2 and 5.6.4, it follows that the ARE of the rank-order test (based on $\mathscr{L}_N$) with respect to the normal-theory likelihood-ratio test (based on λ_N), when the actual d.f. is F, is given by

$$e(\mathscr{L}, \lambda) = \frac{\Delta_{\mathscr{L}}}{\Delta_\lambda}$$

$$= \frac{\sum_{j=1}^{p}\sum_{j'=1}^{p}\sum_{l=1}^{q}\sum_{l'=1}^{q} \beta_{jl}\beta_{j'l'}\tilde{C}_{ll'}\tau^{jj'}(F)}{\sum_{j=1}^{p}\sum_{j'=1}^{p}\sum_{l=1}^{q}\sum_{l'=1}^{q} \beta_{jl}\beta_{j'l'}\tilde{C}_{ll'}\sigma^{jj'}}. \tag{5.7.2}$$

If we define

$$\xi_{jj'} = \sum_{l=1}^{q} \sum_{l'=1}^{q} \beta_{jl}^{0}\beta_{j'l'}^{0}\tilde{C}_{ll'} \qquad \text{for} \quad j, j' = 1, \ldots, p,$$

$$\Xi = \left((\xi_{jj'})\right)_{j,j'=1,\ldots,p}, \tag{5.7.3}$$

then (5.7.2) can be written as

$$e(\mathscr{L}, \lambda) = \frac{\operatorname{Tr}(\Xi \mathbf{T}^{-1}(F))}{\operatorname{Tr}(\Xi \Sigma^{-1})}. \tag{5.7.4}$$

Thus, in the general multivariate case, the ARE not only depends on Σ and $\mathbf{T}(F)$, but also depends on $\boldsymbol{\beta}^0$ and $\tilde{\mathbf{C}}$ through Ξ. The following points are noteworthy in this context.

5.7.1. The Univariate Case, $p = 1$

In this case, all the three matrices Ξ, Σ, and $\mathbf{T}(F)$ are scalar, so that (5.7.4) reduces to

$$e(\mathscr{L}, \lambda) = \frac{\sigma_{11}}{\tau_{11}(F)} = \frac{B^2(F_{[1]}, \phi_1)\sigma_{11}}{\nu_{11}}, \tag{5.7.5}$$

which does not depend on $\boldsymbol{\beta}^0$, $\tilde{\mathbf{C}}$, and q. Equation (5.7.5) depends only on F and the score function $\phi_1(u)$, $0 < u < 1$, and it happens to coincide with the usual ARE expression for the two-sample rank-order statistics with respect to the Student t-statistic (when the two distributions differ in locations only). This has been studied in great detail in the literature (e.g., Section 3.8 of Puri and Sen, 1971); without going into the details, we briefly present here some important cases.

First, suppose $\phi_1(u) = u$ for $0 < u < 1$, i.e., we use the Wilcoxon scores. Then $\nu_{11} = \frac{1}{12}$ and $B(F_{[1]}, \phi_1) = \int_{-\infty}^{\infty} f_{[1]}^2(x)\,dx$, so that (5.7.5) reduces to

$$e(\mathscr{L}_W, \lambda) = 12\sigma_{11}\left(\int_{-\infty}^{\infty} f_{[1]}^2(x)\,dx\right)^2, \tag{5.7.6}$$

where $\mathscr{L}_W$ stands for the Wilcoxon version of the rank-order statistic. It is known that (5.7.6) is bounded from below by 0.864 for all continuous d.f.'s with finite second moment. Also, when the d.f. F is normal, (5.7.6) reduces to $3/\pi = 0.955$, which is quite close to unity. For many nonnormal d.f.'s, particularly the ones with heavy tails (the Cauchy d.f., double-exponential d.f., logistic d.f., etc.), (5.7.6) exceeds one, so that asymptotically we are better off using the rank-order test instead of the normal-theory test.

Second, consider the normal-score statistics where

$$\phi_1(u) = \Phi^{-1}(u), \qquad 0 < u < 1, \tag{5.7.7}$$

and $\Phi(x)$ is the standard normal d.f. whose density function is $\phi(x)$, $-\infty < x < \infty$. Then $\nu_{11} = \int_0^1[\Phi^{-1}(u)]^2\,du - [\int_0^1\Phi^{-1}(u)\,du]^2 = \int_{-\infty}^{\infty}x^2\,d\Phi(x) - [\int_{-\infty}^{\infty}x\,d\Phi(x)]^2 = 1$, and

$$B(F_{[1]}, \phi_1) = \int_{-\infty}^{\infty}\frac{d}{dx}\Phi^{-1}(F_{[1]}(x))\,dF_{[1]}(x)$$

$$= \int_{-\infty}^{\infty}\frac{f_{[1]}^2(x)}{\Phi'(\Phi^{-1}(F(x)))}\,dx. \tag{5.7.8}$$

Thus, denoting by $\mathscr{L}_\Phi$ the rank-order statistic based on the normal scores statistics, we have

$$e(\mathscr{L}_\Phi, \lambda) = \sigma_{11}\left(\int_{-\infty}^{\infty}\frac{f_{[1]}^2(x)}{\phi(\Phi^{-1}(F(x)))}\,dx\right)^2. \tag{5.7.9}$$

It is known (see, for example, Section 3.8 of Puri and Sen, 1971) that (5.7.9) is bounded below by 1 for all F, and the lower bound is attained only when F is the normal distribution. Thus, the normal-score rank-order statistic is asymptotically at least as efficient as the normal-theory test, whether the actual d.f. F is normal or not.

5.7.2. Coordinatewise Independent Multivariate Distributions

In this case, the p variates are mutually stochastically independent, that is, in (5.4.1),

$$F(\mathbf{x}) = \prod_{j=1}^{p} F_{[j]}(x_j) \qquad \forall \mathbf{x} \in R^p. \tag{5.7.10}$$

Then both $\mathbf{\Sigma}$ and $\mathbf{T}(F)$ are diagonal matrices, so that (5.7.4) reduces to

$$e(\mathscr{L}, \lambda) = \left(\sum_{j=1}^{p}\frac{\xi_{jj}}{\tau_{jj}(F)}\right)\Bigg/\left(\sum_{j=1}^{p}\frac{\xi_{jj}}{\sigma_{jj}(F)}\right). \tag{5.7.11}$$

If we denote the efficiency factors for the p marginals (as in (5.7.5)) by

$$e_j(\mathscr{L},\lambda) = \frac{\sigma_{jj}}{\tau_{jj}(F)} = \frac{B^2(F_{[j]},\phi_j)\sigma_{jj}}{\nu_{jj}}, \qquad j = 1,\ldots,p, \tag{5.7.12}$$

then from (5.7.11) and (5.7.12) we obtain that

$$e(\mathscr{L},\lambda) = \sum_{j=1}^{p} \omega_j e_j(\mathscr{L},\lambda), \tag{5.7.13}$$

where

$$\omega_j = \frac{\xi_{jj}/\sigma_{jj}}{\sum_{k=1}^{p}(\xi_{kk}/\sigma_{kk})}, \qquad 1 \le j \le p. \tag{5.7.14}$$

Since, by (5.7.3), $\xi_{jj} \ge 0\ \forall 1 \le j \le p$, the ω_j are nonnegative weights adding up to unity, so that from (5.7.13),

$$\min_{1\le j\le p} e_j(\mathscr{L},\lambda) \le e(\mathscr{L},\lambda) \le e(\mathscr{L},\lambda) \le \max_{1\le j\le p} e_j(\mathscr{L},\lambda) \qquad \forall \boldsymbol{\beta}^0, \tilde{\mathbf{C}}. \tag{5.7.15}$$

Hence, if we use the normal-score statistic, then by (5.7.9) and the discussion following it, $e_j(\mathscr{L},\lambda) \ge 1$ for $1 \le j \le p$ and all F, so that we have by (5.7.15) that $e(\mathscr{L}_\Phi,\lambda) \ge 1$ for all F, where the equality holds when F is normal. Similarly, for the Wilcoxon scores, $e(\mathscr{L}_W,\lambda) \ge 0.864$ for all continuous F.

5.7.3. Multinormal D.F.

If in (5.4.1), $F(\mathbf{x})$ is itself a multivariate normal d.f. with dispersion matrix $\boldsymbol{\Sigma}$, then if one uses the normal scores, it readily follows that $B^2(F_{[j]},\phi_j) = 1/\sigma_{jj}$, $1 \le j \le p$, and hence

$$\mathbf{T}(F) = \boldsymbol{\Sigma}, \tag{5.7.16}$$

so that, by (5.7.4),

$$e(\mathscr{L}_\Phi,\lambda) = 1 \qquad \forall \boldsymbol{\beta}^0, \mathbf{C}^*. \tag{5.7.17}$$

Thus, for parent normal d.f.'s, the normal-score and the normal-theory likelihood-ratio tests are asymptotically power-equivalent. In this case, by (5.7.4), (5.7.16), and the definition of the Wilcoxon scores, we have

$$e(\mathscr{L}_W, \lambda) = e(\mathscr{L}_W, \mathscr{L}_\Phi) = \frac{\operatorname{Tr}(\boldsymbol{\Xi}\mathbf{T}_N^{-1}(\mathbf{F}))}{\operatorname{Tr}(\boldsymbol{\Xi}\boldsymbol{\Sigma}^{-1})}, \tag{5.7.18}$$

where $\mathbf{T}_W(\mathbf{F})$ has the elements

$$\frac{\pi}{3}[\sigma_{jj}\sigma_{j'j'}]^{1/2}\rho^g_{jj'}, \qquad 1 \le j, j' \le p, \tag{5.7.19}$$

and

$$\rho^g_{jj'} = \frac{6}{\pi}\sin^{-1}\left(\tfrac{1}{2}\rho_{jj'}\right), \qquad \text{where} \quad \sigma_{jj'} = [\sigma_{jj}\sigma_{j'j'}]^{1/2}\rho_{jj'} \tag{5.7.20}$$

for $j, j' = 1, \ldots, p$. Thus, if we let

$$\boldsymbol{\Xi}^* = \mathbf{D}^*\boldsymbol{\Xi}\mathbf{D}^*, \qquad \text{where} \quad \mathbf{D}^* = \operatorname{diag}\left(\sigma_{11}^{-1/2}, \ldots, \sigma_{pp}^{-1/2}\right), \tag{5.7.21}$$

and denote

$$e(\mathscr{L}_W, \lambda) = e(\mathscr{L}_W, \mathscr{L}_\Phi) = \frac{3}{\pi}\frac{\operatorname{Tr}\left(\boldsymbol{\Xi}^*\mathbf{P}_g^{-1}\right)}{\operatorname{Tr}(\boldsymbol{\Xi}^*\mathbf{P}^{-1})}, \tag{5.7.22}$$

we have the following.

Lemma 5.7.1. With the definition in (5.7.20),

$$\frac{\pi}{3}\mathbf{P}_g - \mathbf{P} \text{ is p.s.d.} \tag{5.7.23}$$

The proof of the lemma follows by using the usual expansion of $\sin^{-1}x$ and a well-known result on the characteristic roots of product of matrices; it is left as an exercise (Exercise 5.7.1).

Since the trace of a matrix is the sum of the characteristic roots, on using the fact that

$$\frac{\operatorname{Tr}(\mathbf{AB}^{-1})}{\operatorname{Tr}(\mathbf{AC}^{-1})} = \frac{\operatorname{Tr}(\mathbf{AC}^{-1}\mathbf{CB}^{-1})}{\operatorname{Tr}(\mathbf{AC}^{-1})} = \frac{\operatorname{Tr}(\mathbf{A}^*\mathbf{CB}^{-1})}{\operatorname{Tr}(\mathbf{A}^*)}, \tag{5.7.24}$$

where $\mathbf{A}^* = \mathbf{AC}^{-1}$, and the fact that

$$\frac{\operatorname{Tr}(\mathbf{A}^*\mathbf{X})}{\operatorname{Tr}(\mathbf{A}^*)} \leq \operatorname{Ch}_1(\mathbf{X}) \qquad \forall \mathbf{A}^* \text{ p.s.d.}, \tag{5.7.25}$$

we obtain from (5.7.22)–(5.7.25) that for every $\boldsymbol{\beta}^0, \mathbf{C}$,

$$e(\mathscr{L}_W, \lambda) = e(\mathscr{L}_W, \mathscr{L}_\Phi) \leq \operatorname{Ch}_1\left(P\left\{\frac{\pi}{3}\mathbf{P}_g\right\}^{-1}\right) \leq 1. \tag{5.7.26}$$

Consequently, for the family of multinormal d.f.'s, the Wilcoxon-score statistics is never asymptotically better than the normal-score statistic. For specific values of p (≥ 2), (5.7.26) can be evaluated to obtain suitable bounds. For $p = 2$, both $\mathbf{P}$ and $\mathbf{P}_g$ become functions of a single ρ (namely, ρ_{12}), so that (5.7.26) can be easily evaluated. We refer to Chapter 4 of Puri and Sen (1971), where similar bounds are studied in detail.

5.7.4. General Multivariate D.F.

As has been mentioned earlier, (5.7.4), in general, depends on $\mathbf{T}$ and $\boldsymbol{\Sigma}$ as well as $\boldsymbol{\Xi}$ (which depends on $\boldsymbol{\beta}$ and $\mathbf{C}^*$). Now, noting that by (5.7.3),

$$\boldsymbol{\Xi} = \boldsymbol{\beta}\mathbf{C}^*\boldsymbol{\beta}', \tag{5.7.27}$$

(5.7.4) reduces to

$$\frac{\operatorname{Tr}(\boldsymbol{\beta}\tilde{\mathbf{C}}\boldsymbol{\beta}')\mathbf{T}^{-1}(F)}{\operatorname{Tr}(\boldsymbol{\beta}\tilde{\mathbf{C}}\boldsymbol{\beta}')\boldsymbol{\Sigma}^{-1}}, \tag{5.7.28}$$

and this explicitly shows the dependence on $\boldsymbol{\beta}$, $\mathbf{C}$, $\boldsymbol{\Sigma}$, and $\mathbf{T}(F)$. To evaluate bounds for (5.7.29) which do not depend on $\boldsymbol{\beta}$, let us first consider hypersurfaces for which

$$(\boldsymbol{\beta}\tilde{\mathbf{C}}\boldsymbol{\beta}')\boldsymbol{\Sigma}^{-1} = k\mathbf{I}_p, \qquad k \geq 0, \tag{5.7.29}$$

where $\mathbf{I}_p$ is the unit matrix of order p. Note that by varying k ($0 \leq k < \infty$), we scan the entire pq-dimensional plane of $\boldsymbol{\beta}$. Also, subject to (5.7.29), (5.7.28) reduces to

$$\frac{k \operatorname{Tr} \boldsymbol{\Sigma}\mathbf{T}^{-1}(F)}{pk} = p^{-1}\operatorname{Tr}\boldsymbol{\Sigma}\mathbf{T}^{-1}(F) = p^{-1}\sum_{j=1}^{p}\operatorname{Ch}_j(\boldsymbol{\Sigma}\mathbf{T}^{-1}(F)) \tag{5.7.30}$$

which do not depend on k. Further, the upper and lower bounds for the rhs of (5.7.30) are respectively

$$\mathrm{Ch}_1(\mathbf{\Sigma T}^{-1}(F)) \quad \text{and} \quad \mathrm{Ch}_p(\mathbf{\Sigma T}^{-1}(F)), \tag{5.7.31}$$

so that we have

$$\mathrm{Ch}_p(\mathbf{\Sigma T}^{-1}(F)) = \inf_{\boldsymbol{\beta},\tilde{\mathbf{C}}} e(\mathscr{L},\lambda) \leq \sup_{\boldsymbol{\beta},\tilde{\mathbf{C}}} e(\mathscr{L},\lambda) = \mathrm{Ch}_1(\mathbf{\Sigma T}^{-1}(F)). \tag{5.7.32}$$

The bounds in (5.7.32) provide useful information on the directional (over $\boldsymbol{\beta}$) fluctuation of the ARE when the score functions and the underlying d.f. F are held fixed. For particular cases, such as the Wilcoxon scores, and particular d.f., such as the multinormal one, when $p = 2$ or 3, various workers have obtained specific values for these bounds by varying $\mathbf{\Sigma}$ over its entire space; some of these results are presented in Chapter 5 of Puri and Sen (1971).

Next, we consider briefly the case of $\mathscr{L}_N^+$ and λ_N^+. For this, we define

$$\xi_{jj'} = \sum_{l=0}^{q} \sum_{l'=0}^{q} B_{jl}^* B_{j'l'}^* \tilde{C}_{ll'}^*, \qquad j, j' = 1, \ldots, p, \tag{5.7.33}$$

and let

$$\tilde{\mathbf{\Xi}} = ((\xi_{jj'}))_{j,j'=1,\ldots,p}. \tag{5.7.34}$$

Then, by virtue of Theorems 5.5.3 and 5.6.5, we obtain that

$$e(\mathscr{L}^+,\lambda^+) = \frac{\Delta_{\mathscr{L}^+}}{\Lambda_{\lambda^+}} = \frac{\mathrm{Tr}\,\tilde{\mathbf{\Xi}}\mathbf{T}^{-1}(F)}{\mathrm{Tr}\,\tilde{\mathbf{\Xi}}\mathbf{\Sigma}^{-1}}, \tag{5.7.35}$$

which has the same form as in (5.7.4), with the sole change that $\mathbf{\Xi}$ is replaced here by $\tilde{\mathbf{\Xi}}$. Consequently, all that has been shown in (5.7.4)–(5.7.32) remains good for (5.7.35) too. In view of this correspondence, the details are not reproduced here.

5.8. ASYMPTOTIC OPTIMALITY OF RANK-ORDER STATISTICS

We shall study here the asymptotic power equivalence of the likelihood-ratio test and certain rank-order tests when the underlying d.f. is not necessarily normal.

We assume that F in (5.4.1) has an absolutely continuous density function f. Let then $p(\mathbf{X}_{N1},\ldots,\mathbf{X}_{NN};\boldsymbol{\alpha},\boldsymbol{\beta})$ be the joint density function of $\mathbf{E}_N=(\mathbf{X}_{N1},\ldots,\mathbf{X}_{NN})$, and let $\hat{\boldsymbol{\alpha}}_N,\hat{\boldsymbol{\beta}}_N$ be the maximum-likelihood estimates of $\boldsymbol{\alpha}$ and $\boldsymbol{\beta}$ respectively. Also, let $\hat{\boldsymbol{\alpha}}_N^*$ be the maximum-likelihood estimate of $\boldsymbol{\alpha}$ under $H_0^{(1)}:\boldsymbol{\beta}=\mathbf{0}$. Then, for testing $H_0^{(1)}$, the likelihood function is defined by

$$L_N=\frac{p(\mathbf{E}_N;\hat{\boldsymbol{\alpha}}k_N^*,\mathbf{0})}{p(\mathbf{E}_N,\hat{\boldsymbol{\alpha}}_N,\hat{\boldsymbol{\beta}}_N)}. \tag{5.8.1}$$

By the standard results on the asymptotic distribution theory of likelihood-ratio statistics (Wald, 1943; Feder, 1968; and others), it follows that under $H_0^{(1)}$, $-2\log L_N$ has asymptotically a chi-square distribution with pq DF, and under $H_N^{(1)}$ (see Section 5.5), it has noncentral chi-square distribution with pq DF and noncentrality parameter

$$\Delta_L=\sum_{j=1}^{p}\sum_{j'=1}^{p}\sum_{l=1}^{q}\sum_{l'=1}^{q}\beta_{jl}^0\beta_{j'l'}^0C_{ll'}\gamma^{jj'}(F), \tag{5.8.2}$$

where

$$\left(\left(\gamma^{jj'}(F)\right)\right)=\boldsymbol{\Gamma}^{-1}(\mathbf{F})=\left(\left(\gamma_{jj'}(F)\right)\right)^{-1}, \tag{5.8.3}$$

$$\gamma_{jj'}(F)=E\left(\frac{\partial\log f(\mathbf{X}_i)}{\partial X_{ij}}\,\frac{\partial\log f(\mathbf{X}_i)}{\partial X_{ij'}}\right)_{\mathbf{0}} \tag{5.8.4}$$

for $j,j'=1,\ldots,p$, and $\mathbf{0}$ indicates that the derivative is evaluated at $\boldsymbol{\beta}=\mathbf{0}$. Note that by (5.7.3) and (5.8.2)

$$\Delta_L=\operatorname{Tr}\boldsymbol{\Xi}\boldsymbol{\Gamma}^{-1}(F). \tag{5.8.5}$$

Now, for each $\boldsymbol{\beta}\in R^{pq}$, we define a surface by the equation

$$\Delta_L=a,\qquad 0<a<\infty. \tag{5.8.6}$$

Consider now a transformation $\boldsymbol{\beta}\to\bar{\boldsymbol{\beta}}$ chosen in such a way that in terms of $\bar{\boldsymbol{\beta}}$, (5.8.6), reduces to

$$\sum_{j=1}^{p}\sum_{l=1}^{q}\left(\bar{\beta}_{jl}^2\right)=a,\qquad 0<a<\infty. \tag{5.8.7}$$

We denote by $S_a(\boldsymbol{\beta})$ and $S_a'(\bar{\boldsymbol{\beta}})$ the surfaces in (5.8.6) and (5.8.7) respectively. Now, for any point $\boldsymbol{\beta}^0$ and any positive ρ, consider the set $\omega(\boldsymbol{\beta}^0, \rho)$ consisting of all points $\boldsymbol{\beta}$ which lie on the same $S_a(\boldsymbol{\beta})$ as $\boldsymbol{\beta}^0$ and satisfy the condition that $\|\boldsymbol{\beta} - \boldsymbol{\beta}^0\| < \rho$, where $\|\cdot\|$ stands for the Euclidean norm. We denote by $\omega'(\boldsymbol{\beta}, \rho)$ the image of $\omega(\boldsymbol{\beta}, \rho)$ by the transformation from $\boldsymbol{\beta}$ to $\bar{\boldsymbol{\beta}}$, and by $A(\omega)$ the area of the set ω. Then we have the following result, which follows from Theorem VIII of Wald (1943), with further justifications for local alternatives by Feder (1968).

Theorem 5.8.1. *For testing $H_0^{(1)}: \boldsymbol{\beta} = \mathbf{0}$ against $H^{(1)}: \boldsymbol{\beta} \neq \mathbf{0}$, the likelihood-ratio test considered above: (i) has asymptotically best average power with respect to the surfaces $S_a(\boldsymbol{\beta})$ and weight function $\delta(\boldsymbol{\beta}) = \lim_{\rho \to 0}\{A[\omega'(\boldsymbol{\beta}, \rho)]/A[\omega(\boldsymbol{\beta}, \rho)]\}$, (ii) has asymptotically best constant power on the surfaces $S_a(\boldsymbol{\beta})$, and (iii) is an asymptotically most stringent test.*

Thus, when speaking of the asymptotic optimality of the likelihood-ratio test, we keep in mind the regularity conditions of Wald (1943) and regard the optimality in the light of Theorem 5.8.1. For optimality in the nonlocal sense, we may refer to Hoeffding (1965), Bahadur (1967), and others. We shall show that under certain conditions on F and for suitable choice of the ϕ_j, $j = 1, \ldots, p$, the test based on $\mathscr{L}_N$ has also asymptotically the test average power on the same family of ellipsoids.

Let us define

$$g_j(x_j) = \frac{\partial}{\partial x_j} \log f(x_1, \ldots, x_p) = \frac{f_j'(x_j|\mathbf{x})}{f_j(x_j|\mathbf{x})}, \tag{5.8.8}$$

where

$$f_j(x_j|\mathbf{x}) = \frac{f(\mathbf{x})}{\int_{-\infty}^{\infty} f(\mathbf{x})\, dx_j}, \qquad j = 1, \ldots, p, \tag{5.8.9}$$

are the conditional densities of the jth variate, given the others $(1 \le j \le p)$, and

$$f_j'(x_j|\mathbf{x}) = \frac{d}{dx_j} f_j(x_j|\mathbf{x}), \qquad j = 1, \ldots, p. \tag{5.8.10}$$

Consider then the statistics

$$U_{N,jl} = \sum_{i=1}^{N} c_{il} g_j(X_{i1}, \ldots, X_{i,j-1}, X_{i,j+1}, \ldots, X_{ip}) \tag{5.8.11}$$

for $j = 1, \ldots, p$, $l = 1, \ldots, q$, and let

$$L_N^* = \sum_{j=1}^{p} \sum_{j'=1}^{p} \sum_{l=1}^{q} \sum_{l'=1}^{q} U_{N,jl} U_{N,j'l'} C_N^{ll'} \gamma^{jj'}(F), \qquad (5.8.12)$$

where $((C_N^{ll'})) = \mathbf{C}_N^{-1}$ and the $\gamma^{jj'}(F)$ are defined by (5.8.3)–(5.8.4). Then, from the results of Wald (1943), it follows that

$$-2\log L_N - L_N^* \to 0 \text{ in probability} \quad \text{as} \quad N \to \infty. \qquad (5.8.13)$$

Therefore, Theorem 5.8.1 applies to L_N^* also.

Consider first the univariate case, i.e., $p = 1$. Here, the conditional density in (5.8.9) agrees with the density $f_{[1]}(x) = f(x)$ and $f'_{[1]}(x) = f'(x)$. If f and f' are given, we may consider the score function

$$\psi(u) = -\frac{f'(F^{-1}(u))}{f(F^{-1}(u))}, \qquad 0 < u < 1. \qquad (5.8.14)$$

It follows then that $\int_0^1 \psi(u)\, du = 0$, and

$$\int_0^1 \psi^2(u)\, du = I(f) = \int_{-\infty}^{\infty} \left[\frac{f'(x)}{f(x)}\right]^2 dF(x) = \gamma_{11}(F) \qquad (5.8.15)$$

is the Fisher information of the density f. Also, by (5.5.15) and (5.8.14), we have by partial integration

$$B(F, \psi) = \int_0^1 \psi'(u) f(F^{-1}(u))\, du$$

$$= \int_0^1 \psi(u) \left[-\frac{f'(F^{-1}(u))}{f(F^{-1}(u))}\right] du = \int_0^1 \psi^2(u)\, du. \qquad (5.8.16)$$

Consequently,

$$\frac{\int_0^1 \psi^2(u)\, du}{B^2(F, \psi)} = \frac{1}{I(f)}. \qquad (5.8.17)$$

As such, by Theorem 5.5.2, for $p = 1$ and ψ given by (5.8.14), $\mathscr{L}_N$ ($= \mathscr{L}_{N,\psi}$, say) has asymptotically (under $\{H_N^{(1)}\}$) a noncentral chi-square distribution with q DF and noncentrality parameter

$$I(f)\left[\sum_{l=1}^{q}\sum_{l'=1}^{q}\beta_{1l}^{0}\beta_{1l'}^{0}\tilde{C}_{ll'}\right]. \tag{5.8.18}$$

From (5.8.2), (5.8.3), (5.8.4), and (5.8.15), we conclude that under $\{H_N^{(1)}\}$, $-2\log L_N$ has asymptotically the same distribution. That is, under $\{H_N^{(1)}\}$, $\mathscr{L}_{N,\psi}$ and $-2\log L_N$ are asymptotically power-equivalent. Hence, by virtue of Theorem 5.8.1, we have the following.

Theorem 5.8.2. *Under $\{H_N^{(1)}\}$, the likelihood-ratio test and the rank-order test based on $\mathscr{L}_{N,\psi}$ with ψ defined by* (5.8.14) *are asymptotically power-equivalent, and hence they share the asymptotic optimality.*

We may add that if in (5.3.1), $F(x) = F_0((x-\alpha)/\delta)$, where α and δ (>0) are unknown and F_0 is specified, then in (5.8.14) one may work with $\psi(u) = -f_0'(F_0^{-1}(u))/f_0(F_0^{-1}(u))$.

Let us now proceed to the general multivariate case. Here, additional conditions are needed to establish the asymptotic optimality of $\mathscr{L}_N$. The reason is that $\mathscr{L}_N$ involves the coordinatewise rank statistics $S_{N,jl}$, $1\le j\le p$, $1\le l\le q$, whereas L_N or L_N^* involves the functions g, which in general depend on all p coordinates $X_{i1},\dots,X_{ip}$. However, if this dependence is of some special form, the asymptotic optimality follows. For this, we define the jth marginal of F by $F_{[j]}$, the corresponding density function by $f_{[j]}$, and its first derivative by $f_{[j]}'(x)$, $j=1,\dots,p$. Let us then assume that for each j ($=1,\dots,p$)

$$g_j(x_j|\mathbf{x}) = \sum_{k=1}^{p}\frac{h_{jk}f_{[k]}'(x_k)}{f_{[k]}(x_k)}, \tag{5.8.19}$$

where the h_{jk} are real constants not all equal to zero. Note that (5.8.19) holds for (1) multivariate normal d.f.'s and (2) coordinatewise independent distributions, and it may hold for some other d.f.'s.

Let us now define

$$W_{N,jl} = \sum_{i=1}^{N}\frac{c_{il}f_{[j]}'(X_{ij})}{f_{[j]}(X_{ij})}, \qquad 1\le j\le p,\quad 1\le l\le q. \tag{5.8.20}$$

From (5.8.11), (5.8.15), and (5.8.20), we have

$$U_{N,jl} = \sum_{k=1}^{p}h_{jk}W_{N,kl} \qquad \text{for } 1\le j\le p,\quad 1\le l\le q, \tag{5.8.21}$$

so that by (5.8.12) and (5.8.21),

$$L_N^* = \sum_{j=1}^{p}\sum_{j'=1}^{p}\sum_{l=1}^{q}\sum_{l'=1}^{q} C_N^{ll'}\gamma^{jj'}(F)\left\{\sum_{k=1}^{q} h_{jk}W_{N,jk}\right\}\left\{\sum_{k'=1}^{q} h_{j'k'}W_{N,j'k'}\right\}$$

$$= \sum_{l=1}^{q}\sum_{l'=1}^{q}\sum_{k=1}^{p}\sum_{k'=1}^{p} C_N^{ll'}W_{N,kl}W_{N,k'l'}\left(\sum_{j=1}^{p}\sum_{j'=1}^{p}\gamma^{jj'}(F)h_{jk}h_{j'k'}\right)$$

$$= \sum_{k=1}^{p}\sum_{k'=1}^{p}\sum_{l=1}^{q}\sum_{l'=1}^{q} W_{N,kl}W_{N,k'l'}C_N^{ll'}\gamma_*^{kk'}(F), \quad \text{say,} \tag{5.8.22}$$

where for every $k, k' = 1,\ldots,p$.

$$\gamma_*^{kk'}(F) = \sum_{j=1}^{p}\sum_{j'=1}^{p} h_{jk}h_{j'k'}\gamma^{jj'}(F) \tag{5.8.23}$$

Also, by (5.8.19), $\gamma_{jj'}(F) = E(g_j(X_{ij})g_{j'}(X_{ij'})|H_0^{(1)}) = \sum_{k=1}^{p}\sum_{k'=1}^{p} h_{jk}h_{j'k'}E[f'_{[k]}(X_{ik})f_{[k']}(X_{ik'})/f_{[k]}(X_{ik})f_{[k']}(X_{ik'})]$ for $j, j' = 1,\ldots,p$. So we denote

$$\tilde{\gamma}_{kk'}(F) = E\left\{\frac{d}{dX_{ik}}\log f_{[k]}(X_{ik})\;\frac{d}{dX_{ik'}}\log f_{[k']}(X_{ik'})\middle| H_0^{(1)}\right\}; \tag{5.8.24}$$

$$\tilde{\mathbf{\Gamma}}(F) = ((\tilde{\gamma}_{kk'}(F)))_{k,k'=1,\ldots,p}; \tag{5.8.25}$$

we then have

$$\mathbf{\Gamma}(F) = \mathbf{H}\tilde{\mathbf{\Gamma}}(F)\mathbf{H}', \qquad \text{where} \quad \mathbf{H} = ((h_{jk})). \tag{5.8.26}$$

Thus, if we assume that $\mathbf{H}$ is nonsingular, we have

$$\mathbf{\Gamma}^{-1}(F) = (\mathbf{H}')^{-1}\tilde{\mathbf{\Gamma}}^{-1}(F)\mathbf{H}^{-1}, \tag{5.8.27}$$

so that

$$\mathbf{H}'\mathbf{T}^{-1}(\mathbf{F})\mathbf{H} = \tilde{\mathbf{\Gamma}}^{-1}(F). \tag{5.8.28}$$

Thus, by (5.8.23) and (5.8.28), we have

$$\gamma_*^{kk'}(F) = \tilde{\gamma}^{kk'}(F) \qquad \forall k, k' = 1,\ldots,p, \tag{5.8.29}$$

so that by (5.8.22) and (5.8.29) we have

$$L_N^* = \sum_{k=1}^{p} \sum_{k'=1}^{p} \sum_{l=1}^{q} \sum_{l'=1}^{q} W_{N,kl} W_{N,k'l'} C_N^{ll'} \tilde{\gamma}^{kk'}(F). \qquad (5.8.30)$$

We now define for each $j\ (= 1, \ldots, p)$

$$\psi_j(u) = -\frac{f'_{[j]}\left(F_{[j]}^{-1}(u)\right)}{f_{[j]}\left(F_{[j]}^{-1}(u)\right)}, \qquad 0 < u < 1, \qquad (5.8.31)$$

and define $L_{N,jl}$ as in (5.4.4) with the scores based on the score functions in (5.8.31). In this case, by (5.8.31),

$$E\left[L_{N,jl} | H_0^{(1)}\right] = 0 \qquad \forall 1 \le j \le p, 1 \le l \le q, \qquad (5.8.32)$$

while by (5.4.16), (5.4.17) and (5.8.24), $\boldsymbol{\nu}(F) = \tilde{\boldsymbol{\Gamma}}(F)$. Finally, by using (4.2.46), (4.2.47), (4.2.49), (4.2.50), and (4.2.24), it follows that under $H_0^{(1)}$ as well as $\{H_N^{(1)}\}$, $S_{N,jk}$ is asymptotically equivalent in quadratic mean to $W_{N,jk}$ for each $j\ (= 1, \ldots, p)$ and $k\ (= 1, \ldots, q)$. Hence, by Lemmas 4.2.2 and 4.2.3, $\mathbf{S}_N$ and $\mathbf{W}_N$ have the same asymptotic distribution theory. Consequently, by (5.8.29) and (5.8.32), we have on using Theorem 5.4.1 that under $\{H_N^{(1)}\}$,

$$\mathscr{L}_N - L_N^* \xrightarrow{P} 0, \qquad (5.8.33)$$

and this leads to the following.

Theorem 5.8.3. *Under* (5.8.19) *and for a p.d.* $\mathbf{H}$, $\mathscr{L}_N$ *is asymptotically (under* $\{H_N^{(1)}\}$*) power-equivalent to the likelihood-ratio test, and hence is asymptotically optimal.*

In a similar manner, the likelihood-ratio statistic can be worked out for the hypothesis $H_0^{(2)}$, and the asymptotic optimality of $\mathscr{L}_N^+$ (under $\{H_N^{(2)}\}$) can be established under the conditions of the last two theorems, depending on the univariate or the multivariate case.

5.9. RANK TESTS FOR CERTAIN SPECIFIC PROBLEMS

In Section 4.4, we have considered some specific problems of general linear models which are of special practical importance. For these problems, the appropriate rank-order statistics ($\mathscr{L}_N$ or $\mathscr{L}_N^+$) are obtained here in explicit forms and their properties studied.

5.9.1. Multivariate Two-Sample Problem

This relates to the model (4.4.1), which is a special case of (5.4.1) with $q = 1$ and the $\mathbf{c}_i$ satisfying (4.4.4). Thus, defining the vector $\mathbf{L}_N$ as in (4.4.2)–(4.4.3) and the matrix $\mathbf{V}_N$ as in (5.4.10), the corresponding rank-order statistic is given by

$$\mathscr{L}_N = \frac{n_1 + n_2}{n_1 n_2}[\mathbf{L}_N - n_1\bar{\mathbf{a}}_N]'\mathbf{V}_N^{-1}[\mathbf{L}_N - n_1\bar{\mathbf{a}}_N], \tag{5.9.1}$$

where $\bar{\mathbf{a}}_N = (\bar{a}_{N,1}, \ldots, \bar{a}_{N,p})'$ and

$$\bar{a}_{N,j} = \frac{1}{N}\sum_{i=1}^{N} a_{Nj}(i), \qquad 1 \leq j \leq p. \tag{5.9.2}$$

The hypothesis of equality of the two distributions being a special case of $H_0^{(1)}$ (with $q = 1$), we obtain from the results of Section 5.4 that under $H_0^{(1)}$, $\mathscr{L}_N$ has asymptotically chi-square distribution with p DF. The non-null distribution of $\mathscr{L}_N$ follows from the results of Section 5.5, its efficiency from Section 5.7, and its optimality from Section 5.8.

5.9.2. Multivariate Multisample Problem

This corresponds to the models (4.4.5)–(4.4.6), and we define the coordinatewise statistics $\mathbf{T}_{Nk}$, $k = 1, \ldots, c$ (≥ 2), as in (4.4.7)–(4.4.8). Also, let $\mathbf{V}_N$ be defined as in (5.4.10). In this case, (4.4.6) is a particular case of (5.4.2) with $q = c - 1$. The corresponding $\mathscr{L}_N$ (obtained by symmetrizing) comes out as

$$\mathscr{L}_N = \sum_{k=1}^{c} n_k^{-1}[\mathbf{T}_{Nk} - n_k\bar{\mathbf{a}}_N]'\mathbf{V}_N^{-1}[\mathbf{T}_{Nk} - n_k\bar{\mathbf{a}}_N], \tag{5.9.3}$$

where $\bar{\mathbf{a}}_N$ is defined by (5.9.2). Here, under the hypothesis of equality of the c d.f.'s, $\mathscr{L}_N$ has asymptotically chi-square with $p(c-1)$ DF, and the non-null distribution follows again from Section 5.5. The efficiency and the optimality results are parallel to those in the two-sample case and follow from Sections 5.7 and 5.8.

5.9.3. Multivariate One-Sample Problem

This corresponds to the model (5.4.1), where $q = 0$ (i.e., $\mathbf{c}_i = 0\ \forall i \geq 1$), and our hypothesis of interest is

$$H_0: \boldsymbol{\alpha} = \mathbf{0}, \tag{5.9.4}$$

i.e., F is diagonally symmetric about $\mathbf{0}$. Define $S_{N,j}^{+} = S_{N,j0}^{+}$, $j = 1, \ldots, p$, as in (5.4.37), and let

$$\mathbf{S}_N^{+} = \left(S_{N,1}^{+}, \ldots, S_{N,p}^{+}\right)'. \tag{5.9.5}$$

Also, let $\tilde{\mathbf{V}}_N$ be defined as in (5.4.49)–(5.4.51). Since $C_{N,00}^{*} = N$ and $C_{N,ll'}^{*} = 0$, $(l, l') \neq (0,0)$, as a special case of (5.4.56), the test statistic is given by

$$\mathscr{L}_N^{+} = N^{-1}\left(\mathbf{S}_N^{+\prime}\tilde{\mathbf{V}}_N^{-1}\mathbf{S}_N^{+}\right). \tag{5.9.6}$$

Here, under (5.9.4), $\mathscr{L}_N^{+}$ has asymptotically a chi-square distribution with p DF. That it has asymptotically a non-null distribution follows from the results of Section 5.5. The asymptotic efficiency and optimality of $\mathscr{L}_N^{+}$ follow from the results of Sections 5.7 and 5.8.

In passing, we may remark that for the multivariate one- and two-sample location problems, the optimal (normal-theory) invariant test is based on the Hotelling T^2-statistics, while for the multisample problem, the likelihood-ratio test statistic is asymptotically equivalent to the generalized Hotelling-Lawley trace criterion. In their form, $\mathscr{L}_N$ and $\mathscr{L}_N^{+}$ are similar to the T^2 and the generalized T^2 statistics, where the roles of the sample-mean vectors and covariance matrices are replaced by the vector of the average rank scores and the rank-covariance matrix $\mathbf{V}_N$ or $\tilde{\mathbf{V}}_N$.

EXERCISES

5.2.1. Show that for the two-sample location model, where in (5.2.1) the c_i can assume the values 0 and 1, if the two sample sizes are n_1 and n_2 respectively and if $n_1/N \to \lambda$, $0 < \lambda < 1$, then $A_N^{-1}C_N^{-2}L_N$ converges in probability to $\mu(\beta, \lambda)$, which is positive or negative according as β is positive or negative; the score function is assumed to be nondecreasing and square-integrable with a finite number of jumps. Hence, or otherwise, show that for small β, $\mu(\beta, \lambda)$ behaves like a scale multiple of β.

5.2.2. Consider the joint probability function $p_\beta(\mathbf{R})$ of the rank vector $\mathbf{R}_N = (R_{N1}, \ldots, R_{NN})$ when $\beta_0 = 0$ in (5.2.1). Show that as $\beta \to 0$, $p_\beta(\mathbf{R})/p_0(\mathbf{R}) = 1 + \beta\Sigma_{i=1}^{N}(c_i - \bar{c}_N)a_N^{*}(R_{Ni}) + o(\beta)$, where the $a_N^{*}(i)$ are defined after (5.2.14). Specify regularity conditions on the underlying d.f. F under which this expansion is a valid one. Hence, comment on the locally most powerful rank (LMPR) test for $H_0: \beta = 0$ against $\beta > 0$. Obtain explicit forms for the LMPR test statistics when F in (5.2.1) is normal, double-exponential, and logistic.

5.2.3. Verify that under the Noether condition on the c_i and (5.2.23) on the scores $a_N(i)$, the Lindeberg condition in (5.2.18) holds, and hence the asymptotic normality of the standardized form of L_N (under H_0) is in order.

5.2.4. Verify that under (5.2.24) and the Noether condition on the scores $a_N(i)$, the Lindeberg condition in (5.2.18) holds, insuring the asymptotic normality of the standardized form of L_N.

5.2.5. Let X_i, $i \geq 1$, be independent r.v.'s with continuous d.f.'s F_i, $i \geq 1$. Define L_N as in (5.2.5). Also, let $\bar{F}_N = N^{-1}\Sigma_{i=1}^N F_i$, $T_N = (NA_N^2 C_N^2)^{-1/2} L_N$, and $\tau_N = \int_{-\infty}^{\infty} \phi(\bar{F}_N(x))\, dF_N^*(x)$, where $F_N^*(x) = N^{-1/2} A_N^{-1} C_N^{-1} \Sigma_{i=1}^N (c_i - \bar{c}_N) F_i(x)$. Then show that under (5.2.24) and $\int_0^1 |\phi(u)|^r\, du < \infty$, for some $r > 2$, $T_N - \tau_N \to 0$ a.s. as $N \to \infty$. (Sen and Ghosh, 1972.)

5.2.6. For the preceding problem, show that the L_r condition on the score function can be replaced by the L_1 condition provided ϕ is of bounded variation on any closed subinterval of (0,1). (Hájek, 1974; Sen, 1981a, p. 120.)

5.2.7. Corresponding to the test in (5.2.15), construct a randomized test procedure having the exact level of significance α.

5.3.1. Provide a proof of Theorem 5.3.2.

5.4.1. Using (5.4.7), verify (5.4.8) and (5.4.9).

5.4.2. Show that v_{Njj}, defined by (5.4.10), converges to ν_{jj}, defined by (5.5.16).

5.4.3. Provide a proof of (5.4.20) when the scores $a_N(i)$ are taken as $E\phi(U_N^{(i)})$, $i = 1, \ldots, N$, as in (4.2.4).

5.4.4. Verify (5.4.27). (Puri and Sen, 1969b.)

5.4.5. Provide a proof of Theorem 5.4.4.

5.4.6. Provide a proof of Theorem 5.4.5.

5.5.1. Use the definitions of $\mathscr{L}_N$ and $\mathscr{L}_N^*$ in (5.4.11)–(5.4.13) and (5.5.21), and show that by the Courant theorem $\mathrm{Ch}_p(\boldsymbol{\nu}\mathbf{V}_N^{-1}) \leq \mathscr{L}_N/\mathscr{L}_N^* \leq \mathrm{Ch}_1(\boldsymbol{\nu}\mathbf{V}_N^{-1})$. Use (5.5.9), (5.5.19), and the above inequality to show that (5.5.20) holds.

5.6.1. Verify the formulae in (5.6.12).

5.6.2. Provide a proof of Theorem 5.6.5.

5.7.1. Provide a proof of Lemma 5.7.1. (Puri and Sen, 1969a.)

5.8.1. Define L_N as in (5.8.1). Show that $-2\log L_N$ can be expressed as a quadratic form in the MLE of $(\boldsymbol{\alpha}, \boldsymbol{\beta})$ plus a remainder term which is negligible, in probability. For this quadratic form, examine the character of its discriminant and show that it is an idempotent matrix for which the standard Cochran theorem on quadratic forms may be applied. Hence, or otherwise, show that under $H_0: \boldsymbol{\beta} = \mathbf{0}$, $-2\log L_N$ is asymptotically distributed as a central chi-square variable with pq DF. (Wald, 1943; Feder, 1968.)

5.9.1. For the multivariate multisample location problem, for normal F, obtain the likelihood-ratio statistics. Use Lemma 5.6.3 and show that this statistic is asymptotically equivalent to the Hotelling–Lawley trace criterion, even when the underlying d.f. is not necessarily normal (but has a finite second-moment matrix). (Sen and Puri, 1970.)

CHAPTER 6

Rank-Order Estimation Theory in Some Linear Models

6.1. INTRODUCTION

The conventional parametric estimation theory in linear models is based either on the classical least-squares method, or on the maximum-likelihood estimation method. The former method usually yields computationally simpler estimates. However, these estimates are not usually very robust for outliers and gross errors or heavy-tailed distributions. The latter procedure demands the knowledge of the underlying distribution and retains its (asymptotic) optimality so long as this assumption on the distribution is true. In extreme cases the procedure may yield totally inconsistent estimates (e.g., when the actual c.d.f. is Cauchy and we assume it to be normal). For this reason, we shall study in this chapter some alternative robust estimators, based on suitable rank-order statistics, which remain valid for a broad class of c.d.f.'s and retain high efficiency too.

In Section 6.2, we start with the simple regression model in the univariate setup and proceed to estimate the slope and the intercept. Section 6.3 deals with the univariate multiple-regression setup. Section 6.4 deals with the multivariate problem; both the simple and multiple-regression models are considered in this context. In the last section we consider the problem of interval estimation. In each section, results on asymptotic relative efficiency are studied in detail along with other asymptotic properties of the estimators. A few basic results in this context are stated in this chapter and are proved systematically in the Appendix. As in earlier chapters, the theory developed here is based on the assumption of continuous distribution functions and the consequent absence of ties. For the study of some of the corresponding problems in the presence of ties, the reader is referred to Padmanabhan and Puri (1979, 1983).

6.2. ESTIMATION IN SIMPLE LINEAR REGRESSION MODELS

Let $X_1, \ldots, X_n$ be independent r.y.'s with continuous c.d.f.'s $F_1(x), \ldots, F_N(x)$, respectively, all defined on $(-\infty, \infty)$. It is assumed that

$$F_i(x) = F(x - \beta_0 - \beta(c_i - \bar{c}_N)), \qquad -\infty < X < \infty, \tag{6.2.1}$$

for $i = 1, \ldots, N$, where β_0 (intercept) and β (regression coefficients) are unknown parameters; $c_1, \ldots, c_N$ are known (regression) constants (not all equal); and $\bar{c}_N = N^{-1}\Sigma_{i=1}^{N} c_i$. Based on suitable rank statistics, we desire to provide robust and efficient estimators of β_0 and β. Because of the differences of the underlying regularity conditions, we shall find it convenient to develop the theory in the following order: (1) estimation of β, (2) estimation of β_0 treating β as known, and (3) simultaneous estimation of β_0 and β.

6.2.1. Estimation of the Regression Coefficient β

We motivate the procedure through the following alignment principle (Hodges and Lehmann, 1963; Sen, 1963; Adichie, 1967b): For every real b $(-\infty < b < \infty)$, define

$$X_i(b) = X_i - b(c_i - \bar{c}_N), \qquad 1 \le i \le N. \tag{6.2.2}$$

If β were known, then $X_1(\beta), \ldots, X_N(\beta)$ would have been i.i.d.r.v.'s with the c.d.f. $F(x - \beta_0)$, $-\infty < x < \infty$, i.e., $X_i(\beta)$ would have the same c.d.f. as that of X_i under the null hypothesis $H_0 : \beta = 0$. Let $T_N = T_N(X_1, \ldots, X_N)$ be a suitable test statistic for testing H_0 against one- or two-sided alternatives, and let $G_N(t) = P\{T_N \le t \mid H_0\}$. Usually, G_N is specified (under H_0) and we let μ_N be a measure of location of G_N. If G_N is symmetric, we may take μ_N as the point of symmetry; otherwise, we may let μ_N be $E[T_N \mid H_0]$, the median of G_N, or the like. The basic idea is that we have maximum confidence in H_0 when T_N is closest to μ_N. Let us denote

$$T_N(b) = T_N(X_1(b), \ldots, X_N(b)), \qquad -\infty < b < \infty. \tag{6.2.3}$$

Then we proceed to choose $\hat{\beta}_N$ as an estimator of β, for which the aligned statistic $|T_N(\hat{\beta}_N) - \mu_N|$ is minimum. In order to obtain a meaningful and unique estimator in this way, we need to impose some regularity conditions. Specifically, we assume that

$$T_N(b) \text{ is nonincreasing in } b : -\infty < b < \infty. \tag{6.2.4}$$

Then there is either a unique value of b or a (half-open) interval for which $T_N(b)$ is closest to μ_N. Thus, to define $\hat{\beta}_N$ uniquely we let

$$\hat{\beta}_{N,1} = \inf\{b : T_N(b) < \mu_N\}, \qquad \hat{\beta}_{N,2} = \sup\{b : T_N(b) > \mu_N\}, \tag{6.2.5}$$

$$\hat{\beta}_N = \tfrac{1}{2}(\hat{\beta}_{N,1} + \hat{\beta}_{N,2}), \tag{6.2.6}$$

that is, $\hat{\beta}_N$ is the center of gravity of the set of admissible solutions of b for which $|T_N(b) - \mu_N| =$ minimum. The estimator defined in this way includes, among others, the least-squares estimator as a special case where

$$T_n = C_N^{-2} \sum_{i=1}^{N} (c_i - \bar{c}_N)(X_i - \bar{X}_N), \qquad C_N^2 = \sum_{i=1}^{N} (c_i - \bar{c}_N)^2,$$

$$\bar{X}_N = N^{-1} \sum_{i=1}^{N} X_i, \tag{6.2.7}$$

and where $\mu_N = 0$. Here, $T_N(b) = T_N - b$, so that $\hat{\beta}_N = T_N(0)$ is the classical least-squares estimator of β. In this case, $T_N(b)$ is strictly linear in b, and $\hat{\beta}_N$ is unique. Secondly, consider the case of Kendall's tau, where

$$T_N = \sum_{1 \le i \le j \le N} \operatorname{sgn}(c_i - c_j)\operatorname{sgn}(X_i - X_j). \tag{6.2.8}$$

Then $\mu_N = 0$ and (6.2.4) holds, though $T_N(b)$ is not necessarily linear in b. Here we have

$$\beta_N = \operatorname{median}\left\{ \frac{X_i - X_j}{c_i - c_j} : c_i \neq c_j, 1 \le i \le j \le N \right\}; \tag{6.2.9}$$

refer to Sen (1968) for details.

Our main interest is in the estimator based on linear rank statistics of the form

$$L_N = \sum_{i=1}^{N} (c_i - \bar{c}_N) a_N(R_{Ni}), \tag{6.2.10}$$

where $a_N(1) \le \cdots \le a_N(N)$ (not all equal) are suitable scores and R_{Ni} is the rank of X_i among $X_1, \ldots, X_N$, $1 \le i \le N$. Also, we define $L_N(b)$ as in (6.2.3) for $T_N = L_N$. Then we prove first the following theorem relating to (6.2.4) (cf. Sen, 1969a; Jurečková, 1969).

Theorem 6.2.1. *Let* $a_N(1) \le \cdots \le a_N(N)$ *(not all equal) be N scores,* $c_1 \le \cdots \le c_N$ *(not all equal) be known constants, and* $Y_1, \dots, Y_N$ *be independent random variables with continuous c.d.f.'s* $F_1(x), \dots, F_N(x)$ *respectively. Further, let* $R_{Ni}(b)$ *be the rank of* $Y_i(b) = Y_i - bc_i$ *among* $Y_1(b), \dots, Y_N(b)$, $1 \le i \le N$, *and* $L_N(b)$ *be defined then by* (6.2.10) *for* $R_{Ni} = R_{Ni}(b)$, $1 \le i \le N$. *Under the above conditions, there exists* N^* *points* $b_1 < \cdots < b_{N^*}$, *with*

$$N - 1 \le N^* \le \binom{N}{2},$$

such that (1) $L_N(b) = L_N(b_s + 0)$ *for* $b_s < b < b_{s+1}$, $s = 0, 1, \dots, N^*$ $(b_0 = -\infty,\ b_{N^*+1} = +\infty)$, (2) $L_N(b_s - 0) \ge L_N(b_s) \ge L_N(b_s + 0)$, $1 \le s \le N^*$, *and* (3) $L_N(b) > 0$ *for* $b < b_1$ *and* < 0 *for* $b > b_{N^*+1}$. *Thus,* $L_N(b)$ *is nonincreasing in* $b: -\infty < b < \infty$.

[Note that in order to define the $L_N(b)$ properly at b_s, $s = 1, \dots, N^*$, we adopt the usual convention of average scores for tied observations.]

Proof. Let $c_N^* = c_i - c_1$ (≥ 0), $1 \le i \le N$, and rewrite $L_N(b)$ as

$$L_N(b) = \sum_{i=1}^{N} c_i^* a_N(R_{Ni}(b)) + (c_1 - \bar{c}_N) \sum_{i=1}^{N} a_N(i), \qquad (6.2.11)$$

where the last term does not depend on b. Let then

$$Y_i^*(b) = Y_i - bc_i^*, \qquad 1 \le i \le N, \qquad -\infty < b < \infty. \qquad (6.2.12)$$

Now, (6.2.12) represents N straight lines in b. The ith and i'th lines either are parallel (if $c_i = c_{i'}$) or intersect at a single point $b_{ii'}$ (if $c_i \ne c_{c'}$), $1 \le i < i' \le N$. Let $N^* = \{(c_i, c_{i'}) : c_i \ne c_{i'},\ 1 \le i < i' \le N\}$. Then,

$$N - 1 \le N^* \le \binom{N}{2}.$$

We denote the ordered values of the $b_{ii'}$ by $b_1, \dots, b_{N^*}$. As the F_i are all assumed to be continuous, ties among $Y_1, \dots, Y_N$ (and hence, among $b_1, \dots, b_{N^*}$) can be neglected in probability. Thus, $b_1 < \cdots < b_{N^*}$ in probability. Now, for any s $(0 \le s < N^*)$, consider the open interval $b_s < b < b_{s+1}$. Since, in this interval, no two lines in (6.2.12) intersect, the ranks of $Y_1^*(b), \dots, Y_N^*(b)$ are the same as those of $Y_1^*(b_s + 0), \dots, Y_N^*(b_s + 0)$, and hence,

$$L_N(b) = L_N(b_s + 0) \qquad \text{for} \quad b_s < b < b_{s+1}. \qquad (6.2.13)$$

At $b = b_s$ (> 0), let the two intersecting lines be the ith and i'th ones, and let $R_{Ni}(b_s - 0) = q_1$, $R_{Ni'}(b_s - 0)] = q_2$, $i' > i$. As $c_{i'}^* > c_i^*$ (otherwise, the two lines do not intersect), we must have (1) $q_1 = q_2 - 1 = q - 1$ (say), (2) $R_{Ni}(b_s + 0) = q = 1 + R_{Ni'}(b_s + 0)$, and (3) $R_{Nj}(b_s + 0) = R_{Nj}(b_s - 0)$ for the remaining $n - 2$ values of j. Thus, for $b_s > 0$,

$$\begin{aligned} L_N(b_s + 0) &= L_N(b_s - 0) - (c_{i'}^* - c_i^*)[a_N(q) - q_N(q - 1)] \\ &\le L_N(b_s - 0). \end{aligned} \tag{6.2.14}$$

At $b = b_s$, $Y_i^*(b_s) = Y_{i'}^*(b_s)$, so that by the convention of average scoring, we have

$$\begin{aligned} L_N(b_s) &= L_N(b_s - 0) - \tfrac{1}{2}(c_{i'}^* - c_i^*)[a_N(q) - a_N(q - 1)] \\ &= \tfrac{1}{2}\{L_N(b_s - 0) + L_N(b_s + 0)\}, \end{aligned} \tag{6.2.15}$$

and hence, by (6.2.14), $L_N(b_s - 0) \ge L_N(b_s) \ge L_N(b_s + 0)$, $1 \le s \le N^*$. A similar proof holds for $b_s < 0$. Hence, $L_N(b)$ is nonincreasing in $b: -\infty < b < \infty$. Finally, we rewrite $L_N(b)$ as

$$\frac{1}{2N} \sum_{1 \le i < i' \le N} (c_{i'}^* - c_i^*)[a_N(R_{Ni'}(b)) - a_N(R_{Ni}(b))], \tag{6.2.16}$$

where for $b < b_1$, $Y_{i'}^*(b) > Y_i^*(b)$ for every $i' > i$, so that $a_N(R_{Ni'}(b) - a_N(R_{Ni}(b)) \ge 0$ for every $1 \le i < i' \le N$, with at least one strict inequality for some (i, i'): $c_i \ne c_{i'}$. Hence, $L_N(b) > 0$ for $b < b_1$. Similarly, $L_N(b) < 0$ for $b > b_{N^*+1}$. The above proof is adapted from Sen (1969a); Jurečková (1969) incorporated Lehmann's (1966) concept of dependence and provided an alternative proof.

Note that under $H_0: \beta = 0$, $(R_{N1}, \ldots, R_{NN})$ has the uniform distribution over the $N!$ permutations of $(1, \ldots, N)$, so that $E(L_N | H_0) = 0$. Hence, for L_N, (6.2.4) holds, and in (6.2.5) we have $\mu_N = 0$. In passing, we remark that we may have the c.d.f. of L_N, under $H_0: \beta = 0$, symmetric about 0, in which case μ_N ($= 0$) is also the point of symmetry of this distribution. To this end, we state the following lemma, whose proof is left as an exercise (see Exercise 6.2.1).

Lemma 6.2.2. If either F is symmetric or $c_j + c_{N-j+1} = 2\bar{c}_N$, $1 \le j \le N$, then under $H_0: \beta = 0$, L_N is distributed symmetrically about 0.

Unlike the case of (6.2.9) or the least-squares estimator, for $\hat{\beta}_N$ based on L_N, in general, an explicit expression in terms of the x_i and c_i may not be possible. However, an interative procedure may be employed to compute $\hat{\beta}_N$ up to any desired level of accuracy. One may start with the trial solution (6.2.9), compute the corresponding $L_N(b)$, and then, if $L_N(b) > 0$ (< 0), choose a higher (lower) value of b, and by applying the Newton–Raphson or other methods, iterate the estimator until $L_N(b) = 0$ or is closest to it. To illustrate this, we consider the following data, where we assume that the model (6.2.1) holds.

Example 6.2.1. The following table relates to the amount of fertilizer used (c) and the yield of potatotes (X), both in kilograms, in seven equal plots:

c	3	3	4	5.5	6.5	6.5
X	127.6	131.9	132.5	140.0	145.0	150.0

By (6.2.7), the classical least-squares estimate of β is $\hat{\beta}_{N,\mathrm{LS}} = 5.007$. Also, here there are 5 distinct values of c and 19 values of the divided differences entering into (6.2.9). These are 4.90, 5.96, 4.96, 4.98, 6.40, 0.60, 3.80, 3.24, 3.74, 5.18, 7.00, 5.00, 5.00, 7.00, 0.50, 3.66, 7.00, 5.00, and 6.67. Hence, the estimate based on Kendall's tau statistic is the median of this set, $\hat{\beta}_{N,K} = 5.00$. Next, let us consider the estimator $\hat{\beta}_{N,W}$ based on L_N for the Wilcoxon-type scores, that is, $L_N = \sum_{i=1}^{N}(c_i - \bar{c}_N)[R_{Ni} - (N+1)/2] = \sum_{i=1}^{N} c_i[R_{Ni} - (N+1)/2]$. For $b = 5.00$, we have

$X(b)$	112.6	116.9	112.5	114.5	112.5	112.5 117.5
$R(b)$	4	6	2	5	2	2 7

Hence, $L_N(5.00) = 0 + 6 - 8 + 5 - 11 - 13 + 19.5 = -1.5$. Thus, $\hat{\beta}_{N,W} \le 5.00$. We consider the trial value 4.999 and obtain similarly $L_N(4.999) = 30.5 - 29.5 = 1$. Actually, for every $\varepsilon > 0$, $L_N(5 - \varepsilon) > 0$. Thus, by (6.2.5),

$$\hat{\beta}_{N,W} = \tfrac{1}{2}(5.00 + 5.00^{-}) = 5.00.$$

In the special case of the two-sample location problem where $c_1 = \cdots = c_{n_1} = 0$, $c_{n_1+1} = \cdots = c_{n_1+n_2} = 1$, $N = n_1 + n_2$, $n_i \ge 1$, $i = 1, 2$, one can write

$$\hat{\beta}_{N,W} = \text{median}\{X_i - X_j : 1 \le j \le n_1,\ n_1 + 1 \le i \le N\}; \quad (6.2.17)$$

see Hodges and Lehmann (1963) and Sen (1968). But for general scores and/or general c_i, such a simple expression may not be available. However, as in the case of $\hat{\beta}_{N,W}$, the computations may be made for general $\hat{\beta}_N$ by the same iteration method.

6.2.2. Estimation of β_0 When β Is Specified

The statistics T_N, L_N and so on of Section 6.2.1 are invariant under shift, and as a result, the estimation $\hat{\beta}_N$ based on them is also invariant under shift. Thus, we were able to estimate β by $\hat{\beta}_N$, treating β_0 as a nuisance parameter. The situation is somewhat different when we want to estimate β_0; usually the statistics are not invariant under translation by the concomitant variates. Thus, we proceed to consider first the case when β is specified; in the next section, we shall extend it to the case where β is unknown.

Since β is specified, we work with $X_i^* = X_i - \beta(c_i - \bar{c}_N)$, $1 \le i \le N$, where we note that $X_1^*, \ldots, X_N^*$ are i.i.d.r.v.'s with a continuous c.d.f. $F(x - \beta_0)$. Let $S_N = S_N(X_1^*, \ldots, X_N^*)$ be a suitable test statistic for testing $H_0^*: \beta_0 = 0$; S_N may be parametric or nonparametric. We assume that (1) under H_0^*, the distribution of S_N has a specified location μ_N^* [it may be $E(S_N | H_0^*)$ or the point of symmetry of the distribution, if it is assumed to be symmetric], and (2)

$$S_N(a) = S_N(X_1^* - a, \ldots, X_N^* - a) \text{ is nonincreasing in } a \in R. \tag{6.2.18}$$

Then, by an appeal to the same alignment principle as in Section 6.2.1, we define the estimator $\hat{\beta}_{0,N}$ by

$$\tilde{\beta}_{0,N}^{(1)} = \sup\{a: S_N(a) > \mu_N^*\}, \quad \tilde{\beta}_{0,N}^{(2)} = \inf\{a: S_N(a) < \mu_N^*\}; \tag{6.2.19}$$

$$\tilde{\beta}_{0,N} = \tfrac{1}{2}\left(\tilde{\beta}_{0,N}^{(1)} + \tilde{\beta}_{0,N}^{(2)}\right). \tag{6.2.20}$$

First, we consider two particular cases. If we let $S_N(a) = \sum_{i=1}^N (X_i^* - a)$, then under H_0^*, $S_N(0)$ has expectation 0 and also (6.2.18) holds. Hence, by (6.2.19) and (6.2.20), we have here $\tilde{\beta}_{0,N} = (1/N)\sum_{i=1}^N X_i^* = (1/N)\sum_{i=1}^N X_i^* = \bar{X}_N^*$, the classical least-squares estimator. We need to assume that the c.d.f. F has a finite first moment, so that $E(X_1^* | H_0^*) = 0$. Second, let

$$S_N(a) = \sum_{i=1}^N \operatorname{sgn}(X_i^* - a). \tag{6.2.21}$$

Then, for continuous F with $F(0) = \frac{1}{2}$, $E[\operatorname{sgn}(X_i^*)|H_0^*] = \frac{1}{2}(1 - 1) = 0$ and also (6.2.18) holds. Thus, the corresponding estimator is

$$\tilde{\beta}_{0,N}^{(M)} = \text{median}\left[X_i^* : 1 \leq i \leq N\right]. \tag{6.2.22}$$

Here, we need not assume that F has a finite first moment, but only that $F(0) = \frac{1}{2}$.

Our main interest lies in estimating β_0 based on a class of signed rank statistics, and for this purpose, *we assume that F is symmetric about* 0. Consider then

$$S_N = \sum_{i=1}^{N} \operatorname{sgn} X_i^* a_N^*(R_{Ni}^+), \tag{6.2.23}$$

where $a_N^*(1) \leq \cdots \leq a_N^*(N)$ (not all equal) and R_{Ni}^+ is the rank of $|X_i^*|$ among $|X_1^*|, \ldots, |X_N^*|$, $1 \leq i \leq N$. In (6.2.23), on replacing X_i^* by $X_i^* - a$, $1 \leq i \leq N$, we define the corresponding ranks by $R_{Ni}^+(a)$, $1 \leq i \leq N$ and the statistic by $S_N(a)$. Note that $R_{Ni}^+(a)$ is nonincreasing in a if $X_i^* - a > 0$, and nondecreasing in a if $X_i^* - a < 0$, while $\operatorname{sgn}(X_i^* - a)$ is nonincreasing in a. Hence, we have the following lemma, the proof of which is left as an exercise.

Lemma 6.2.3. For nondecreasing $a_N^*(i)$, $S_N(a)$ is nonincreasing in $a \in R$.

Also, note that under $H_0^* : \beta_0 = 0$, we have $\operatorname{sgn} X_i^* = \pm 1$ with probability $\frac{1}{2}$, $(R_{N1}^+, \ldots, R_{NN}^+)$ assumes all possible permutations of $(1, \ldots, N)$ with the common probability $(N!)^{-1}$, and $(\operatorname{sgn} X_1^*, \ldots, \operatorname{sgn} X_N^*)$ and $(R_{N1}^+, \ldots, R_{NN}^+)$ are stochastically independent. Therefore, under H_0^*, $S_N(X_1^*, \ldots, X_N^*)$ and $S_N(-X_1^*, \ldots, -X_N^*)$ have the same distribution, that is, S_N is distributed symmetrically about 0. Hence, we can use the estimator $\hat{\beta}_{0,N}$ defined by (6.2.19)–(6.2.20), where $\mu_N^* = 0$.

In particular, if we let $a_N^*(i) = i/(N + 1)$, $1 \leq i \leq N$, (i.e., we use the Wilcoxon signed rank statistic), we may rewrite S_N as

$$\frac{2}{N+1} \sum_{1 \leq i \leq j \leq N} \operatorname{sgn}\left(X_i^* + X_j^*\right), \tag{6.2.24}$$

so that we obtain from (6.2.19), (6.2.20), and (6.2.24) that

$$\tilde{\beta}_{0,N}^{(W)} = \text{median}\left\{\tfrac{1}{2}\left(X_i^* + X_j^*\right) : 1 \leq i \leq j \leq N\right\}. \tag{6.2.25}$$

For general scores, an explicit expression in terms of the X_i^*, as in (6.2.25), may not be possible, and we may have to apply an iterative procedure as we did in Section 6.2.1.

6.2.3. Estimation of β_0 When β Is Unknown

We estimate β as in Section 6.2.1 and denote the estimator by $\hat{\beta}_N$. Consider then the aligned observations and the corresponding statistic, defined by

$$\tilde{X}_i = X_i - \hat{\beta}_N(c_i - \bar{c}_N), \quad 1 \le i \le N, \qquad \tilde{S}_N = S_N(\tilde{X}_1, \dots, \tilde{X}_N). \quad (6.2.26)$$

We consider the estimator $\hat{\beta}_{0,N}$ defined by

$$\hat{\beta}_{0,N}^{(1)} = \sup\{a : \tilde{S}_N(a) > \mu_N^*\}, \qquad \hat{\beta}_{0,N}^{(2)} = \inf\{a : \tilde{S}_N(a) < \mu_N^*\}, \quad (6.2.27)$$

$$\hat{\beta}_{0,N} = \tfrac{1}{2}(\hat{\beta}_{0,N}^{(1)} + \hat{\beta}_{0,N}^{(2)}), \quad (6.2.28)$$

$\tilde{S}_N(a) = S_N(\tilde{X}_1 - a, \dots, \tilde{X}_N - a)$ being assumed to be nonincreasing in a. The rest of the discussion follows as in Section 6.2.2. We may note that while the least-squares estimator $\overline{X}_N$ does not depend on β (or $\hat{\beta}_N$), the other ones usually do.

For the data of Example 6.2.1, we note that the least-squares estimator of β_0 is 114.14, while the median estimator (6.2.22) is 112.6. Let us consider the 28 midranges $\frac{1}{2}(X_i + X_j)$, $1 \le i \le j \le 7$. These are 112.60, 114.75, 112.55, 113,55, 112.55, 112.55, 115.05, 116.9, 114.70, 115.70, 114, 70, 114.70, 117.2, 112.5, 113.0, 112.5, 112.5, 115.0, 114.5, 113.5, 113.5, 116.0, 112.5, 112.5, 115.0, 112.5, 115.0, 117.5. Thus $\hat{\beta}_0^{(w)} = \frac{1}{2}(113.55 + 114.5) = 114.025$. For general scores, again an iterative procedure has to be employed.

6.2.4. Invariance and Symmetry of the Estimators

For every real a, b, and d, let $\mathbf{X}_N = (X_1, \dots, X_N)'$, and

$$\mathbf{X}_N(a, b, d) = d[\mathbf{X}_N + a\mathbf{1}_N + b(\mathbf{c}_N - \bar{c}_N\mathbf{1}_N)]. \quad (6.2.29)$$

The estimators $\hat{\beta}_N$ and $\hat{\beta}_{0,N}$, defined by (6.2.6) and (6.2.28), but based on $\mathbf{X}_N(a, b, d)$, are denoted by $\hat{\beta}_N(a, b, d)$ and $\hat{\beta}_{0,N}(a, b, d)$, respectively.

Lemma 6.2.5. For every real a, b, and d, for the estimators based on L_N and S_N in (6.2.10) and (6.2.23),

$$\hat{\beta}_N(a, b, d) = d\hat{\beta}_N(0, 0, 1) + db, \quad (6.2.30)$$

$$\hat{\beta}_{0,N}(a, b, d) = d\hat{\beta}_{0,N}(0, 0, 1) + ad. \quad (6.2.31)$$

Proof. The proof is trivial for $d = 0$. Let $d > 0$. Then, by (6.2.10), for every a, b, b_1, and d,

$$\begin{aligned} L_N(\mathbf{X}_N(a,b,d) - b_1(\mathbf{c}_N - \bar{c}_N\mathbf{1}_N)) &= L_N\big(\mathbf{X}_N\big(a, b - d^{-1}b_1, d\big)\big) \\ &= L_N\big(\mathbf{X}_N\big(0, b - d^{-1}b_1, 1\big)\big) \\ &= L_N\big(d^{-1}b_1 - b\big), \qquad (6.2.32) \end{aligned}$$

as $R_{N1}, \ldots, R_{NN}$ remain invariant when the X_i are all multiplied by a common factor d (> 0) or translated by a common shift a. Then (6.2.30) follows directly from (6.2.5), (6.2.6), and (6.2.32) when $d > 0$. The case of $d < 0$ follows analogously. Also, by (6.2.26) and (6.2.30), for every real a, b, and d,

$$\begin{aligned} \hat{X}_N(a,d) &= \mathbf{X}_N(a,b,d) - \hat{\beta}_N(a,b,d)(\mathbf{c}_N - \bar{c}_N\mathbf{1}_N) \\ &= d\{\mathbf{X}_N + a\mathbf{1}_N - \hat{\beta}_N(0,0,1)(\mathbf{c}_N - \bar{c}_N\mathbf{1}_N)\} \\ &= d\hat{\mathbf{X}}_N + ad\mathbf{1}_N, \\ \hat{\mathbf{X}}_N &= \big(X_1 - \hat{\beta}_N(c_1 - \bar{c}_N), \ldots, X_N - \hat{\beta}_N(c_N - \bar{c}_N)\big)'. \end{aligned} \qquad (6.2.33)$$

Further, for every a, a_1, and d ($\neq 0$),

$$\begin{aligned} S_N(\hat{\mathbf{X}}_N(a,d) - a_1\mathbf{1}_N) &= S_N\big(\hat{\mathbf{X}}_N\big(a - d^{-1}a_1, d\big)\big) \\ &= S_N\big(\hat{\mathbf{X}}_N\big(a - d^{-1}a_1, 1\big)\big) = \hat{S}_N\big(d^{-1}a_1 - a\big), \end{aligned} \qquad (6.2.34)$$

as the ranks $R^+_{N1}, \ldots, R^+_{NN}$ remain invariant when all the observations are multiplied by a common factor d ($\neq 0$). Thus, (6.2.31) follows from (6.2.27), (6.2.28), (6.2.33), and (6.2.34). The proof follows.

In passing, we may remark that (6.2.30)–(6.2.31) also hold for the classical least-squares estimator.

In the next theorem, we establish, under appropriate regularity conditions, the symmetry of the distributions of $\hat{\beta}_N$ and $\hat{\beta}_{0,N}$ around β and β_0, respectively.

Theorem 6.2.5. *For L_N in (6.2.10), assume that $a_N(1) \leq \cdots \leq a_N(N)$ (not all equal) and that either of the following holds: (1) $c_i + c_{N-i+1} = 2\bar{c}_N$, $1 \leq i \leq N$, or (2) $a_N(i) + a_N(N-i+1) = \text{const}$, $1 \leq i \leq N$; for S_N in*

(6.2.23), *assume that* $a_N^*(1) \leq \cdots \leq a_N^*(N)$ (*not all equal*). *Then the symmetry of* F *insures that the distributions of* $\hat{\beta}_N$ *and* $\hat{\beta}_{0,N}$ *are also symmetric around* β *and* β_0, *respectively*.

Proof. By virtue of Lemma 6.2.4, without any loss of generality, we may assume that $\beta_0 = \beta = 0$, so that F is symmetric about 0. Then both $\mathbf{X}_N$ and $(-1)\mathbf{X}_N$ have the same joint distribution, so that $\hat{\beta}_N(\mathbf{X}_N)$ and $\hat{\beta}_N((-1)\mathbf{X}_N)$ both have the same d.f., and the same holds for $\hat{\beta}_{0,N}(\mathbf{X}_N)$ and $\hat{\beta}_{0,N}((-1)\mathbf{X}_N)$. Thus to prove the theorem, it suffices to show that (1) $\hat{\beta}_N((-1)\mathbf{X}_N) = -\hat{\beta}_N(\mathbf{X}_N)$ and (2) $\hat{\beta}_{0,N}((-1)\mathbf{X}_N) = -\hat{\beta}_{0,N}(\mathbf{X}_N)$. Now, under the hypotheses of the theorem, on noting that if X_i has rank R_i among $X_1, \dots, X_N$, then among $-X_1, \dots, -X_N$, the rank of $-X_i$ is $N + 1 - R_i$, $1 \leq i \leq N$, we have $L_N(\mathbf{X}_N) = -L_N((-1)\mathbf{X}_N)$. Hence, by (6.2.5), $\hat{\beta}_N((-1)\mathbf{X}_N) = -\hat{\beta}_N(\mathbf{X}_N)$. Similarly, $S_N((-1)\mathbf{X}_N) = -S_N(\mathbf{X}_N)$, as the signs are inverted but the ranks R_{Ni}^+, $1 \leq i \leq N$, remain the same. Hence, here also $\hat{\beta}_{0,N}((-1)\mathbf{X}_N) = -\hat{\beta}_{0,N}(\mathbf{X}_N)$.

6.2.5. Asymptotic Normality of the Estimators

First consider the case of $\hat{\beta}_N$. We assume that

$$\max_{1 \leq i \leq N} N^{-1}(c_i - \bar{c}_N)^2 \to 0 \quad \text{as} \quad N \to \infty, \tag{6.2.35}$$

$$0 < \lim_{N \to \infty} N^{-1} \sum_{i=1}^{N} (c_i - \bar{c}_N)^2 = C_0^2 < \infty. \tag{6.2.36}$$

Also, let $P_N^{(b)}$ denote the probability under the hypothesis that $\beta = N^{-1/2}b$, so that $P_0 = P_N^{(0)}$ represents the null situation. Then, by (6.25) and (6.2.10),

$$\begin{aligned} P\{N^{1/2}[\hat{\beta}_{N,2} - \beta] \leq b\} &= P_0\{\hat{\beta}_{N,2} \leq N^{-1/2}b\} \\ &= P_0\{L_N(N^{-1/2}b) < 0\} \\ &\leq P_N^{(-b)}\{L_N(0) \leq 0\} \\ &= P_N^{(-b)}\{N^{-1/2}L_N(0) \leq 0\}, \end{aligned} \tag{6.2.37}$$

as $L_N(N^{-1/2}b)$, under $H_0: \beta = 0$, has the same distribution as of $L_N(0)$ under $\beta = N^{-1/2}b$. As in Chapter 2, we assume that $a_N(i) = E\phi(U_{Ni})$ or $\phi(i/(N + 1))$, $1 \leq i \leq N$, where ϕ is absolutely continuous and square-integrable inside (0, 1). Without any loss of generality, we let $\bar{\phi} = \int_0^1 \phi(u)\, du = 0$.

Let then

$$\psi(u) = -\frac{f'(F^{-1}(u))}{f(F^{-1}(u))}, \qquad 0 < u < 1,$$

$$A_\psi^2 = \int_0^1 \psi^2(u)\,du, \tag{6.2.38}$$

and

$$\rho^2(\phi,\psi) = \frac{\left(\int_0^1 \phi(u)\psi(u)\,du\right)^2}{A_\psi^2 A_\phi^2}, \qquad A_\phi^2 = \int_0^1 \phi^2(u)\,du, \tag{6.2.39}$$

where we assume that F is absolutely continuous with a finite Fisher information A_ψ^2. Then, under $\beta = -N^{-1/2}b$, $N^{1/2}L_N(0)$ is asymptotically normal with mean $bC_0^2\rho(\phi,\psi)A_\phi A_\psi$ and variance $C_0^2A_\phi^2$. Thus, if $\Phi(x)$ is the standard normal c.d.f., the rhs of (6.2.37) converges to $\Phi(bC_0\rho(\phi,\psi)A_\psi)$ for every real b. The same limiting result holds for $\hat{\beta}_{N,1}$. Hence, from (6.2.5), (6.2.6), and the above we arrive at the following theorem.

Theorem 6.2.6. *Under the assumptions* (6.2.35) *and* (6.2.36), *for every real* b,

$$\lim_{N\to\infty} P\{N^{1/2}(\hat{\beta}_N - \beta) \le b\} = \Phi\big(bC_0\rho(\phi,\psi)A_\psi\big), \tag{6.2.40}$$

that is, $N^{1/2}(\hat{\beta}_N - \beta)$ *is asymptotically normal with* 0 *mean and variance*

$$\left\{A_\psi^2\rho^2(\phi,\psi)C_0^2\right\}^{-1}. \tag{6.2.41}$$

Consider then the estimator $\tilde{\beta}_{0,N}$, defined by (6.2.20). Here, for S_N, defined by (6.2.23), we let $a_N^*(i) = E\phi^*(U_{Ni})$ or $\phi^*(i/(N+1))$, $1 \le i \le N$, and assume that $\phi^*(u) = \phi((1+u)/2)$, $0 < u < 1$; $\phi(u) + \phi(1-u) = 0$, $0 < u < 1$; and ϕ satisfies the conditions of Theorem 6.2.6. Then, for symmetric F, the following result holds: for every real a,

$$\lim_{N\to\infty} P\{N^{1/2}(\tilde{\beta}_{0,N} - \beta_0) \le a\} = \Phi\big(a\rho(\phi,\psi)A_\psi\big), \tag{6.2.42}$$

that is, $N^{1/2}(\tilde{\beta}_{0,N} - \beta_0)$ is asymptotically normal with 0 mean and variance

$$\left\{A_\psi^2\rho^2(\phi,\psi)\right\}^{-1}. \tag{6.2.43}$$

The proof is very similar to that of the preceding theorem and hence is left as an exercise (see Exercise 6.2.7).

Finally, we proceed to study the asymptotic normality of $\hat{\beta}_{0,N}$ as well as the joint normality of $(\hat{\beta}_N, \hat{\beta}_{0,N})$. In this context, first we present the following results, which are proved in somewhat more generality in the Appendix (see Sections A.4.1–A.4.2):

1. For every real K $(0 < K < \infty)$, under (6.2.1), (6.2.35) and (6.2.36), with $\beta = 0$, as $N \to \infty$,

$$\sup_{b:|b| \le K} \left| L_N(N^{-1/2}b) - L_N(0) + N^{1/2} b C_0^2 \rho(\phi, \psi) A_\phi A_\psi \right| = o_p(N^{1/2}). \tag{6.2.44}$$

2. Consider the usual signed rank statistic S_N based on the observations $X_i - a - b(c_i - \bar{c}_N)$, $1 \le i \le N$, and denote it by $S_N(a, b)$. Assume that (1) F is symmetric, (2) $\phi^*(u) = \phi((1+u)/2)$, $0 < u < 1$, and ϕ satisfies the conditions of Theorem 6.2.6. Then, for every real K $(0 < K < \infty)$, under (6.2.1) with $\beta_0 = \beta = 0$, as $N \to \infty$,

$$\sup\Big\{ \left| N^{-1/2} \{ S_N(N^{-1/2}a, N^{-1/2}b) - S_N(0,0) \} + a\rho(\phi, \psi) A_\phi A_\psi \right| :$$
$$|a| \le K, |b| \le K \Big\} \to 0 \quad \text{in probability.} \tag{6.2.45}$$

Then we have the following theorem.

Theorem 6.2.7. *Under the above assumptions, $N^{1/2}(\hat{\beta}_N - \beta, \hat{\beta}_{0,N} - \beta_0)$ has asymptotically a bivariate normal distribution with null mean vector and dispersion matrix*

$$\left\{ A_\psi^2 \rho^2(\phi, \psi) \right\}^{-1} \operatorname{diag}\left[C_0^{-2}, 1 \right]. \tag{6.2.46}$$

Proof. Note that if F is symmetric and $\beta_0 = \beta = 0$ in (6.2.1), then $N^{-1/2}(L_N(0), S_N(0,0))$ has asymptotically a bivariate normal distribution with null mean vector and dispersion matrix $A_\phi^2 \operatorname{diag}[C_0^2, 1]$. Also, for every real a, b,

$$P\{N^{1/2}(\hat{\beta}_{N,2} - \beta) \le b,\ N^{1/2}(\hat{\beta}^{(2)}_{0,N} - \beta_0) \le a\}$$

$$= P\{\hat{\beta}_{N,2} \le N^{-1/2}b,\ \hat{\beta}^{(2)}_{0,N} \le N^{-1/2}a | \beta_0 = \beta = 0\}$$

$$= P\{N^{-1/2}L_N(N^{-1/2}b) \le 0,$$

$$N^{-1/2}S_N(N^{-1/2}a, \hat{\beta}_N) \le 0 | \beta_0 = \beta = 0\}. \tag{6.2.47}$$

Further, by Theorem 6.2.6, $N^{1/2}|\hat{\beta}_N - \beta| = O_p(1)$, so that when $\beta = 0$, $\hat{\beta}_N = O_p(N^{-1/2})$, Hence, by (6.2.44), (6.2.45), and (6.2.47), the rhs of (6.2.47) reduces (as $N \to \infty$) to

$$P\{N^{-1/2}L_N(0) \le bC_0^2\rho(\phi, \psi)A_\phi A_\psi,$$

$$N^{-1/2}S_N(0,0) \le a\rho(\phi, \psi)A_\phi A_\psi | \beta = \beta_0 = 0\}, \tag{6.2.48}$$

which converges to

$$\Phi(bC_0\rho(\phi, \psi)A_\psi)\Phi(a\rho(\phi, \psi)A_\psi), \tag{6.2.49}$$

and hence the result follows.

Remark. If in (6.2.1), $F_i(x) = F(x - \beta^0 - \beta c_i)$, $i \ge 1$, and $\lim_{N\to\infty} N^{-1}\sum_{i=1}^N c_i = \bar{c}$ exists, then in (6.2.26) we define $\tilde{X}_i = X_i - \hat{\beta}_N c_i$, $1 \le i \le N$, so that $\hat{\beta}_{0,N}$ in (6.2.28) will estimate β^0. In this case, the only change in Theorem 6.2.7 is in (6.2.46), which should be

$$\{A_\psi^2\rho^2(\phi, \psi)\}^{-1}\begin{pmatrix} C_0^{-2} & -\bar{c}C_0^{-2} \\ -\bar{c}C_0^{-2} & 1 + \bar{c}^2/C_0^2 \end{pmatrix}. \tag{6.2.50}$$

6.2.6. Asymptotic Relative Efficiency of the Estimators

We shall now study the ARE of the estimators of (β_0, β). For this purpose, we make use of the following definitions.

Definition 6.2.6.1. Let $\{\mathbf{U}_N\}$ and $\{V_N\}$ be two sequences of estimators of a common parameter (vector) $\boldsymbol{\theta}$, $\psi(N)$ be a nondecreasing sequence of positive numbers with $\lim_{N\to\infty}\psi(N) = \infty$, and $\{N^* = N^*(N)\}$ be a sequence of positive integers such that

$$\psi(N^*)[\mathbf{U}_{N^*} - \boldsymbol{\theta}] \xrightarrow{\mathscr{L}} \mathbf{Z} \quad \text{and} \quad \psi(N)[\mathbf{V}_N - \boldsymbol{\theta}] \xrightarrow{\mathscr{L}} \mathbf{Z}, \tag{6.2.51}$$

where $\mathbf{Z}$ has a nondegenerate c.d.f. $G(\mathbf{z})$, $\mathbf{z} \in R^p$, $p \geq 1$. Then the ARE of $\{\mathbf{V}_N\}$ with respect to $\{\mathbf{U}_N\}$ is defined by

$$e_1 = \lim_{N\to\infty} \frac{N^*(N)}{N}, \tag{6.2.52}$$

provided the limit exists.

An immediate consequence of this definition is the following proposition.

Proposition 6.2.6.1. If for some k, $0 \leq k < \infty$,

$$\mathscr{L}\big(N^{1/2}(\mathbf{U}_N - \boldsymbol{\theta})\big) \to N_p(\mathbf{0}, \boldsymbol{\Sigma}) \quad \text{and} \quad \mathscr{L}\big(N^{1/2}(\mathbf{V}_N - \theta)\big) \to N_p(\mathbf{0}, k\boldsymbol{\Sigma}), \tag{6.2.53}$$

then $e_1 = e(\mathbf{V}, \mathbf{U}) = k^{-1}$.

The proof is left as an exercise (Exercise 6.2.9). In particular, if $p = 1$, both the dispersion matrices are scalar, so that e_1 reduces to the reciprocal of the ratio of the asymptotic variances. For $p \geq 2$, the proportionality of the two asymptotic dispersion matrices may not be generally true. In such a case, we make use of the following definition; here, for a nonsingular dispersion matrix $\boldsymbol{\Sigma}$, the *generalized variance* is defined as the determinant $|\boldsymbol{\Sigma}|$ of $\boldsymbol{\Sigma}$ (Cramér 1946).

Definition 6.2.6.2. Suppose that in the setup of Definition 6.2.6.1,

$$\psi(N^*)[\mathbf{U}_{N^*} - \boldsymbol{\theta}] \xrightarrow{\mathscr{L}} \mathbf{Z}_1 \quad \text{and} \quad \psi(N)[\mathbf{V}_N - \boldsymbol{\theta}] \xrightarrow{\mathscr{L}} \mathbf{Z}_2, \tag{6.2.54}$$

where $\mathbf{Z}_j$ has a nonsingular dispersion matrix $\boldsymbol{\Sigma}_j$, $j = 1, 2$, such that

$$|\boldsymbol{\Sigma}_1| = |\boldsymbol{\Sigma}_2|. \tag{6.2.55}$$

Then the ARE of $\{\mathbf{V}_N\}$ with respect to $\{\mathbf{U}_N\}$ is defined by (6.2.52).

An immediate consequence of this definition is the following proposition.

Proposition 6.2.6.2. If for some nonsingular $\boldsymbol{\Sigma}_1$ and $\boldsymbol{\Sigma}_2$

$$N^{1/2}[\mathbf{U}_N - \boldsymbol{\theta}] \xrightarrow{\mathscr{L}} \mathscr{N}_p(\mathbf{0}, \boldsymbol{\Sigma}_1) \quad \text{and} \quad N^{1/2}[\mathbf{V}_N - \boldsymbol{\theta}] \xrightarrow{\mathscr{L}} \mathscr{N}_p(\mathbf{0}, \boldsymbol{\Sigma}_2), \tag{6.2.56}$$

then the ARE of $\mathbf{V}_N$ *with respect to* $\mathbf{U}_N$ *(in the sense of Definition* 6.2.6.2) *is given by*

$$e_2 = e_2(V, U) = \left(\frac{|\boldsymbol{\Sigma}_1|}{|\boldsymbol{\Sigma}_2|} \right)^{1/p} = |\boldsymbol{\Sigma}_1 \boldsymbol{\Sigma}_2^{-1}|^{1/p}. \tag{6.2.57}$$

The proof is left as an exercise (see Exercise 6.2.10).

For the model (6.2.1), the classical least-squares estimator of (β_0, β) is given by $(\check{\beta}_{0,N}, \check{\beta}_N)$, where as in (6.2.7),

$$\check{\beta}_{0,N} = \overline{X}_N \quad \text{and} \quad \check{\beta}_N = C_N^{-2} \sum_{i=1}^{N} (c_i - \bar{c}_N)(X_i - \overline{X}_N). \tag{6.2.58}$$

At this stage, we assume that F in (6.2.1) possesses a finite and positive variance σ^2. Then, if we write

$$X_i = \beta_0 + \beta(c_i - \bar{c}_N) + e_i, \qquad 1 \le i \le N, \tag{6.2.59}$$

the e_i are i.i.d.r.v.'s with mean 0 and variance σ^2. From (6.2.58) and (6.2.59), we obtain that

$$(\check{\beta}_{0,N} - \beta_0) = N^{-1} \sum_{i=1}^{N} e_i \quad \text{and} \quad (\check{\beta}_N - \beta) = \sum_{i=1}^{N} c_{Ni}^* e_i, \tag{6.2.60}$$

where $c_{Ni}^* = (c_i - \bar{c}_N) / \sum_{i=1}^{N} (c_i - \bar{c}_N)^2$, $1 \le i \le N$.

In this context, we make use of the following special central limit theorem, proved in Hájek and Šidák (1967, p. 153): If $\{a_{Ni}, 1 \le i \le N\}$, $N \ge 1$ is a double sequence of real numbers such that $\max_{1 \le i \le N} a_{Ni}^2 / \sum_{i=1}^{N} a_{Ni}^2 \to 0$ as $N \to \infty$, then, for a sequence $\{e_i, i \ge 1\}$ of i.i.d.r.v.'s with mean 0 and variance σ^2,

$$\left(\sum_{i=1}^{N} a_{Ni}^2 \right)^{-1/2} \sum_{i=1}^{N} a_{Ni} e_i \xrightarrow{\mathscr{L}} \mathscr{N}(0, \sigma^2). \tag{6.2.61}$$

By assuming (6.2.35) and (6.2.36), and considering a linear combination $N^{1/2}[\lambda_1(\check{\beta}_{0,N} - \beta_0) + \lambda_2(\check{\beta}_N - \beta)]$ [where $(\lambda_1, \lambda_2) \neq \mathbf{0}$], we obtain by using the above theorem and (6.2.60) that

$$N^{1/2}((\check{\beta}_{0,N} - \beta_0), (\check{\beta}_N - \beta)) \xrightarrow{\mathscr{L}} \mathscr{N}_2(\mathbf{0}, \sigma^2 \operatorname{diag}(1, C_0^{-2})). \tag{6.2.62}$$

Using Theorem 6.2.7, (6.2.62), and Proposition 6.2.6.1, we find that the ARE of $(\hat{\beta}_{0,\mathrm{N}}, \hat{\beta}_N)$ with respect to $(\check{\beta}_{0,N}, \check{\beta}_N)$ is given by

$$e(F) = \sigma^2\rho^2(\phi, \psi)A_\psi^2, \tag{6.2.63}$$

where A_ψ^2 and $\rho^2(\phi, \psi)$ are defined by (6.2.38) and (6.2.39).

In (5.7.5), we have discussed the ARE of rank-order tests with respect to the classical analysis of variance tests, and it agrees with our (6.2.63). Thus, in this case, the ARE results are shared by both the tests and the derived estimates. However, this is not true in general, and the ARE may be more complicated to express. For this situation, we consider the following estimate, classically known as the Brown–Mood (1950) estimator.

Corresponding to the set of points (c_i, X_i), $1 \le i \le N$, let $\tilde{c}_N$ be the median of $(c_1, \ldots, c_N)$, defined in the usual way. Then, the estimator $(\dot{\beta}_N, \dot{\beta}_{0N})$ of (β, β_0) is the solution of the equations

$$\operatorname*{median}_{\{i:\, c_i \le \tilde{c}_N\}} \{X_i - \dot{\beta}_{0N} - \dot{\beta}_N(c_i - \tilde{c}_N)\} = 0 = \operatorname*{median}_{\{i:\, c_i > \tilde{c}_N\}} \{X_i - \dot{\beta}_{0N} - \dot{\beta}_N(c_i - \tilde{c}_N)\}. \tag{6.2.64}$$

As in the proof of Theorem 6.2.1, we may let $c_1 \le \cdots \le c_N$ with at least one strict inequality sign and note that given the (c_i, X_i), $1 \le i \le n$, $X_i - a - b(c_i - \tilde{c}_N)$ is decreasing in a and is increasing or decreasing in b according as $c_i \le \tilde{c}_N$ or $> \tilde{c}_N$. We rewrite $X_i - a - b(c_i - \bar{c}_N)$ as $X_i - a' - b(c_i - \tilde{c}_N)$, where $a' = a + b(\tilde{c}_N - \overline{C}_N)$. Note that $X_i - a' - b(c_i - \tilde{c}_N)$ is decreasing in a' and is increasing or decreasing in b according as $c_i \le \tilde{c}_N$ or $> \tilde{c}_N$, and it is a continuous function of (a', b). For any given a', $X_i - a' - b(c_i - \tilde{c}_N)$ continuously and monotonically varies from $-\infty$ to $+\infty$ (as b varies from $-\infty$ to $+\infty$) if $c_i < \tilde{c}_N$, and from $+\infty$ to $-\infty$ if $c_i > \tilde{c}_N$. Hence, the two functions median$\{X_i - a' - b(c_i - \tilde{c}_N) : c_i \le \tilde{c}_N\}$

and median$\{X_i - a' - b(c_i - \tilde{c}_N) : c_i > \tilde{c}_N\}$ intersect at a point b_0, where

$$\text{median}\{X_i - a' - b_0(c_i - \tilde{c}_N) : c_i \le \tilde{c}_N\} = h(a')$$
$$= \text{median}\{X_i - a' - b_0(c_i - \tilde{c}_N) : c_i > \tilde{c}_N\} = h(a'). \quad (6.2.65)$$

Letting then $a = : a_0 + b_0(\tilde{c}_N - \bar{c}_N) = h(a')$, we note that for (a_0, b_0), (6.2.64) holds. Hence, a solution for $(\dot{\beta}_N, \dot{\beta}_{0N})$ always exists.

Hill (1962) has established the asymptotic normality of $N^{1/2}(\dot{\beta}_N - \beta, \dot{\beta}_{0N} - \beta_0)$ and has shown that the asymptotic mean vector is $\mathbf{0}$ and the dispersion matrix is $\mathbf{T} = ((\tau_{\alpha\beta}))$, where

$$\mathbf{T} = [f(0)(\gamma_1 - \gamma_2)]^{-2} \begin{pmatrix} \frac{1}{4} & -\frac{1}{4}(\gamma_1 + \gamma_2) \\ -\frac{1}{4}(\gamma_1 + \gamma_2) & \frac{1}{2}(\gamma_1^2 + \gamma_2^2) \end{pmatrix} \quad (6.2.66)$$

and $\gamma_1 = \int_{1/2}^{1} h(t)\,dt$, $\gamma_2 = \int_0^{1/2} h(t)\,dt$. We assume that $c_i = h(i/N)$, $1 \le i \le N$, where $h(t)$ is a strictly monotonic and continuous function on $(0, 1)$ with $h(0) = h_0 < h_1 = h(1)$. We leave the details as an exercise (see Exercise 6.2.12).

For comparing $(\dot{\beta}_N, \dot{\beta}_{0N})$ with the other competing estimators, Definition 6.2.6.1 with (6.2.51)–(6.5.52) is no longer applicable, and hence, we propose to use (6.2.57). Since $c_i = h(i/N)$, $1 \le i \le N$, here we have $C_0^2 = \int_0^1 h^2(t)\,dt - (\int_0^1 h(t)\,dt)^2$, so that by (6.2.46), the generalized variance for $N^{1/2}(\hat{\beta}_N - \beta, \hat{\beta}_{0N} - \beta_0)$ is

$$\{A_\psi^2 \rho^2(\phi, \psi)\}^2 \left[\int_0^1 h^2(t)\,dt - \left(\int_0^1 h(t)\,dt\right)^2\right]^{-1}. \quad (6.2.67)$$

On the other hand, by (6.2.66),

$$\|\mathbf{T}\| = \frac{1}{16 f^4(0)(\gamma_1 - \gamma_2)^2}. \quad (6.2.68)$$

Hence, the ARE of $(\dot{\beta}_N, \dot{\beta}_{0N})$ with respect to $(\hat{\beta}_N, \hat{\beta}_{0N})$ is given by

$$e_2([\dot{\beta}, \dot{\beta}_0], [\hat{\beta}, \hat{\beta}_0]) = \left[\frac{4 f_{(0)}^2}{A_\psi^2 \rho^2(\phi, \psi)}\right] \left[\frac{(\gamma_1 - \gamma_2)^2}{\int_0^1 h^2(t)\,dt - \left(\int_0^1 h(t)\,dt\right)^2}\right]^{1/2}$$

(6.2.69a)

Similarly,

$$e_2([\dot{\beta}, \dot{\beta}_0], [\bar{\beta}, \bar{\beta}_0]) = 4\sigma^2 f^2(0) \left[\frac{(\gamma_1 - \gamma_2)^2}{\int_0^1 h^2(t)\,dt - \left(\int_0^1 h(t)\,dt\right)^2} \right]^{1/2}. \tag{6.2.69b}$$

In the special case of the Wilcoxon scores (i.e., $\phi(u) \equiv u$), (6.2.69a) reduces to

$$\frac{f^2(0)}{3\left(\int_{-\infty}^{\infty} f^2(x)\,dx\right)^2} \left\{ \frac{(\gamma_1 - \gamma_2)^2}{\int_0^1 h^2(t)\,dt - \left(\int_0^1 h(t)\,dt\right)^2} \right\}^{1/2} \tag{6.2.70}$$

and in particular when $h(t) = a + bt$, $0 \le t \le 1$, (6.2.70) equals $\frac{1}{2}\sqrt{3}\,\{f^2(0)/3(\int_{-\infty}^{\infty} f^2(x)\,dx)^2\}$, which is $< f^2(0)/3(\int_{-\infty}^{\infty} f^2(x)\,dx)^2$, which is the ARE of the sign test with respect to the signed rank test. The loss in the ARE (due to the additional factor $\frac{1}{3}\sqrt{3}$ is due to the fact that the partitioning into two ordered subsets may not utilize the full information in the combined set.

In computing the ARE (6.2.63), we have assumed that (6.2.36) holds. The latter assumption may be replaced by a weaker one if we are interested only in the marginal ARE of $\hat{\beta}_N$ with respect to $\check{\beta}_N$. Suppose we assume that $C_N^2 = \sum_{i=1}^N (c_i - \bar{c}_N)^2 \to \infty$ as $N \to \infty$ and that for every $a: 0 < a < \infty$,

$$C_N^2 = Q(N) \quad \text{and} \quad \lim_{N \to \infty} \frac{Q(aN)}{Q(N)} = s(a) \text{ exists}, \tag{6.2.71}$$

where $s(a)$ is increasing in a and $s(1) = 1$. For example, if $c_i = b_0 + ib$, $i \ge 1$, $b > 0$, then $s(a) = a^3$. One can then virtually repeat the proof of Theorem 6.2.6 and show that as $N \to \infty$,

$$C_N(\hat{\beta}_N - \beta) \xrightarrow{\mathscr{L}} \mathscr{N}\left(0, \left[A_\psi^2 \rho^2(\phi, \psi)\right]^{-1}\right) \tag{6.2.72}$$

and

$$C_N(\check{\beta}_N - \beta) \xrightarrow{\mathscr{L}} \mathscr{N}(0, \sigma^2); \tag{6.2.73}$$

the details are left as an exercise (see Exercise 6.2.14). Then, by (6.2.51), (6.2.52), (6.2.71), (6.2.72), and (6.2.73), we have

$$e(\hat{\beta}, \check{\beta}) = s^{-1}\left(\sigma^2 A_\psi^2 \rho^2(\phi, \psi)\right). \tag{6.2.74}$$

For convex $s(a)$ with $s(1) = 1$, (6.2.74) usually moves closer to unity.

6.3. ESTIMATION IN MULTIPLE REGRESSION MODELS

As a natural extension of the model (6.2.1) we consider the following. Let $X_1, \ldots, X_N$ be independent r.v.'s with continuous c.d.f.'s $F_1(x), \ldots, F_N(x)$, respectively, all defined on $(-\infty, \infty)$, where

$$F_i(x) = F(x - \beta_0 - \boldsymbol{\beta}'(\mathbf{c}_i - \bar{\mathbf{c}}_N)), \qquad 1 \le i \le N, \tag{6.3.1}$$

β_0 and $\boldsymbol{\beta} = (\beta_1, \ldots, \beta_q)'$ are unknown parameters, $\mathbf{c}_i' = (c_{i1}, \ldots, c_{iq})$, $i = 1, \ldots, N$ are known q-vectors ($q \ge 1$) and $\bar{\mathbf{c}}_N = N^{-1}\Sigma_{i=1}^N \mathbf{c}_i$. Here also, based on suitable rank statistics, we want to estimate β_0 and $\boldsymbol{\beta}$.

First, consider the estimation of $\boldsymbol{\beta}$. To motivate, we start with the classical least-squares estimator $\check{\boldsymbol{\beta}}_N$ of $\boldsymbol{\beta}$. As in (6.2.7), we now consider the q-vector

$$\mathbf{T}_N(\mathbf{b}) = \sum_{i=1}^N (\mathbf{c}_i - \bar{\mathbf{c}}_N)\left[X_i - (\mathbf{c}_i - \bar{\mathbf{c}}_N)'\mathbf{b}\right] \tag{6.3.2}$$

where $\mathbf{b}' = (b_1, \ldots, b_q) \in R^q$. For $\mathbf{b} = \boldsymbol{\beta}$, $X_i - (\mathbf{c}_i - \bar{\mathbf{c}}_N)'\boldsymbol{\beta}$, $1 \le i \le N$, are i.i.d.r.v.'s with the c.d.f. $F(x - \beta_0)$, and hence $E\mathbf{T}_N(\boldsymbol{\beta}) = \mathbf{0}$. Thus, by the same alignment procedure as in Section 6.2, we are inclined to choose an estimator $\check{\boldsymbol{\beta}}_N$ of $\boldsymbol{\beta}$ for which $\mathbf{T}_N(\check{\boldsymbol{\beta}}_N)$ is closest to $\mathbf{0}$. Note that by (6.3.2),

$$\mathbf{T}_N(\mathbf{b}) = \mathbf{T}_N(0) - \mathbf{C}_N\mathbf{b}; \qquad \mathbf{C}_N = \sum_{i=1}^N (\mathbf{c}_i - \bar{\mathbf{c}}_N)(\mathbf{c}_i - \bar{\mathbf{c}}_N)'. \tag{6.3.3}$$

Thus, if $\mathbf{C}_N$ is nonsingular and $\mathbf{C}_N^{-1}$ is the inverse of $\mathbf{C}_N$, then

$$\check{\boldsymbol{\beta}}_N = \mathbf{C}_N^{-1}\mathbf{T}_N(\mathbf{0}) \quad \text{and} \quad \mathbf{T}_N(\check{\boldsymbol{\beta}}_N) = \mathbf{0}. \tag{6.3.4}$$

We now adopt the same alignment procedure, but replacing $\mathbf{T}_N$ by the linear rank statistics

$$\mathbf{L}_N(\mathbf{b}) = \sum_{i=1}^N (\mathbf{c}_i - \bar{\mathbf{c}}_N)a_N(R_{Ni}(\mathbf{b})) = \left(L_{N1}(\mathbf{b}), \ldots, L_{Nq}(\mathbf{b})\right)', \tag{6.3.5}$$

where $R_{Ni}(\mathbf{b})$ is the rank of $\mathbf{X}_i - \mathbf{b}'(\mathbf{c}_i - \bar{\mathbf{c}}_N)$ among $X_\alpha - \mathbf{b}'(\mathbf{c}_\alpha - \bar{\mathbf{c}}_N)$, $1 \leq \alpha \leq {}_N$, for $i = 1, \dots, N$, and the scores $a_N(1), \dots, a_N(N)$ are (nondecreasing and) defined as in (6.2.10). Unfortunately, unlike (6.3.3), $\mathbf{L}_N(\mathbf{0}) - \mathbf{L}_N(\mathbf{b})$ may not be a very smooth function of $\mathbf{b}$, so that an explicit solution, as in (6.3.4), may not exist. Further, unlike (6.3.4), for the estimator $\hat{\boldsymbol{\beta}}_N$ (based on $\mathbf{L}_N$), $\mathbf{L}_N(\hat{\boldsymbol{\beta}}_N)$ may not be exactly equal to $\mathbf{0}$, and hence we need to define a *norm* measuring the closeness to $\mathbf{0}$. For a q-vector $\mathbf{x}$, we consider the norm

$$\|\mathbf{x}\| = \max_{1 \leq j \leq q} |x_j| \text{ or } \sum_{j=1}^{q} |x_j| \text{ or } (\mathbf{x}'\mathbf{x})^{1/2}, \tag{6.3.6}$$

though there may be other possibilities. Thus, we are inclined to propose $\hat{\boldsymbol{\beta}}_N$ as an estimator of $\boldsymbol{\beta}$ if

$$\|\mathbf{L}_N(\hat{\boldsymbol{\beta}}_N)\| \text{ is minimum for } \left\{\|\mathbf{L}_N(\mathbf{b})\| : \mathbf{b} \in R^q\right\}. \tag{6.3.7}$$

For $q = 1$, (6.3.7) holds for a (possibly) open interval of $\mathbf{b}$, and we considered the midpoint of that interval as an estimator. For $q \geq 1$, we define

$$D_N = \left\{\mathbf{b} : \|\mathbf{L}_N(\mathbf{b})\| \text{ is minimum}\right\}. \tag{6.3.8}$$

If D_N is a convex set, then a natural choice for $\hat{\boldsymbol{\beta}}_N$ is the center of gravity of D_N. For $q = 1$, the monotonicity of $L_N(b)$ (in b) enables us to show that D_N is convex. However, for $q > 1$, the situation is somewhat different. For example, for $q = 2$, for a fixed b_2, $L_{N1}(\mathbf{b})$ is decreasing in b_1, and similarly, for a fixed b_1, $L_{N2}(\mathbf{b})$ is decreasing in b_2. But, for a fixed b_2, we may not know precisely how $L_{N2}(\mathbf{b})$ varies with b_1, and similarly for $L_{N1}(\mathbf{b})$ with fixed b_1. Thus, for every fixed b_2, proceeding as in (6.2.5)–(6.2.6), we may obtain $\hat{\beta}_{1N}(= \hat{\beta}_{1N}(b_2))$ for which $|L_{N1}(\mathbf{b})|$ is a minimum. However, if this solution does not behave regularly with b_2, we may end up with either multiple solutions for (6.3.8) (which are not adjacent) or no solution.

The existence of a closed convex hull for D_N can be proved under additional regularity conditions on the $\mathbf{c}_i$, $i \geq 1$. But this may localize the scope of applicability of the theory to rather restricted situations. On the other hand, for large N, under fairly general conditions, $\mathbf{L}_N(\mathbf{0}) - \mathbf{L}_N(\mathbf{b})$ behaves as a smooth function of $\mathbf{b}$, and we shall see that this naturally yields a solution with a meaningful physical interpretation. In this way, the situation is comparable to the existence of the maximum-likelihood estimators, where for large sample sizes, in the neighborhood of the true parameter point, the solution exists, in probability, under fairly general regularity conditions. Thus, we shall assume for the time being that D_n is a closed convex set and take

$$\hat{\boldsymbol{\beta}}_N = \text{center of gravity of } D_N. \tag{6.3.9}$$

Once defined this way, the invariance property (6.2.30) extends readily to the vector case. Namely, if we set

$$\mathbf{X}_N(a, \mathbf{b}, d) = d\{\mathbf{X}_N + a\mathbf{1}_N + \mathbf{b}'[(\mathbf{c}_1 - \bar{\mathbf{c}}_N), \ldots, (\mathbf{c}_N - \bar{\mathbf{c}}_N)]\}, \tag{6.3.10}$$

and if $\hat{\boldsymbol{\beta}}_N(a, \mathbf{b}, d)$ denotes the estimator based on $\mathbf{X}_N(a, \mathbf{b}, d)$, then

$$\hat{\boldsymbol{\beta}}_N(a, \mathbf{b}, d) = d\hat{\beta}(0, \mathbf{0}, 1) + d\mathbf{b} \qquad \forall \mathbf{b} \in R^q,\ a, d \in R. \tag{6.3.11}$$

Our main interest lies in studying the nature of D_N for large N and the properties of $\hat{\boldsymbol{\beta}}_N$. For this, we define

$$\mathbf{c}_{(j)} = (c_{ij}, \ldots, c_{Nj}), \qquad j = 1, \ldots, q, \tag{6.3.12}$$

and assume that it is possible to write

$$\mathbf{c}_{(j)} = \mathbf{c}_{(j)}^{(1)} + \mathbf{c}_{(j)}^{(2)} \qquad \left(\text{i.e.,}\quad c_{ij} = c_{ij}^{(1)} + c_{ij}^{(2)}\ \forall i, j\right), \tag{6.3.13}$$

where for every N $(> q)$ and $1 \le j, l \le q$, $1 \le i, i' \le N$,

$$\left(c_{ij}^{(k)} - c_{i'j}^{(k)}\right)\left(c_{il}^{(k)} - c_{i'l}^{(k)}\right) \ge 0 \qquad \text{for}\quad k = 1, 2, \tag{6.3.14}$$

$$\left(c_{ij}^{(1)} - c_{i'j}^{(1)}\right)\left(c_{il}^{(2)} - c_{i'l}^{(2)}\right) \le 0. \tag{6.3.15}$$

Further, writing $\bar{c}_{j)}^{(k)} = N^{-1}\Sigma_{i=1}^{N} c_{ij}^{(k)}$, $k = 1, 2$, $j = 1, \ldots, q$, we assume that either

$$\left(\mathbf{c}_{(j)}^{(k)} - \bar{\mathbf{c}}_{(j)}^{(k)}\right)\left(\mathbf{c}_{(j)}^{(k)} - \bar{\mathbf{c}}_{(j)}^{(k)}\right)' = 0 \tag{6.3.16}$$

for all but a finite number of N, or

$$\left(\mathbf{c}_{(j)}^{(k)} - \bar{\mathbf{c}}_{(j)}^{(k)}\right)\left(\mathbf{c}_{(j)}^{(k)} - \bar{\mathbf{c}}_{(j)}^{(k)}\right)' > 0 \tag{6.3.17}$$

for all but a finite number of N, and in the latter case,

$$\lim_{N\to\infty} \frac{\max\limits_{1 \le i \le N} \left(c_{ij}^{(k)} - \bar{c}_j^{(k)}\right)^2}{\sum\limits_{i=1}^{N} \left(c_{ij}^{(k)} - \bar{c}_j^{(k)}\right)^2} = 0. \tag{6.3.18}$$

Further, for every k (= 1, 2) and j (= 1, ..., q),

$$N^{-1}\left(\mathbf{c}_{(j)}^{(k)} - \bar{\mathbf{c}}_{(j)}^{(k)}\right)\left(\mathbf{c}_{(j)}^{(k)} - \bar{c}_{(j)}^{(k)}\right)' \leq M\ (< \infty) \qquad \forall N. \tag{6.3.19}$$

Finally, we assume that

$$N^{-1}\mathbf{C}_N \to \boldsymbol{\Lambda} = (\boldsymbol{\lambda}^{(1)}, \ldots, \boldsymbol{\lambda}^{(q)}) \text{ is p.d.} \tag{6.3.20}$$

First, we consider the following result (due to Jurečková, 1971a), which forms the basis for our subsequent results [here P_0 denotes the probability under $H_0: \boldsymbol{\beta} = \mathbf{0}$, and $\rho(\phi, \psi)$, A_ϕ^2, and A_ψ^2 are defined by (6.2.38) and (6.2.39)]:

Under (6.3.1) *and the assumptions made above, for every* $\varepsilon > 0$ *and* $0 < K < \infty$,

$$\lim_{N\to\infty} P_0\Big\{ \sup_{\mathbf{u}: \|\mathbf{u}\| < K} N^{-1/2}\Big\| \mathbf{L}_N(N^{-1/2}\mathbf{u}) - \mathbf{L}_N(\mathbf{0}) +$$

$$N^{1/2}\{\rho(\phi, \psi) A_\phi A_\psi\}\boldsymbol{\Lambda}\mathbf{u}\Big\| > \varepsilon\Big\} = 0. \tag{6.3.21}$$

The proof of (6.3.21) is considered in the Appendix (see Section A.4.3). For a suitably large K, let us define a subset of D_N,

$$\tilde{D}_N = \left\{\mathbf{b} : |\mathbf{L}_N(\mathbf{b})| \text{ is minimum and } N^{1/2}\|\boldsymbol{\beta} - \mathbf{b}\| \leq K\right\}, \tag{6.3.22}$$

and suppose that the center of gravity of $\tilde{D}_N$ is $\tilde{\boldsymbol{\beta}}_N$. Then, by virtue of (6.3.21), it follows that $\tilde{D}_N$ is convex, in probability, and further, by (6.3.11) and (6.3.21),

$$N^{-1/2}\|\mathbf{L}_N(\tilde{\boldsymbol{\beta}}_N)\| \xrightarrow{P} 0 \quad \text{as} \quad N \to \infty, \tag{6.3.23}$$

$$N^{1/2}\sup\{\|\mathbf{b} - \mathbf{b}'\| : \mathbf{b}, \mathbf{b} \in \tilde{D}_N\} \xrightarrow{P} 0 \quad \text{as} \quad N \to \infty. \tag{6.3.24}$$

Moreover (see Theorem 5.3.1), under P_0,

$$N^{1/2}\mathbf{L}_N(\mathbf{0}) \xrightarrow{\mathscr{L}} \mathscr{N}_q\left(\mathbf{0}, A_\phi^2\boldsymbol{\Lambda}\right), \tag{6.3.25}$$

and hence, from (6.3.21) through (6.3.25), it follows that

$$N^{1/2}(\tilde{\boldsymbol{\beta}}_N - \boldsymbol{\beta}) \xrightarrow{\mathscr{L}} \mathscr{N}_q\big(\mathbf{0}, \{\rho(\phi,\psi)A_\psi\}^{-2}\boldsymbol{\Lambda}^{-1}\}. \tag{6.3.26}$$

To obtain the asymptotic distribution of $\hat{\boldsymbol{\beta}}_N$ defined by (6.3.9), we need to show that

$$P\{\tilde{\boldsymbol{\beta}}_N \neq \hat{\boldsymbol{\beta}}_N | \boldsymbol{\beta}\} \to 0 \qquad \text{as} \quad N \to \infty, \tag{6.3.27}$$

so that the asymptotic distribution in (6.3.26) applies then as well to $\hat{\boldsymbol{\beta}}_N$. Now, by virtue of (6.3.23), it suffices to show that for every $\varepsilon > 0$ and $\eta > 0$,

$$\lim_{N\to\infty} P_{\boldsymbol{\beta}}\left\{\inf\left[N^{-1/2}\|\mathbf{L}_N(\mathbf{b})\| : \mathbf{b} \in D_N \setminus \tilde{D}_N\right] < \eta\right\} < \varepsilon, \tag{6.3.28}$$

where $A \setminus B$ stands for $A - B$. The proof of (6.3.28) is left as an exercise. Thus, we have as $N \to \infty$,

$$N^{1/2}(\hat{\boldsymbol{\beta}}_N - \boldsymbol{\beta}) \xrightarrow{\mathscr{L}} \mathscr{N}_q\left(\mathbf{0}, \{\rho(\phi,\psi)A_\psi\}^{-2}\boldsymbol{\Lambda}^{-1}\right). \tag{6.3.29}$$

Consider now the least-squares estimator $\check{\boldsymbol{\beta}}_N$, defined by (6.3.4). By the same technique as in (6.2.61)–(6.2.62), it follows that as $N \to \infty$,

$$N^{1/2}(\boldsymbol{\beta}_N - \boldsymbol{\beta}) \xrightarrow{\mathscr{L}} \mathscr{N}_q(\mathbf{0}, \sigma^2\boldsymbol{\Lambda}^{-1}), \tag{6.3.30}$$

where $\sigma^2 = \mathrm{Var}(X_1)$ (see Exercise 6.3.5). If the form of the density function f (corresponding to the c.d.f. F) were known, one could have used the method of maximum likelihood for the estimation of $\boldsymbol{\beta}$. If the density f satisfies the usual regularity conditions (Cramér, 1946, pp. 500–503), then it follows by standard arguments (see Exercise 6.3.31) that for $\boldsymbol{\beta}_N^*$, the maximum-likelihood estimator of $\boldsymbol{\beta}$, we have

$$N^{1/2}(\boldsymbol{\beta}_N^* - \boldsymbol{\beta}) \xrightarrow{\mathscr{L}} \mathscr{N}_q\left(\mathbf{0}, \frac{1}{I(f)}\boldsymbol{\Lambda}^{-1}\right), \tag{6.3.31}$$

where

$$I(f) = \int_{-\infty}^{\infty} (f'(x)/f(x))^2\, dF(x) = A_\psi^2 \tag{6.3.32}$$

is the Fisher information.

Comparing (6.3.29), (6.3.30), and (6.3.31), it follows from Proposition 6.2.6.1 that

$$e(\hat{\boldsymbol{\beta}}, \check{\boldsymbol{\beta}}) = \sigma^2\rho^2(\phi, \psi)A_\psi^2, \tag{6.3.33}$$

$$e(\hat{\boldsymbol{\beta}}, \boldsymbol{\beta}^*) = \rho^2(\phi, \psi). \tag{6.3.34}$$

Since these familiar expressions have already been studied, we shall not enter into the details again.

We may remark that the computation of $\hat{\boldsymbol{\beta}}_N$, in general, requires some iteration procedures. In the same spirit as the *method of scoring* dealing with the solution of *maximum likelihood estimators*, we may employ the following linearized estimators of $\boldsymbol{\beta}$, based on rank statistics, due to Kraft and van Eeden (1972). (See also Hettmansperger, 1984.)

Define $\boldsymbol{\beta}_N$ as in (6.3.4), $\mathbf{L}_N(\mathbf{b})$ as in (6.3.2), and $\mathbf{C}_N$ as in (6.3.3). Let then

$$\bar{\bar{\boldsymbol{\beta}}}_N = \boldsymbol{\beta}_N + \left[\mathbf{C}_N^{-1}\mathbf{L}_N(\boldsymbol{\beta}_N)\right]\left[A_\phi A_\psi \rho(\phi, \psi)\right]^{-1}. \tag{6.3.35}$$

Then, by (6.3.21) and (6.3.35), it follows that

$$N^{1/2}\|\bar{\bar{\boldsymbol{\beta}}}_N - \hat{\boldsymbol{\beta}}_N\| \xrightarrow{P} 0 \quad \text{as} \quad N \to \infty, \tag{6.3.36}$$

and hence $\hat{\boldsymbol{\beta}}_N$ and $\bar{\bar{\boldsymbol{\beta}}}_N$ are asymptotically equivalent. Also, recall that

$$A_\phi A_\psi \rho(\phi, \psi) = L(\phi, \psi) = \int_0^1 \phi(u)\psi(u)\, du$$

$$= \int_{-\infty}^{\infty} \frac{d}{dx}\phi(F(x))\, dF(x) = D(F), \quad \text{say}, \tag{6.3.37}$$

and hence, if D_n is any consistent estimator of $D(F)$, then

$$\bar{\bar{\boldsymbol{\beta}}}_N = \check{\boldsymbol{\beta}}_N + D_N^{-1}\left[\mathbf{C}_N^{-1}\mathbf{L}_N(\check{\boldsymbol{\beta}}_N)\right] \tag{6.3.38}$$

is an asymptotically equivalent form of $\hat{\boldsymbol{\beta}}_N$. In particular, if the form of F is known and it only involves an unknown scale parameter γ, then apart from γ we have $\phi(u) \equiv \psi(u)$, so that $D(F)$ reduces to $1/\gamma$, and hence, a consistent estimator of γ suffices; this case has been dealt with in detail by Kraft and van Eeden (1972). For unknown F, we note that by (6.3.21) and (6.3.30), for every (fixed) $\mathbf{b} \neq \mathbf{0}$,

$$N^{-1/2}\left\|\mathbf{L}_N(\check{\boldsymbol{\beta}}_N) - \mathbf{L}_N(\check{\boldsymbol{\beta}}_N - N^{-1/2}\mathbf{b}) + N^{1/2}D(F)\boldsymbol{\Lambda}\mathbf{b}\right\| \xrightarrow{P} 0,$$

so that as a consistent estimator of $D(F)$, we have

$$D_N = \left(\frac{[\mathbf{L}_N(\check{\boldsymbol{\beta}}_N) - \mathbf{L}_N(\check{\boldsymbol{\beta}}_N - N^{-1/2}\mathbf{b})]'[\mathbf{L}_N(\check{\boldsymbol{\beta}}_N) - \mathbf{L}_N(\check{\boldsymbol{\beta}}_N - N^{-1/2}\mathbf{b})]}{(\mathbf{b}'\mathbf{C}_N^2\mathbf{b})/N^2} \right)^{1/2}, \tag{6.3.39}$$

For some adaptive estimators, see Hušková and Sen (1984).

So far, we have considered the problem of estimating $\boldsymbol{\beta}$, treating β_0 as a nuisance parameter. Now, we proceed to estimate β_0. If $\boldsymbol{\beta}$ is known, then, writing $X_i^* = X_i - \boldsymbol{\beta}'(c_i - \bar{\mathbf{c}}_N)$, $1 \le i \le N$, and noting that the X_i^* are i.i.d.r.v.'s with a c.d.f. $F(X - \beta_0)$ symmetric about β_0, we can proceed as in (6.2.18)–(6.2.20) and derive a parallel estimator of β_0. We consider the general case when $\boldsymbol{\beta}$ is unknown, and we estimate it by $\hat{\boldsymbol{\beta}}_n$, as earlier in this section. Let then

$$\hat{X}_i = X_i - \hat{\boldsymbol{\beta}}_N'(\mathbf{c}_i - \bar{\mathbf{c}}_N), \qquad 1 \le i \le N; \tag{6.3.40}$$

define $\hat{S}_N = S_N(\hat{X}_1, \ldots, \hat{X}_N)$ as in (6.2.23), with the X_i^* being replaced by $\hat{X}_i$, $1 \le i \le N$; and also let

$$\hat{S}_N(a) = S_N(\hat{X}_1 - a, \ldots, \hat{X}_N - a), \qquad -\infty < a < \infty. \tag{6.3.41}$$

Then, as in (6.2.27)–(6.2.28), we define

$$\hat{\beta}_{0,N} = \tfrac{1}{2}(\hat{\beta}_{0,N}^{(1)} + \hat{\beta}_{0,N}^{(2)}), \tag{6.3.42}$$

where

$$\hat{\beta}_{0,N}^{(1)} = \sup\{a : \hat{S}_N(a) > 0\}, \quad \hat{\beta}_{0,N}^{(2)} = \inf\{a : \hat{S}_N(a) < 0\}. \tag{6.3.43}$$

Consider now the setup of (6.2.45) with $b(c_i - \bar{c}_N)$ replaced by $\mathbf{b}'(\mathbf{c}_i - \bar{\mathbf{c}}_N)$, $1 \le i \le N$, and $S_N(N^{-1/2}a, N^{-1/2}b)$ replaced by $S_N(N^{-1/2}a, N^{-1/2}\mathbf{b})$. Then the following direct generalization of (6.2.45) holds, and its proof is given in the Appendix (see Section A.4.4):

Under $H_0: \beta_0 = 0$, $\boldsymbol{\beta} = \mathbf{0}$, *for every real* K $(0 < K < \infty)$, *as* $N \to \infty$,

$$\sup\Big\{\Big|N^{-1/2}[S_N(N^{-1/2}a, N^{-1/2}\mathbf{b}) - S_N(0, \mathbf{0})] + a\rho(\phi, \psi)A_\phi A_\psi\Big| : |a| \le K, \|b\| \le K\Big\} \xrightarrow{P} 0. \tag{6.3.44}$$

As a result, by a straightforward generalization of the method of proving Theorem 6.2.7, we have the following theorem.

Theorem 6.3.1. *Under the assumptions* (6.3.13)–(6.3.20), *as* $N \to \infty$,

$$N^{1/2}\left(\hat{\beta}_{0,N} - \beta_0\right) \xrightarrow{\mathscr{L}} \mathscr{N}\left(0, \left\{A_{\psi}^2 \rho^2(\phi, \psi)\right\}^{-1}\right) \tag{6.3.45}$$

and

$$N^{1/2}\left(\beta_{0,N} - \beta_0, \hat{\beta}_N' - \boldsymbol{\beta}'\right) \xrightarrow{\mathscr{L}} \mathscr{N}_{q+1}\left(\mathbf{0}, \left\{A_{\psi}^2 \rho^2(\phi, \psi)\right\}^{-1} \boldsymbol{\Sigma}^*\right) \tag{6.3.46}$$

where

$$\boldsymbol{\Sigma}^* = \begin{pmatrix} 1 & \mathbf{0} \\ \mathbf{0} & \boldsymbol{\Lambda} \end{pmatrix}^{-1}, \tag{6.3.47}$$

$\boldsymbol{\Lambda}$ *being defined by* (6.3.20).

In passing, we may remark that if $\check{\beta}_{0,N} = \overline{X}_N$ is the classical least-squares estimator of β_0, then (6.3.45` holds for $\check{\beta}_{0,N}$ with $\{A_{\psi}^2 \rho^2(\phi, \psi)\}^{-1}$ replaced by $\sigma^2 = \operatorname{Var} X_1$, and a similar argument holds for (6.3.46). Hence, the ARE results studied earlier remain valid in this case too.

We conclude this section with a remark on the median estimator of $(\beta_0, \boldsymbol{\beta}')$. The procedure in (6.2.64) can be extended to the case of $q + 1$ ordered subsets involving $q + 1$ unknowns β_0 and $\boldsymbol{\beta}$, and simultaneously equating them, one may derive estimates of β_0 and $\boldsymbol{\beta}$. But for $q > 1$, the asymptotic treatment in (6.2.65) through (6.2.67) obviously becomes more involved, and here also the computation of the ARE becomes quite involved.

6.4. ESTIMATION IN MULTIVARIATE LINEAR MODELS

We consider now some multivariate generalizations of the theory studied in the previous two sections, and for this purpose we consider the following model. Let $\mathbf{X}_i = (X_{i1}, \ldots, X_{ip})'$, $i = 1, \ldots, N$ $(p \geq 1)$ be N independent random vectors with continuous c.d.f.'s $F_1(\mathbf{x}), \ldots, F_N(x)$, respectively, all defined on $E^p = (-\infty, \infty)^p$, and assume that

$$F_i(\mathbf{x}) = F(\mathbf{x} - \boldsymbol{\beta}_0 - \boldsymbol{\beta}(\mathbf{c}_i - \bar{\mathbf{c}}_N)), \qquad 1 \leq i \leq N, \tag{6.4.1}$$

where $\boldsymbol{\beta}_0 = (\beta_{01}, \ldots, \beta_{0p})'$ and $\boldsymbol{\beta} = ((\beta_{jk}))_{j=1,\ldots,p,\,k=1,\ldots,q}$ $(q \geq 1)$ are unknown parameters; $\mathbf{c}_i = (c_{i1}, \ldots, c_{iq})'$, $i = 1, \ldots, N$, are vectors of known (regression) constants; $\bar{\mathbf{c}}_N = N^{-1}\Sigma_{i=1}^N \mathbf{c}_i$; and F is (unknown but) continuous. The simple regression model (6.2.1) is a particular case of (6.4.1) with $p = q = 1$, and similarly, the univariate multiple regression model (6.3.1) is a special case of (6.4.1) with $p = 1$, $q \geq 1$. We are basically interested here in the estimation of $\boldsymbol{\beta}_0$ and $\boldsymbol{\beta}$.

In passing, we may remark that the model (6.4.1) includes various common models of special interest as particular cases. For example, if the $\mathbf{c}_i$ are all equal i.e., $\mathbf{c}_1 = \cdots = \mathbf{c}_N = \bar{\mathbf{c}}_N$, we have

$$F_i(\mathbf{x}) = F(\mathbf{x} - \boldsymbol{\beta}_0), \qquad i = 1, \dots, N, \tag{6.4.2}$$

which relates to the classical multivariate one-sample location problem. For the model (6.4.2), the $\mathbf{X}_i$ are i.i.d., and if we assume that the p marginal c.d.f.'s are all symmetric, $\boldsymbol{\beta}_0$ is the vector of marginal medians. This problem has been treated in a fairly detailed manner in Chapter VI of Puri and Sen (1971). Secondly, suppose that $N = n_1 + n_2$ where both n_1 and n_2 are positive integers, and further, $\mathbf{c}_1 = \cdots = \mathbf{c}_{n_1} = 0$ and $\mathbf{c}_{n_1+1} = \cdots = \mathbf{c}_{n_1+n_2} = 1$, $q = 1$. Then $\mathbf{X}_1, \dots, \mathbf{X}_{n_1}$ are i.i.d. with the c.d.f. $F(\mathbf{x} - \boldsymbol{\beta}_0)$, while $\mathbf{X}_{n_1+1}, \dots, \mathbf{X}_N$ are so with the c.d.f. $F(\mathbf{x} - \beta_0 - \boldsymbol{\beta})$, where $\boldsymbol{\beta}$ is the difference of locations of the two distributions. Thus, in this case, the problem of estimation of $\boldsymbol{\beta}$ reduces to the multivariate two-sample location problem, also treated in detail in Chapter 6 of Puri and Sen (1971). More generally, suppose that $N = n_1 + \cdots + n_c$, $c \geq 2$, where the n_k are all positive integers, and suppose that $\mathbf{c}_1 = \cdots = \mathbf{c}_{n_1} \neq \mathbf{c}_{n_1+1} = \cdots = \mathbf{c}_{n_1+n_2} \neq \cdots \neq c_{n_1+\cdots+n_{c-1}} = \cdots = \mathbf{c}_N$, $q = 1$. Then $\mathbf{X}_{n_1+\cdots+n_{l-1}+j}$, $j = 1, \dots, n_l$ (where $n_0 = 0$) are i.i.d. with the c.d.f.

$$F_l(\mathbf{x}) = F(\mathbf{x} - \boldsymbol{\beta}_0 - \boldsymbol{\beta}_l), \qquad l = 1, \dots, c, \tag{6.4.3}$$

where the $\boldsymbol{\beta}_l$, $l = 1, \dots, c$, are unknown parameters (vectors). Writing $\boldsymbol{\theta}_l = \boldsymbol{\beta}_0 + \boldsymbol{\beta}_l$, $1 \leq l \leq c$, we have $F_l(\mathbf{x}) = F(\mathbf{x} - \boldsymbol{\theta}_l)$, $1 \leq l \leq c$, so that the model refers to the classical multivariate multisample location problem, also treated in detail in Chapter 6 of Puri and Sen (1971). We suggest the reader have a look into these simple models first for a better understanding of the theory to follow.

Let us denote the marginal c.d.f. of X_{ij} by $F_{ij}(x)$, $x \in E$, so that by (6.4.1), we have for every $i (= 1, \dots, N)$,

$$F_{ij}(x) = F\big(x - \beta_{0j} - \boldsymbol{\beta}_j(\mathbf{c}_i - \bar{\mathbf{c}}_N)\big), \quad x \in E, \tag{6.4.4}$$

for every j $(1 \leq j \leq p)$, where $\boldsymbol{\beta}_j = \boldsymbol{\beta}_j = (\beta_{j1}, \dots, \beta_{jq})$, $j = 1, \dots, p$ [so that $\boldsymbol{\beta}' = (\boldsymbol{\beta}_1', \dots, \boldsymbol{\beta}_p')$]. Thus, for every fixed j, (6.4.4) agrees with the model (6.3.1) and, in the particular case of $q = 1$, with (6.2.1). Hence, the theory developed in the preceeding two sections can be incorporated here (coordinatewise) to estimate $(\beta_{0j}, \boldsymbol{\beta}_j)$ for every $j = 1, \ldots, p$, and adjoining these estimators in appropriate vector or matrix forms, we arrive at the estimators of $(\boldsymbol{\beta}_0, \boldsymbol{\beta})$. With this coordinatewise reduction, the operational procedure for estimating $(\boldsymbol{\beta}_0, \boldsymbol{\beta})$ remains the same as in Sections 6.2 and 6.3, and hence we need not elaborate it. However, to study the various properties of the estimators, especially the asymptotic properties, we need to consider some further details which we discuss below. It will be more convenient to discuss the simple case ($q = 1$) of multivariate simple regression first, and then present the case $q > 1$.

6.4.1. Multivariate Simple Regression Model

Consider the vector of linear rank statistics

$$L_N = \left(L_N^{(1)}, \ldots, L_N^{(p)}\right)', \tag{6.4.5}$$

$$L_N^{(j)} = \sum_{i=1}^{N} (c_i - \bar{c}_N) a_N^{(j)}\left(R_{ij}^{(N)}\right), \qquad j = 1, \ldots, p, \tag{6.4.6}$$

where for each j ($1 \le j \le p$), $a_N^{(j)}(1) \le \cdots \le a_N^{(j)}(N)$ are suitable rank scores, and $R_{ij}^{(N)}$ is the rank of X_{ij} among $X_{ij}, \ldots, X_{Nj}$ for $1 \le i \le N$. Also, let

$$X_{ij}(b) = X_{ij} - b(c_i - \bar{c}_N), \qquad 1 \le i \le N, \quad b \in E, \tag{6.4.7}$$

and let $R_{ij}^{(N)}(b)$ be the rank of $X_{ij}(b)$ among $X_{ij}(b), \ldots, X_{Nj}(b)$ for $1 \le i \le N$. Finally, in (6.4.6), on replacing the ranks $R_{ij}^{(N)}$ by $R_{ij}^{(N)}(b)$, we denote the resulting quantities by $L_N^{(j)}(b)$ for $j = 1, \ldots, p$. Note that by Theorem 6.2.1, $L_N^{(j)}(b)$ is decreasing in b, and hence, on proceeding as in (6.2.5)–(6.2.6) with $\mu_N = \mu_N^{(j)} = [N^{-1}\sum_{i=1}^N a_N^{(j)}(i)]\sum_{i=1}^N (c_i - \bar{c}_N) = 0$, we obtain the estimator

$$\hat{\beta}_{j,N} = \tfrac{1}{2}\left(\hat{\beta}_{j,N}^{(1)} + \hat{\beta}_{i,N}^{(2)}\right). \tag{6.4.8}$$

$$\hat{\beta}_{j,N}^{(1)} = \inf\left\{b \colon L_N^{(j)}(b) < 0\right\}, \; \hat{\beta}_{j,N}^{(2)} = \sup\left\{b \colon L_N^{(j)}(b) > 0\right\}, \tag{6.4.9}$$

for $1 \le j \le p$. Let then

$$\hat{\boldsymbol{\beta}}_N = \left(\hat{\beta}_{1,N}, \ldots, \hat{\beta}_{p,N}\right)', \tag{6.4.10}$$

which is our estimator of $\boldsymbol{\beta}$. Similarly, keeping in mind the developments of

Section 6.2.2, we define

$$\mathbf{S}_N = \left(S_N^{(1)}, \dots, S_N^{(p)}\right)', \tag{6.4.11}$$

$$S_N^{(j)} = \sum_{i=1}^{N} \operatorname{sgn} X_{ij} a_N^{*(j)}\left(R_{ij,N}^{+}\right), \qquad 1 \le j \le p, \tag{6.4.12}$$

where $a_N^{*(j)}(1) \le \cdots \le a_N^{*(j)}(N)$ are scores, and $R_{ij,N}^{+}$ is the rank of $|X_{ij}|$ among $|X_{ij}|, \dots, |X_{Nj}|$, $1 \le i \le N$, for $j = 1, \dots, p$. Replacing the X_{ij} by $X_{ij} - a$, we denote the resulting $R_{ij,N}^{+}$ by $R_{ij,N}^{+}(a)$ and $S_N^{(j)}$ by $S_N^{(j)}(a)$, for $j = 1, \dots, p$. Then, for the case of known $\boldsymbol{\beta}$, the estimator of $\boldsymbol{\beta}_0$ is $\tilde{\boldsymbol{\beta}}_{0,N} = (\tilde{\beta}_{0,N}^{(1)}, \dots, \tilde{\beta}_{0,N}^{(p)})'$, where $\tilde{\beta}_{0,N}^{(j)}$ is defined by (6.2.19)–(6.2.20) with $S_N(a)$ replaced by $S_n^{(j)}(a)$ and $X_{ij} - a$ by $X_{ij} - \beta_j(c_i - \bar{c}_N) - a = X_{ij}^{*} - a$, $1 \le i \le N$, $j = 1, \dots, p$. For the case of unknown $\boldsymbol{\beta}$, we define $\hat{X}_{ij} = X_{ij} - \hat{\beta}_{j,N}(c_i - \bar{c}_N)$, $1 \le i \le N$ $(1 \le j \le p)$ and $\hat{X}_{ij}(a) = \hat{X}_{ij} - a$, $\hat{R}_{ij,N}^{+}(a) =$ rank of $|\hat{X}_{ij}(a)|$ among $|\hat{X}_{ij}(a)|, \dots, |\hat{X}_{Nj}(a)|$, for $1 \le i \le N$; and in (6.4.12), on replacing X_{ij} and $R_{ij,N}^{+}$ by $\hat{X}_{ij}(a)$ and $\hat{R}_{ij,N}^{+}(a)$, respectively, we define $\hat{S}_N^{(j)}(a)$. Then, the estimator $\hat{\beta}_{0,N}^{(j)}$ of β_j is defined as in (6.2.27)–(6.2.28), where $\hat{S}_N(a)$ is replaced by $\hat{S}_N^{(j)}(a)$ and $\mu_N^{*} = 0$.

For the study of the asymptotic distribution and efficiency of these estimators, we assume that

$$a_N^{(j)}(i) = E\varphi_j(U_{Ni}) \text{ or } \varphi_j\left(\frac{i}{N+1}\right),$$

$$a_N^{*(j)}(i) = E\varphi_j^{*}(U_{Ni}) \text{ or } \varphi_j^{*}\left(\frac{i}{N+1}\right) \tag{6.4.13}$$

for $i = 1, \dots, N$, where for each j $(= 1, \dots, p)$

φ_j (or φ_j^{*}) is absolutely continuous and square-integrable inside $(0, 1)$ and monotonic for $j = 1, \dots, p$. (6.4.14)

Let then for every $j(= 1, \dots, p)$,

$$\psi_j(u) = -\frac{f_j'\left(F_j^{-1}(u)\right)}{f_j\left(F_j^{-1}(u)\right)}, \qquad 0 < u < 1, \tag{6.4.15}$$

where F_j is the jth marginal c.d.f. corresponding to F and is assumed to have an absolutely continuous pdf $f_j(x)$, a.e. Also, let

$$\nu_{jl} = \int_{-\infty}^{\infty}\int_{-\infty}^{\infty} \phi_j(F_j(x))\phi_l(F_l(y))\, dF_{jl}(x, y)$$

$$-\left(\int_0^1 \phi_j(u)\, du\right)\left(\int_0^1 \phi_l(v)\, dv\right) \tag{6.4.16}$$

for $j, l = 1, \ldots, p$, whose F_{jl} is the bivariate joint c.d.f. of the (j, l)th variates corresponding to the c.d.f. F, $\boldsymbol{\nu} = ((\nu_{il}))$; and let

$$\gamma_j = \gamma_j(\phi_j, \psi_j) = \int_0^1 \phi_j(u)\psi_j(u)\, du, \qquad j = 1, \ldots, p, \tag{6.4.17}$$

$$\tau_{jl} = \frac{\nu_{jl}}{\gamma_j \gamma_l}, \quad j, l = 1, \ldots, p, \qquad \mathbf{T} = ((\tau_{jl})) \tag{6.4.18}$$

Then we have the following theorem.

Theorem 6.4.1. *Under the assumptions made above,*

$$N^{1/2}[\hat{\boldsymbol{\beta}}_N - \boldsymbol{\beta}] \xrightarrow{\mathscr{L}} \mathscr{N}_p(\mathbf{0}, C_0^{-2}\mathbf{T}), \tag{6.4.19}$$

where C_0^2 is defined by (6.2.36).

Proof. Let $\mathbf{L}_N(\mathbf{b}) = (L_N^{(1)}(b_1), \ldots, L_N^{(p)}(l_p))'$, $\mathbf{b} \in E^p$, where the $L_N^{(j)}(b_j)$ are defined earlier. Also, let $\mathbf{x} \le \mathbf{y}$ be the coordinatewise inequality $x_j \le y_j$, $1 \le j \le p$. Then, by (6.4.9), we have, for every $\mathbf{b} \in E^p$,

$$P\{N^{1/2}(\tilde{\boldsymbol{\beta}}_N^{(2)} - \boldsymbol{\beta}) \le \mathbf{b}|\boldsymbol{\beta}\} = P\{N^{1/2}\tilde{\boldsymbol{\beta}}_N^{(2)} \le \mathbf{b}|\boldsymbol{\beta} = \mathbf{0}\}$$

$$= P\{\mathbf{L}_N(N^{-1/2}\mathbf{b}) \le \mathbf{0}|\boldsymbol{\beta} = \mathbf{0}\}$$

$$= P\{\mathbf{L}_N(\mathbf{0}) \le \mathbf{0}|\boldsymbol{\beta} = -N^{-1/2}\mathbf{b}\}$$

$$= P\{N^{-1/2}\mathbf{L}_N(\mathbf{0}) \le \mathbf{0}|\boldsymbol{\beta} = -N^{-1/2}\mathbf{b}\}. \tag{6.4.20}$$

Now, from Theorem 5.5.2, it follows that under $H_N^{(-\mathbf{b})}: \boldsymbol{\beta} = -N^{-1/2}\mathbf{b}$, $N^{-1/2}\mathbf{L}_N(\mathbf{0})$ is asymptotically normally distributed with mean vector

$$-\boldsymbol{\lambda} = -(\lambda_1, \ldots, \lambda_p), \text{ where } \lambda_j = C_0^2 b_j \gamma_j, \quad 1 \le j \le p, \tag{6.4.21}$$

and dispersion matrix $C_0^2\boldsymbol{\nu}$, where $\boldsymbol{\lambda}$ is defined by (6.4.16). Hence, the rhs of

(6.4.20) converges to

$$(2\pi)^{-p/2}|\boldsymbol{\nu}|^{-1/2}C_0^{-p}\int_{-\infty}^{\lambda_1}\cdots\int_{-\infty}^{\lambda_p}\exp\left\{\frac{-1}{2C_0^2}\mathbf{t}'\boldsymbol{\nu}^{-1}\mathbf{t}\right\}d\mathbf{t}$$

$$=(2\pi)^{-p/2}|\mathbf{T}|^{-1/2}C_0^{p}\int_{-\infty}^{b_1}\cdots\int_{-\infty}^{b_p}\exp\{-\tfrac{1}{2}C_0^2\mathbf{u}'\mathbf{T}^{-1}\mathbf{u}\}\,d\mathbf{u},\qquad(6.4.22)$$

where $\mathbf{u}=(u_1,\ldots,u_p)'$ with $u_j=t_j/C_0^2\gamma_j$, $j=1,\ldots,p$. The same limiting result holds for $\hat{\boldsymbol{\beta}}_N^{(1)}$. Hence, the theorem follows from (6.4.8), (6.4.9), (6.4.20), and (6.4.22).

For the estimator $\tilde{\boldsymbol{\beta}}_{0,N}$ of $\hat{\boldsymbol{\beta}}_{0,N}$, we need to assume that the p marginal c.d.f.'s $F_1,\ldots,F_p$ are all symmetric. Further, in (6.4.13), we set for each j $(=1,\ldots,p)$ and $u\in(0,1)$,

$$\varphi_j^*(u)=\varphi_j\left(\frac{1+u}{2}\right),\quad \varphi_j(u)+\varphi_j(1-u)=0.\qquad(6.4.23)$$

Then, for the particular case of known $\boldsymbol{\beta}$ (say, equal to $\mathbf{0}$), the proof of the following theorem follows along the lines of the proof of the preceeding theorem (and hence is omitted).

Theorem 6.4.2. *For diagonally symmetric F, under the hypotheses of Theorem* 6.4.1,

$$N^{1/2}[\tilde{\boldsymbol{\beta}}_{0,N}-\tilde{\beta}_0]\xrightarrow{\mathscr{L}}\mathscr{N}_p(\mathbf{0},\mathbf{T}).\qquad(6.4.24)$$

Let us finally consider the asymptotic distribution theory of $N^{1/2}[(\hat{\boldsymbol{\beta}}_{0,N}-\boldsymbol{\beta}_0)',(\hat{\boldsymbol{\beta}}_N-\boldsymbol{\beta})']'$. We have the following theorem.

Theorem 6.4.3. *For diagonally symmetric F, under the assumptions made above,*

$$N^{1/2}\left[(\hat{\boldsymbol{\beta}}_{0,N}-\boldsymbol{\beta}_0)',(\hat{\boldsymbol{\beta}}_N-\boldsymbol{\beta})'\right]'\xrightarrow{\mathscr{L}}\mathscr{N}_{2p}(\mathbf{0},\mathbf{D})\qquad(6.4.25\text{a})$$

where

$$\mathbf{D}=\mathbf{T}\otimes\operatorname{diag}(1,C_0^{-2}).\qquad(6.4.25\text{b})$$

Proof. We make use here of both (6.2.44) and (6.2.45), for each of the coordinates. Thus, in (6.2.44), we replace $L_N(N^{-1/2}b)$, $L_N(0)$, and $\rho(\varphi, \psi)A_\varphi A_\psi$ by $L_N^{(j)}(N^{-1/2}b)$, $L_N^{(j)}(0)$, and γ_j, respectively, for $j = 1, \ldots, p$, and obtain that under $H_0: \boldsymbol{\beta} = \mathbf{0}$,

$$\sup_{b:|b|\le K} \left| L_N^{(j)}(N^{-1/2}b) - L_N^{(j)} + N^{1/2}bC_0^2\gamma_j \right| = o_p(N^{1/2}), \qquad 1 \le j \le p. \tag{6.4.26}$$

Similarly, in (6.2.45), we replace $S_N(N^{-1/2}a, N^{-1/2}b)$, $S_N(0,0)$, and $\rho(\varphi, \psi)A_\varphi A_\psi$ by $S_N^{(j)}(N^{-1/2}a, N^{-1/2}b)$, $S_N^{(j)}(0,0)$, and γ_j, respectively, and obtain that under $H_0^*: \boldsymbol{\beta}_0 = \mathbf{0}$, $\boldsymbol{\beta} = \mathbf{0}$,

$$\sup\Big\{ \Big| N^{-1/2}\big\{ S_N^{(j)}(N^{-1/2}a, N^{-1/2}b) - S_N^{(j)}(0,0) + a\gamma_j \Big| : \\ |a| \le K, |b| \le K \Big\} \xrightarrow{P} 0, \qquad 1 \le j \le p. \tag{6.4.27}$$

Further, for F diagonally symmetric, it follows by some standard steps (along the lines of Sections 4.2 and 4.3) that

$$\mathscr{L}_0\big(N^{-1/2}[\mathbf{L}_N(\mathbf{0}), \mathbf{S}_N(\mathbf{0})]\big) \to N_{2p}\big(0, \boldsymbol{\nu} \otimes \operatorname{diag}(C_0^2, 1)\big), \tag{6.4.28}$$

where $\mathbf{L}_N(\mathbf{0}) = (L_N^{(1)}(0), \ldots, L_N^{(b)}(0))'$ and $\mathbf{S}_N(\mathbf{0}) = (S_N^{(1)}(0,0), \ldots, S_N^{(p)}(0,0))'$, and $\mathscr{L}_0$ stands for the law under the hypothesis $H_0^*: \boldsymbol{\beta}_0 = \boldsymbol{\beta} = 0$; we leave the proof as an exercise (viz. Exercise 6.4.3). The rest of the proof follows virtually as a direct multivariate extension of the proof of Theorem 6.2.7 along the lines of the proof of Theorem 6.4.1, and hence is not reproduced.

Let us next consider the classical least-squares estimators of $\boldsymbol{\beta}_0$ and $\boldsymbol{\beta}$. These are given by

$$\check{\boldsymbol{\beta}}_{0,N} = \overline{\mathbf{X}}_N\check{\boldsymbol{\beta}}_N = C_N^{-2}\sum_{i=1}^{N}(c_i - \bar{c}_N)(\mathbf{X}_i - \overline{\mathbf{X}}_N) \tag{6.4.29}$$

By the same technique as in (6.2.59)–(6.2.62), along with a direct multivariate extension, we obtain that

$$N^{1/2}\big[(\boldsymbol{\beta}_{0,N} - \boldsymbol{\beta}_0)', (\boldsymbol{\beta}_N - \boldsymbol{\beta})'\big]' \xrightarrow{\mathscr{L}} \mathscr{N}_{2p}\big(\mathbf{0}, \boldsymbol{\Sigma} \otimes \operatorname{diag}(1, C_0^{-1})\big), \tag{6.4.30}$$

where

$$\mathbf{\Sigma} = \left(\left(\mathrm{Cov}(X_{ij}, X_{il})\right)\right)_{j,l=1,\ldots,p}; \tag{6.4.31}$$

we leave the details of the proof as an exercise (see Exercise 6.4.4).

Finally, let $\boldsymbol{\beta}^*_{0,N}$ and $\boldsymbol{\beta}^*_N$ be the maximum-likelihood estimators of $\boldsymbol{\beta}_0$ and $\boldsymbol{\beta}$, respectively. Also let

$$\mathbf{I}(f) = \left(\int_{-\infty}^{\infty}\int_{-\infty}^{\infty} \psi_j(F_j(x))\psi_l(F_l(y))\, dF_{jl}(x, y)\right)_{j,l=1,\ldots,p} \tag{6.4.32}$$

be the information matrix, where the ψ_j are defined by (6.4.15). Then (See Exercise 6.4.5), by well-known results on the asymptotic distribution of the maximum-likelihood estimators, we have

$$N^{1/2}\left[(\boldsymbol{\beta}^*_{0,N} - \boldsymbol{\beta}_0)', (\boldsymbol{\beta}^*_N - \boldsymbol{\beta})'\right]' \xrightarrow{\mathscr{L}} \mathscr{N}_{2p}\left(\mathbf{0}, \mathbf{I}^{-1}(f) \otimes \mathrm{diag}(1, C_0^{-2})\right). \tag{6.4.33}$$

We are now in a position to study the asymptotic relative efficiencies of the competing estimators. For this purpose, we recall the definitions of ARE in Section 6.2.6 and note that, in our case, we are not, in general, able to apply Definition 6.2.6.1, but Definition 6.2.6.2 along with Proposition 6.2.6.2 can be used to define and interpret suitable measures. In this way, by (6.2.57), Theorem 6.4.3, and (6.4.30), we obtain that the ARE of the rank-order estimator with respect to the least-squares estimator is given by

$$e_{\hat{\boldsymbol{\beta}},\check{\boldsymbol{\beta}}} = e_{(\hat{\boldsymbol{\beta}}_0,\hat{\boldsymbol{\beta}}),(\beta_0,\beta)} = \left\{\frac{|\boldsymbol{\Sigma}|}{|T|}\right\}^{1/p}. \tag{6.4.34}$$

If we rewrite

$$\mathbf{\Sigma} = ((\sigma_{jl})) = ((\sigma_j\sigma_l\rho_{jl})),$$

$$\boldsymbol{\nu} = ((\nu_{jl})) = ((\nu_{jj}^{1/2}\nu_{ll}^{1/2}\rho_{jl}^*)), \tag{6.4.35}$$

where $\sigma_j^2 = V(X_{ij})$ and ρ_{jl} is the Pearsonian correlation coefficient between the jth and lth variates, and if we let

$$\mathbf{R} = ((\rho_{jl})) \quad \text{and} \quad \mathbf{R}^* = ((\rho_{jl}^*)), \tag{6.4.36}$$

then from (6.4.34)–(6.4.36), we have

$$e_{\hat{\boldsymbol{\beta}},\check{\boldsymbol{\beta}}} = \left\{ \prod_{j=1}^{p} \frac{\sigma_j^2 \gamma_j^2}{\nu_{jj}} \right\}^{1/p} \left\{ \frac{|\mathbf{R}|}{|\mathbf{R}^*|} \right\}^{1/p}$$

$$= \left\{ \prod_{j=1}^{p} e_{\hat{\beta}_j,\check{\beta}_j} \right\}^{1/p} \left\{ \frac{|\mathbf{R}|}{|\mathbf{R}^*|} \right\}^{1/p}. \tag{6.4.37}$$

The first factor on the rhs of (6.4.37) is the geometric mean of the marginal AREs, while the second factor depends on the joint c.d.f. F, through $\mathbf{R}$ and $\mathbf{R}^*$. In general, the relationship between $\mathbf{R}$ and $\mathbf{R}^*$ is not so explicit as to enable us to derive suitable bounds for $\{|\mathbf{R}|/|\mathbf{R}^*|\}^{1/p}$. However, for some specific cases, such bounds are available and provide reasonable ideas about the ARE. We consider the following specific cases.

(a) ***Wilcoxon scores.*** $\varphi_j(u) = (u - \frac{1}{2})$, $0 \le u \le 1$, $j = 1, \ldots, p$. In this case,

$$e_{\hat{\beta}_j,\check{\beta}_j} = 12\sigma_j^2 \left(\int_{-\infty}^{\infty} f_j^2(x)\, dx \right)^2, \qquad j = 1, \ldots, p, \tag{6.4.38}$$

and it is well known that for every $j = 1, \ldots, p$

$$e_{\hat{\beta}_j,\check{\beta}_j} \begin{cases} \ge 0.864 & \text{for all } F_j, \\ = 3/\pi & \text{when } \quad F_j \equiv \text{normal}. \end{cases} \tag{6.4.39}$$

Further, in this case, $\rho_{jl}^* = \rho_{jl}^g$ is the grade correlation given by

$$\rho_{jl}^g = 12 \int_{-\infty}^{\infty} \int_{-\infty}^{\infty} \left[F_j(x) - \tfrac{1}{2} \right] \left[F_l(y) - \tfrac{1}{2} \right] dF_{jl}(x, y), \qquad j, l = 1, \ldots, p. \tag{6.4.40}$$

Note that if the X_{ij}, X_{il} are pairwise uncorrelated, $\rho_{jl} = 0$ for $j \ne l = 1, \ldots, p$, but $|\rho_{jl}^g| \le 1$, and hence $|\mathbf{R}|/|\mathbf{R}^g| \ge 1$. As a result, from (6.4.39) and the above, we arrive at the following. Let $\mathscr{F}^* = \{F\colon \rho_{jl} = 0\ \forall 1 \le j < l \le p\}$. Then, for the Wilcoxon-score estimators,

$$e_W = e_{\hat{\boldsymbol{\beta}},\check{\boldsymbol{\beta}}} \ge 0.864 \qquad \text{for all} \quad F \in \mathscr{F}^*. \tag{6.4.41}$$

Another important case often arising in practice is the equicorrelation (intraclass correlation) model where $X_{i1}, \ldots, X_{ip}$ are interchangeable or symmetric dependent (apart from the location–regression variations). In this case, $\rho_{jl} = \rho$ for every $1 \le j < l \le p$ where $-(p-1)^{-1} \le \rho < 1$. Thus, here,

$$|\mathbf{R}| = (1-\rho)^{p-1}[1+(p-1)\rho], \tag{6.4.42}$$

$$|\mathbf{R}^g| = (1-\rho_g)^{p-1}[1+(p-1)\rho_g], \tag{6.4.43}$$

where ρ_g is the common value of ρ_{jl}^g, $1 \le j < l \le p$. Hence,

$$\frac{|\mathbf{R}|}{|\mathbf{R}^g|} = \left[\frac{1-p}{1-p_g}\right]^{p-1} \frac{1+(p-1)\rho}{1+(p-1)\rho_g}. \tag{6.4.44}$$

Thus, if the relationship between ρ and ρ^g is known, suitable bounds for (6.4.44) can be derived. For example, if ρ and ρ_g satisfy

$$\rho \le 0 \;\Rightarrow\; \rho \le \rho_g \le 0 \quad \text{and} \quad \rho \ge 0 \;\Rightarrow\; \rho \ge \rho_g \ge 0, \tag{6.4.45}$$

then $|\mathbf{R}|/|\mathbf{R}^g| \le 1$, and $-(p-1)^{-1} < \rho < 1$, and hence

$$e_{\hat{\boldsymbol{\beta}}, \check{\boldsymbol{\beta}}} \le \left\{ \prod_{j=1}^{p} e_{\hat{\beta}_j, \check{\beta}_j} \right\}^{1/p}. \tag{6.4.46}$$

In particular, if $\lim_{\rho \to -(p-1)^{-1}} \rho_g \ne -(p-1)^{-1}$, then it follows from (6.6.44) that $\lim_{\rho \to -(p-1)^{-1}} |R|/|R^g| = 0$, so that (6.4.37) converges to 0 as $\rho \to -(p-1)^{-1}$. However, this is a degenerate case and ceases to be of much practical interest.

Let us next consider the case of F being a multinormal c.d.f. In this case,

$$\rho_{il}^g = \frac{6}{\pi} \sin^{-1}\left(\tfrac{1}{2}\rho_{jl}\right), \qquad 1 \le j \le l \le p, \tag{6.6.47}$$

so that $|\mathbf{R}|/|\mathbf{R}^g|$ becomes a function of ρ_{jl}, $1 \le j < l \le p$. For the particular case of $p = 2$, we have

$$\left(\frac{|\mathbf{R}|}{|\mathbf{R}^g|}\right)^{1/2} = \left(\frac{1-\rho_{12}^2}{1-(36/\pi^2)\left(\sin^{-1}\frac{1}{2}\rho_{12}\right)^2}\right)^{1/2}, \tag{6.6.48}$$

and, minimizing and maximizing (6.6.48) with respect to ρ_{12} and using

(6.4.39), we obtain

$$0.91 \le e_W \le 0.95 \qquad \text{for every} \quad \rho_{12} \in (-1, 1), \tag{6.4.49}$$

where the lower bound is attained when $\rho_{12} \to \pm 1$. For $p > 2$, the dependence of $|\mathbf{R}|/|\mathbf{R}^g|$ on ρ_{jl}, $1 \le j < l \le p$, becomes more complicated; it may even tend to 0 in some degenerate cases.

(b) ***Normal scores.*** The normal scores were $\varphi_j(u) = \Phi^{-1}(u)$, $0 < u < 1$, $j = 1, \ldots, p$, where Φ is the standard normal d.f. In this case,

$$e_N^{(j)} = e_{\hat{\beta}_j, \check{\beta}_j} = \sigma_j^2 \left(\int_0^1 \Phi^{-1}(u) \psi_j(u)\, du \right)^2, \tag{6.4.50}$$

and it is known (Puri and Sen, 1971, Chapters 3, 6) that

$$e_N^{(j)} \ge 1 \qquad \forall F_j, \quad j = 1, \ldots, p, \tag{6.4.51}$$

where the equality sign holds iff F_j is a normal c.d.f., $j = 1, \ldots, p$. Further, in this case, $\mathbf{R}^*$ has the element

$$\rho_{jl}^* = \int_{-\infty}^{\infty} \int_{-\infty}^{\infty} \Phi^{-1}(F_j(x)) \Phi^{-1}(F_l(y))\, dF_{jl}(x, y), \qquad 1 \le j \le l \le p. \tag{6.4.52}$$

Hence, as in the case of the Wilcoxon scores, if the variates are pairwise uncorrelated (i.e., $\rho_{jl} = 0\ \forall j \ne l$), then $|\mathbf{R}| = 1$, $|R^*| \le 1$, so that

$$e_N \ge 1 \qquad \forall F \in \mathscr{F}^*. \tag{6.4.53}$$

Secondly, if F itself is a multinormal c.d.f., then $\mathbf{R}^* = \mathbf{R}$, and hence

$$e_N = 1 \qquad \text{for all multinormal } F. \tag{6.4.54}$$

Finally, for the intraclass correlation model too, under (6.4.45) (with ρ_g replaced by $\rho^* \equiv$ the common value of the ρ_{jl}^*), $e_N \le (\prod_{j=1}^p e_N^{(j)})^{1/p}$.

For the ARE of $\hat{\boldsymbol{\beta}}_N$ with respect to the maximum-likelihood estimator $\boldsymbol{\beta}^*$, we have

$$e_{\hat{\boldsymbol{\beta}}, \boldsymbol{\beta}^*} = e_{(\hat{\boldsymbol{\beta}}_0, \hat{\boldsymbol{\beta}}), (\boldsymbol{\beta}_0^*, \boldsymbol{\beta}^*)} = \left\{ \frac{|\mathbf{I}(f)|}{|\mathbf{T}|} \right\}^{-1/p}. \tag{6.4.55}$$

Now, by the multivariate extension of the classical Cramér–Rao inequality, it can be shown that $\mathbf{T} - \mathbf{I}^{-1}(f)$ is a p.s.d. matrix, and hence $\mathbf{I}(f)\mathbf{T} - \mathbf{I}$ is also p.s.d., where $\mathbf{I} = \text{diag}(1, \ldots, 1)$. Thus, the characteristic roots of $\mathbf{I}(f)\mathbf{T}$ are all ≥ 1, and hence $\{|\mathbf{I}(f)||\mathbf{T}|\}^{-1/p} = \{$product of the characteristic roots of $\mathbf{I}(f)\mathbf{T}\}^{-1/p} < 1$, where the equality sign holds iff $\hat{\boldsymbol{\beta}}_n \equiv \boldsymbol{\beta}^*$.

A final remark: as in (6.2.71)–(6.2.74), we need not confine ourselves to the condition (6.2.36), if we need to study the ARE of $\hat{\boldsymbol{\beta}}_N$ with respect to $\breve{\boldsymbol{\beta}}_N$ (or $\boldsymbol{\beta}^*$). The only change here will be to replace (6.2.74) by

$$e_{\hat{\boldsymbol{\beta}}, \breve{\boldsymbol{\beta}}} = s^{-1}\left(\left\{\frac{|\boldsymbol{\Sigma}|}{|\mathbf{T}|}\right\}^{1/p}\right), \tag{6.4.56}$$

where $s(\cdot)$ is defined by (6.2.71).

6.4.2. Multivariate Multiple Regression Model

Consider the matrix

$$\mathbf{L}_N = \left(\left(L_{Nk}^{(j)}\right)\right)_{j=1,\ldots,p,\, k=1,\ldots,q}, \tag{6.4.57}$$

$$L_{Nk}^{(j)} = \sum_{i=1}^{N} (c_{ik} - \bar{c}_{Nk}) a_N^{(j)}\left(R_{ij}^{(N)}\right), \qquad 1 \leq j \leq p, \quad 1 \leq k \leq q, \tag{6.4.58}$$

where $a_N^{(j)}(\cdot)$ and $R_{ij}^{(N)}$ are defined as in (6.4.6). Also, let

$$\mathbf{B} = \begin{pmatrix} \mathbf{b}_1 \\ \vdots \\ \mathbf{b}_p \end{pmatrix}, \qquad \text{where} \quad \mathbf{b}_j = \left(b_1^{(j)}, \ldots, b_q^{(j)}\right), \quad j = 1, \ldots, p, \tag{6.4.59}$$

$$X_{ij}(\mathbf{b}_j) = X_{ij} - \mathbf{b}_j(\mathbf{c}_i - \bar{\mathbf{c}}_N), \qquad 1 \leq j \leq p, \quad 1 \leq i \leq N. \tag{6.4.60}$$

Replacing the X_{ij} by $X_{ij}(\mathbf{b}_j)$ in the ranking process, we denote the ranks by $R_{ij}(\mathbf{b}_j)$ and the resulting $L_{Nk}^{(j)}$ by $L_{Nk}^{(j)}(\mathbf{b}_j)$ for $1 \leq k \leq q$, $1 \leq j \leq p$. Finally, let

$$\mathbf{L}_N^{(j)}(\mathbf{b}_j) = \left(L_{N1}^{(j)}(\mathbf{b}_j), \ldots, L_{Nq}^{(j)}(\mathbf{b}_j)\right), \qquad j = 1, \ldots, p. \tag{6.4.61}$$

If we write

$$\boldsymbol{\beta} = \begin{pmatrix} \boldsymbol{\beta}_1 \\ \vdots \\ \boldsymbol{\beta}_p \end{pmatrix}, \qquad \boldsymbol{\beta}_j = \left(\beta_1^{(j)}, \ldots, \beta_q^{(j)}\right), \quad j = 1, \ldots, p, \tag{6.4.62}$$

then, proceeding as in (6.3.7)–(6.3.9), we define the estimator $\hat{\beta}_{N,j}$ of $\boldsymbol{\beta}_j$, based on $\mathbf{L}_N^{(j)}(\mathbf{b}_j)$, for $j = 1, \ldots, p$, and then let

$$\hat{\boldsymbol{\beta}}_N = \begin{pmatrix} \hat{\boldsymbol{\beta}}_{N,1} \\ \vdots \\ \hat{\boldsymbol{\beta}}_{N,p} \end{pmatrix}, \tag{6.4.63}$$

which is the desired estimator of $\boldsymbol{\beta}$.

To estimate $\boldsymbol{\beta}_0$, we consider, for each j,

$$\hat{X}_{ij}(a) = X_{ij} - a - \hat{\boldsymbol{\beta}}_{N,j}(\mathbf{c}_i - \bar{\mathbf{c}}_N), \qquad 1 \le i \le N, \tag{6.4.64}$$

and define $\hat{S}_N^{(j)}(a)$ as we did after (6.4.12), for $j = 1, \ldots, p$. Then the procedure for estimating $\boldsymbol{\beta}_0$, on the basis of $\hat{\mathbf{S}}_N(\mathbf{a})$, is the same as after (6.4.12). We denote the resulting estimator by $\hat{\boldsymbol{\beta}}_{0,N}$. Thus, in the multivariate multiple regression model, we treat each of the p variates separately, apply the univariate technique developed in Section 6.3, and then complete the matrix $\hat{\boldsymbol{\beta}}_N$ [or $(\hat{\boldsymbol{\beta}}_{0,N}, \hat{\boldsymbol{\beta}}_N)$].

For the study of the asymptotic properties of this estimator, we make the assumptions (6.3.12) through (6.3.20). Then, parallel to (6.3.21) and (6.4.26), we have for every $j(= 1, \ldots, p)$,

$$\sup_{\mathbf{u}:\|\mathbf{u}\|<K} \left\{ N^{-1/2} \left\| \mathbf{L}_N^{(j)}(N^{-1/2}\mathbf{u}) - \mathbf{L}_N^{(j)}(\mathbf{0}) + N^{1/2}\gamma_j \boldsymbol{\Lambda}\mathbf{u} \right\| \right\} \xrightarrow{P} 0, \tag{6.4.65}$$

when $H_0: \boldsymbol{\beta} = \mathbf{0}$ holds. Then (6.3.23)–(6.3.24) hold coordinatewise, and (6.3.25) extends, under $H_0: \boldsymbol{\beta} = \mathbf{0}$, to

$$N^{-1/2}\mathbf{L}_N(\mathbf{0}) \xrightarrow{\mathcal{L}} \mathcal{N}_{pq}(\mathbf{0}, \boldsymbol{\nu} \otimes \boldsymbol{\Lambda}), \tag{6.4.66}$$

where $\boldsymbol{\nu}$ is defined by (6.4.16). Hence, it follows by some standard steps that

$$N^{1/2}(\hat{\boldsymbol{\beta}}_N - \boldsymbol{\beta}) \xrightarrow{\mathcal{L}} \mathcal{N}_{pq}(\mathbf{0}, \mathbf{T} \otimes \boldsymbol{\Lambda}^{-1}); \tag{6.4.67}$$

we leave the proof as an exercise (see Exercise 6.4.10).

Also, for each j $(= 1, \ldots, p)$, (6.3.44) insures that

$$\sup\left\{ \left| N^{-1/2}\left\{ \mathbf{S}_N^{(j)}\left(N^{-1/2}a, N^{-1/2}b_j\right) - S_N^{(j)}(0,0) + a\gamma_j \right\} \right| : \right.$$

$$\left. |a| \le K, \ \|\mathbf{b}\| \le K \right\} \xrightarrow{P} 0 \tag{6.4.68}$$

when $H_0: \boldsymbol{\beta}_0 = 0$, $\boldsymbol{\beta} = 0$ holds. Hence, Theorem 6.3.1 can be extended (in the same manner as in Theorems 6.4.1–6.4.3) to prove the following theorem.

Theorem 6.4.4. *Under the hypotheses of Theorem* 6.3.1 *and Theorem* 6.4.3, *as* $N \to \infty$,

$$N^{1/2}(\hat{\boldsymbol{\beta}}_{0,N} - \boldsymbol{\beta}_0) \xrightarrow{\mathcal{L}} \mathcal{N}_p(\mathbf{0}, \mathbf{T}), \tag{6.4.69}$$

$$N^{1/2}[\hat{\boldsymbol{\beta}}_{0,N} - \boldsymbol{\beta}_0), (\hat{\boldsymbol{\beta}}_N - \boldsymbol{\beta})] \xrightarrow{\mathcal{L}} N_{p(q+1)}(\mathbf{0}, \mathbf{D}^*), \tag{6.4.70}$$

where

$$\mathbf{D}^* = \mathbf{T} \otimes \begin{bmatrix} 1 & \mathbf{0} \\ \mathbf{0} & \Lambda \end{bmatrix}^{-1}, \tag{6.4.71}$$

Λ *being defined by* (6.3.20).

The proof is left as an exercise (see Exercise 6.4.11).

It may be remarked that the classical least-squares estimators of $\boldsymbol{\beta}_0$ and $\boldsymbol{\beta}$ are

$$\check{\boldsymbol{\beta}}_{0,N} = \bar{\mathbf{X}}_N \tag{6.4.72}$$

$$\check{\boldsymbol{\beta}}_N = \left[\sum_{i=1}^{N} (\mathbf{X}_i - \bar{\mathbf{X}}_N)(\mathbf{c}_i - \bar{\mathbf{c}}_N)'\right]\mathbf{C}_N^{-1} \tag{6.4.73}$$

Since these are linear estimators, by considering a linear combination of all these $p(q+1)$ estimators, we are in a position to make use of (6.2.61) and prove the asymptotic normality; the Cramér–Wold technique ensures that

$$N^{1/2}[(\boldsymbol{\beta}_{0,N} - \boldsymbol{\beta}_0), (\boldsymbol{\beta}_N - \boldsymbol{\beta})] \xrightarrow{\mathcal{L}} \mathcal{N}_{p(q+1)}(\mathbf{0}, \check{\mathbf{D}}), \tag{6.4.74}$$

where

$$D = \boldsymbol{\Sigma} \otimes \begin{bmatrix} 1 & \mathbf{0} \\ \mathbf{0} & \Lambda \end{bmatrix}^{-1}, \tag{6.4.75}$$

$\boldsymbol{\Sigma}$ being defined by (6.4.31). In a similar manner, it can be shown (see Exercise 6.4.12) that if $(\boldsymbol{\beta}_{0,N}^*, \boldsymbol{\beta}_N^*)$ is the maximum-likelihood estimator of $(\boldsymbol{\beta}_0, \boldsymbol{\beta})$, then

$$N^{1/2}[(\boldsymbol{\beta}_{0,N}^* - \boldsymbol{\beta}_0), (\boldsymbol{\beta}_N^* - \boldsymbol{\beta})] \xrightarrow{\mathcal{L}} \mathcal{N}_{p(q+1)}(\mathbf{0}, \mathbf{D}_0), \tag{6.4.76}$$

where

$$\mathbf{D}_0 = \left(\mathbf{I}(f) \otimes \begin{bmatrix} 1 & \mathbf{0} \\ \mathbf{0} & \mathbf{\Lambda} \end{bmatrix}\right)^{-1}, \tag{6.4.77}$$

$\mathbf{I}(f)$ being defined by (6.4.32).

By virtue of Theorem 6.4.4, (6.4.74), (6.4.75), (6.4.76), and (6.4.77), we are in a position to study the ARE of the different estimators. In this case, also we adopt Definition 6.2.6.2 and Proposition 6.2.6.2, and the results are precisely the same as in (6.4.34) through (6.4.56).

We conclude this section with the remark that as the multivariate multiple regression model relates to estimators which are coordinatewise the estimators for the univariate multiple regression model for the p marginals, the process of obtaining efficient linearized estimators, discussed in detail in (6.3.35) through (6.3.39), extends in a straightforward manner. For each coordinate, we derive the linearized estimators and then complete the matrix by augmentation.

6.5. RANK-BASED INTERVAL ESTIMATION IN LINEAR MODELS

Based on suitable rank statistics, the theory of point estimation of parameters in univariate and multivariate linear models has been systematically developed in Section 6.2 through 6.4. The corresponding theory of interval estimation will be considered in this section. For the specific single-sample as well as several-sample location models (in the univariate as well as multivariate setups), this theory has been presented in Chapter 6 of Puri and Sen (1971). Therefore we shall mainly consider here the case of linear models, and cite some of these special cases as exercises.

Consider the simple regression model in (6.2.1). In (6.2.5), the rank (point) estimator of the regression slope β was obtained by equating the aligned rank $T_N(b)$ to 0, the central value of the null-hypothesis distribution of $T_N(0)$. Note that under the null hypothesis $H_0: \beta = 0$, $T_N(0)$ has a distribution independent of the underlying F, and hence, for every α $(0 < \alpha < 1)$ and $c_1, \dots, c_N$, there exist two real values t_{NL} and t_{NU} and an α_N such that

$$P\{t_{NL} \leq T_N(0) \leq t_{NU} | H_0\} = 1 - \alpha_N \ (\geq 1 - \alpha), \tag{6.5.1}$$

where α_N does not depend on F and converges to α as $N \to \infty$. Thus, if we define

$$\hat{\beta}_{N,L} = \inf\{b: T_N(b) \leq t_{NU}\}, \tag{6.5.2}$$

$$\hat{\beta}_{N,U} = \sup\{b: T_N(b) \geq t_{NL}\}, \tag{6.5.3}$$

then, by using the monotonicity property of $T_N(b)$ in b (see Theorem 6.2.1) along with (6.5.1) through (6.5.3), we conclude that

$$P\{\hat{\beta}_{N,L} \le \beta \le \hat{\beta}_{N,U} | \beta\} = 1 - \alpha_N, \tag{6.5.4}$$

and this provides a distribution-free confidence interval for β. Note that in this context, we have made use of the linear rank statistic $L_N(b)$ for $T_N(b)$ along with the null distribution of $L_N(0)$ to arrive at the interval $[\hat{\beta}_{N,L}, \hat{\beta}_{N,U}]$ in (6.5.4).

A very similar case holds for the Kendall tau statistic, where in (6.2.9), instead of the median, we need to take appropriate percentiles of these (divided) differences $\{(X_j - X_i)/(c_j - c_i) : c_i \neq c_j, 1 \le i < j \le N\}$ (see Exercises 6.5.1 and 6.5.2). Similarly, if we consider the simple location model for which in (6.2.1) $\beta = 0$, then, by using Lemma 6.2.3, we may employ the aligned signed rank statistic $S_N(a)$ in (6.2.18) and (6.2.23), and obtain a confidence interval for the location parameter β_0 by equating $S_N(a)$ to the upper and lower $50\alpha_N\%$ points of the null-hypothesis distribution of $S_N(0)$; see Exercises 6.5.3 and 6.5.4. Here also, α_N does not depend on the underlying F and converges to a specified α as $N \to \infty$. For the particular case of the Wilcoxon signed rank statistics, this alignment procedure was first considered by Moses (1953), and for general rank statistics it is discussed in detail in Chapter 6 of Puri and Sen (1971). For both the uniparameter models [relating to the regression slope β (β_0 nuisance) and location β_0 (β specified)], we have thus a genuinely distribution-free confidence interval based on appropriate rank statistics.

Note that by Theorem 5.2.3, defining A_N^2 and C_N^2 as in (5.2.11) and (5.2.12), we obtain that

$$\frac{t_{NU}}{A_N C_N} \to \tau_{\alpha/2} \text{ and } \frac{t_{NL}}{A_N C_N} \to -\tau_{\alpha/2} \quad \text{as} \quad N \to \infty, \tag{6.5.5}$$

where τ_ε is the upper $100\varepsilon\%$ point of the standard normal d.f. Φ. Hence, we may virtually repeat the proof of Theorem 6.2.6 and conclude that for every real t, as $N \to \infty$,

$$P\{C_N(\hat{\beta}_{N,U} - \beta)A_\psi \rho(\psi,\phi) \le t\} \to \Phi(t - \tau_{\alpha/2}), \tag{6.5.6}$$

$$P\{C_N(\hat{\beta}_{N,L} - \beta)A_\psi \rho(\psi,\phi) \le t\} \to \Phi(t + \tau_{\alpha/2}), \tag{6.5.7}$$

where A_ψ and $\rho(\psi,\phi)$ are defined by (6.2.38)–(6.2.39). Exercise 6.5.5 is to give a formal proof of (6.5.6) and (6.5.7). By virtue of (6.5.6), (6.5.7), and the

Jurečková linearity of $L_N(b)$ in (6.2.44), we conclude that

$$C_N(\hat{\beta}_{N,U} - \hat{\beta}_{N,L})A_\psi \rho(\psi, \phi) \xrightarrow{P} 2\tau_{\alpha/2} \quad \text{as} \quad N \to \infty. \tag{6.5.8}$$

Let $\delta_N = \hat{\beta}_{N,U} - \hat{\beta}_{N,L}$ be the width of the confidence interval in (6.5.4). For two score functions ϕ and ϕ^*, we denote the corresponding δ_N by $\delta_N(\phi)$ and $\delta_N(\phi^*)$, respectively. Then, by virtue of (6.5.8), we conclude that

$$\frac{\delta_N(\phi^*)}{\delta_N(\phi)} \xrightarrow{P} \frac{\rho(\psi, \phi)}{\rho(\psi, \phi^*)} \quad \text{as} \quad N \to \infty. \tag{6.5.9}$$

Thus, if we employ the squared length of the confidence interval as a measure of the efficacy (cf. Theorem 3 of Sen, 1966a), then, from the above, we conclude that the ARE of the confidence interval based on the score function ϕ^* with respect to the one based on the score function ϕ is given by $\{\rho(\psi, \phi)/\rho(\psi, \phi^*)\}^2$, and this agrees with the Piman ARE of the corresponding point estimators, discussed in detail in Section 6.2. In passing, we may remark that for the normal-theory model, the width of the confidence interval for β based on the least-squares (maximum-likelihood) estimators is equal to $2C_N^{-1}s_e t_{N-2,\alpha}$, where s_e^2 is the mean square due to error and $t_{N-2,\alpha}$ is the upper $50\alpha\%$ point of the Student t-distribution with $N-2$ DF. Now, $t_{N-2,\alpha} \to \tau_{\alpha/2}$ as $N \to \infty$. Also, for F not necessarily normal but admitting a finite second moment, $s_e^2 \to \sigma^2$, in probability, as $N \to \infty$. Hence, in this case also (6.5.8) holds, with $A_\psi \rho(\psi, \phi)$ replaced by σ^{-1}. Thus, the ARE of the rank procedure in (6.5.4) with respect to the parametric procedure is given by

$$\sigma^2 \rho^2(\psi, \phi) A_\psi^2, \tag{6.5.10}$$

which agrees with (6.2.63). Hence, the discussions following (6.2.63) also apply to the confidence-interval problem.

For the simple location model, (6.5.6)–(6.5.10) all hold, where C_N and $t_{N-2,\alpha}$ are to be replaced by $N^{1/2}$ and $t_{N-1,\alpha}$, respectively. In view of the fact that the details are given in Chapter 6 of Puri and Sen (1971), we put these as Exercises 6.5.6 and 6.5.7.

Consider next the multiple regression model in (6.3.1). For the point estimation of $\boldsymbol{\beta}$, $\mathbf{L}_N(\mathbf{b})$ was employed in (6.3.7)–(6.3.8). Note that under $H_0: \boldsymbol{\beta} = \mathbf{0}$, $A_N^{-2}(\mathbf{L}_N(\mathbf{0}))'\mathbf{C}_N^{-1}(\mathbf{L}_N(\mathbf{0}))$ is a genuinely distribution-free statistic, and hence there exist a real $l_{\alpha N}$ and an $\alpha_n (\le \alpha)$ such that

$$P\left\{ A_N^{-2}(\mathbf{L}_N(\mathbf{0}))'\mathbf{C}_N^{-1}(\mathbf{L}_N(\mathbf{0})) \le l_{\alpha N} | H_0 \right\} = 1 - \alpha_N, \tag{6.5.11}$$

where α_N (independent of F) converges to α as $N \to \infty$. Further, $l_{\alpha N} \to \chi^2_{q,\alpha}$ as N increases. Thus, parallel to (6.3.8), if we define

$$I_N = \left\{\mathbf{b}: A_N^{-2}(\mathbf{L}_N(\mathbf{b}))'\mathbf{C}_N^{-1}(\mathbf{L}_N(\mathbf{b})) \le l_{\alpha N}\right\}, \tag{6.5.12}$$

then I_N provides a confidence region for $\boldsymbol{\beta}$ with converage probability $1 - \alpha_N$ provided that I_N is a closed convex region.

There are, however, some problems in verifying the convexity of I_N. Note that by (6.3.5), $\mathbf{L}_N(\mathbf{b}) = (L_{N1}(\mathbf{b}), \dots, L_{Nq}(\mathbf{b}))'$, where for each j $(= 1, \dots, q)$, $L_{Nj}(\mathbf{b})$ is nonincreasing in b_j when the other $q - 1$ components of $\mathbf{b}$ are held fixed, while this may not be so in b_r for $r \neq j$. As in the case of the point estimates in Section 6.3, one may need more stringent regularity conditions to insure that I_N is a closed convex region, and this in turn makes it difficult to construct genuinely distribution-free confidence regions for $\boldsymbol{\beta}$ in the multiple regression model (6.3.1). Even if I_N can be characterized as a closed convex region, exact determination of this set may become prohibitively laborious as N increases.

For this reason, we make use of the basic linearity results in (6.3.21) and provide some asymptotically distribution-free (ADF) confidence regions which are easier to construct. Note that using (6.3.7), (6.3.8), and (6.3.21), we conclude that for large N, I_N becomes equivalent (in probability) to the following region:

$$I_N^0 = \left\{\mathbf{b}: (\mathbf{b} - \hat{\boldsymbol{\beta}}_N)'\mathbf{C}_N(\mathbf{b} - \hat{\boldsymbol{\beta}}_N) \le \frac{l_\alpha}{A_\psi^2 \rho^2(\psi, \phi)}\right\}, \tag{6.5.13}$$

where $\hat{\boldsymbol{\beta}}_N$ is the point estimator of $\boldsymbol{\beta}$, derived in (6.3.7)–(6.3.8). To derive an ADF confidence region from (6.5.3), we replace the unknown $\gamma = A_\psi \rho(\psi, \phi)$ by a consistent estimator $\hat{\gamma}_N$ which we provide below. Let $\mathbf{e}_j$ be the q-vector with 1 at the jth position and 0 elsewhere, for $j = 1, \dots, q$. For some positive a $(< \infty)$, we define then, for each j $(= 1, \dots, q)$,

$$\hat{\gamma}_{Nj} = (\mathbf{C}_{Nj}\mathbf{e}_j)^{-1} a^{-1} N^{1/2} \frac{L_{Nj}(\hat{\boldsymbol{\beta}}_N - N^{-1/2}a\mathbf{e}_j) - L_{Nj}(\hat{\boldsymbol{\beta}}_N - N^{-1/2}a\mathbf{e}_j)}{2A_N}, \tag{6.5.14}$$

where $\mathbf{C}_{Nj}$ is the jth row (vector) of $\mathbf{C}_N$, defined by (6.3.3). Note that by (6.3.21), for each j $(= 1, \dots, q)$, $\hat{\gamma}_{Nj}$ converges in probability to γ, and hence, as an estimator of γ, we choose

$$\hat{\gamma}_N = q^{-1}(\hat{\gamma}_{N1} + \cdots + \hat{\gamma}_{Nq}). \tag{6.5.15}$$

Let us then consider the set

$$I_N^* = \left\{\mathbf{b} : (\mathbf{b} - \hat{\boldsymbol{\beta}}_N)'\mathbf{C}_N(\mathbf{b} - \hat{\boldsymbol{\beta}}_N) \le \hat{\gamma}_N^{-2}\chi^2_{q,\alpha}\right\}. \qquad (6.5.16)$$

Then I_N^* is the desired (ADF) confidence region for $\boldsymbol{\beta}$ in the model (6.3.1). Note that for the normal-theory model, the confidence region for $\boldsymbol{\beta}$ based on the classical least-squares estimators ($\bar{\boldsymbol{\beta}}_N$) is of the form

$$\bar{I}_N = \left\{\mathbf{b} : (\mathbf{b} - \bar{\boldsymbol{\beta}}_N)'\mathbf{C}_N(\mathbf{b} - \bar{\boldsymbol{\beta}}_N) \le s_e^2 q \mathscr{F}_{q, N-q-1; \alpha}\right\}, \qquad (6.5.17)$$

where the mean square due to error (s_e^2) converges to σ^2, and the critical value $\mathscr{F}_{q, N-q-1; \alpha}$ of the variance ratio statistic converges to $q^{-1}\chi^2_{q,\alpha}$ as N increases. Hence, comparing (6.5.16) and (6.5.17), we conclude that even in the case of the multiple regression model (6.3.1), the ARE of the rank procedure with respect to the classical parametric procedure is given by (6.5.10).

Let us next consider the case of $\boldsymbol{\beta}^* = (\beta_0, \boldsymbol{\beta}')'$ for the model (6.3.1). If we let $\mathbf{b}^* = (b_0, \mathbf{b}')'$ and if $\hat{\boldsymbol{\beta}}_N^* = (\hat{\beta}_{0,N}, \hat{\boldsymbol{\beta}}_N')'$ is the point estimator of $\boldsymbol{\beta}^*$, defined as in (6.3.7)–(6.3.8) and (6.3.42)–(6.3.43), then on letting

$$I_N^{**} = \left\{\mathbf{b} : (\mathbf{b}^* - \hat{\boldsymbol{\beta}}_N^*)'\mathbf{C}_N^{*-1}(\mathbf{b}^* - \hat{\boldsymbol{\beta}}_N^*) \le \hat{\gamma}_N^{-2}\chi^2_{q+1,\alpha}\right\}, \qquad (6.5.18)$$

where $\hat{\gamma}_N$ is defined as in (6.5.15) and

$$\mathbf{C}_N^* = \begin{pmatrix} N & \mathbf{0} \\ \mathbf{0} & \mathbf{C}_N \end{pmatrix},$$

we conclude on using (6.3.21) and (6.3.44) that I_N^{**} is an ADF confidence region for $\boldsymbol{\beta}^*$ with asymptotic coverage probability equal to $1 - \alpha$. Exercise 6.5.10 is to verify these details. The ARE results remain the same in this case too.

Finally, let us comment on the general multivariate regression model in (6.4.1). With the aid of the point estimators developed in Section 6.4, an ADF confidence region can be constructed as in (6.5.16) or (6.5.18), where $\chi^2_{q,\alpha}$ (or $\chi^2_{q+1,\alpha}$) needs to be replaced by $\chi^2_{pq,\alpha}$ (or $\chi^2_{p(q+1),\alpha}$), and also we need to estimate the matrix $\boldsymbol{\nu}$ defined in (6.4.16). These are worked out in detail in Section 7.5 [see (7.5.11) through (7.5.25)]. For the case of the multivariate one-sample model [for which in (6.4.1) $\boldsymbol{\beta} = \mathbf{0}$], some of the details are given in Chapter 6 of Puri and Sen (1971).

EXERCISES

6.2.1. Provide a formal proof of Lemma 6.2.2.

6.2.2. Provide a formal proof of the assertion (6.2.17).

6.2.3. Prove that $S_N(a)$ defined in (6.2.21) is nonincreasing in a.

6.2.4. Provide a formal proof of Lemma 6.2.3.

6.2.5. Provide a formal proof of the assertion (6.2.25).

6.2.6. Provide the missing details in the proof of Theorem 6.2.6.

6.2.7. Provide a formal proof of the assertion (6.2.42).

6.2.8. Prove that $N^{1/2}|\hat{\beta}_N - \beta| = O_p(1)$, where $\hat{\beta}_N$ is as in Theorem 6.2.7.

6.2.9. Provide a formal proof of Proposition 6.2.6.1.

6.2.10. Provide a formal proof of Proposition 6.2.6.2.

6.2.11. Provide a formal proof of the assertion (6.2.64).

6.2.12. Prove that $N^{1/2}(\dot{\beta}_N - \beta, \dot{\beta}_{0N} - \beta)$ has asymptotically (as $N \to \infty$) the $\mathcal{N}(0, T)$ distribution, where T is given by (6.2.65). (Hill, 1962.)

6.2.13. Verify the assertion (6.2.66).

6.2.14. Provide formal proofs of the assertions (6.2.72), (6.2.73), and (6.2.74).

6.3.1. Verify the assertion (6.3.11).

6.3.2. Provide formal proofs of the assertions (6.3.23) and (6.3.24).

6.3.3. Provide a formal proof of the assertion (6.3.26).

6.3.4. Provide a formal proof of the assertion (6.3.28). (Jurečková, 1971a.)

6.3.5. Provide a formal proof of the assertion (6.3.30).

6.3.6. Provide a formal proof of the assertion (6.3.31).

6.3.7. Verify the assertions (6.3.33) and (6.3.34).

6.3.8. Provide a formal proof of the assertion (6.3.36).

6.3.9. Provide a formal proof of Theorem 6.3.1.

6.3.10. Verify the assertion (A.4.15) in Appendix A.4.3.

6.3.11. Provide a formal proof of the assertion (A.4.17).

6.3.12. Provide a formal proof of the assertion (A.4.18).

6.3.13. Provide a formal proof of the assertion (A.4.19).

6.3.14. Provide a formal proof of the assertion (A.4.20).

6.4.1. Provide the missing details in the proof of Theorem 6.4.1.

6.4.2. Provide a formal proof of Theorem 6.4.2.

6.4.3. Verify the assertions (6.4.26), (6.4.27), and (6.4.28).

6.4.4. Provide a formal proof of the assertion (6.4.30).

6.4.5. Provide a formal proof of the assertion (6.4.33).

6.4.6. Verify the assertion (6.4.37).

6.4.7. Verify the assertion (6.4.46).

6.4.8. Verify the assertion (6.4.49).

6.4.9. Provide formal proofs of the assertions (6.4.65) and (6.4.66).

6.4.10. Provide formal proofs of the assertions (6.4.67) and (6.4.68).

6.4.11. Provide a formal proof of Theorem 6.4.4.

6.4.12. Provide formal proofs of the assertions (6.4.74) and (6.4.76).

In the following, let X_i, $1 \le i \le N$, be an independent random sample from a distribution with finite Fisher information. Consider the statistics $S_{\Delta N} = N^{-1}\Sigma_{i=1}^N c_{Ni} R_{Ni}^{\Delta}$, where R_{Ni}^{Δ} is the rank of $X_i + \Delta d_{Ni}$ among $\{X_j + \Delta d_{Nj}, 1 \le j \le N\}$, $\Delta \in [0,1]$, and d_{Ni} and c_{Ni}, $1 \le i \le N$, are real constants. Assume that $\Sigma c_{Ni} = \Sigma d_{Ni} = 0$, $\Sigma c_{Ni}^2 = \Sigma d_{Ni}^2 = 1$, $\lim_{N\to\infty} \max_{1\le i\le N} c_{Ni}^2 = \lim_{N\to\infty} \max_{1\le i\le N} d_{Ni}^2 = 0$, $\lim_{N\to\infty} \Sigma c_{Ni} d_{Ni} = b^2 > 0$, and $\lim_{N\to\infty}[\max_{1\le i\le N}(c_{Ni}^2 d_{Ni}^2)(\Sigma_{i=1}^N c_{Ni}^2 d_{Ni}^2)^{-1}] = 0$. For fixed Δ and N, let $\hat{S}_{\Delta N} = \Sigma_{i=1}^N E(S_{\Delta N} | X_i) - (N-1)ES_{\Delta N}$, $Z_N = \Sigma_{i=1}^k \lambda_i (S_{\Delta_i} - S_{0N})$, and $\hat{Z}_N = \Sigma_{i=1}^k \lambda_i(\hat{S}_{\Delta_i} - \hat{S}_{0N})$, where λ_i, $1 \le i \le k$, are any fixed real numbers and $0 \le \Delta_1 < \Delta_2 < \cdots < \Delta_k \le 1$ are parameter values.

6.4.13. Show that $E\hat{Z}_N = EZ_N = \Sigma_{i=1}^k \lambda_i ES_{\Delta_i N}$, $E(Z_N - \hat{Z}_N)^2 = \operatorname{Var} Z_N - \operatorname{Var} \hat{Z}_N$, and $\lim_{N\to\infty}[(\operatorname{Var} Z_N - \operatorname{Var} \hat{Z}_N)(\operatorname{Var} \hat{Z}_N)^{-1}] = 0$. (Jurečková 1973.)

6.4.14. (continued.) Show that $\hat{Z}_N$ (and consequently Z_N) has asymptotically, as $N \to \infty$, the normal $(E\hat{Z}_N, \operatorname{Var} \hat{Z}_N)$ distribution. (Jurečková, 1973.)

6.4.15. (continued.) Consider the random process $\mathscr{S}_{\Delta N} = A_N^{-1}[S_{\Delta N} - S_{0N} - ES_{\Delta N}]$ where $A_N^2 = \Sigma_{i=1}^N c_{Ni}^2 d_{Ni}^2 + 3N^{-1}(\Sigma_{i=1}^N c_{Ni} d_{Ni})^2$. Show that the random vector $(S_{\Delta_1 N}, \dots, S_{\Delta_k N})$ has asymptotically, as $N \to \infty$, the normal distribution with mean vector zero and covariance matrix $(\sigma_{ij})_{i,j=1,\dots,k}$, where $\sigma_{ij} = C\Delta_i\Delta_j$ and $C = \int f^3(x)\,dx - [\int f^2(x)\,dx]^2$. (Jurečková, 1973.)

6.4.16. (continued.) Prove that the sequence of distributions of $\mathscr{S}_{\Delta N}$ is tight. (Jurečková, 1973.) [The sequence of distributions is tight if for each $\varepsilon > 0$, $\eta > 0$, there exists a δ, $0 < \delta < 1$ and an integer N_0 such that $P\{\omega''(S_N, \delta) \geq \varepsilon\} \leq \eta$, $N \geq N_0$, where $\omega''(\mathscr{S}_N, \delta) = \sup \min\{|S_{\Delta N} - S_{\Delta_1 N}, S_{\Delta N} - S_{\Delta_2 N}|\}$, where the supremum is over Δ_1, Δ_2, and Δ satisfying $\Delta_1 \leq \Delta \leq \Delta_2$ and $\Delta_2 - \Delta_1 \leq \delta$. (Billingsley, 1968.)]

For results corresponding to Exercises 6.4.13–6.4.16 for a wider class of score generating functions see Carlson and Puri (1985), Puri and Wu (1984, 1985a), Hušková and Jurečková (1981), and Hušková (1982).

6.5.1. Show that for the model (6.2.1), under $H_0: \beta = 0$, the Kendall tau statistic T_N in (6.2.8) is distribution-free with mean 0 and variance $v_N^2 = \{N(N-1)(2N+5) - \sum_{j=1}^{t} u_j(u_j - 1)(2u_j + 5)\}/18$, where $u_1, \dots, u_t$ ($t \geq 2$) are the frequencies with which the c_i assume the distinct values $c_{(1)} < \cdots < c_{(t)}$. Hence, or otherwise, show that (6.5.1) holds for T_N in (6.2.8), where $t_{NU}/v_N \to \tau_{\alpha/2}$ and $t_{NL}/v_N \to -\tau_{\alpha/2}$, as $N \to \infty$. Show further that the cardinality of the set $S_N = \{(i, j): c_i \neq c_j, 1 \leq i < j \leq N\}$ is equal to

$$\sum_{1 \leq i < j \leq t} u_i u_j = \binom{N}{2} - \sum_{j=1}^{t} \binom{u_j}{2} = N^*, \quad \text{say.}$$

Denote the ordered values of the N^* divided differences $\{(X_j - X_i)/(c_j - c_i): (i, j) \in S_N\}$ by $b_{(1)} \leq \cdots \leq b_{(N^*)}$. Then, show that for the Kendall tau statistic, (6.5.4) reduces to $b_{(M_1)} \leq \beta \leq b_{(M_2)}$, where $M_1 = (N^* - t_{NU})/2$ and $M_2 = (N^* - t_{NL})/2$. (Sen, 1968.)

6.5.2. (continued.) Define $C_N^2 = \sum_{i=1}^{N}(c_i - \bar{c}_N)^2$, $A_N^2 = \{N(N^2 - 1) - \sum_{j=1}^{t} u_j(u_j^2 - 1)\}/12$, and $r_N = \sum_{i=1}^{N}[i - (N+1)/2](c_i - \bar{c}_N)/(C_N A_N)$. Then, show that for the Kendall tau statistic, for $\hat{\beta}_{N,L}$ and $\hat{\beta}_{N,U}$ as obtained in the preceeding problem, we have $r_N C_N(\hat{\beta}_{N,U} - \hat{\beta}_{N,L}) \to \tau_{\alpha/2}(3^{1/2}\int_{-\infty}^{\infty} f^2(x)\,dx)^{-1}$ in probability as $N \to \infty$. Hence, or otherwise, comment on the ARE of this procedure, with especial attention to the possible range of r_N. Can r_N be equal to 1? Obtain the confidence interval for the difference of locations in the two-sample model as a special case of the Kendall tau, and comment on r_N in this case. (Sen, 1968.)

6.5.3. Consider the model (6.2.1) with $\beta = 0$, so that we have the one-sample location model. Let $Y_{ij} = (X_i + X_j)/2$ for $1 \leq i \leq j \leq N$, and denote the ordered values of these $M = \binom{N+1}{2}$ midranges by $Y_{(1)} \leq \cdots \leq Y_{(M)}$. Show that for $S_N(a)$, if we use the Wilcoxon signed rank statistic, then the confidence interval for β_0 is given by $Y_{(M^*)} \leq \beta_0 \leq Y_{(M+1-M^*)}$, where M^* is related to the critical value of the test statistic. (Moses, 1953.)

6.5.4. Use the sign statistic instead of the Wilcoxon signed rank statistic, and show that the confidence limits are given by two order statistics of the X_i and the binomial law determines their ranks.

6.5.5. Use the identity [by (6.5.3)] that for every real u, $P\{\hat{\beta}_{N,U} < \beta + u|\beta\} = P\{T_N(u) < t_{NL}|\beta = 0\}$, along with the linearity result in (6.2.44) and the asymptotic normality of $T_N(0)$ under $H_0: \beta = 0$, to verify (6.5.6). Provide a parallel proof of (6.5.7).

6.5.6. For the signed rank statistic $S_N(a)$, show that under (6.2.1) with $\beta_0 = \beta = 0$, $S_N(0)$ has a distribution symmetric about 0, so that $t_{NL} = -t_{NU}$. Hence, use the asymptotic normality of $S_N(0)$ to verify (6.5.5) with C_N^2 replaced by N. Use the same technique as in the preceeding exercise to verify that (6.5.6) and (6.5.7) also hold for the location model, where again C_N^2 has to be replaced by N.

6.5.7. Use (6.3.44) and verify that (6.5.8) holds for the confidence interval based on the signed rank statistic, where C_N^2 needs to be replaced by N.

6.5.8. Show that the estimator $\hat{\gamma}_N$ in (6.6.16) converges to $\gamma = A_\psi \rho(\psi, \phi)$, in probability, as $N \to \infty$. Hence, or otherwise, use (6.5.12) and (6.3.21) and verify that (6.5.16) is stochastically equivalent to (6.5.13), which in turn is stochastically equivalent to (6.5.12).

6.5.9. With reference to (6.5.17), show that (1) $q\mathcal{F}_{q, N-q; \alpha} \to \chi^2_{q,\alpha}$ as $N \to \infty$, (2) $s_e^2 \to \sigma^2$ in probability as $N \to \infty$, whenever the underlying d.f. has a finite second moment, and (3) the linear estimates $\bar{\boldsymbol{\beta}}_N$ have jointly (asymptotically) a multinormal distribution under the usual assumptions of the Noether condition and finite second moments. Hence, show that (6.5.17) provides an asymptotic confidence region for $\boldsymbol{\beta}$ with (asymptotic) coverage probability $1 - \alpha$. (Sen and Puri, 1970.)

6.5.10. Verify that (6.5.18) provides an asymptotically distribution-free confidence region for $\boldsymbol{\beta}^*$ with asymptotic coverage probability $1 - \alpha$.

CHAPTER 7

Asymptotically Distribution-Free Aligned Rank-Order Tests for Some General Linear Hypotheses

7.1. INTRODUCTION

In Chapter 5, for some univariate as well as multivariate linear models, under suitable hypotheses of invariance (of the joint distribution of the sample observations under appropriate groups of transformations mapping the sample space onto itself), some genuinely distribution-free tests were considered. For a general linear model, particularly, in the context of subhypothesis testing, this invariance does not generally hold, and hence, it may not be possible to formulate genuinely distribution-free tests based on the ranks of the original observations. In some cases, it may be possible to eliminate some of the nuisance parameters by suitable transformations (data reduction) and then apply the theory developed in Chapter 5 to such a reduced model. But such a transformation may introduce some arbitrariness in the choice of the reduced data set (leading to possibly different conclusions from the same original data set) and involve some loss of information due to data reduction. The theory developed in Chapter 6 enables us to provide an alternative procedure where the nuisance parameters are eliminated through estimation and the arbitrariness mentioned above is avoided by the use of *aligned rank-order statistics*. These aligned statistics based on the ranks of the residuals are not in general strictly distribution-free. Nevertheless, they are robust and asymptotically distribution-free (ADF).

In this chapter, tests for general linear hypotheses based on aligned rank-order statistics are considered, and their various (asymptotic) properties (including efficiency and optimality) are studied. In Section 7.2, we start with the simple regression model. Section 7.3 deals with the univariate multiple regression setup, and their multivariate counterparts are considered in Sections 7.4 and 7.5. Simultaneous inference procedures are also considered here. Rank tests for restricted alternatives based on Roy's (1953) union–intersection (UI) principle are considered in Section 7.6. The concluding section is devoted to the theory of *preliminary-test inference* (both estimation and testing) procedures based on suitable rank statistics. Rank tests for linear models in the case of stochastic predictors with relevance to the so-called *growth-curve models* and for grouped data problems will be considered in Chapter 8.

7.2. ALIGNED RANK-ORDER TESTS FOR THE INTERCEPT IN A SIMPLE REGRESSION MODEL

As in Section 5.2, let $X_1, \ldots, X_N$ be independent real-valued r.v.'s with

$$F_i(x) = P\{X_i \le x\} = F(x - \beta_0 - \beta(c_i - \bar{c}_N)),$$

$$1 \le i \le N, \quad -\infty < x < \infty, \tag{7.2.1}$$

where β_0 (intercept) and β (slope) are unknown parameters, $c_1, \ldots, c_N$ are known constants, $\bar{c}_N = N^{-1}\Sigma_{i=1}^N c_i$, and F is an unspecified continuous d.f. As in (5.2.4), consider the null hypothesis

$$H_0: \beta_0 = 0 \text{ vs. } H_1: \beta_0 > 0 \qquad (\text{on} \quad H_2: \beta_0 \ne 0), \tag{7.2.2}$$

where β is treated as a nuisance parameter. Note that for $\beta \ne 0$, under $H_0: \beta_0 = 0$, the X_i are not necessarily i.i.d.r.v.'s, and hence, no hypothesis of invariance holds. In some cases (e.g. $c_i = a + b_i$, $1 \le i \le N$), one may match X_i (viz., $Y_i = (X_i + X_{N-i+1})/2$, $1 \le i \le [(N+1)/2]$), and on the reduced variables ($Y_1, \ldots, Y_{N^*}$, $N^* = [(N+1)/2]$) use the usual signed rank statistics for testing $H_0: \beta_0 = 0$. In general, for arbitrary $c_1, \ldots, c_N$, such a matching poses the following problems: (1) the reduced r.v.'s may not be i.i.d.r.v.'s, (2) there may not be any unique choice of the pairs to form the reduced variables, and (3) some loss of information may occur due to this data reduction. For these reasons, we proceed to estimate first the nuisance parameter β. As in Section 6.2, we denote such an estimator of β by $\hat{\beta}_N$ and consider the residuals $\hat{X}_i = X_i - \hat{\beta}_N(c_i - \bar{c}_N)$, $1 \le i \le N$. The main idea is to construct suitable (aligned) rank-order statistics based on these residuals, to show that such statistics are asymptotically distribution-free, and to employ them for testing H_0 in (7.2.2).

With the above motivation, we define, as in (6.2.2), $X_i(b) = X_i - b(c_i - \bar{c}_N)$, $1 \le i \le N$, $-\infty < b < \infty$, and let

$$L_N(b) = \sum_{i=1}^{N} (c_i - \bar{c}_N) a_N(R_{Ni}(b)), \tag{7.2.3}$$

where the scores $a_N(1), \dots, a_N(N)$ are defined as in (6.2.10) and

$$R_{Ni}(b) = \sum_{j=1}^{N} u(X_i - bc_i - X_j + bc_j), \qquad 1 \le i \le N. \tag{7.2.4}$$

As in (6.2.5) and (6.2.6), let

$$\hat{\beta}_{N,1} = \inf\{b: L_N(b) < 0\}, \quad \hat{\beta}_{N,2} = \sup\{b: L_N(b) > 0\};$$

$$\hat{\beta}_N = \tfrac{1}{2}(\hat{\beta}_{N,1} + \hat{\beta}_{N,2}). \tag{7.2.5}$$

Having obtained the estimate $\hat{\beta}_N$, consider the residuals $\hat{X}_i = X_i - \hat{\beta}_N(c_n - \bar{c}_N)$, $1 \le i \le N$, and the aligned signed rank statistic

$$\hat{S}_N = \sum_{i=1}^{N} \operatorname{sgn}(\hat{X}_i)\, a_N^*(\hat{R}_{Ni}^+), \tag{7.2.6}$$

where $\hat{R}_{Ni}^+ = \sum_{j=1}^{N} u(|X_i| - X_j|)$, $1 \le i \le N$, and the scores $a_N^*(1), \dots, a_N^*(N)$ are defined as in (6.2.23). Also, let

$$T_N = N^{-1/2} A_N^{*-1} \hat{S}_N, \qquad \text{where} \quad A_N^{*2} = N^{-1} \sum_{i=1}^{N} \left[a_N^*(i)\right]^2. \tag{7.2.7}$$

For testing $H_0: \beta_0 = 0$, we use T_N as a test statistic. Towards this, first, we consider the following.

Theorem 7.2.1. *Suppose that (1) F is symmetric about 0 and has an absolutely continuous p.d.f. f with finite Fisher information $I(f)$, (2) $N^{-1}\sum_{i=1}^{N}(c_i - \bar{c}_N)^2 \to C^2$ $(0 < C < \infty)$ and $N^{-1}\{\max_{1 \le i \le N}(c_i - \bar{c}_N)^2\} \to 0$ as $N \to \infty$, (3) the scores are defined by $a_N(i) = E\phi(U_{Ni})$ or $\phi(i/(N+1))$, and $a_N^*(i) = E\phi^*(U_{Ni})$ or $\phi^*(i/(N+1))$, $1 \le i \le N$, where $U_{N1} < \cdots < U_{NN}$ are the ordered r.v.'s of a sample of size N from the uniform (0, 1) d.f., and (4) $\phi^*(u) = \phi((1+u)/2)$, $0 < u < 1$, where $\phi(t)$ is nondecreasing in t, $\phi(t) + \phi(1-t) = 0$ $\forall t \in (0,1)$, and $0 < A_\phi^2 = \int_0^1 \phi^2(u)\,du < \infty$. Then under $H_0: \beta_0 = 0$, T_N has asymptotically a normal distribution with 0 mean and unit variance.*

Proof. Note that under (3) and (4), $A_N^* \to A_\phi$ as $N \to \infty$. Also, by Theorem 6.2.7, $N^{1/2}(\hat{\beta}_N - \beta)$ is $O_p(1)$, and hence, by (6.2.45) [where we let

$a = 0$ and $b = N^{1/2}(\hat{\beta}_N - \beta)$], under H_0, as $N \to \infty$,

$$N^{-1/2}\left[\hat{S}_N - S_N(\beta)\right] \xrightarrow{P} 0, \tag{7.2.8}$$

where $S_N(\beta) = \sum_{i=1}^N \operatorname{sgn} X_i(\beta)\, a_N^*(R_{Ni}^+(\beta))$, $X_i(\beta) = X_i - \beta(c_i - \bar{c}_N)$, and $R_{Ni}^+(\beta) = \sum_{j=1}^N u(|X_i(\beta)| - |X_j(\beta)|)$, $1 \le i \le N$. Note that under $H_0: \beta_0 = 0$, $X_i(\beta)$, $i = 1, \ldots, N$ are i.i.d.r.v.'s with d.f. F symmetric about 0, and hence, by Theorem 5.2.4, $N^{-1/2}S_N(\beta) \sim \mathcal{N}(0, A_\phi^2)$. The proof of the theorem now follows from (7.2.8).

By virtue of Theorem 7.2.1, we have the following asymptotically distribution-free test: For testing $H_0: \beta_0 = 0$ vs. $H_1: \beta_0 > 0$ (or $H_2: \beta_0 \neq 0$), reject the null hypothesis when $T_N \ge \tau_\alpha$ (or $|T_N| \ge \tau_{\alpha/2}$), where τ_ε is the upper $100\varepsilon\%$ point of the standard normal d.f. and α ($0 < \alpha < 1$) is the desired level of significance of the test.

For the study of the asymptotic power properties of this aligned rank test, as in Section 5.5, we consider a sequence of local alternatives (for which the power is bounded away from 1). Assume

$$H_{1(N)}: (7.2.1) \text{ holds for } \beta_0 = N^{-1/2}\lambda, \text{ for some (fixed) } \lambda. \tag{7.2.9}$$

Also, let $\psi(u)$, $0 < u < 1$, A_ψ^2, A_ϕ^2 and $\rho(\phi, \psi)$ be as defined in (6.2.38)–(6.2.39). Then, by an appeal to (6.2.45), (7.2.8), and the contiguity of the probability measure under $\{H_{1(N)}\}$ with respect to the measure under H_0 [insured by (7.2.9) and assumptions (1), (2), (3), and (4) of Theorem 7.2.1], we arrive at the following.

Theorem 7.2.2. *Under* $\{H_{1(N)}\}$ *in* (7.2.9) *and assumptions* (1), (2), (3) *and* (4) *of Theorem* 7.2.1, T_N *is asymptotically normal with mean* $\lambda A_\psi \rho(\phi, \psi)$ *and variance* 1.

It follows from Theorem 7.2.2 that under $\{H_{1(N)}\}$, the asymptotic power of the aligned rank test is equal to

$$1 - \Phi\left(\tau_\alpha - \lambda A_\psi \rho(\phi, \psi)\right) \qquad \text{(one-sided case)}, \tag{7.2.10}$$

$$1 - \Phi\left(\tau_{\alpha/2} - \lambda A_\psi \rho(\phi, \psi)\right) + \Phi\left(-\tau_{\alpha/2} - \lambda A_\psi \rho(\phi, \psi)\right) \qquad \text{(two-sided case)}, \tag{7.2.11}$$

where $\Phi(x)$ is the standard normal d.f.

To study the asymptotic efficiency and optimality of the aligned rank test, we consider some alternative tests for $H_0: \beta_0 = 0$. In the classical (normal F) case, the least-squares estimator of β_0 is

$$\tilde{\beta}_{0N} = \overline{X}_N = N^{-1} \sum_{i=1}^{N} X_i, \tag{7.2.12}$$

and the likelihood-ratio statistic reduces to

$$\lambda_N = N^{1/2} \overline{X}_N / s_N, \tag{7.2.13}$$

where $s_N^2 = (N-2)^{-1} \Sigma_{i=1}^N [X_i - \overline{X}_N - \tilde{\beta}_N(c_i - \bar{c}_N)]^2$, and $\tilde{\beta}_N = \Sigma_{i=1}^N (X_i - \overline{X}_N)(c_i - \bar{c}_N) / \Sigma_{i=1}^N (c_i - \bar{c}_N)^2$ is the least-squares estimator of β. One can express

$$s_N^2 = \frac{N}{N-2} \left\{ \frac{1}{N} \sum_{i=1}^{N} e_i^2 - (\overline{X}_N - \beta_0)^2 - (\tilde{\beta}_N - \beta)^2 \frac{1}{N} \sum_{i=1}^{N} (c_i - \bar{c}_N)^2 \right\}, \tag{7.2.14}$$

where $e_i = X_i - \beta_0 - \beta(c_i - \bar{c}_N)$, $i = 1, \ldots, N$, are i.i.d.r.v.'s with mean 0 and variance σ^2 (assumed to be finite). Thus, by the Khinchin law of large numbers,

$$N^{-1} \sum_{i=1}^{N} e_i^2 \xrightarrow{P} \sigma^2 \quad \text{as} \quad N \to \infty, \tag{7.2.15}$$

while by the consistency of the least-squares estimators,

$$\overline{X}_N - \beta_0 \xrightarrow{P} 0 \quad \text{and} \quad \tilde{\beta}_N - \beta \xrightarrow{P} 0.$$

Hence, under (2) of Theorem 7.2.1,

$$s_N^2 \xrightarrow{P} \sigma^2 \quad \text{as } N \to \infty \qquad \text{whenever} \quad \sigma^2 < \infty. \tag{7.2.16}$$

Using the central limit theorem on $N^{1/2}\overline{X}_N$ and the Slutzky theorem, we obtain from (7.2.15) that under $\{H_{1(N)}\}$,

$$\lambda_N \xrightarrow{\mathcal{L}} \mathcal{N}(\lambda/\sigma, 1) \tag{7.2.17}$$

and hence, under $H_0: \beta_0 = 0$, λ_N is asymptotically $\mathcal{N}(0,1)$. We leave the details to an exercise (Exercise 7.2.1).

In view of (7.2.12) and (7.2.17), the ARE of $\{T_N\}$ with respect to $\{\lambda_N\}$ [for $\{H_{1(N)}\}$ in (7.2.9)] is given by

$$e(T_N, \lambda_N) = \sigma^2 A_\psi^2 \rho^2(\phi, \psi)$$

$$= \frac{\sigma^2 \left(\int_0^1 \phi(u) \psi(u) \, du \right)^2}{\int_0^1 \phi^2(u) \, du}. \tag{7.2.18}$$

We have encountered this familiar expression in Section 5.7 [see (5.7.5)], and therefore we omit detailed discussion of it.

Instead of the normal-theory likelihood-ratio test (based on the least-squares estimators), one may also use the likelihood-ratio test when the form of F is assumed to be specified. In that case (see Exercise 7.2.2), parallel to (7.2.17), we would have an asymptotic normal distribution with mean $\lambda I(f)$ and variance 1. Hence, the ARE would be equal to

$$\frac{A_\psi^2 \rho^2(\phi, \psi)}{I(f)} = \rho^2(\phi, \psi), \qquad \text{as} \quad A_\psi^2 = I(f). \tag{7.2.19}$$

Hence, an optimal choice of the score function, for a given F, is

$$\phi(u) \equiv -\frac{f'(F^{-1}(u))}{f(F^{-1}(u))} = \psi(u), \qquad 0 < u < 1. \tag{7.2.20}$$

In Section 6.2.3 [see (6.2.27) and (6.2.28)], we have considered $\hat{\beta}_{0,N}$ as an estimator of β_0 when β is unknown. Also, in Theorem 6.2.7, it has been shown that as $N \to \infty$,

$$N^{1/2}(\hat{\beta}_{0,N} - \beta_0) \xrightarrow{\mathcal{L}} \mathcal{N}\left(0, \left[A_\psi \rho(\phi, \psi)\right]^{-2}\right). \tag{7.2.21}$$

Moreover, if we define $\hat{R}_{Ni}^+(a) = \sum_{j=1}^N u(|\hat{X}_i - a| - |\hat{X}_j - a|)$, $1 \le i \le N$,

$$\hat{S}_N(a) = \sum_{i=1}^N \operatorname{sgn}(\hat{X}_i - a) \, a_N^*(\hat{R}_{Ni}^+(a)), \tag{7.2.22}$$

and

$$\hat{\beta}_{0L,N} = \inf\{a: N^{-1/2}\hat{S}_N(a) < A_\phi \tau_{\alpha/2}\}, \tag{7.2.23}$$

$$\hat{\beta}_{0U,N} = \sup\{a: N^{-1/2}\hat{S}_N(a) > -A_\phi \tau_{\alpha/2}\}, \tag{7.2.24}$$

then it follows by using (6.2.45) that as $N \to \infty$,

$$B_N = \frac{2\tau_{\alpha/2}}{N^{1/2}(\hat{\beta}_{0U,N} - \hat{\beta}_{0L,N})} \xrightarrow{P} A_\psi \rho(\phi, \psi). \tag{7.2.25}$$

Consequently, by (7.2.21), (7.2.25), and the Slutzky theorem,

$$2\tau_{\alpha/2}\frac{\hat{\beta}_{0,N} - \beta_0}{\hat{\beta}_{0U,N} - \hat{\beta}_{0L,N}} \xrightarrow{\mathscr{L}} \mathscr{N}(0,1). \tag{7.2.26}$$

Thus, a large-sample test for $H_0: \beta_0 = 0$ can be based on the statistic

$$T_N^* = \frac{2\tau_{\alpha/2}\hat{\beta}_{0,N}}{\hat{\beta}_{0U,N} - \hat{\beta}_{0L,N}}. \tag{7.2.27}$$

By virtue of (7.2.25) and (7.2.26), the asymptotic power of the test based on T_N^* [for $\{H_{1(N)}\}$ in (7.2.9)] is again given by (7.2.12) or (7.2.13), so that if T_N and T_N^* both involve the same score function ϕ, then the tests based on T_N and T_N^* are asymptotically power-equivalent. From computational aspect, T_N^* involves the additional computation of $\hat{\beta}_{0,N}$, $\hat{\beta}_{0U,N}$, and $\hat{\beta}_{0L,N}$ (requiring usually trial-and-error solution) and hence we recommend the use of T_N.

We conclude this section with the remark that (6.2.45) underlies both (7.2.8) and (7.2.25). But (6.2.45) is not dependent on b as far as the displacement [viz. $a\rho(\phi,\psi)A_\phi A_\psi$] is concerned. This shows that for defining the residuals $\hat{X}_i$ (needed for both T_N and T_N^*), one may replace $\hat{\beta}_N$ in (7.2.5) by any other estimator $\tilde{\beta}_N$ such that $N^{1/2}(\tilde{\beta}_N - \beta) = O_p(1)$. For example, we may take $\tilde{\beta}_N$ as the least-squares estimator, defined before (7.2.16) Other possibilities also remain open. For the least-squares estimator, we need that F has a finite second moment; alternative conditions may be needed for other estimators. One advantage of using $\hat{\beta}_N$ in (7.2.5) is that we do not need any additional restriction on the score function ϕ or the d.f. F, besides the ones needed to justify (6.2.45).

7.3. ALIGNED RANK TESTS FOR SUBHYPOTHESES IN MULTIPLE LINEAR REGRESSION MODELS

As in (6.3.1), let $X_1, \ldots, X_N$ be independent r.v.'s such that

$$F_i(x) = P\{X_i \le x\} = F\big(x - \beta_0 - \boldsymbol{\beta}'(\mathbf{c}_i - \bar{\mathbf{c}}_N)\big),$$
$$1 \le i \le N, \quad -\infty < x < \infty, \tag{7.3.1}$$

where $\mathbf{c}_i = (c_{i1}, \ldots, c_{iq})'$, $1 \le i \le N$, are known vectors of regressors, $\bar{\mathbf{c}}_N = (1/N)\Sigma_{i=1}^N \mathbf{c}_i$, and $\boldsymbol{\beta}^{*\prime} = (\beta_0, \boldsymbol{\beta}')$ is an unknown parametric vector [where $\boldsymbol{\beta}' = (\beta_1, \ldots, \beta_q)$] and $q \ge 1$. A direct generalization of (7.2.2) would be to consider

$$H_0^{(1)}: \beta_0 = 0 \quad \text{vs.} \quad H_1^{(1)}: \beta_0 > 0 \quad \big(\text{or } H_2^{(1)}: \beta_0 \ne 0\big), \tag{7.3.2}$$

where $\boldsymbol{\beta}$ is unspecified. We may also partition $\boldsymbol{\beta}$ as

$$\underset{1\times q}{\boldsymbol{\beta}'} = (\underset{1\times q_1}{\boldsymbol{\beta}_1'}, \underset{1\times q_2}{\boldsymbol{\beta}_2'}), \tag{7.3.3}$$

where $q_1 \ge 0$, $q_2 \ge 0$, $q_1 + q_2 = q \ge 1$, and need to test

$$H_0^{(2)}: \boldsymbol{\beta}_2 = \mathbf{0} \quad \text{vs.} \quad H_1^{(2)}: \boldsymbol{\beta}_2 \ne \mathbf{0}, \tag{7.3.4}$$

where $\beta_0, \boldsymbol{\beta}_1$ are not specified. We may also need to test

$$H_0^{(3)}: \beta_0 = 0, \boldsymbol{\beta}_2 = \mathbf{0} \quad \text{vs.} \quad H_1^{(3)}: (\beta_0, \boldsymbol{\beta}_2')' \ne \mathbf{0}, \tag{7.3.5}$$

where $\boldsymbol{\beta}_1$ is not specified. Note that for $q_2 = 0$, (7.3.5) reduces to (7.3.2), and hence, in the sequel, we shall consider the case of (7.3.4) and (7.3.5) in detail, and later on briefly present the case of (7.3.2).

7.3.1. Aligned Rank Tests for $H_0^{(2)}$

We refer back to Section 6.3 for notation and define

$$\mathbf{L}_N(\mathbf{b}) = \big(L_{N1}(\mathbf{b}), \ldots, L_{Nq}(\mathbf{b})\big)', \qquad \mathbf{b} \in R^q, \tag{7.3.6}$$

as in (6.3.5). Let then

$$[\mathbf{L}_N(\mathbf{b})]' = \big([\mathbf{L}_{N(1)}(\mathbf{b})]', [\mathbf{L}_{N(2)}(\mathbf{b})]'\big), \tag{7.3.7}$$

where

$$\mathbf{L}_{N(1)}(\mathbf{b}) = \left(L_{N1}(\mathbf{b}), \ldots, L_{Nq_1}(b)\right)',$$

$$\mathbf{L}_{N(2)}(\mathbf{b}) = \left(L_{Nq_1+1}(\mathbf{b}), \ldots, L_{Nq}(\mathbf{b})\right)'. \tag{7.3.8}$$

We also define $\mathbf{C}_N$ as in (6.3.3), and assume that (6.3.12) through (6.3.20) all hold. In addition, we assume that F in (7.3.1) belongs to the class of absolutely continuous d.f.'s with absolutely continuous pdf f having a finite Fisher information $I(f) = \int (f'/f)^2 \, dF$. Furthermore, $\phi(u)$, the score function, is nondecreasing in u,

$$\int_0^1 |\phi(u)| \{u(1-u)\}^{-1/2} \, du < \infty, \tag{7.3.9}$$

and

$$0 < A_\phi^2 = \int_0^1 \phi^2(u) \, du - \left(\int_0^1 \phi(u) \, du\right)^2 < \infty. \tag{7.3.10}$$

To test $H_0^{(2)}$, we first estimate $\boldsymbol{\beta}_2$ under $H_0^{(2)}$. For this, we partition the $\mathbf{c}_i$ as

$$\mathbf{c}_i = \begin{pmatrix} \mathbf{c}_{i(1)} \\ \mathbf{c}_{i(2)} \end{pmatrix} \begin{matrix} q_1 \times 1 \\ q_2 \times 1 \end{matrix}, \qquad 1 \le i \le n, \tag{7.3.11}$$

so that under $H_0^{(2)}$, we have, by (7.3.1) and (7.3.4),

$$F_i(x) = F\left(x - \beta_0 - \boldsymbol{\beta}_1' \mathbf{c}_{i(1)}\right), \qquad 1 \le i \le N. \tag{7.3.12}$$

We again appeal to Section 6.3 and, as in (6.3.7) and (6.3.8), define

$$D_N = \left\{\mathbf{b}_1 \in R^{q_1}: \sum_{k=1}^{q_1} |L_{Nk}(\mathbf{b}_1, \mathbf{0})| = \text{minimum}\right\}. \tag{7.3.13}$$

For each N, we choose a unique $\boldsymbol{\beta}_{1,N} \in D_N$ (viz. the centroid of D_N), so that as in (6.3.26) we have, under $H_0^{(2)}$,

$$N^{1/2}(\hat{\boldsymbol{\beta}}_{1,N} - \boldsymbol{\beta}_1) \xrightarrow{\mathscr{L}} \mathscr{N}_{q_1}\left(\mathbf{0}, \{A_\psi \rho(\phi, \psi)\}^{-2} \boldsymbol{\Lambda}_{11}^{-1}\right), \tag{7.3.14}$$

where A_ψ and $\rho(\phi, \psi)$ are defined by (7.2.10) and (7.2.11), and where

$$\lim_{N\to\infty} N^{-1}\mathbf{C}_N = \underset{q\times q}{\mathbf{\Lambda}} = \begin{pmatrix} \mathbf{\Lambda}_{11}, & \mathbf{\Lambda}_{12} \\ \mathbf{\Lambda}_{21}, & \mathbf{\Lambda}_{22} \end{pmatrix} \tag{7.3.15}$$

and $\mathbf{\Lambda}_{11}^{-1}$ is the inverse of $\mathbf{\Lambda}_{11}$. Now, defining $\mathbf{L}_{N(2)}(\mathbf{b})$ as in (7.3.8), we let

$$\hat{\mathbf{L}}_{N(2)} = \mathbf{L}_{N(2)}(\hat{\boldsymbol{\beta}}_{1,N}, \mathbf{0}) = \left(\hat{L}_{N,q_1+1}, \dots, \hat{L}_{N,q}\right)'; \tag{7.3.16}$$

$$\hat{L}_{N,j} = \sum_{i=1}^{N} (c_{ij} - \bar{c}_{Nj}) a_N(\hat{R}_{Ni}), \qquad q_1 + 1 \le j \le q; \tag{7.3.17}$$

$$\hat{R}_{Ni} = \text{rank of } X_i - \hat{\boldsymbol{\beta}}_{1,N}'\mathbf{c}_{i(1)} \text{ among } X_\alpha - \hat{\boldsymbol{\beta}}_{1,N}'\mathbf{c}_{\alpha(1)}, \quad 1 \le \alpha \le N,$$

$$i = 1, \dots, N. \tag{7.3.18}$$

Also, parallel to (7.3.15), we partition $\mathbf{C}_N$ as

$$\mathbf{C}_N = \begin{pmatrix} \mathbf{C}_{N11}, & \mathbf{C}_{N12} \\ \mathbf{C}_{N21}, & \mathbf{C}_{N22} \end{pmatrix} \tag{7.3.19}$$

and let

$$\mathbf{C}_{N22\cdot 1} = \mathbf{C}_{N22} - \mathbf{C}_{N21}\mathbf{C}_{N11}^{-1}\mathbf{C}_{N12}. \tag{7.3.20}$$

Then, the proposed test for $H_0^{(2)}: \boldsymbol{\beta}_2 = \mathbf{0}$ is based on the statistic

$$\mathscr{L}_N^{(2)} = A_N^{-2}\left(\hat{\mathbf{L}}_{N(2)}'\mathbf{C}_{N22\cdot 1}^{-1}\mathbf{L}_{N(2)}\right), \tag{7.3.21}$$

where

$$A_N^2 = (N-1)^{-1}\sum_{i=1}^{N} [a_N(i) - \bar{a}_N]^2 \qquad (N \ge 2). \tag{7.3.22}$$

We show first that under $H_0^{(2)}: \boldsymbol{\beta}_2 = \mathbf{0}$, $\mathscr{L}_N^{(2)}$ has a limiting central chi-square distribution. Towards this, we have the following theorem.

Theorem 7.3.1. *Under $H_0^{(2)}: \boldsymbol{\beta}_2 = \mathbf{0}$ and the assumptions made earlier, $\mathscr{L}_N^{(2)}$ has asymptotically the (central) chi-square distribution with q_2 DF.*

Proof. By virtue of (6.3.21), (6.3.23), (6.3.24), (7.3.13), and (7.3.14), we obtain that under the hypothesis of Theorem 7.3.1, as $N \to \infty$,

$$N^{-1/2}\left\|\hat{\mathbf{L}}_{N(2)} - \mathbf{L}_{N(2)}(\boldsymbol{\beta}_1, \mathbf{0}) + A_\phi A_\psi \rho(\phi, \psi)\mathbf{C}_{N21}(\hat{\boldsymbol{\beta}}_{1,N} - \boldsymbol{\beta}_1)\right\| \xrightarrow{P} 0, \tag{7.3.23}$$

$$N^{-1/2}\big\|\mathbf{L}_{N(1)}(\hat{\boldsymbol{\beta}}_{1,N}, \mathbf{0}) - \mathbf{L}_{N(1)}(\boldsymbol{\beta}_1, 0) + A_\phi A_\psi \rho(\phi, \psi)\mathbf{C}_{N11}(\hat{\boldsymbol{\beta}}_{1,N} - \boldsymbol{\beta}_1)\big\| \xrightarrow{P} 0. \tag{7.3.24}$$

As a result, from (7.3.23) and (7.3.24), we have

$$N^{-1/2}\left\|\hat{\mathbf{L}}_{N(2)} - \left\{\mathbf{L}_{N(2)}(\boldsymbol{\beta}_1, \mathbf{0}) - \mathbf{C}_{N21}\mathbf{C}_{N11}^{-1}\mathbf{L}_{N(1)}(\boldsymbol{\beta}_1, \mathbf{0})\right\}\right\| \xrightarrow{P} 0. \tag{7.3.25}$$

Now, under $H_0^{(2)}: \boldsymbol{\beta}_2 = \mathbf{0}$, $[\mathbf{L}_{N(1)}(\boldsymbol{\beta}_1, \mathbf{0}), \mathbf{L}_{N(2)}(\boldsymbol{\beta}_1, 0)]$ has the same (joint) distribution as of $[\mathbf{L}_{N(1)}(\mathbf{0}), \mathbf{L}_{N(2)}(\mathbf{0})]$ under $H_0^*: \boldsymbol{\beta} = 0$, where, by Theorem 5.3.1 [see (5.3.11)], under H_0^*,

$$N^{-1/2}\left(\mathbf{L}_{N(1)}(\mathbf{0}), \mathbf{L}_{N(2)}(\mathbf{0})\right) \xrightarrow{\mathscr{L}} \mathscr{N}_q\left(\mathbf{0}, A_\phi^2 \boldsymbol{\Lambda}\right), \tag{7.3.26}$$

so that by (7.3.25) and (7.3.26), under $H_0^{(2)}$,

$$N^{-1/2}\hat{\mathbf{L}}_{N(2)} \xrightarrow{\mathscr{L}} \mathscr{N}_q\left(\mathbf{0}, \{\boldsymbol{\Lambda}_{22} - \boldsymbol{\Lambda}_{21}\boldsymbol{\Lambda}_{11}^{-1}\boldsymbol{\Lambda}_{12}\}A_\phi^2\right). \tag{7.3.27}$$

Finally, by (7.3.15), (7.3.19), and (7.3.20),

$$N^{-1}\mathbf{C}_{N22\cdot 1} \to \boldsymbol{\Lambda}_{22\cdot 1} = \boldsymbol{\Lambda}_{22} - \boldsymbol{\Lambda}_{21}\boldsymbol{\Lambda}_{11}^{-1}\boldsymbol{\Lambda}_{12}. \tag{7.3.28}$$

Therefore, the desired result follows from (7.3.21), (7.3.27), (7.3.28), and the fact that $A_N^2 \to A_\phi^2$ as $N \to \infty$.

Let $\chi_t^2(\alpha)$ be the upper $100\alpha\%$ point of the central chi-square d.f. with t DF. Then we have the following asymptotic test: Reject or accept $H_0^{(2)}: \boldsymbol{\beta}_2 = \mathbf{0}$ according as

$$\mathscr{L}_N^{(2)} \geq \text{ or } < \chi_{q_2}^2(\alpha), \tag{7.3.29}$$

where α $(0 < \alpha < 1)$ is the desired level of significance of the test.

For the study of the asymptotic power properties of the test based on $\mathscr{L}_N^{(2)}$, as in (7.2.9), we confine ourselves to a class of local alternatives:

$$H_N^{(2)}: (7.3.1) \text{ holds for } \boldsymbol{\beta}_2 = N^{-1/2}\boldsymbol{\delta}_2, \tag{7.3.30}$$

where $\boldsymbol{\delta}_2' = (\delta_{q_1+1}, \ldots, \delta_q) \in R^{q_2}$.

Note that the joint density of the X_{Ni} $(1 \le i \le N)$ under $H_0^{(2)}: \boldsymbol{\beta}_2 = \mathbf{0}$ is

$$p_N = \prod_{i=1}^{N} f\left(x_i - \beta_0 - \boldsymbol{\beta}_i'\mathbf{c}_{i(1)}\right) \tag{7.3.31}$$

and under $H_N^{(2)}$, it is

$$q_N = \prod_{i=1}^{N} f\left(x_i - \beta_0 - \boldsymbol{\beta}_1'\mathbf{c}_{i(1)} - N^{-1/2}\boldsymbol{\delta}_2'\mathbf{c}_{i(2)}\right). \tag{7.3.32}$$

Thus, if we let $y_i = x_i - \beta_0 - \boldsymbol{\beta}_1\mathbf{c}_{i(1)}$, $1 \le i \le N$, and $d_{Ni} = N^{-1/2}\boldsymbol{\delta}_2'\mathbf{c}_{i(2)}$, $1 \le i \le N$, we may rewrite

$$\frac{q_N}{p_N} = \prod_{i=1}^{N} \frac{f(y_i - d_{Ni})}{f(y_i)}. \tag{7.3.33}$$

[Actually, without any loss of generality we may set $\bar{c}_{Nj} = 0$, $1 \le j \le q$, and hence, in (7.3.33), $f(y_i)$ may also be written as $f(y_i - \bar{d}_N)$ where $\bar{d}_N = N^{-1}\Sigma_{i=1}^N d_{Ni} = 0$.] We are naturally tempted to use the powerful tool of continguity, displayed in Chapter 2, Section 7. By an appeal to Theorem 2.7.5, and leaving the details as an exercise (viz., Exercise 7.3.1), we conclude that

$$\{q_N\} \text{ is continguous to } \{p_N\}. \tag{7.3.34}$$

A direct consequence of (7.3.34) is that if $\{B_N\}$ is any sequence of events such that $P(B_N|p_N) \to 0$, then $P(B_N|q_N) \to 0$, as $N \to \infty$. As a result, (7.3.25) continues to hold under $\{H_N^{(2)}\}$ as well. On the other hand, under $\{H_N^{(2)}\}$ in (7.3.30), $N^{-1/2}\{\mathbf{L}_{N(1)}(\boldsymbol{\beta}_1, \mathbf{0}), \mathbf{L}_{N(2)}(\boldsymbol{\beta}_1, 0)\}$ [having the same joint distribution as of $N^{-1/2}\{\mathbf{L}_{N(1)}(0,0), \mathbf{L}_{N(2)}(\mathbf{0},0)\}$ under $H_N^{*(2)}: \boldsymbol{\beta}_1 = \mathbf{0}$, $\boldsymbol{\beta}_2 = N^{-1/2}\boldsymbol{\delta}$] has asymptotically a multinormal distribution with dispersion matrix $A_\phi^2\boldsymbol{\Lambda}$ and mean vector

$$\left\{\rho(\phi, \psi)A_\phi A_\psi\right\}\left(\boldsymbol{\Lambda}_{12}\boldsymbol{\delta}_2, \boldsymbol{\Lambda}_{22}\boldsymbol{\delta}_2\right); \tag{7.3.35}$$

this follows from Theorems 2.5.3 and 2.6.1 and the fact that under (7.3.30) and the assumed regularity conditions, for μ_N defined by (2.6.7), $N^{-1/2}\mu_N$ converges to (7.3.35) when we take c_i as $\mathbf{c}_i$ and d_{Ni} as in the discussion after (7.3.22). Consequently, under $\{H_N^{(2)}\}$, parallel to (7.3.27), we have

$$N^{-1/2}\hat{\mathbf{L}}_{N(2)} \xrightarrow{\mathcal{L}} \mathcal{N}_{q_2}\big(\{\rho(\phi,\psi)A_\phi A_\psi\}(\boldsymbol{\Lambda}_{22} - \boldsymbol{\Lambda}_{21}\boldsymbol{\Lambda}_{11}^{-1}\boldsymbol{\Lambda}_{12})\boldsymbol{\delta}_2,$$

$$(\boldsymbol{\Lambda}_{22} - \boldsymbol{\Lambda}_{21}\boldsymbol{\Lambda}_{11}^{-1}\boldsymbol{\Lambda}_{12})A_\phi^2\big). \quad (7.3.36)$$

Hence, by (7.3.21), (7.3.28), and (7.3.36), we arrive at the following theorem.

Theorem 7.3.2. *Under* $\{H_N^{(2)}\}$ *in* (7.3.30) *and the regularity conditions of Theorem* 7.3.1, $\mathcal{L}_N^{(2)}$ *has asymptotically a noncentral chi-square distribution with* q_2 *DF and noncentrality parameter*

$$\Delta_{\mathcal{L}}^{(2)} = \{A_\psi^2\rho^2(\phi,\psi)\}\boldsymbol{\delta}_2'(\boldsymbol{\Lambda}_{22} - \boldsymbol{\Lambda}_{21}\boldsymbol{\Lambda}_{11}^{-1}\boldsymbol{\Lambda}_{12})\boldsymbol{\delta}_2. \quad (7.3.37)$$

As in Section 5.6, we consider the usual normal-theory test for $H_0^{(2)}$. Define $\mathbf{A}_N$ and $\mathbf{H}_N$ as in (5.6.1) and (5.6.2) respectively, and note that here A_N^2 is a scalar $[= \sum_{i=1}^N (X_i - \bar{X}_N)^2/N]$, while we have

$$\underset{q\times 1}{\mathbf{H}_N} = N^{-1}\sum_{i=1}^{N} X_i\big[\mathbf{c}_{i(1)}' - \bar{\mathbf{c}}_{(1)}', \mathbf{c}_{i(2)}' - \bar{\mathbf{c}}_{(2)}'\big]' = (\underset{1\times q_1}{H_{N1}'}, \underset{1\times q_2}{H_{N(2)}'})'. \quad (7.3.38)$$

Also, let

$$\mathbf{C}_N^* = N^{-1}\mathbf{C}_N = \begin{pmatrix} \mathbf{C}_{N11}^* & \mathbf{C}_{N12}^* \\ \mathbf{C}_{N21}^* & \mathbf{C}_{N22}^* \end{pmatrix}, \quad (7.3.39)$$

where $\mathbf{C}_N$ is defined by (6.3.3) and partitioned as in (7.3.19). Then the least-squares estimator of $\boldsymbol{\beta}$ in (7.3.1) is $\tilde{\boldsymbol{\beta}}_N = \mathbf{C}_N^{*-1}\mathbf{H}_N$ [see (5.6.2)], and under $H_0^{(2)}: \boldsymbol{\beta}_2 = \mathbf{0}$, the least-squares estimator of $\boldsymbol{\beta}_1$ is

$$\check{\boldsymbol{\beta}}_{1,N} = \mathbf{C}_{N11}^{*-1}\mathbf{H}_{N1}. \quad (7.3.40)$$

Thus, the normal-theory likelihood-ratio test statistic is

$$\lambda_N^{(2)} = \left[\frac{NA_N - \tilde{\boldsymbol{\beta}}_N'\mathbf{C}_N^*\tilde{\boldsymbol{\beta}}_N}{NA_N - \check{\boldsymbol{\beta}}_{1,N}'C_{N11}^*\check{\boldsymbol{\beta}}_{1,N}}\right]^{N/2}. \quad (7.3.41)$$

Note that $\tilde{\boldsymbol{\beta}}_N$ and $\check{\boldsymbol{\beta}}_{1,N}$ are both linear in $(X_1, \ldots, X_N)$, and hence, proceeding as in Lemmas 5.6.1, 5.6.2, and 5.6.3, we obtain that

$$A_N \xrightarrow{P} \sigma^2 = \operatorname{Var}(X_i) \qquad \text{as} \quad N \to \infty, \tag{7.3.42}$$

$$(\tilde{\boldsymbol{\beta}}_N - \boldsymbol{\beta})' \mathbf{C}_N^* (\boldsymbol{\beta}_N - \boldsymbol{\beta}) = O_p(N^{-1}), \tag{7.3.43}$$

$$\left(\check{\boldsymbol{\beta}}_{1,N} - \boldsymbol{\beta}_1\right)' \mathbf{C}_{N11}^* \left(\check{\boldsymbol{\beta}}_{1,N} - \boldsymbol{\beta}_1\right) = O_p(N^{-1}) \qquad \left(\text{under } H_0^{(2)}\right), \tag{7.3.44}$$

$$-2 \log \lambda_N^{(2)} \xrightarrow{\mathscr{L}} \chi_{q_2}^2 \qquad \text{under } H_0^{(2)}, \tag{7.3.45}$$

and under $\{H_N^{(2)} : \boldsymbol{\beta}_2 = N^{-1/2}\boldsymbol{\delta}_2\}$,

$$-2 \log \lambda_N^{(2)} \xrightarrow{\mathscr{L}} \chi_{q_2, \Delta_L^{(2)}}^2, \tag{7.3.46}$$

where

$$\Delta_L^{(2)} = \sigma^{-2} \boldsymbol{\delta}_2' \boldsymbol{\Lambda}_{22 \cdot 1} \boldsymbol{\delta}_2 \tag{7.3.47}$$

In view of the similarity with Theorem 5.6.4, the proof of (7.3.46) is left as an exercise (see Exercise 7.3.2). Note that for (7.3.46) to hold, F in (7.3.1) does not need to be normal—it needs to have a finite and positive σ^2. If, however, F *is* normal, then $\sigma^2 = 1/I(f) = 1/A_\psi^2$, so that (7.3.47) reduces to

$$A_\psi^2 \left[\boldsymbol{\delta}_2' (\boldsymbol{\Lambda}_{22} - \boldsymbol{\Lambda}_{21} \boldsymbol{\Lambda}_{11}^{-1} \boldsymbol{\Lambda}_{12}) \boldsymbol{\delta}_2 \right] \qquad (\text{when } F \text{ is normal}). \tag{7.3.48}$$

As in Section 5.7, the ARE of the aligned rank test with respect to the normal-theory likelihood-ratio test (when the underlying d.f. is F) is given by the ratio of the two noncentrality parameters in (7.3.37) and (7.3.47), viz.,

$$e\left(\mathscr{L}_n^{(2)}, -2 \log \lambda_N^{(2)} | F\right) = \frac{\Delta_{\mathscr{L}}^{(2)}}{\Delta_L^{(2)}} = \sigma^2 A_\psi^2 \rho^2(\phi, \psi)$$

$$= \frac{\sigma^2 \left(\int_0^1 \phi(u) \psi(u) \, du \right)^2}{A_\phi^2}. \tag{7.3.49}$$

Now, (7.3.49) agrees with (5.7.5), and hence, the discussion following (5.7.5) applies to $\mathscr{L}_n^{(2)}$ as well. The asymptotic optionality of $\mathscr{L}_n^{(2)}$ will be studied in a later section.

7.3.2. Rank-Order Tests for $H_0^{(3)}$

As in Section 7.2, we need in this case that F in (7.3.1) is symmetric about 0, so that the signs of the residuals may be employed for testing $\beta_0 = 0$. Here also, the first step is to eliminate the nuisance parameter $\boldsymbol{\beta}_1$, and for this we proceed as in (7.3.6) through (7.3.13) and define $\hat{\boldsymbol{\beta}}_{1,N}$ as we did after (7.3.13). Let then

$$\hat{X}_i = X_i - \hat{\boldsymbol{\beta}}_{1,N}\mathbf{c}_{i(1)} \qquad \text{for} \quad i = 1, \ldots, N, \tag{7.3.50}$$

and standardize the $\mathbf{c}_i$ so that $\bar{\mathbf{c}}_N = N^{-1}\Sigma_{i=1}^N \mathbf{c}_i = \mathbf{0}$. Define then $\hat{S}_N$ as in (7.2.6) and $\hat{L}_{N(2)}$ as in (7.3.16)–(7.3.17), where $a_N^*(i)$ and $a_N(i)$ satisfy condition (3) in Theorem 7.2.1, so that $A_\phi^* = A_\phi$. Let then

$$\mathscr{L}_N^{(3)} = \mathscr{L}_N^{(2)} + \left(NA_N^{*2}\right)^{-1}\hat{S}_N^2, \tag{7.3.51}$$

where A_N^{*2} is defined by (7.2.7). Then we have the following.

Theorem 7.3.3. *For F symmetric about 0, under $H_0^{(3)}$ and the regularity conditions in Theorem 7.2.1, $\mathscr{L}_N^{(3)}$ has asymptotically the central chi-square distribution with $q_2 + 1$ DF.*

Proof. First, we show that under the hypothesis of the theorem, $(N^{-1/2}\hat{S}_N, N^{-1/2}\hat{\mathbf{L}}_{N(2)})$ has asymptotically a $(q_2 + 1)$-variate normal distribution with mean vector $\mathbf{0}$ and dispersion matrix

$$A_\phi^2 \begin{bmatrix} 1 & \mathbf{0}' \\ \mathbf{0} & \boldsymbol{\Lambda}_{22} - \boldsymbol{\Lambda}_{21}\boldsymbol{\Lambda}_{11}^{-1}\boldsymbol{\Lambda}_{12} \end{bmatrix} \tag{7.3.52}$$

Note that by (6.3.26), under $H_0^{(3)}$, $\boldsymbol{\beta}' = (\boldsymbol{\beta}_1', \mathbf{0}')$ and $N^{1/2}(\hat{\boldsymbol{\beta}}_{1,N} - \boldsymbol{\beta}_1)$ is asymptotically normal with mean $\mathbf{0}$ and dispersion matrix $\{\rho(\phi, \psi) A_\psi\}^{-2}\Lambda_{11}^{-1}$, and hence, $N^{1/2}\|\hat{\boldsymbol{\beta}}_{1,N} - \boldsymbol{\beta}_1\| = O_p(1)$. Also, we note that under $H_0^{(3)}$,

$$X_i^0 = X_i - \boldsymbol{\beta}_2'\mathbf{c}_{i(1)}, \quad i = 1, \ldots, N, \quad \text{are i.i.d.r.v.'s} \tag{7.3.53}$$

with d.f. F, symmetric about 0. Let S_N^0 be the signed rank statistic in (7.2.6), but based on the X_i^0 in (7.3.53), that is,

$$S_N^0 = \sum_{i=1}^{N} \operatorname{sgn} X_i^0 a_N^*\left(R_{Ni}^{0+}\right), \tag{7.3.54}$$

where R_{Ni}^{0+} is the rank of $|X_i^0|$ among $|X_1^0|, \ldots, |X_N^0|$ for $i = 1, \ldots, N$. Then, by an appeal to (6.3.44), we conclude that under $H_0^{(3)}$,

$$N^{-1/2}|\hat{S}_N - S_N^0| \xrightarrow{P} 0 \quad \text{as} \quad N \to \infty. \tag{7.3.55}$$

Similarly, for $\hat{\mathbf{L}}_{N(2)}$ we use (7.3.25), (7.3.26), and (7.3.27). Further, as in the proof of Theorem 6.2.7, we have under $\beta_0 = 0$, $\boldsymbol{\beta} = \mathbf{0}$ that $N^{-1/2}\{S_N^0, \mathbf{L}_N(0)\}$ is asymptotically normal with null mean vector and dispersion matrix

$$A_\phi^2 \begin{bmatrix} 1 & \mathbf{0}' \\ \mathbf{0} & \boldsymbol{\Lambda} \end{bmatrix}, \tag{7.3.56}$$

so that $N^{-1/2}S_N^0$ is asymptotically independent of $N^{-1/2}\mathbf{L}_N(0)$. Thus, by (7.3.25), (7.3.26), (7.3.27), and (7.3.56), we conclude that (7.2.52) holds, and this, along with (7.3.51) and Theorem 7.3.1, shows that $\mathscr{L}_N^{(3)}$ is asymptotically chi-square with $q_2 + 1$ DF, Q.E.D.

Thus, parallel to (7.3.29), we propose the following asymptotic test: Reject or accept $H_0^{(3)}$ according as

$$\mathscr{L}_N^{(3)} \geq \text{ or } \leq \chi^2_{q_2+1}(\alpha), \tag{7.3.57}$$

where α $(0 < \alpha < 1)$ is the desired level of significance of the test.

For the study of the asymptotic power properties of the test based on $\mathscr{L}_N^{(3)}$, we consider $\{H_N^{(3)}\}$, given by

$$H_N: \beta_0 = N^{-1/2}\delta_0, \quad \boldsymbol{\beta}_2 = N^{-1/2}\boldsymbol{\delta}_2, \quad \left(\delta_0, \boldsymbol{\delta}_2'\right) \in R^{q_2+1}. \tag{7.3.58}$$

As in (7.3.31) through (7.3.34), the contiguity of the probability measure under $\{H_N^{(3)}\}$ with respect to that under $H_0^{(3)}$ holds. Therefore, we may repeat the proof of Theorem 7.3.2 and additionally use (6.3.44) and claim that under $\{H_N^{(3)}\}$, (a) (7.3.35) holds and (b) $N^{-1/2}\hat{S}_N$ is asymptotically independent of $N^{-1/2}\hat{\mathbf{L}}_{N(2)}$ and is normally distributed with mean $\delta_0 A_\psi \rho(\phi, \psi)$ and variance A_ϕ^2. Thus, combining this with (7.3.36), we obtain from (7.3.51) and the Cochran theorem that under $\{H_N^{(3)}\}$, $\mathscr{L}_N^{(3)}$ has asymptotically the noncentral chi-square distribution with $q_2 + 1$ DF and the noncentrality parameter

$$\Delta_{\mathscr{L}}^{(3)} = \Delta_{\mathscr{L}}^{(2)} + \delta_0^2 A_\psi^2 \rho^2(\phi, \psi), \tag{7.3.59}$$

where $\Delta_{\mathscr{L}}^{(2)}$ is defined by (7.3.37).

As in Section 5.6 and (7.3.38)–(7.3.47), we may consider the normal-theory likelihood-ratio statistic for $H_0^{(3)}$. This will be similar to (7.3.41), viz.,

$$\lambda_N^{(3)} = \left\{ \frac{\sum_1^N (X_i - \bar{X}_N)^2 - \tilde{\boldsymbol{\beta}}_N' \mathbf{C}_N^* \boldsymbol{\beta}_N}{\sum_1^N X_i^2 - \tilde{\boldsymbol{\beta}}_{1,N}' \mathbf{C}_{N11}^* \tilde{\boldsymbol{\beta}}_{1,N}'} \right\}^{N/2}, \tag{7.3.60}$$

where $\tilde{\boldsymbol{\beta}}_N$ and $\check{\boldsymbol{\beta}}_{1,N}$ are defined by (7.3.40). Parallel to (7.3.45), we have

$$-2\log \lambda_N^{(3)} \xrightarrow{\mathscr{L}} \chi_{q_2+1}^2 \qquad \text{under } H_0^{(3)}; \tag{7.3.61}$$

and under $\{H_N^{(3)}\}$ in (7.3.58), we have

$$-2\log \lambda_N^{(3)} \xrightarrow{\mathscr{L}} \chi_{q_2+1,\,\Delta_L^{(3)}}^2, \tag{7.3.62}$$

where

$$\Delta_L^{(3)} = \Delta_L^{(2)} + \sigma^{-2}\delta_0^2. \tag{7.3.63}$$

and $\Delta_L^{(2)}$ is defined in (7.3.47). Hence, by (7.3.59), (7.3.62), and (7.3.63),

$$e\left(\mathscr{L}_N^{(3)}, -2\log \lambda_N^{(3)} | F\right) = \frac{\sigma^2 \left(\int_0^1 \phi(u)\psi(u)\, du \right)^2}{A_\phi^2}$$

$$= e\left(\mathscr{L}_N^{(2)}, -2\log \lambda_N^{(2)} | F\right). \tag{7.3.64}$$

7.3.3. Rank-Order Tests for $H_0^{(1)}$

These follow as a particular case of $\mathscr{L}_N^{(3)}$ where we put $q_1 = q$ (or $q_2 = 0$). Here, (7.3.50) relates to $\hat{X}_i = X_i - \boldsymbol{\beta}_N \mathbf{c}_i$, $i = 1, \ldots, N$, so that $\hat{\mathbf{L}}_N \xrightarrow{P} \mathbf{0}$, and, parallel to (7.3.51), we take

$$\mathscr{L}_N^{(1)} = \left(NA_N^{*2}\right)^{-1} \hat{S}_N^2. \tag{7.3.65}$$

Theorem 7.3.3 holds with 1 DF, and in (7.3.59) we have

$$\mathscr{L}_N^{(1)} = \delta_0^2 A_\psi^2 \rho^2(\phi, \psi). \tag{7.3.66}$$

In (7.3.61) we have χ_1^2, and in (7.3.63)

$$\Delta_L^{(1)} = \sigma^{-2}\delta_0^2, \tag{7.3.67}$$

so that (7.3.64) continues to hold for $\mathscr{L}_N^{(1)}$ vs. $-2\log\lambda_N^{(1)}$ as well.

7.3.4. Some Further Rank-Order Tests in Some Specific Models

The model (7.3.1) includes a broad class of linear models, and the solutions in Sections 7.3.1 through 7.3.3 work out well for these models. There are, however, some specific models where a variant form of aligned rank-order tests work out nicely. As an important case, we consider the case of a several-sample regression model, considered by Sen (1969a, 1972a, b). Consider k (≥ 2) independent samples of sizes $n_1, \dots, n_k$, respectively, and let $X_{i1}, \dots, X_{in_i}$ be n_i independent r.v.'s (representing the ith sample) following the model

$$X_{ij} = \beta_{0i} + \beta_{(i)}c_{ij} + e_{ij}, \qquad 1 \leq j \leq n_i, \quad 1 \leq i \leq k, \tag{7.3.68}$$

where $N = n_1 + \cdots + n_k$ and all the N r.v.'s $\{e_{ij}\}$ are i.i.d. with the d.f. F. We may characterize this model as a special case of (7.3.1) as follows. Let us write

$$\beta_{0i} = \beta_0 + \beta_i^0, \quad 1 \leq i \leq k, \qquad \beta_{(i)} = \beta_{(1)} + \beta_i^*, \quad 1 \leq i \leq k, \tag{7.3.69}$$

where $\beta_1^0 = \beta_1^* = 0$ and $\boldsymbol{\beta}' = (\beta_2^0, \dots, \beta_k^0, \beta_{(1)}, \beta_2^*, \dots, \beta_k^*)$ (so that $p = 2k - 1$), and let $\mathbf{X}_N' = (X_{11}, \dots, X_{1n_1}, \dots, X_{k1}, \dots, X_{kn_k})$, so that (7.3.68) corresponds to (7.3.1) with $p = 2k - 1$ and with $\mathbf{c}_r$ [$-(2k-1)$ vectors] having the elements 1 and c_{ij} in the $(i-1)$th and $[k + (i-1)]$th rows when $r = n_0 + \cdots + n_{i-1} + j$, for $j = 1, \dots, n_i$, $1 \leq i \leq k$ (where $n_0 = 0$). Several problems of real interest arise in this context (see Sen, 1971a, b, 1972b):

1. Testing the parallelism of the regression lines, viz.,

$$H_0: \beta_2^* = \cdots = \beta_k^* = 0 \quad \text{vs.} \quad H_1: \beta_j^* \neq 0 \text{ for at least one } 2 \leq j \leq k. \tag{7.3.70}$$

In this context, we may or may not make the assumption that

$$H_0^*: \beta_2^0 = \cdots = \beta_k^0 = 0. \tag{7.3.71}$$

2. Testing the identity of the intercepts $\beta_{01}, \dots, \beta_{0k}$, viz., the hypothesis in (7.3.71), where again the lines may or may not be parallel.

If we desire to test for (7.3.70) assuming that (7.3.71) holds, then in (7.3.69) we have $\boldsymbol{\beta}' = (\beta_{(1)}, \beta_2^*, \dots, \beta_k^*)$, so that the problem reduces to that of testing $H_0^{(2)}$ in (7.3.4) where $q_2 = k - 1$. Thus, we may use the test statistic $\mathscr{L}_N^{(2)}$ in (7.3.21), and under H_0 in (7.3.70), given (7.3.71), it will have asymptotically the central chi-square distribution with $k-1$ DF (by Theorem 7.3.1). The asymptotic power of the test for local alternatives of the type (7.3.30) is again given by the tail probability of a noncentral chi-square d.f. with $k-1$ DF and noncentrality parameter $\Delta_{\mathscr{L}}^{(2)}$ given by (7.3.37). Similarly, if one intends to test for the equality of the intercepts [i.e., (7.3.71)] assuming the equality of the slopes, one has $\boldsymbol{\beta}' = (\beta_2^0, \dots, \beta_k^0, \beta_{(1)})$ $(p = k)$, so that the problem is similar and the test in Section 7.3.1 applies. The problem becomes more complicated when we want to test for H_0 in (7.3.70) without assuming (7.3.71) or vice versa.

First, we consider tests for H_0 in (7.3.70) without imposing (7.3.71). For the model (7.3.68), we let

$$\bar{c}_i = \frac{1}{n_i} \sum_{j=1}^{n_i} c_{ij}, \qquad C_{(i)}^2 = \sum_{j=1}^{n_i} (c_{ij} - \bar{c}_i)^2, \quad 1 \le i \le k, \tag{7.3.72}$$

$$C_N^2 = \sum_{i=1}^{k} C_{(i)}^2 \qquad \gamma_{n_i} = \frac{C_{(i)}^2}{C_N^2}, \qquad 1 \le i \le k, \tag{7.3.73}$$

and assume that there exist positive numbers γ_0 $(\le 1/k)$, $\gamma_1, \dots, \gamma_k$ such that

$$0 < \gamma_0 \le \lim_{n \to \infty} \gamma_{n_i} = \gamma_i \le 1 - \gamma_0 < 1 \qquad \forall 1 \le i \le k. \tag{7.3.74}$$

Based on the n_i observations in the ith sample, define $L_{n_i}^{(i)}(b)$ as in (7.2.3) and (7.2.4), and let

$$L_N(b) = \sum_{i=1}^{k} L_{n_i}^{(i)}(b) \qquad \forall -\infty < b < \infty. \tag{7.3.75}$$

Note that under the regularity conditions of Theorem 7.2.1, the $L_{n_i}^{(i)}(b)$ as well as $L_N(b)$ are decreasing in b, and as in (7.2.5), under H_0 in (7.3.70), we estimate $\beta_{(1)}$ by $\hat{\beta}_N$ based on $L_N(b)$. Let then

$$\hat{L}_{n_i} = L_{n_i}^{(i)}(\hat{\beta}_N) \qquad \text{for} \quad i = 1, \ldots, k, \tag{7.3.76}$$

$$\hat{\mathscr{L}}_N = A_N^{-2} \sum_{i=1}^{k} \left(\hat{L}_{n_i}^2 / C_{(i)}^2\right), \tag{7.3.77}$$

where A_N^2 is defined by (7.3.22). We use $\hat{\mathscr{L}}_N$ as the desired test statistic.

Let $\hat{\beta}_{(i), n_i}$ be the estimator of $\beta_{(i)}$ based on $L_{n_i}^{(i)}(b)$ for $i = 1, \ldots, k$, so that as $N \to \infty$,

$$\left(C_{(i)} A_n\right)^{-1} L_{n_i}^{(i)}\left(\hat{\beta}_{(i), n_i}\right) = o_p(1) \quad \forall i = 1, \ldots, k. \tag{7.3.78}$$

Then, by Theorem 6.2.7, whatever β_{0i} are,

$$C_{(i)} \left| \hat{\beta}_{(i), n_i} - \beta_{(i)} \right| = O_p(1) \qquad \forall 1 \leq i \leq k. \tag{7.3.79}$$

Similarly, using (6.3.21)–(6.3.24), (7.2.5), and (7.3.75), we have under H_0 in (7.3.70) and the regularity conditions of Theorem 7.2.1,

$$C_N |\hat{\beta}_N - \beta_{(1)}| = O_p(1), \qquad \hat{L}_N = L_N(\hat{\beta}_N) = o_p(C_N). \tag{7.3.80}$$

By (7.3.78), (7.3.79), (7.3.80), (7.3.73), and the linearity of the $L_{n_i}^{(i)}(b)$ (in b), we obtain that under H_0 in (7.3.70),

$$\left| C_N \left(\hat{\beta}_N - \sum_{i=1}^{k} \gamma_{Ni} \hat{\beta}_{(i), n_i} \right) \right| \xrightarrow{P} 0 \qquad \text{as} \quad N \to \infty; \tag{7.3.81}$$

$$\hat{\mathscr{L}}_N \underset{P}{\sim} A_\psi^2 \rho^2(\phi, \psi) \left\{ \sum_{i=1}^{k} C_{(i)}^2 \left(\hat{\beta}_{(i), n_i} - \hat{\beta}_N \right)^2 \right\}, \tag{7.3.82}$$

where for different i $(= 1, \ldots, k)$, $C_{(i)}(\hat{\beta}_{(i), n_i} - \beta_i)$ are independently and asymptotically normally distributed with means 0 and variances equal to $1/A_\psi^2 \rho^2(\psi, \phi)$. By (7.3.81), (7.3.82), and the Cochran theorem, we conclude that under H_0 in (7.3.70) and the regularity conditions of Theorem 7.2.1, $\hat{\mathscr{L}}_N$ has asymptotically the central chi-square distribution with $k - 1$ DF, and for local alternatives its noncentral (chi-square) distribution follows along the lines of Theorem 7.3.2. Hence, we may proceed as in (7.3.29) with

$q_2 = k - 1$. Exercise 7.3.3 is to show that under $\{H_N\}$ where $H_N: \beta_j^* = C_N^{-1}\delta_j$, $1 \le j \le k$ (with $\delta_1 = 0$), the non-centrality parameter for $\mathscr{L}_N$ is

$$\Delta_{\mathscr{L}} = A_\psi^2 \rho^2(\phi, \psi) \sum_{i=1}^{k} \gamma_i (\delta_i - \bar{\delta})^2, \quad \text{where} \quad \bar{\delta} = \sum_{i=1}^{k} \gamma_i \delta_i. \tag{7.3.83}$$

The normal-theory likelihood-ratio test, discussed in (7.3.38)–(7.3.47), here corresponds to the test statistic

$$\{(k-1)s_e^2\}^{-1} \sum_{i=1}^{k} C_{(i)}^2 \left(\tilde{\beta}_{(i)n_i} - \tilde{\tilde{\beta}}_N\right)^2 = \mathscr{F}_{k-1}, \quad \text{say}, \tag{7.3.84}$$

where $\tilde{\beta}_{(i)n_i}$ is the classical least-squares estimator of $\beta_{(i)}$ based on the ith sample, $\tilde{\tilde{\beta}}_N = \Sigma_{i=1}^k \gamma_{Ni} \tilde{\beta}_{(i)n_i}$, and $\mathscr{S}_e^2$ is the (pooled) within-sample mean square due to error. Exercise 7.3.4 is to show that under $\{H_N\}$ and for F admitting a finite and positive variance σ^2, $(k-1)\mathscr{F}_{k-1}$ has asymptotically a noncentral chi-square distribution with $k-1$ DF and the non-centrality parameter

$$\Delta_{\mathscr{F}} = \sigma^{-2} \sum_{i=1}^{k} \gamma_i (\delta_i - \bar{\delta})^2. \tag{7.3.85}$$

(The null distribution follows by letting $\Delta_{\mathscr{F}} = 0$.) Hence, the ARE of $\hat{\mathscr{L}}_N$ with respect to the variance-ratio test $\mathscr{F}_{k-1}$ agrees with (7.3.49).

Let us now consider rank tests for the identity of the intercepts in (7.3.71) without assuming that H_0 in (7.3.70) holds. In addition to the notation in (7.3.72)–(7.3.73), we let

$$q_{(i)}^2 = \left(n_i \bar{c}_i^2\right) / C_{(i)}^2, \qquad i = 1, \dots, k, \tag{7.3.86}$$

and assume that as $N \to \infty$,

$$\liminf n_i^{-1} C_{(i)}^2 \ge C_0^2 > 0, \qquad \limsup q_{(i)}^2 \le q_0^2 < \infty. \tag{7.3.87}$$

As in the discussion (7.3.78), let $\hat{\beta}_{(i), n_i}$ be the estimator of $\beta_{(i)}$ based on $L_{n_i}^{(i)}(b)$ [as in (7.2.5)] for $1 \le i \le k$, and as in (7.2.6), based on $X_{ij} - \hat{\beta}_{(i), n_i} c_{ij} - a$, $1 \le j \le n_i$, define the signed rank statistic by

$$\hat{S}_{n_i}^{(i)}(a), \qquad i = 1, \dots, k, \quad -\infty < a < \infty. \tag{7.3.88}$$

Let then

$$\hat{S}_N^{*(a)} = \sum_{i=1}^{N} n_i^{-1} \omega_{Ni} S_{n_i}^{(i)}(a); \tag{7.3.89}$$

$$\omega_{Ni} = n_i \left(1 + q_{(i)}^2\right)^{-1} \Big/ \sum_{j=1}^{k} n_j \left(1 + q_{(j)}^2\right)^{-1}, \qquad 1 \le i \le k. \tag{7.3.90}$$

Now, under (7.3.71), we estimate the (common value) parameter β_0 by

$$\hat{\beta}_{0,N} = \tfrac{1}{2}\left(\hat{\beta}_{0,N}^{(1)} + \hat{\beta}_{0,N}^{(2)}\right), \tag{7.3.91}$$

where

$$\hat{\beta}_{0,N}^{(1)} = \sup\{a : \hat{S}_N(a) > 0\}, \quad \hat{\beta}_{0,N}^{(2)} = \inf\{a : \hat{S}_N^{*(a)} < 0\}. \tag{7.3.92}$$

Then, as in Sen (1972a), the proposed test statistic is

$$\hat{\mathscr{L}}_N^0 = \frac{1}{A_N^{*2}} \sum_{i=1}^{k} \frac{\left\{\hat{S}_{n_i}^{(i)}(\hat{\beta}_{0,N})\right\}^2}{n_i\left(1 + q_{(i)}^2\right)}, \tag{7.3.93}$$

where A_N^{*2} is defined by (7.2.7).

Now proceeding as in the case of $\hat{\mathscr{L}}_N$ [but using (6.3.44) for the individual-sample $\hat{S}_{n_i}^{(i)}(a)$, $1 \le i \le k$], it follows (Exercise 7.3.5) that under H_0^* in (7.3.71), $\hat{\mathscr{L}}_N^0$ has asymptotically a chi-square distribution with $k - 1$ DF. Also, under $\{H_N^*\}$, where

$$H_N^* : \beta_i^0 = N^{-1/2}\delta_i^0, \quad 1 \le i \le k \quad \left(\text{with} \quad \delta_1^0 = 0\right), \tag{7.3.94}$$

and the regularity conditions of Theorem 7.2.1, $\hat{\mathscr{L}}_N$ has asymptotically a noncentral chi-square distribution with $k - 1$ DF and noncentrality parameter

$$\Delta_{\mathscr{L}0} = A_\psi^2 \rho^2(\phi, \psi) \sum_{i=1}^{k} \frac{\lambda^{(i)}\left(\delta_i^0 - \bar{\delta}^0\right)^2}{1 + q_{(i)}^2}, \tag{7.3.95}$$

where we assume that $n_i/N \to \lambda^{(i)}$ and $q_{(i)}$ also converges to a limit as $N \to \infty$, $\forall i \le n \le k$.

The construction of the normal-theory likelihood ratio is similar to that in (7.3.85), and we leave the details as an exercise (see Exercise 7.3.6). Also, the asymptotic relative efficiency in this case agrees again with (7.3.69) (see Exercise 7.3.7). This leads us to the study of the asymptotic optimality of aligned rank tests for general linear model, which we do in the next subsection.

7.3.5. Asymptotic Optimality of Aligned Rank Tests

In this subsection, we study the asymptotic optimality of the aligned rank tests for local alternatives. Basically, as in Chapter 5, we show that these aligned rank-order tests are asymptotically power-equivalent to the likelihood-ratio tests (for local alternatives) when certain regularity conditions hold, and this enables us to transmit the local optimality properties of the likelihood-ratio tests to the aligned rank-order tests.

We let $f_i(x; \boldsymbol{\beta}^*) = f(x - \beta_0 - \Sigma_{j=1}^q \beta_j c_{ij})$, $1 \le i \le N$, and assume that for every i $(1 \le i \le N)$, there exists a non-null set B^* such that

$$\left| \frac{\partial}{\partial \beta^*} f_i(x; \boldsymbol{\beta}^*) \right| \le \mathbf{U}_1(x) \qquad \forall \boldsymbol{\beta}^* \in B^*,\ x \in R, \quad (7.3.96)$$

$$\left| \frac{\partial^2}{\partial \boldsymbol{\beta}^* \, \partial \boldsymbol{\beta}^{*\prime}} f_i(x; \boldsymbol{\beta}^*) \right| \le \mathbf{U}_2(x) \qquad \forall \boldsymbol{\beta}^* \in B^*,\ x \in R, \quad (7.3.97)$$

where $\mathbf{U}_1$ and $\mathbf{U}_2$ are both integrable [with respect to $F(x, \boldsymbol{\beta}^*)$]. Writing

$$\mathbf{h}_i(x; \beta^*) = \frac{\partial^2}{\partial \boldsymbol{\beta}^* \, \partial \boldsymbol{\beta}^{*\prime}} \log f_i(x; \boldsymbol{\beta}^*), \qquad 1 \le i \le N, \quad (7.3.98)$$

we assume that

$$\left\| E\left\{ h_i\left(X_i; \boldsymbol{\beta}^* \right) | \beta^* \right\} \right\| < \infty \quad (7.3.99)$$

and

$$\lim_{\delta \downarrow 0} E\left\{ \sup_{\|\boldsymbol{\beta}^* - \tilde{\boldsymbol{\beta}}^*\| < \delta} \left\| h_i\left(X_i; \boldsymbol{\beta}^* \right) - \mathbf{h}_i\left(X_i; \tilde{\boldsymbol{\beta}}^* \right) \right\| \right\} = 0 \quad (7.3.100)$$

uniformly in $1 \le i \le N$, where $\| \cdot \|$ stands for the "sup norm". Finally, we assume that the maximum-likelihood estimator of $\boldsymbol{\beta}^*$ exists and is uniformly consistent. We denote this (unique) estimator by $\tilde{\boldsymbol{\beta}}_N^*$.

Let us consider the likelihood-ratio test statistic for testing $H_0^{(2)}$ in (7.3.4) and denote it by $\lambda_N^{(2)}$. Then, proceeding as in Section 5.6, it follows that under $H_N^{(2)}$ in (7.3.30), $-2\log\lambda_N^{(2)}$ has asymptotically a noncentral chi-square distribution with q_2 DF and noncentrality parameter

$$\Delta_\lambda^{(2)} = I(f)\{\boldsymbol{\delta}_2'(\boldsymbol{\Lambda}_{22} - \boldsymbol{\Lambda}_{21}\boldsymbol{\Lambda}_{11}^{-1}\boldsymbol{\Lambda}_{12})\boldsymbol{\delta}_2\}; \tag{7.3.101}$$

we leave the details as an exercise (see Exercise 7.3.7). Note that $A_\psi^2 = I(f)$ = Fisher information, so that by Theorem 7.3.2 and the above, we conclude that the aligned rank test $\mathscr{L}_N^{(2)}$ is asymptotically power-equivalent to the likelihood-ratio test $\lambda_N^{(2)}$ [when $\{H_N\}$ in (7.3.30) holds], provided that

$$\rho^2(\phi,\psi) = 1 \qquad [\text{i.e.,}\quad \phi(u) = \psi(u),\quad 0 < u < 1]. \tag{7.3.102}$$

On the other hand, the likelihood-ratio test is an asymptotically most stringent test, so that under (7.3.102), when $\{H_N\}$ in (7.3.30) holds, $\mathscr{L}_N^{(2)}$ is also an asymptotically most stringent test. Note that both the likelihood-ratio test and the aligned rank-order test have asymptotically constant power over ellipsoids in the parametric space specified by

$$\boldsymbol{\delta}_2'(\boldsymbol{\Lambda}_{22} - \boldsymbol{\Lambda}_{21}\boldsymbol{\Lambda}_{11}^{-1}\boldsymbol{\Lambda}_{12})\boldsymbol{\delta}_2 = \text{const}, \tag{7.3.103}$$

and hence, under (7.3.102), $\mathscr{L}_N^{(2)}$ also has asymptotically the best average power with respect to surfaces on these ellipsoids (and a uniform distribution $q\delta_2$ on these surfaces) as well as best constant power on these surfaces. Thus, (7.3.102) characterizes the asymptotic optimality of the aligned rank test based on $\mathscr{L}_N^{(2)}$. A very similar case holds for $\mathscr{L}_N^{(3)}$, $\mathscr{L}_N^{(1)}$, and T_N. Note that when F is a normal d.f., we have $\psi(u) = F^{-1}(u)$, $0 < u < 1$, so that when $\phi(u)$ is the inverse of the standard normal d.f. (i.e., we use normal-score statistics), the corresponding $\mathscr{L}_N^{(2)}$ is asymptotically optimal. If F is a logistic d.f., when $\psi(u) \equiv 2u - 1$, so that the use of Wilcoxon scores leads to an asymptotically optimal aligned rank test. Similar cases hold for other common d.f.'s. Note that if F is a location-scale family of d.f.'s, then though F may depend on unknown location or scale parameters, $\psi(u)$ does not depend on them, and hence, for such a specified F, the corresponding $\psi(u)$ can be obtained to yield an asymptotically optimal test based on the corresponding $\mathscr{L}_N^{(2)}$ with $\phi \equiv \psi$.

7.4. ALIGNED RANK TESTS FOR SUBHYPOTHESES IN MULTIVARIATE LINEAR MODELS

We shall now consider the multivariate extensions of the results considered in the preceding two subsections. For this purpose, we generalize the model in (7.2.1) and (7.3.1) as follows. Let $\mathbf{X}_1, \ldots, \mathbf{X}_N$ be N independent random vectors, where $\mathbf{X}_i = (X_{i1}, \ldots, X_{ip})'$ has a continuous $p(\geq 1)$-variate d.f.

$$F_i(\mathbf{x}) = F(\mathbf{x} - \boldsymbol{\beta}_0 - \boldsymbol{\beta}(\mathbf{c}_i - \bar{\mathbf{c}}_N)), \qquad 1 \leq i \leq N, \quad \mathbf{x} \in R^p, \quad (7.4.1)$$

where $\boldsymbol{\beta}_0' = (\beta_{01}, \ldots, \beta_{0p})'$ (the vector of intercepts) and $\boldsymbol{\beta} = ((\beta_{jk}))$ (the $p \times q$ matrix of partial regression coefficients) are unknown parameters; $\mathbf{c}_i' = (c_1, \ldots, c_{iq})'$, $i = 1, \ldots, N$, are known regression constants (vectors); $q \geq 1$; and the form of F is not specified. As in Section 7.3, we partition $\boldsymbol{\beta}$ as

$$\underset{p \times q}{\boldsymbol{\beta}} = (\underset{p \times q_1}{\boldsymbol{\beta}_1}, \underset{p \times q_2}{\boldsymbol{\beta}_2}); \quad q_1 \geq 0, \quad q_2 \geq 0, \quad q_1 + q_2 = q, \quad (7.4.2)$$

and consider the hypotheses

$$H_0^{(1)}: \boldsymbol{\beta}_0 = \mathbf{0} \quad \text{vs.} \quad H_1^{(1)}: \boldsymbol{\beta}_0 \neq \mathbf{0}, \quad (7.4.3)$$

where $\boldsymbol{\beta}$ is not specified;

$$H_0^{(2)}: \boldsymbol{\beta}_2 = \mathbf{0} \quad \text{vs.} \quad H_1^{(2)}: \boldsymbol{\beta}_2 \neq \mathbf{0}, \quad (7.4.4)$$

where $\boldsymbol{\beta}_0, \boldsymbol{\beta}_1$ are not specified;

$$H_0^{(3)}: \boldsymbol{\beta}_0 = \mathbf{0}, \boldsymbol{\beta}_2 = 0 \quad \text{vs.} \quad H_1^{(3)}: (\boldsymbol{\beta}_0, \boldsymbol{\beta}_2) \neq \mathbf{0}, \quad (7.4.5)$$

where $\boldsymbol{\beta}_1$ is not specified. Actually, $H_0^{(1)}$ is a special case of $H_0^{(3)}$ when $q_2 = 0$, so we shall consider only the case of $H_0^{(2)}$ and $H_0^{(3)}$.

Note that for $\boldsymbol{\beta} = \mathbf{0}$, $\mathbf{X}_1, \ldots, \mathbf{X}_N$ are i.i.d.r.v.'s having the d.f. $F(\mathbf{x} - \boldsymbol{\beta}_0)$, $\mathbf{x} \in R^p$, and in Chapter 5, some conditionally distribution-free (as well as asymptotically distribution-free) tests for $H_0: \boldsymbol{\beta} = \mathbf{0}$ vs. $H_1: \boldsymbol{\beta} \neq \mathbf{0}$ were studied. Similarly, under $H_0^*: \boldsymbol{\beta}_0 = 0$, $\boldsymbol{\beta} = \mathbf{0}$, when F is diagonally symmetric about $\mathbf{0}$, conditionally (as well as asymptotically) distribution-free rank tests have been studied in Chapter 5. However, for $H_0^{(j)}$, $j = 1, 2, 3$, if $q_1 > 0$, the $\mathbf{X}_i$ are not identically distributed and genuinely distribution-free tests may not generally exist. For this reason, we formulate some aligned rank tests which are asymptotically distribution free.

7.4.1. Aligned Rank Tests for $H_0^{(2)}$

As in Section 5.4.1, we define

$$\mathbf{L}_N = ((L_{N,jl}))_{j=1,\ldots,p,\, l=1,\ldots,q}, \quad (7.4.6)$$

where for every $1 \le j \le p$, $1 \le l \le q$,

$$L_{N,jl} = \sum_{i=1}^{N} (c_{il} - \bar{c}_{Nl}) a_{Nj}(R_{Ni}^{(j)}), \tag{7.4.7}$$

in which $R_{Ni}^{(j)}$ is the rank of X_{ij} among $X_{1j}, \ldots, X_{Nj}$ for $1 \le i \le N$, $j = 1, \ldots, p$, and the scores $a_{Nj}(1), \ldots, a_{Nj}(N)$ are defined by (4.2.4). As in (5.4.16)–(5.6.17), we let for $j, j' = 1, \ldots, p$

$$\nu_{jj'}(F) = \int_{-\infty}^{\infty}\int_{-\infty}^{\infty} \phi_j(F_{[j]}(x))\phi_{j'}(F_{[j']}(y))\, dF_{[jj']}(x, y) - \bar{\phi}_j\bar{\phi}_{j'}, \tag{7.4.8}$$

$$\bar{\phi}_j = \int_0^1 \phi_j(u)\, du, \qquad 1 \le j \le p, \tag{7.4.9}$$

and assume that

$$\boldsymbol{\nu}(F) = ((\nu_{jj'}(F))) \text{ is p.d. and finite.} \tag{7.4.10}$$

Also, as in (4.2.9), we let

$$\mathbf{C}_N^* = N^{-1} \sum_{i=1}^{N} (\mathbf{c}_i - \bar{\mathbf{c}}_N)(\mathbf{c}_i - \bar{\mathbf{c}}_N)'. \tag{7.4.11}$$

Then, as in Section 7.3, we assume that (6.3.3) through (6.3.12) hold and, in addition, (7.3.9) holds for each ϕ_j $(1 \le j \le p)$ and for every j $(1 \le j \le p)$, the marginal d.f. $F_{[j]}$ belongs to the class of absolutely continuous d.f.'s with absolutely continuous pdf $f_{[j]}$ having a finite Fisher information.

Let $\mathbf{B} = ((b_{jl}))$ be a $p \times q$ matrix of real elements, and write

$$\mathbf{X}_i(\mathbf{B}) = \mathbf{X}_i - \mathbf{B}\mathbf{c}_i = (X_{i1}(\mathbf{b}_1), \ldots, X_{ip}(\mathbf{b}_p))', \qquad 1 \le i \le N, \tag{7.4.12}$$

where $\mathbf{b}_j$ is the jth row of $\mathbf{B}$, for $j = 1, \ldots, p$. Also, let

$$R_{Ni}^{(j)}(\mathbf{B}) = R_{Ni}^{(j)}(\mathbf{b}_j) = \text{rank of } X_{ij}(\mathbf{b}_j) \text{ among } X_{1j}(\mathbf{b}_j), \ldots, X_{Nj}(\mathbf{b}_j),$$
$$\text{for} \quad i = 1, \ldots, N; \; j = 1, \ldots, p. \tag{7.4.13}$$

Finally, in (7.4.7), we replace $R_{Ni}^{(j)}$ by $R_{Ni}^{(j)}(\mathbf{B})$ and denote the corresponding rank-order statistics as

$$\mathbf{L}_N(\mathbf{B}) = \left(\left(L_{N,jl}(\mathbf{b}_j)\right)\right). \tag{7.4.14}$$

Note that for each j, l, $S_{N,jl}(\mathbf{b}_j)$ is a stochastic process in R^q (generated by $\mathbf{b}_j \in R^q$).

Note that under $H_0^{(2)}: \boldsymbol{\beta}_2 = \mathbf{0}$, (7.4.1) reduces to

$$F_i(\mathbf{x}) = F\left(\mathbf{x} - \boldsymbol{\beta}_0 - \boldsymbol{\beta}_1(\mathbf{c}_{i(1)} - \bar{\mathbf{c}}_{N(1)})\right), \qquad 1 \le i \le N, \quad \mathbf{x} \in R^p, \tag{7.4.15}$$

where $\mathbf{c}_{i(1)}$, $1 \le i \le N$, are defined by (7.3.11) and $\bar{\mathbf{c}}_{N(1)} = N^{-1}\sum_{i=1}^N \mathbf{c}_{i(1)}$. Consider the statistic (matrix)

$$\underset{p \times q_1}{\mathbf{L}_{N(1)}}(\mathbf{B}_1, \mathbf{0}) = \left(\left(L_{N,jl}(\mathbf{b}_{j(1)}, \mathbf{0})\right)\right)_{j=1,\ldots,p,\, l=1,\ldots,q_1}, \tag{7.4.16}$$

where $\mathbf{0}$ is a q_2 null vector, and

$$\mathbf{b}_{j(1)} = \left(b_{j1}, \ldots, b_{jq_1}\right), \quad 1 \le j \le p. \tag{7.4.17}$$

For each j, proceeding as in (7.3.13)–(7.3.14), we define $\hat{\boldsymbol{\beta}}_{j(1),N}$ $(1 \le j \le p)$ and let

$$\hat{\boldsymbol{\beta}}_{1,N} = \left(\hat{\boldsymbol{\beta}}_{1(1),N}, \ldots, \hat{\boldsymbol{\beta}}_{p(1),N}\right)', \tag{7.4.18}$$

$$\begin{aligned}\hat{\mathbf{L}}_{N(2)} &= \left(\left(\hat{L}_{N,jl}\right)\right)_{1 \le j \le p,\, q_1 < l \le q} \\ &= \left(\left(L_{N,jl}\left(\hat{\boldsymbol{\beta}}_{1,N}, \mathbf{0}\right)\right)\right),\end{aligned} \tag{7.4.19}$$

$$\underset{pq_2 \times pq_2}{\hat{\mathbf{H}}_N} = \left(\left(\hat{L}_{N,jl}\hat{L}_{N,j'l'}\right)\right)_{j,j'=1,\ldots,p,\, l,l'=q_1+1,\ldots,q_2}, \tag{7.4.20}$$

$$\hat{R}_{Ni}^{(j)} = R_{Ni}^{(j)}\left(\hat{\boldsymbol{\beta}}_{1,N}\right), \qquad 1 \le j \le N, \quad 1 \le j \le p, \tag{7.4.21}$$

$$\hat{\mathbf{M}}_N = \left(\left(\hat{m}_{Njj'}\right)\right), \tag{7.4.22}$$

where for every $j, j' = 1, \ldots, p$

$$\hat{m}_{Njj'} = \frac{1}{N-1}\left[\sum_{i=1}^N a_{Nj}\left(R_{Ni}^{(j)}\right)a_{Nj'}\left(R_{Ni}^{(j')}\right) - N\bar{a}_{Nj}\bar{a}_{Nj'}\right], \tag{7.4.23}$$

and $\bar{a}_{Nj} = N^{-1}\Sigma_{i=1}^{N} a_{Nj}(i)$, $1 \le j \le p$. We define $\mathbf{C}_N^0$ as in (7.3.20) and let

$$\hat{\mathbf{G}}_N = \hat{M}_N \otimes \mathbf{C}_N^0, \tag{7.4.24}$$

$$\hat{\mathscr{L}}_N^{(2)} = \operatorname{Tr}\left(\hat{\mathbf{H}}_N \hat{\mathbf{G}}_N^{-1}\right)$$

$$= \sum_{j=1}^{p} \sum_{j'=1}^{p} \sum_{l=q_1+1}^{q} \sum_{l'=q_1+1}^{q} \hat{L}_{N,jl} \hat{L}_{N,j'l'} \hat{m}_N^{jj'} c_N^{0ll'}, \tag{7.4.25}$$

where

$$\hat{\mathbf{M}}_N^{-1} = \left(\left(\hat{m}_N^{jj'}\right)\right) \quad \text{and} \quad \left(\mathbf{C}_N^0\right)^{-1} = \left(\left(c_N^{0ll'}\right)\right). \tag{7.4.26}$$

We propose $\hat{\mathscr{L}}_N^{(2)}$ as a test statistic for testing $H_0: \boldsymbol{\beta}_2 = \mathbf{0}$. In particular, if $p = 1$, then $\hat{m}_{N11} = A_N^2$, defined by (7.3.22), so that (7.4.22) reduces to (7.3.21). Thus, $\hat{\mathscr{L}}_N^{(2)}$ may be regarded as a direct multivariate extension of $\mathscr{L}_N^{(2)}$ in (7.3.21). We may note the analogy of $\hat{\mathscr{L}}_N^{(2)}$ and the classical Lawley–Hotelling (trace) statistic for the MANOVA (multivariate analysis of variance) problem. The latter is based on the least-squares estimators (maximum-likelihood estimators under the assumption that F is multinormal) and the estimator of the dispersion matrix of F, while $\mathscr{L}_N^{(2)}$ is based on aligned rank-order statistics and their estimated dispersion matrix. In the case of normal-theory MANOVA, it is known (Anderson, 1959, Chapter 8) that the Lawley–Hotelling trace statistic and the (Wilks) likelihood-ratio statistic are asymptotically equivalent. For the nonparametric case, a similar equivalence theorem is proved in Chapter 8 of Puri and Sen (1971), dealing with the problem of multivariate independence. By virtue of the results in Chapter 6, such a result is also true in our case here (see Exercise 7.4.1). Thus, we could have proposed the alternative test statistic

$$\tilde{\mathscr{L}}_N^{(2)} = \left\{ \frac{\|\hat{\mathbf{G}}_N\|}{\|\hat{\mathbf{H}}_N + \hat{\mathbf{G}}_N\|} \right\}^{N/2}, \tag{7.4.27}$$

where $\|\mathbf{A}\|$ stands for the determinant of $\mathbf{A}$, and it follows that

$$\hat{\mathscr{L}}_N^{(2)} + 2 \log \tilde{\mathscr{L}}_N^{(2)} \xrightarrow{P} 0 \quad \text{as} \quad N \to \infty \tag{7.4.28}$$

(under $H_0^{(2)}$ as well as under local alternatives). On the other hand, $\hat{\mathscr{L}}_N^{(2)}$ is computationally simpler, and we shall now proceed to show that it has a simple asymptotic distribution too. Towards this we consider first the following lemma.

Lemma 7.4.1. Under the assumed regularity conditions on the d.f. F, the score functions $\phi_1, \ldots, \phi_p$, and the constants $\{c_i\}$, when $H_0^{(2)}$ holds,

$$\hat{\mathbf{M}}_N \to \boldsymbol{\nu}(F) \quad \text{in probability} \qquad \text{as} \quad N \to \infty. \tag{7.4.29}$$

Proof. By virtue of Theorem 5.4.1, it suffices to show that under $H_0 : \boldsymbol{\beta} = \mathbf{0}$,

$$\hat{\mathbf{M}}_N - \mathbf{M}_N \xrightarrow{P} 0 \qquad \text{as} \quad N \to \infty, \tag{7.4.30}$$

where $\mathbf{M}_N$ $(= \mathbf{V}_N)$ is defined by (5.4.10). Now, $\hat{m}_{Njj} = m_{Njj}$, $1 \le j \le p$, so it suffices to consider the case of $j \neq j'$ $(= 1, \ldots, p)$, and show that as $N \to \infty$, under $H_0 : \boldsymbol{\beta} = \mathbf{0}$,

$$\frac{1}{N-1} \sum_{i=1}^{N} \left[a_{Nj}\left(\hat{R}_{Ni}^{(j)}\right) a_{Nj'}\left(\hat{R}_{Ni}^{(j')}\right) - a_{Nj}\left(R_{Ni}^{(j)}\right) a_{Nj'}\left(R_{Ni}^{(j')}\right) \right] \xrightarrow{P} 0. \tag{7.4.31}$$

Now, we may virtually repeat the steps in (5.4.21)–(5.4.30) and show that

$$\frac{1}{N-1} \sum_{i=1}^{N} a_{Nj}\left(R_{Ni}^{(j)}\right) a_{Nj'}\left(R_{Ni}^{(j')}\right)$$

$$= \int_{-\infty}^{\infty} \int_{-\infty}^{\infty} \psi_j\left(\frac{N}{N+1} H_{N[j]}(x)\right)$$

$$\psi_{j'}\left(\frac{N}{N+1} H_{N[j']}(y)\right) dH_{N[j,j']}^{*}(x, y) + o_p(1), \tag{7.4.32}$$

$$\frac{1}{N-1} \sum_{i=1}^{N} a_{Nj}\left(\hat{R}_{Ni}^{(j)}\right) a_{Nj'}\left(\hat{R}_{Ni}^{(j')}\right)$$

$$= \int_{-\infty}^{\infty} \int_{-\infty}^{\infty} \psi_j\left(\frac{N}{N+1} \hat{H}_{N[j]}(x)\right)$$

$$\psi_{j'}\left(\frac{N}{N+1} \hat{H}_{N[j']}(y)\right) d\hat{H}_{N[j,j']}^{*}(x, y) + o_p(1), \tag{7.4.33}$$

where $\psi_j(u)$, $0 < u < 1$, is a polynomial [and hence, a bounded and continuous function of $u \in (0, 1)$] for every j $(= 1, \dots, p)$; $H_{N[j]}(x) = N^{-1}\sum_{\alpha=1}^{N} c(x - X_{\alpha j})$, $1 \le j \le p$; $H^*_{N[i,j']}(x, y) = N^{-1}\sum_{i=1}^{N} c(x - X_{ij}) \cdot c(y - X_{ij'})$, $j \neq j' = 1, \dots, p$; $\hat{H}_{N[j]}(x) = N^{-1}\sum_{i=1}^{N} c(x - \hat{X}_{ij})$; and $\hat{H}^*_{N[j,j']}(x, y) = N^{-1}\sum_{i=1}^{N} c(x - \hat{X}_{ij}) c(y - \hat{X}_{ij'})$, where $\hat{\mathbf{X}}_i = \mathbf{X}_i - \hat{\boldsymbol{\beta}}_{1,N}\mathbf{c}_{i(1)}$ for $1 \le i \le N$. Since (7.3.14) holds for each j $(= 1, \dots, p)$ and $\max\{N^{-1/2}\|c_i - \bar{c}_N\| : 1 \le i \le N\} \to 0$, by the fact that

$$\max\{\|\mathbf{X}_i - \hat{X}_i\| : 1 \le i \le N\} \xrightarrow{P} 0 \tag{7.4.34}$$

and the continuity of $F_{[j]}, F_{[j,j']}$, we obtain as $N \to \infty$,

$$\sup\left\{\left|H^*_{N[j,j']}(x, y) - H^*_{N[j,j']}(x, y)\right| : (x, y) \in R^2\right\} \xrightarrow{P} 0. \tag{7.4.35}$$

[Note that (7.4.35) insures that $\sup\|H_{N[j]} - \hat{H}_{N[j]}\| \xrightarrow{P} 0$ for every $1 \le j \le p$.] Thus, by (7.4.32), (7.4.33), (7.4.35), and the continuity (as well as boundedness) of the ψ_j's, we conclude that (7.4.31) holds].

Now, note that in (7.4.19), $L_{N,jl}(\hat{\boldsymbol{\beta}}_{1,N}, 0) = L_{N,jl}(\hat{\boldsymbol{\beta}}_{j(1),N}, 0)$ for every $1 \le j \le p$, so that using (7.3.23)–(7.3.25), coordinatewise, we arrive at the following.

Lemma 7.4.2. Under $H_0^{(2)} : \boldsymbol{\beta}_2 = \mathbf{0}$ and the assumed regularity conditions,

$$N^{-1/2}\left\{\hat{\mathbf{L}}_{N(2)} - \mathbf{L}_{N(2)}(\boldsymbol{\beta}_1, \mathbf{0}) + \mathbf{L}_{N(1)}(\boldsymbol{\beta}_1, \mathbf{0}) \cdot \mathbf{C}_{N11}^{-1}\mathbf{C}_{N12}\right\} \xrightarrow{P} \mathbf{0}, \tag{7.4.36}$$

as $N \to \infty$.

Now, under $H_0^{(2)}$, $(\mathbf{L}_{N(1)}(\boldsymbol{\beta}_1, \mathbf{0}), \mathbf{L}_{N12}(\boldsymbol{\beta}_1, \mathbf{0})) = L_N(\boldsymbol{\beta}_1, \mathbf{0})$ has the same distribution as of $\mathbf{L}_N(\mathbf{0})$ under $H_0 : \boldsymbol{\beta} = \mathbf{0}$. For the latter, we appeal to Theorems 5.4.1 and 5.4.2, and then utilizing (7.4.26), we arrive at the following lemma.

Lemma 7.4.3. Under $H_0^{(2)} : \boldsymbol{\beta}_2 = \mathbf{0}$ and the assumed regularity conditions,

$$N^{-1/2}\hat{\mathbf{L}}_{N(2)} \xrightarrow{\mathscr{L}} \mathscr{N}_{pq_2}(\mathbf{0}, \boldsymbol{\nu}(F) \otimes \boldsymbol{\Lambda}^*), \tag{7.4.37}$$

where Λ^* is defined by (7.3.28).

Finally, by Lemma 7.4.1 and (7.3.28).

$$N^{-1}\hat{\mathbf{G}}_N \xrightarrow{P} \boldsymbol{\nu}(F) \otimes \boldsymbol{\Lambda}^* \quad \text{as} \quad N \to \infty. \tag{7.4.38}$$

By (7.4.20), (7.4.24), (7.4.25), (7.4.37), and (7.4.38), we arrive at the following theorem.

Theorem 7.4.4. *Under $H_0^{(2)}: \boldsymbol{\beta}_2 = \mathbf{0}$ and the assumed regularity conditions, $\hat{\mathscr{L}}_N^{(2)}$ has asymptotically the (central) chi-square distribution with pq_2 DF.*

Thus, as in (7.3.29), we consider the following test: Reject or accept $H_0^{(2)}$ according as

$$\hat{\mathscr{L}}_N^{(2)} \geq \text{ or } < \chi^2_{pq_2}(\alpha), \tag{7.4.39}$$

where α $(0 < \alpha < 1)$ is the desired level of significance of the test.

For the study of the asymptotic power properties of the test based on $\hat{\mathscr{L}}_N^{(2)}$, as in (7.3.30), we confine ourselves to a class of local alternative hypotheses $\{H_N^{(2)}\}$, where

$$H_N^{(2)}: (7.4.1) \text{ holds for } \boldsymbol{\beta}_2 = N^{-1/2}\boldsymbol{\delta}_2, \tag{7.4.40}$$

where $\boldsymbol{\delta}_2$ is a $p \times q_2$ matrix of real constants. If we let

$$\mathbf{d}_{Ni} = N^{-1/2}\boldsymbol{\delta}_2\mathbf{c}_{i(2)}, \quad \mathbf{Y}_i = \mathbf{X}_i - \boldsymbol{\beta}_0 - \mathbf{d}_{Ni}, \qquad 1 \leq i \leq N, \tag{7.4.41}$$

then we may define the likelihood ratio as in (7.3.33) and establish (7.3.34) on parallel lines; we leave the details as an exercise (see Exercise 7.4.2). A direct consequence of this contiguity is that (7.4.30) holds under $\{H_N^{(2)}\}$ as well. On the other hand, from Theorem 5.4.1 and the fact that under $\{H_N^{(2)}\}$ in (7.4.40) we have $\overline{H}_N \to F$ a.e. as $N \to \infty$, we conclude that $\boldsymbol{\nu}(\overline{H}_N) \to \boldsymbol{\nu}(F)$ as $N \to \infty$. Hence, (7.4.29) holds under $\{H_N^{(2)}\}$ as well. Contiguity also insures that (7.4.36) continues to hold for $\{H_N^{(2)}\}$ in (7.4.40). On the other hand, by an appeal to the asymptotic normality results derived in Section 5.5 [see (5.5.13)–(5.5.17)], we conclude that under $\{H_N^{(2)}\}$, $N^{-1/2} \cdot (\mathbf{L}_{N(1)}(\boldsymbol{\beta}_1, 0), \mathbf{L}_{N(2)}(\boldsymbol{\beta}_1, \mathbf{0}))$ has asymptotically a multinormal distribution with dispersion matrix $\boldsymbol{\nu}(F) \otimes \boldsymbol{\Lambda}$ and mean vector

$$\mathbf{B}[\mathbf{0}, \boldsymbol{\delta}_2]\boldsymbol{\Lambda}, \tag{7.4.42}$$

where

$$\mathbf{B} = \operatorname{diag}\big(B(F_{[j]}, \phi_j), 1 \leq j \leq p\big) \tag{7.4.43}$$

and the $B(F_{[j]}, \phi_j)$ are defined by (5.5.15). From the above results, we arrive at the following.

Theorem 7.4.5. *Under* $\{H_N^{(2)}\}$ *in* (7.4.40) *and the assumed regularity conditions,* $\hat{\mathscr{L}}_N^{(2)}$ *has asymptotically a noncentral chi-square distribution with* pq_2 *DF and noncentrality parameter*

$$\Delta^{(2)} = \operatorname{Tr}\left(\boldsymbol{\delta}_2^* \left[\mathbf{T}(F) \otimes \boldsymbol{\Lambda}^*\right]^{-1}\right), \tag{7.4.44}$$

where $\boldsymbol{\delta}_2^* = ((\delta_{jl}\delta_{j'l'}))_{j,j'=1,\ldots,p_j,\ l,l'=q_1+1,\ldots,p}$, $\mathbf{T}(F)$ *is defined in* (5.5.24)–(5.5.25), *and* $\boldsymbol{\Lambda}^*$ *in* (7.3.28).

Now, referring to the model (7.4.1), we may write

$$\mathbf{X} = (\mathbf{X}_1, \ldots, \mathbf{X}_N) = \boldsymbol{\beta}^*\mathbf{A} + \mathbf{e}, \tag{7.4.45}$$

where $\mathbf{e} = (\mathbf{e}_1, \ldots, \mathbf{e}_N)$ and

$$\underset{p\times(q+1)}{\boldsymbol{\beta}^*} = (\boldsymbol{\beta}_0, \boldsymbol{\beta}), \qquad \underset{(q+1)\times N}{\mathbf{A}} = \begin{bmatrix} 1 & \cdots & 1 \\ \mathbf{c}_1' & \cdots & \mathbf{c}_N' \end{bmatrix}, \tag{7.4.46}$$

with the assumption that

$$E\mathbf{e} = \mathbf{0} \quad \text{and} \quad V(\mathbf{e}) = \boldsymbol{\Sigma}(F) \otimes \mathbf{I}_N, \tag{7.4.47}$$

where $\boldsymbol{\Sigma}(F)$ is the dispersion matrix of F and

$$\boldsymbol{\Sigma}(F) \text{ is p.d. and finite.} \tag{7.4.48}$$

Then, the least-squares estimator of $\boldsymbol{\beta}^*$ is

$$\tilde{\boldsymbol{\beta}}_N^* = (\mathbf{XA}')(\mathbf{AA}')^{-1}\left[= \boldsymbol{\beta}^* + (\mathbf{eA}')(\mathbf{AA}')^{-1}\right]. \tag{7.4.49}$$

Also, the estimator of $\boldsymbol{\Sigma}(F)$ is

$$\begin{aligned} \mathbf{S}_e &= (n-q-1)^{-1}\left\{\mathbf{XX}' - \mathbf{XA}'(\mathbf{AA}')^{-1}\mathbf{X}'\right\} \\ &= (n-q-1)^{-1}\left\{\mathbf{ee}' - \mathbf{eA}'(\mathbf{AA}')^{-1}\mathbf{e}'\right\}. \end{aligned} \tag{7.4.50}$$

Therefore, using the fact that the $\mathbf{e}_i$ are i.i.d.r.v.'s satisfying (7.4.47)–(7.4.48), it follows by some routine computations that

$$\mathbf{S}_e \xrightarrow{P} \boldsymbol{\Sigma}(F) \quad \text{as} \quad N \to \infty, \tag{7.4.51}$$

irrespective of any hypothesis on the $\boldsymbol{\beta}^*$. On the other hand, $\tilde{\boldsymbol{\beta}}_N^*$ is linear in $\mathbf{e}$, and the (multivariate) central limit theorem yields that under the assumptions made on the $\{\mathbf{c}_i\}$ and (7.4.47)–(7.4.48),

$$N^{-1/2}(\tilde{\boldsymbol{\beta}}_N^* - \tilde{\boldsymbol{\beta}}^*) \xrightarrow{\mathscr{L}} \mathscr{N}_{P(q+1)}\left(\mathbf{0}, \boldsymbol{\Sigma}(F) \otimes \begin{pmatrix} 1 & \bar{\mathbf{c}}' \\ \bar{\mathbf{c}} & \boldsymbol{\Lambda} \end{pmatrix}\right), \qquad (7.4.52)$$

where

$$\bar{\mathbf{c}} = \lim_{N\to\infty} \bar{\mathbf{c}}_N \quad \text{and} \quad \boldsymbol{\Lambda} = \lim_{N\to\infty} N^{-1}\mathbf{C}_N = \lim_{N\to\infty} \mathbf{C}_N^*. \qquad (7.4.53)$$

Thus, from (7.4.51), (7.4.52), and the definition of the Lawley–Hotelling trace statistic $T_{N(2)}$ for testing $H_0^{(2)}: \boldsymbol{\beta}_2 = \mathbf{0}$, we conclude that under $\{H_N^{(2)}\}$, $T_{N(2)}$ has asymptotically a noncentral chi-square distribution with pq_2 DF and noncentrality parameter

$$\Delta_\lambda^{(2)} = \operatorname{Tr}\left(\boldsymbol{\delta}_2^*\left[\boldsymbol{\Sigma}(F) \otimes \boldsymbol{\Lambda}^*\right]^{-1}\right), \qquad (7.4.54)$$

where $\boldsymbol{\delta}_2^*$ is defined after (7.4.44) and $\boldsymbol{\Lambda}^*$ in (7.3.28). (As a particular case, under $H_0^{(2)}$, $T_{N(2)}$ has asymptotically the central chi-square distribution with pq_2 DF.) Thus, the asymptotic relative efficiency of $\hat{\mathscr{L}}_N^{(2)}$ with respect to $T_{N(2)}$ is given by

$$e\left(\hat{\mathscr{L}}_N^{(2)}, T_{N(2)}\right) = \frac{\Delta_{\mathscr{L}}^{(2)}}{\Delta_\lambda^{(2)}} = \frac{\operatorname{Tr}\left(\boldsymbol{\delta}_2^*\left[\mathbf{T}(F) \otimes \boldsymbol{\Lambda}^*\right]^{-1}\right)}{\operatorname{Tr}\left(\boldsymbol{\delta}_2^*\left[\boldsymbol{\Sigma}(F) \otimes \boldsymbol{\Lambda}^*\right]^{-1}\right)}, \qquad (7.4.55)$$

which depends, in general, on $\boldsymbol{\delta}_2$, $\mathbf{T}(F)$, $\boldsymbol{\Sigma}(F)$, and $\boldsymbol{\Lambda}^*$.

If F, the underlying d.f., is itself multinormal [with dispersion matrix $\boldsymbol{\Sigma}(F)$] and we use the normal scores for the aligned rank statistics [i.e., $\phi_j(u) = \Phi^{-1}(u)$, $0 < u < 1$, $1 \le j \le p$], then as in Chapter 5, we have $\mathbf{T}(F) = \boldsymbol{\Sigma}(F)$, so that (7.4.55) reduces to 1 for all $\boldsymbol{\delta}_2^*$, $\boldsymbol{\Lambda}^*$, and $\boldsymbol{\Sigma}(F)$. This insures the asymptotic optimality of the aligned rank tests based on normal-score statistics when the underlying d.f. F is normal. For possibly nonnormal F, one has to consider the likelihood-ratio test criterion to derive an asymptotically optimal test, and we shall comment on that later on. But we may note that by (7.4.55) and the Courant theorem,

$$\operatorname{Ch}_p\left(\boldsymbol{\Sigma}(F)\mathbf{T}^{-1}(F)\right) \le e\left(\hat{\mathscr{L}}_N^{(2)}, T_{N(2)}\right) \le \operatorname{Ch}_1\left(\boldsymbol{\Sigma}(F)\mathbf{T}^{-1}(\mathbf{F})\right), \qquad (7.4.56)$$

Thus, the bounds studied in Chapter 5 are also applicable in the present case.

7.4.2. Aligned Rank-Order Tests for $H_0^{(3)}$ in (7.4.5)

Since $\boldsymbol{\beta}_0$ stands for the location of F (when the $\mathbf{c}_i$ are all $\mathbf{0}$ or when $\boldsymbol{\beta} = \mathbf{0}$), as in Section 7.3.2, we need to assume here that F is (diagonally) symmetric about $\mathbf{0}$, so that signed rank statistics may be employed for our testing purpose. First, to eliminate the nuisance parameter $\boldsymbol{\beta}_1$, we employ the estimator $\hat{\boldsymbol{\beta}}_{1,N}$ defined by (7.4.18) and consider the residuals

$$\hat{\mathbf{X}}_i = \mathbf{X}_i - \hat{\boldsymbol{\beta}}_{1,N}\mathbf{c}_{i(1)}, \qquad 1 \le i \le N. \tag{7.4.57}$$

Let $\hat{R}_{Ni}^{+(j)}$ be the rank of $|\hat{X}_{ij}|$ among $|\hat{X}_{1j}|, \ldots, |X_{Nj}|$ for $1 \le i \le N$ and $j = 1, \ldots, p$. Define then [as in (5.4.46)] for each j $(= 1, \ldots, p)$,

$$\hat{S}_{N,j\ell} = \sum_{i=1}^{N} c_{il} s_{gn}\left(\hat{X}_{ij}\right) a_{Nj}^{*}\left(\hat{R}_{Ni}^{+(j)}\right), \quad l = 0, q_1 + 1, \ldots, p \tag{7.4.58}$$

(where $c_{i0} = 1$, $1 \le i \le N$), and, parallel to (5.4.52)–(5.4.53), let

$$\hat{v}_{Njj'} = N^{-1} \sum_{i=1}^{N} sgn\hat{X}_{ij} s_{gn} \hat{X}_{ij'} a_{Nj}^{*}\left(\hat{R}_{Ni}^{+(j)}\right) a_{Nj'}\left(\hat{R}_{Ni}^{+(j')}\right), \qquad j, j' = 1, \ldots, p, \tag{7.4.59}$$

$$\hat{\mathbf{V}}_N = \left(\left(\hat{v}_{Njj'}\right)\right). \tag{7.4.60}$$

Also, we let

$$\mathbf{D}_N = \sum_{i=1}^{N} \left(1, \mathbf{c}_i'\right)'\left(1, \mathbf{c}_i\right) = \begin{pmatrix} N & N\bar{\mathbf{c}}_{(1)}' & N\bar{\mathbf{c}}_{(2)}' \\ N\bar{c}_{(1)} & \mathbf{D}_{N11} & \mathbf{D}_{N12} \\ N\bar{c}_{(2)} & \mathbf{D}_{N21} & \mathbf{D}_{N22} \end{pmatrix}, \tag{7.4.61}$$

where $\mathbf{D}_{N11}$ is $q_1 \times q_1$, $\mathbf{D}_{N22}$ is $q_2 \times q_2$, and $\mathbf{D}_{N12} = \mathbf{D}_{N21}'$ is $q_1 \times q_2$. We rearrange the blocks in $\mathbf{D}_N$ and write it as

$$\tilde{\mathbf{D}}_N = \begin{pmatrix} N & N\bar{\mathbf{c}}_{(2)}' & N\bar{\mathbf{c}}_{(1)}' \\ N\bar{\mathbf{c}}_{(2)} & \mathbf{D}_{N22} & \mathbf{D}_{N21} \\ N\bar{\mathbf{c}}_{(1)} & \mathbf{D}_{N12} & \mathbf{D}_{N11} \end{pmatrix} \tag{7.4.62}$$

and let

$$\mathbf{D}_N^0 = \begin{pmatrix} N & N\bar{\mathbf{c}}'_{(2)} \\ N\bar{\mathbf{c}}_{(2)} & \mathbf{D}_{N22} \end{pmatrix} - \begin{pmatrix} N\bar{\mathbf{c}}_{(1)} \\ \mathbf{D}_{N21} \end{pmatrix} \mathbf{D}_{N11}^{-1}\left(N\bar{\mathbf{c}}_{(1)}, \mathbf{D}_{N12}\right). \quad (7.4.63)$$

Then, parallel to (7.4.20), we let

$$\hat{\mathbf{H}}_N = \left(\left(\hat{S}_{N,jl}\hat{S}_{N,j'l'}\right)\right)_{j,j'=1,\dots,p,\, l,l'=p,q_1+,\dots,\, q)'} \quad (7.4.64)$$

$$\hat{\mathbf{G}}_N = \hat{\mathbf{V}}_N \otimes \hat{\mathbf{D}}_N^0, \quad (7.4.65)$$

and

$$\hat{\mathscr{L}}_N^{(3)} = \operatorname{Tr}\left(\hat{\mathbf{H}}_N \hat{\mathbf{G}}_N^{-1}\right)$$

$$= \sum_{j=1}^{p} \sum_{j'=1}^{p} \sum_{l=0,q_1+1}^{q} \sum_{l'=0,q_1+1}^{q} \hat{S}_{N,jl}\hat{S}_{N,j'l'}\hat{v}_n^{jj'}d_N^{0ll'}. \quad (7.4.66)$$

We propose the use of $\hat{\mathscr{L}}_N^{(3)}$ as a test statistic for testing $H_0^{(3)}$. By using (6.4.65) and (6.4.68), we may virtually repeat the line of proofs of Theorem 4.4.4 and Lemmas 7.4.1, 7.4.2, and 7.4.3 to arrive at the following theorem.

Theorem 7.4.6. *Under $H_0^{(3)}: \boldsymbol{\beta}_0 = \mathbf{0}$, $\boldsymbol{\beta}_2 = \mathbf{0}$ and the assumed regularity conditions, $\hat{\mathscr{L}}_N^{(3)}$ has asymptotically the (central) chi-square distribution with $p(q_2 + 1)$ DF.*

Thus, parallel to (7.4.39), we have the following asymptotic test procedure: Reject or accept $H_0^{(2)}$ according as

$$\hat{\mathscr{L}}_N^{(3)} \geq \text{ or } < \chi^2_{p(q_2+1)}(\alpha), \quad (7.4.67)$$

where α $(0 < \alpha < 1)$ is the desired level of significance of the test.

For the study of the asymptotic power properties, we consider here the class $\{H_N^{(3)}\}$ of local alternative hypotheses, where

$$H_N^{(3)}: (7.4.1) \text{ holds with } \boldsymbol{\beta}_0 = N^{-1/2}\boldsymbol{\delta}_0,\ \boldsymbol{\beta}_2 = N^{-1/2}\boldsymbol{\delta}_2 \quad (7.4.68)$$

and $\boldsymbol{\delta}_0 \in R^p$, $\boldsymbol{\delta}_2 \in R^{pq_2}$.

We can prove contiguity as in the case of $\hat{\mathscr{L}}_N^{(2)}$, and hence, proceeding as in the proof of Theorem 7.4.5, we arrive at the following theorem.

Theorem 7.4.7. *Under $\{H_N^{(3)}\}$ in (7.4.68) and the assumed regularity conditions, $\hat{\mathscr{L}}_N^{(3)}$ has asymptotically a noncentral chi-square distribution with $p(q_2+1)$ DF and noncentrality parameter*

$$\Delta_{\mathscr{L}}^{(3)} = \operatorname{Tr}\left(\boldsymbol{\delta}^* \left[\mathbf{T}(F) \otimes \boldsymbol{\Lambda}^0\right]^{-1}\right), \tag{7.4.69}$$

where $\mathbf{T}(F)$ *is defined by* (5.5.24)–(5.5.25) *and*

$$\boldsymbol{\Lambda}^0 = \lim_{N\to\infty} N^{-1}\mathbf{D}_N^0, \qquad \boldsymbol{\delta}^* = \left(\left(\delta_{jl}\delta_{j'l'}\right)\right)_{j,j'=1,\dots,p,\,l,l'=0,q_2+1,\dots,q} \tag{7.4.70}$$

By an analysis entirely similar to (7.4.45)–(7.4.54), we may construct the Lawley–Hotelling trace statistic $T_{N(3)}$ for testing $H_0^{(3)}$ and show that under $H_N^{(3)}$, $T_{N(3)}$ has asymptotically a noncentral chi-square distribution with $p(q_2+1)$ DF and noncentrality parameter

$$\Delta_{\lambda}^{(3)} = \operatorname{Tr}\left(\boldsymbol{\delta}^* \left[\boldsymbol{\Sigma}(F) \otimes \boldsymbol{\Lambda}^0\right]^{-1}\right), \tag{7.4.71}$$

so that under $H_0^{(3)}$, it has the central $\chi^2_{p(q_2+1)}$ d.f. Thus, in this case, the ARE of $\hat{\mathscr{L}}_N^{(3)}$ with respect to $T_{N(3)}$ is

$$e\left(\hat{\mathscr{L}}_N^{(3)}, T_{N(3)}\right) = \frac{\Delta_{\mathscr{L}}^{(3)}}{\Delta_{\lambda}^{(3)}} = \frac{\operatorname{Tr}\left(\boldsymbol{\delta}^*\left[\mathbf{T}(F)\otimes\boldsymbol{\Lambda}^0\right]^{-1}\right)}{\operatorname{Tr}\left(\boldsymbol{\delta}^*\left[\boldsymbol{\Sigma}(F)\otimes\boldsymbol{\Lambda}^0\right]^{-1}\right)}, \tag{7.4.72}$$

and the comments made between (7.4.55) and (7.4.56) remain applicable to (7.4.72) as well.

For testing $H_0^{(1)}: \boldsymbol{\beta}_0 = \mathbf{0}$ vs. $H^{(1)}: \boldsymbol{\beta}_0 \neq \mathbf{0}$, in (7.3.57) we let $q_1 = q$, so that in (7.4.58) we use only $\hat{S}_{N,j0}$, $1 \le j \le p$ (with $q_1 = q$), and in (7.4.61)–(7.4.62) we take $\mathbf{D}_N$ a scalar equal to N; the resulting statistic in (7.4.66) is $\hat{\mathscr{L}}_N^{(1)}$ that is, $\hat{\mathscr{L}}_N^{(1)} = N^{-1}\Sigma_{j=1}^p\Sigma_{j'=1}^p \hat{S}_{N,j0}\hat{S}_{N,j'0}\hat{v}_N^{jj'}$. In (7.4.67), we replace $\hat{\mathscr{L}}_N^{(3)}$ by $\hat{\mathscr{L}}_N^{(1)}$ and $\chi^2_{p(q_2+1),\alpha}$ by $\chi^2_{p,\alpha}$. Similarly, in Theorem 7.6.7, we replace $\hat{\mathscr{L}}_N^{(3)}$ by $\hat{\mathscr{L}}_N^{(1)}$, and in (7.4.69), $\boldsymbol{\delta}^*$ by $\boldsymbol{\delta}_0$ and $\boldsymbol{\Lambda}^0$ by 1. The ARE is analogous to (7.4.72) and the bounds in (7.4.56) remain intact.

7.4.3. Asymptotic Optimality of Aligned Rank Tests

In this subsection, we shall extend the results of Section 7.3.5 (under additional regularity conditions) to the multivariate case. We let $\boldsymbol{\beta}^* =$

$(\boldsymbol{\beta}_0, \boldsymbol{\beta})$, a $p \times (q+1)$ matrix, write $f_i(\mathbf{x}; \boldsymbol{\beta}^*) = f(x - \boldsymbol{\beta}_0 - \boldsymbol{\beta} c_i)$, $1 \leq i \leq N$, and assume that for every i $(= 1, \ldots, N)$,

$$\left| \frac{\partial}{\partial \boldsymbol{\beta}^*} f_i(\mathbf{x}; \boldsymbol{\beta}^*) \right| \leq \mathbf{U}_1(\mathbf{x}) \quad \forall \boldsymbol{\beta}^* \in B^*, \mathbf{x} \in R^p, \tag{7.4.73}$$

$$\left| \frac{\partial^2}{\partial \boldsymbol{\beta}^* \, \partial \boldsymbol{\beta}^{*\prime}} f_i(\mathbf{x}; \boldsymbol{\beta}^*) \right| \leq \mathbf{U}_2(\mathbf{x}) \quad \forall \boldsymbol{\beta}^* \in B^*, \mathbf{x} \in R^p, \tag{7.4.74}$$

where $\mathbf{U}_1$ and $\mathbf{U}_2$ are both integrable [with respect to $F(x; *)$], B^* is a nonempty subset in $R^{p(q+1)}$ containing $\mathbf{0}$ as an inner point $(\boldsymbol{\beta}^* \in B^*)$. Furthermore, writing

$$\mathbf{H}_i(\mathbf{x}; \boldsymbol{\beta}^*) = \frac{\partial^2}{\partial \boldsymbol{\beta}^* \, \partial \boldsymbol{\beta}^{*\prime}} \log f_i(\mathbf{x}; \boldsymbol{\beta}^*), \qquad 1 \leq i \leq N, \tag{7.4.75}$$

we assume that for every $1 \leq i \leq N$,

$$\left\| E\{\mathbf{H}_i(\mathbf{X}_i; \boldsymbol{\beta}^*) | \boldsymbol{\beta}^*\} \right\| < \infty \tag{7.4.76}$$

and

$$\lim_{\delta \downarrow 0} E\left\{ \sup_{\|\boldsymbol{\beta}^* - \tilde{\boldsymbol{\beta}}^*\| < \delta} \left\| \mathbf{H}_i(\mathbf{X}_i; \boldsymbol{\beta}^*) - H_i(\mathbf{X}_i; \tilde{\boldsymbol{\beta}}^*) \right\| \Big| \boldsymbol{\beta}^* \right\} = 0, \tag{7.4.77}$$

where $\| \cdot \|$ stands for the sup norm. Finally, we assume that the maximum-likelihood estimator of $\boldsymbol{\beta}^*$ exists and is uniformly consistent. We denote this (unique) estimator by $\tilde{\boldsymbol{\beta}}^*$. Also, we let

$$\mathscr{I} = E\left\{ - \frac{\partial^2}{\partial \boldsymbol{\beta}_0 \, \partial \boldsymbol{\beta}_0'} \log f_i(\mathbf{X}_i; \boldsymbol{\beta}^*) \Big|_{\boldsymbol{\beta}^* = \mathbf{0}} \right\}; \tag{7.4.78}$$

$\mathscr{I}$ exists by (7.3.73)–(7.3.76). Further, by (7.3.76),

$$\sum_{i=1}^{N} E\{ H_i(\mathbf{X}_i, \boldsymbol{\beta}^*) | \boldsymbol{\beta}^* = \mathbf{0} \} = \mathscr{I} \otimes \mathbf{D}_N, \tag{7.4.79}$$

where $\mathbf{D}_N$ is defined by (7.4.61). Consider then the likelihood-ratio test $(\lambda_N^{(2)})$ for testing $H_0^{(2)}$ in (7.4.4) when the underlying d.f. satisfies the abovementioned regularity conditions. Then proceeding as in Section 5.8, it follows that under $H_0^{(2)}$, $-2\log\lambda_N^{(2)}$ has asymptotically the central chi-square distribution with pq_2 DF, and under $\{H_N^{(2)}\}$ in (7.4.40), it has asymptotically a noncentral chi-square distribution with pq_2 DF and noncentrality parameter

$$\Delta_\lambda^{(2)} = \operatorname{Tr}\left(\boldsymbol{\delta}_2^*\left[\mathscr{I}^{-1}\otimes\boldsymbol{\Lambda}^*\right]^{-1}\right) = \operatorname{Tr}\left(\boldsymbol{\delta}_2^*\left[\mathscr{I}\otimes\boldsymbol{\Lambda}^{*-1}\right]\right), \quad (7.4.80)$$

where $\boldsymbol{\delta}_2^*$ and $\boldsymbol{\Lambda}^*$ are defined as in (7.6.44). Moreover, under the abovementioned regularity conditions, this likelihood-ratio test has asymptotically best average power with respect to ellipsoidal surfaces (specified by $\Delta_\lambda^{(2)} =$ const) and is an asymptotically most stringent test. Thus, by Theorem 7.4.5 and the above discussion, we conclude that $\hat{\mathscr{L}}_N^{(2)}$ is also asymptotically an optimal test (for testing $H_0^{(2)}$ against $\{H_N^{(2)}\}$) when

$$\Delta_{\mathscr{L}}^{(2)} = \Delta_\lambda^{(2)} \qquad \forall\boldsymbol{\delta}_2^*, \quad (7.4.81)$$

or, in other words, when

$$\mathbf{T}(F) = \mathscr{I}^{-1}. \quad (7.4.82)$$

Recalling the results in Chapter 6, we may note that $\mathbf{T}(F)$ appears in the dispersion matrix of the rank-order estimates, while $\mathscr{I}^{-1}$ appears in the dispersion matrix of the maximum-likelihood estimator. Thus, the condition (7.4.82) is equivalent to saying that if the rank-order estimators (based on $\mathbf{L}_N$ or $\mathbf{S}_N$) are fully efficient, then the aligned rank-order tests are also.

The treatment of optimality of $\hat{\mathscr{L}}_N^{(3)}$ or $\hat{\mathscr{L}}_N^{(1)}$ follows analogously.

7.4.4. Aligned Rank Tests for Parallelism of Regression Surfaces

In this subsection, as a generalization of the model (7.3.68) to the multivariate case, we consider the following model and study appropriate rank-order tests (considered in Sen and Puri, 1977).

Let there be k (≥ 2) independent samples of sizes $n_1, \ldots, n_k$ respectively, where the random vectors $\mathbf{X}_{i1}, \ldots, \mathbf{X}_{in_i}$, constituting the ith sample, are independent with continuous p-variate d.f.'s $F_{i1}, \ldots, F_{in_i}$, all defined on R^p for $i = 1, \ldots, k$, and where

$$F_{ij}(\mathbf{x}) = F(\mathbf{x} - \boldsymbol{\beta}_{0i} - \boldsymbol{\beta}_i\mathbf{c}_{ij}), \qquad 1 \leq j \leq n_i, \quad 1 \leq i \leq k, \quad (7.4.83)$$

$\boldsymbol{\beta}_{0i} = (\beta_{0i}^{(1)}, \ldots, \beta_{0i}^{(p)})'$ and $\boldsymbol{\beta}_i = ((\beta_{is}^{(t)}))_{1\leq s\leq q, 1\leq t\leq p}$ are unknown parameters, $\mathbf{c}_{ij} = (c_{ij}^{1(1)}, \ldots, c_{ij}^{(q)})'$, $1 \leq j \leq n_i$, are known regression constants (vectors) for $i = 1, \dots, k$, and F is a continuous (unknown) d.f. defined on R^p. We are interested in testing

$$H_0: \boldsymbol{\beta}_1 = \cdots = \boldsymbol{\beta}_k = \boldsymbol{\beta} \quad \text{(unknown)} \quad \text{vs.}$$

$$H_1: \boldsymbol{\beta}_i \neq \boldsymbol{\beta}_{i'} \quad \text{for at least one } i \neq i'. \tag{7.4.84}$$

As we are dealing with k-sample analogues of the multivariate problem treated in Section 7.4.1, we assume that for each i $(1 \le i \le k)$ and t $(1 \le t \le p)$, the $\mathbf{c}_{ij}^{(t)}$, $1 \le j \le n_i$, satisfy (6.3.12) through (6.3.20). Let then

$$\mathbf{C}_{n_i} = \sum_{j=1}^{n_i} (\mathbf{c}_{ij} - \bar{\mathbf{c}}_i)(\mathbf{c}_{ij} - \bar{\mathbf{c}}_i)', \qquad 1 \le i \le k, \tag{7.4.85}$$

where $\bar{\mathbf{c}}_i = n_i^{-1}\sum_{j=1}^{n_i}\mathbf{c}_{ij}$, $1 \le i \le k$, and assume that for each i $(1 \le i \le k)$,

$$\mathbf{C}_{n_i}^* = n_i^{-1}\mathbf{C}_{n_i} \to \mathbf{C}_i^* \qquad \text{as} \quad n_i \to \infty, \tag{7.4.86}$$

where, for each $1 \le i \le k$,

$$\mathbf{C}_i^* \text{ is p.d. and has finite elements.} \tag{7.4.87}$$

Also, let $N = n_1 + \cdots + n_k$ and $\lambda_{Ni} = n_i/N$, $1 \le i \le k$. Then we assume that there exists an $\lambda_0 \in (0, k^{-1}]$ such that

$$0 < \lambda_0 \le \lim_{N \to \infty} \lambda_{Ni} = \lambda_i < 1 - \lambda_0 < 1 \qquad \forall 1 \le i \le k. \tag{7.4.88}$$

Finally, as in Section 7.4.1, we assume that all the p marginal d.f.'s of F are absolutely continuous and possess absolutely continuous pdf with finite Fisher information.

As in (7.4.12), we let $\mathbf{X}_{ij}(\mathbf{B}) = \mathbf{X}_{ij} - \mathbf{B}\mathbf{c}_{ij} = (X_{ij}^{(1)}(\mathbf{b}_1), \dots, X_{ij}^{(p)}(\mathbf{b}_p))$, where $\mathbf{b}_j$ is the jth row of $\mathbf{B}$ $(1 \le j \le p)$, and as in (7.4.13), we let $R_{n_ij}^{(t)}(B) = R_{n_ij}^{(t)}(\mathbf{b}_t) = \text{rank of } X_{ij}^{(t)}(\mathbf{b}_t)$ among $X_{i1}^{(t)}(\mathbf{b}_t), \dots, X_{in_i}^{(t)}(\mathbf{b}_t)$, $1 \le j \le n_i$, $1 \le t \le p$, for $i = 1, \dots, k$. Let then for each $i (= 1, \dots, k)$,

$$\mathbf{L}_{n_i}^{(i)}(\mathbf{B}) = \left(\left(L_{n_i(s)}^{(t)(i)}(\mathbf{B})\right)\right), \tag{7.4.89}$$

$$L_{n_i(s)}^{(t)(i)}(\mathbf{B}) = L_{n_i(s)}^{(t)}(\mathbf{b}_t) = \sum_{j=1}^{n_i} \left(c_{ij}^{(s)} - \bar{c}_i^{(s)}\right) a_{n_i t}\left(R_{n_ij}^{(t)}(\mathbf{b}_t)\right),$$

$$1 \le s \le q, \quad 1 \le t \le p, \tag{7.4.90}$$

where the scores $a_{n_i t}(1), \ldots, a_{n_i t}(n_i)$ for $t = 1, \ldots, p$ are defined by (4.2.4). Let then

$$\mathbf{L}_N(\mathbf{B}) = \sum_{i=1}^{k} \mathbf{L}_{n_i}^{(i)}(\mathbf{B}) \qquad \forall \mathbf{B} \in R^{pq}. \tag{7.4.91}$$

To construct an aligned rank statistic for testing H_0 in (7.4.84), we need to eliminate the nuisance parameter $\boldsymbol{\beta}$ through estimation. For this, we let

$$\mathbf{B}_N = \left\{ \mathbf{B} \in R^{pq} : \sum_{t=1}^{p} \sum_{s=1}^{q} \left| L_{N(s)}^{(t)}(\mathbf{B}) \right| = \text{minimum} \right\} \tag{7.4.92}$$

and choose a unique $\hat{\boldsymbol{\beta}}_N \in \mathbf{B}_N$ (e.g. the centroid of $\mathbf{B}_N$) as the estimator of $\boldsymbol{\beta}$ [under H_0 in (7.4.84)]. Consider then the aligned rank statistics

$$\hat{\mathbf{L}}_{n_i}^{(i)} = \mathbf{L}_{n_i}^{(i)}(\hat{\boldsymbol{\beta}}_N), \qquad 1 \le i \le k, \tag{7.4.93}$$

to be employed in the construction of the test statistic. If we let $\hat{R}_{n_i j}^{(t)} = R_{n_i j}^{(t)}(\hat{\boldsymbol{\beta}}_N)$, $1 \le j \le n_i$, $1 \le t \le p$, then as in (7.4.23) we let

$$\hat{m}_{n_i tt'}^{(i)} = \frac{1}{n_i - 1} \left[\sum_{j=1}^{n_i} a_{n_i t}\left(\hat{R}_{n_i j}^{(t)}\right) a_{n_i t'}\left(\hat{R}_{n_i j}^{(t')}\right) - n_i \bar{a}_{n_i t} \bar{a}_{n_i t'} \right] \tag{7.4.94}$$

for $t, t' = 1, \ldots, p$ and $1 \le i \le k$, and let

$$\hat{\mathbf{M}}_N = ((\hat{M}_{Ntt'})) = \sum_{i=1}^{k} \frac{n_i - 1}{N - k} \left(\left(\hat{m}_{n_i tt'}^{(i)} \right) \right), \tag{7.4.95}$$

$$\mathbf{G}_{n_i} = \hat{\mathbf{M}}_N \otimes \mathbf{C}_{n_i}, \qquad i = 1, \ldots, k, \tag{7.4.96}$$

$$\hat{\mathbf{H}}_{n_i} = \left(\left(\hat{L}_{n_i(s)}^{(t)(i)} \hat{L}_{n_i(s')}^{(t')(i)} \right) \right)_{s, s'=1, \ldots, q, t, t'=1, \ldots, p}, 1 \le i \le k. \tag{7.4.97}$$

Then, as in (7.4.25), we consider

$$\hat{\mathscr{L}}_N = \sum_{i=1}^{k} \operatorname{Tr}\left(\hat{\mathbf{H}}_{n_i} \mathbf{G}_{n_i}^{-1} \right). \tag{7.4.98}$$

The statistic $\hat{\mathscr{L}}_N$ is proposed for testing $H_0 : \boldsymbol{\beta}_1 = \cdots = \boldsymbol{\beta}_k$. We may remark that if $q = p = 1$, then $\hat{\mathscr{L}}_N$ in (7.4.98) reduces to $\hat{\mathscr{L}}_N$ in (7.3.77), so that (7.4.98) may be regarded as an extension of (7.3.77) to the case of multivariate observations dealing with multiple regressors. For $p \geq 1$ but $q = 1$, $\mathbf{C}_{n_i}$ are scalar quantities, so that $\hat{\mathscr{L}}_N$ reduces to

$$\sum_{i=1}^{k} C_{n_i}^{-1} \left\{ \sum_{t=1}^{p} \sum_{t'=1}^{p} \hat{L}_{n_i(1)}^{(t)(i)} \hat{L}_{n_i(1)}^{(t')(i)} \hat{M}_N^{tt'} \right\}. \tag{7.4.99}$$

Similarly, for $p = 1$ but $q \geq 1$, (7.4.98) reduces to

$$M_{N11}^{-1} \sum_{i=1}^{k} \hat{\mathbf{L}}_{n_i}^{(i)\prime} \mathbf{C}_{n_i}^{-1} \hat{\mathbf{L}}_{n_i}^{(i)}, \tag{7.4.100}$$

where $\hat{\mathbf{L}}_{n_i}^{(c)\prime} = (\hat{L}_{n_i(1)}^{(1)(i)}, \ldots, \hat{L}_{n_i(q)}^{(1)(i)})'$, $i = 1, \ldots, k$. In the sequel, we proceed to consider the general case of (7.4.98), and later on we shall touch briefly the special cases in (7.4.99) and (7.4.100).

To study the properties of the test based on $\hat{\mathscr{L}}_N$, we first consider some properties of $\hat{\mathbf{L}}_{n_i}^{(i)}$, $1 \leq i \leq k$. As in Section 6.4.2, consider (from the ith sample) the rank-order estimator $\hat{\boldsymbol{\beta}}_{n_i}^{(i)}$ of $\boldsymbol{\beta}_i$. For this, let

$$B_{n_i}^{(i)} = \left\{ B \in R^{pq} : \sum_{t=1}^{p} \sum_{s=1}^{q} \left| L_{n_i(s)}^{(t)(i)}(B) \right| = \text{minimum} \right\}, \tag{7.4.101}$$

$$\hat{\boldsymbol{\beta}}_{ni}^{(i)} = \text{centroid of } B_{n_i}^{(i)}, \qquad 1 \leq i \leq k. \tag{7.4.102}$$

Then, from Theorem 6.4.4. we have for every i $(1 \leq i \leq k)$,

$$n_i^{1/2}\left(\hat{\boldsymbol{\beta}}_{n_i}^{(i)} - \boldsymbol{\beta}_i\right) \overset{\mathscr{L}}{\sim} \mathscr{N}_{pq}(0, \mathbf{T}(F) \otimes \mathbf{C}_i^{*-1}), \tag{7.4.103}$$

where $\mathbf{T}(F)$ is defined by (5.5.24)–(5.5.25) and $\mathbf{C}_i^*$ by (7.4.86), and

$$n_i^{-1/2} \mathbf{L}_{n_i}^{(i)}\left(\hat{\boldsymbol{\beta}}_{n_i}^{(i)}\right) = \mathbf{O}_p(1) \qquad \text{for} \quad i = 1, \ldots, k. \tag{7.4.104}$$

In a similar manner, under H_0 in (7.4.84),

$$N^{1/2}(\hat{\boldsymbol{\beta}}_N - \boldsymbol{\beta}) \overset{\mathscr{L}}{\to} \mathscr{N}_{pq}(\mathbf{O}, \mathbf{T}(F) \otimes \mathbf{C}^{*-1}); \tag{7.4.105}$$

$$N^{-1/2} \mathbf{L}_N(\hat{\boldsymbol{\beta}}_N) = \mathbf{O}_p(1), \tag{7.4.106}$$

where

$$\mathbf{C}^* = \sum_{i=1}^{k} \lambda_i \mathbf{C}_i^*. \tag{7.4.107}$$

Thus, by using (6.4.65) for each t $(=1,\ldots,p)$ and i $(=1,\ldots,k)$, we obtain from (7.4.103) through (7.4.106) that under H_0 in (7.4.84),

$$\begin{aligned}\hat{\mathbf{L}}_{n_i}^{(i)} &= \hat{\mathbf{L}}_{n_i}^{(i)}(\hat{\boldsymbol{\beta}}_N) - \mathbf{L}_{n_i}^{(i)}(\hat{\boldsymbol{\beta}}_{n_i}^{(i)}) + \mathbf{o}_p(n_i^{1/2}) \\ &= \mathbf{B}\mathbf{C}_{n_i}(\hat{\boldsymbol{\beta}}_{n_i}^{(i)} - \hat{\boldsymbol{\beta}}_N) + \mathbf{o}_p(n_i^{1/2}), \qquad 1 \le i \le k, \qquad (7.4.108)\end{aligned}$$

where the diagonal matrix $\mathbf{B}$ is defined by (7.4.43) and the C_{n_i} by (7.4.85). On the other hand, by Lemma 7.4.1 and (7.4.86),

$$n_i^{-1}\mathbf{G}_{n_i} \xrightarrow{P} \boldsymbol{\nu}(F) \otimes \mathbf{C}_i^*, \qquad 1 \le i \le k. \qquad (7.4.109)$$

Further, by (7.4.91), (7.4.104), (7.4.106), and (7.4.108), under H_0 in (7.4.84),

$$\mathbf{B}\left(\sum_{i=1}^{k} \mathbf{C}_{n_i}\right)\hat{\boldsymbol{\beta}}_N = \mathbf{B}\left(\sum_{i=1}^{k} \mathbf{C}_{n_i}\hat{\beta}_{n_i}^{(i)}\right) + \mathbf{o}_p(N^{1/2}). \qquad (7.4.110)$$

Hence, by (7.4.95)–(7.4.98) and (7.4.107)–(7.4.110), $\hat{\mathscr{L}}_N$ in (7.4.98) is asymptotically equivalent (in probability) to a quadratic form in the estimators $\hat{\boldsymbol{\beta}}_{n_i}^{(i)}$, $1 \le i \le k$, and a direct application of the Cochran theorem along with (7.4.103) leads us to the following theorem.

Theorem 7.4.8. *Under H_0 in (7.4.84) and the assumptions made at the beginning of this subsection, $\hat{\mathscr{L}}_N$ in (7.8.98) has asymptotically the central chi-square distribution with $pq(k-1)$ DF.*

Thus, parallel to (7.4.39) or (7.4.67), we prescribe the following asymptotic test procedure for H_0 in (7.4.84): Reject or accept H_0 according as

$$\hat{\mathscr{L}}_N \ge \text{ or } \le \chi^2_{pq(k-1)}(\alpha), \qquad (7.4.111)$$

where α $(0 < \alpha < 1)$ is the desired level of significance of the test.

For the study of the asymptotic power properties of the test based on $\hat{\mathscr{L}}_N$, we consider a sequence $\{H_N\}$ of local alternatives, where

$$H_N: \boldsymbol{\beta}_i = \boldsymbol{\beta}_i^{(N)} = \boldsymbol{\beta} + N^{-1/2}\boldsymbol{\gamma}_i, \quad 1 \le i \le k; \ \sum_{i=1}^{k} \lambda_i \mathbf{C}_i^* \boldsymbol{\gamma}_i = \mathbf{0}. \quad (7.4.112)$$

Now, the contiguity of the probability measures under $\{H_N\}$ with respect to those under H_0 in (7.4.84) follows as in the case of $\hat{\mathscr{L}}_N^{(2)}$ in Section 7.4.1, and this insures that (7.4.108), (7.4.109), and (7.4.110) all hold under $\{H_N\}$ as well. Thus, the asymptotic equivalence (in probability) of $\hat{\mathscr{L}}_N$ to the same quadratic form in the $\hat{\boldsymbol{\beta}}_{n_i}^{(i)}$, $1 \le i \le k$, holds (as in the case of H_0), and therefore, by (7.4.103), (7.4.112), and the Cochran theorem, we arrive at the following theorem.

Theorem 7.4.9. *Under* $\{H_N\}$ *in* (7.4.112) *and the assumptions made at the beginning of this subsection* $\hat{\mathscr{L}}_N$ *in* (7.4.98) *has asymptotically a noncentral chi-square distribution with* $pq(k-1)$ *DF and noncentrality parameter*

$$\Delta_{\mathscr{L}} = \sum_{i=1}^{k} \lambda_i^{-1} \operatorname{Tr}\left(\boldsymbol{\Gamma}_i (\mathbf{T}(F) \otimes \mathbf{C}_i^*)^{-1}\right), \tag{7.4.113}$$

where the $\mathbf{C}_i^*$ *are defined by* (7.4.86), $\mathbf{T}(F)$ *by* (5.5.24)–(5.5.25), *and*

$$\boldsymbol{\Gamma}_i = \left(\left(\gamma_{ts}^{(i)} \gamma_{t's'}^{(i)}\right)\right)_{s,s'=1,\dots,q,\,t,t'=1,\dots,p}, \quad i = 1,\dots,k. \tag{7.4.114}$$

Analogously to (7.4.72), the ARE of $\hat{\mathscr{L}}_N$ with respect to the normal-theory likelihood-ratio test is given by

$$\frac{\sum_{i=1}^{k} \lambda_i^{-1} \operatorname{Tr}\left(\boldsymbol{\Gamma}_i (\mathbf{T}(F) \otimes \mathbf{C}_i^*)^{-1}\right)}{\sum_{i=1}^{k} \lambda_i^{-1} \operatorname{Tr}\left(\boldsymbol{\Gamma}_i (\boldsymbol{\Sigma}(F) \otimes \mathbf{C}_i^*)^{-1}\right)}, \tag{7.4.115}$$

and it depends on the $\boldsymbol{\Gamma}_i$, $\mathbf{C}_i^*$, λ_i as well as $\mathbf{T}(F)$ and $\boldsymbol{\Sigma}(F)$. The lower and upper bounds for the ARE in (7.4.115) are again the smallest and the largest roots of $\boldsymbol{\Sigma}(F)\mathbf{T}^{-1}(F)$, and hence the discussion following (7.4.55) remains applicable here. Finally, the asymptotic optimality condition in (7.4.82) remains intact here also.

For the special case of $q = 1$ or $p = 1$, the adjustment for the number of DF in (7.4.111) is obvious, while in (7.4.113), $\Delta_{\mathscr{L}}$ implies to $\sum_{i=1}^{k} \lambda_i^{-1} C_i^{*-1} [\gamma_i' T^{-1}(F) \gamma_i]$ or $\sum_{i=1}^{k} \lambda_i^{-1} (\gamma_i' C_*^{-1} \gamma_i)/\tau_{111}(F)$, and the same change appears in (7.4.116).

7.5. SOME ADDITIONAL REMARKS ON ALIGNED RANK TESTS FOR SUBHYPOTHESES

The results of this chapter are based on an alignment principle under which the alignment is made by estimating the nuisance parameters and rank statistics are used for the estimation problem (so that the theory developed in Chapter 6 applies). We may remark that in this setup, the critical technique is the choice of congruent scores for the estimation of nuisance parameters and for the test of the parameters under the hypotheses. If different score functions were used for these two different purposes, we would have additional unknown quantities appearing in the linearity results for the aligned rank statistics, and that would create additional problems. On the other hand, under diverse regularity conditions such a restriction can be avoided. We shall briefly discuss here some developments on this line.

One of the nice properties of aligned signed rank statistics is that the linearity result in (6.4.68) (or its univariate counterpart) insures that the influence of the regression residuals can be neglected (asymptotically) under fairly general conditions provided we assume that F is symmetric. Note that for testing $H_0^{(1)}$ or $H_0^{(3)}$ in Sections 7.3 and 7.4, we have assumed that F is symmetric about 0, so that in such a case, we do not need any additional assumption. This shows that in testing $H_0^{(1)}$ or $H_0^{(3)}$, for the alignment procedure, instead of the rank-based estimator $\hat{\boldsymbol{\beta}}_N$, we may use any convenient estimator $\tilde{\boldsymbol{\beta}}_N$ of $\boldsymbol{\beta}$ provided

$$N^{1/2}\|\tilde{\boldsymbol{\beta}}_{2,N} - \boldsymbol{\beta}_2\| = O_p(1) \tag{7.5.1}$$

and the $\mathbf{c}_i$ satisfy the same conditions as in Sections 7.3 and 7.4. Now, (7.5.1) holds, for example, if one uses the classical-least squares estimator, provided the d.f. F has finite second moments (which we do not need for rank estimators). There are other estimators too which satisfy (7.5.1). For testing $H_0^{(2)}$ in Sections 7.3 and 7.4, we do not require F to be symmetric. If, however, we are willing to make the additional assumption that F is symmetric, then instead of using the linear rank statistics (aligned), we may use the signed rank statistics [as in (7.4.58)], and for this, we do not need to use rank estimators—any consistent estimator satisfying (7.5.1) will suffice. The theory is again based on the basic linearity result in (6.4.68) and runs parallel to the one developed in Sections 7.3 and 7.4. For more details in the univariate setup, we refer to Adichie (1978), and the same theory holds for the multivariate case as well.

Among other possibilities, an approach initiated by Jaeckel (1972) and later extended by McKean and Hettmansperger (1976) also deserves to be mentioned in this context. Consider, for example, the univariate multiple regression model (7.3.1), and as in the discussion after (6.3.5), let $R_{N_i}(\mathbf{b})$ be the rank of $X_i(\mathbf{b}) = X_i - \mathbf{b}'\mathbf{c}_i$ among $X_1(\mathbf{b}), \ldots, X_N(\mathbf{b})$ for $i = 1, \ldots, N$, where $\mathbf{b} \in R^q$. Then Jaeckel's dispersion function is defined by

$$D_N(\mathbf{b}) = \sum_{i=1}^{N} X_i(\mathbf{b})\, a_N(R_{Ni}(\mathbf{b})), \qquad \mathbf{b} \in R^q, \tag{7.5.2}$$

where the scores $a_N(1) \leq \cdots \leq a_N(N)$ are defined as in (6.2.10). Consider then $H_0^{(2)}: \boldsymbol{\beta}_2 = 0$ vs. $H^{(2)}: \boldsymbol{\beta}_2 \neq \mathbf{0}$ [as in (7.3.4)], where $\boldsymbol{\beta}' = (\boldsymbol{\beta}_1', \boldsymbol{\beta}_2')$. Let then

$$B_N^0 = \{\mathbf{b} : \mathbf{D}_N(\mathbf{b}) = \text{minimum}\}, \tag{7.5.3}$$

$$\check{\boldsymbol{\beta}}_N = \text{centroid of } B_N^0. \tag{7.5.4}$$

Also, let

$$\mathbf{B}_N^* = \{\mathbf{b}' = (\mathbf{b}_1', \mathbf{0}) : \mathbf{D}_N((\mathbf{b}_1', \mathbf{0}')') = \text{minimum}\}, \tag{7.5.5}$$

$$\check{\boldsymbol{\beta}}_N^* = \text{centroid of } \mathbf{B}_N^*. \tag{7.5.6}$$

Then an aligned test is based on the statistic

$$T_n = D_N((\check{\boldsymbol{\beta}}_N^{*\prime}, \mathbf{0}')') - D_N(\check{\boldsymbol{\beta}}_N), \tag{7.5.7}$$

rejecting $H_0^{(2)}$ for large values of T_N. Also, let us define $\rho(\psi, \phi)$ as in (6.2.39) and let

$$\gamma = \gamma(\phi, \psi) = A_\psi A_\phi \rho(\phi, \psi) = \int_0^1 \phi(u) \psi(u)\, du. \tag{7.5.8}$$

Note that γ is an unknown parameter, but, as in our Chapter 6, we have no problem in estimating γ from the confidence intervals for $\boldsymbol{\beta}$. We denote such an estimator by $\hat{\gamma}_N$, so that

$$\hat{\gamma}_N \to \gamma \quad \text{in probability} \quad \text{as} \quad N \to \infty. \tag{7.5.9}$$

Let then

$$T_n^* = 2A_N^{-2} T_N \hat{\gamma}_N. \tag{7.5.10}$$

By using similar linearity results, it has been shown by McKean and Hettmansperger (1976) that under $H_0^{(2)}$, T_N^* has asymptotically the (central) chi-square distribution with q_2 DF, and they suggested a test procedure similar to (7.3.29) where we replace $\mathscr{L}_N^{(2)}$ by T_N^*. A similar procedure can also be worked out for the multivariate case. This procedure has the same asymptotic properties as the ones considered in Sections 7.3 and 7.4. However, we may note that by (7.5.2), $\mathbf{D}_N(\mathbf{b})$ is not a pure rank statistic—it is a mixed statistic, and the factor $X_i(\mathbf{b})$, $1 \leq i \leq N$, may cause some concern when there are outliers.

Robust tests for linear models based on M-estimators and related functionals have also been considered by various workers. Hettmansperger and Schrader (1980) have considered an analogue of the classical likelihood-ratio test wherein the role of "squared error" appearing in the normal-theory likelihood functions is replaced by some robust competitors $Q(\cdot)$, where Q is a nonnegative function having a well-behaved first derivative ψ (score function) which provides the M-estimators (both under null and alternative hypotheses). A standardized form of the difference of the sum of these residual based Q's under the null and alternative hypotheses has asymptotically chi-square distribution with appropriate DF, and this provides ADF tests. A more simplified (but equally efficient) version of such tests is due to Sen (1982b). These tests are, however, not (aligned) rank tests, and we shall not enter in the details of their properties. For the sake of completeness, we refer to Hettmensperger (1984) and Singer and Sen (1985a, b).

A variety of subhypothesis-testing problems have been treated earlier in this chapter. In a general MANOVA model, in view of the multitude of the parameters, often the overall tests considered earlier fail to provide detailed conclusions on various possible component hypotheses. Detailed testing for these components involves multiple testing on the same set of data, and thereby may lead to a higher overall significance level of these simultaneous tests. Indeed, if a large number of component hypotheses are to be tested simultaneously, the overall significance level may be quite large compared to the ones for the individual tests when done separately. Thus, some care needs to be taken to control the significance level of the simultaneous tests. In the dual problem of simultaneous confidence regions, similarly, the simultaneous convergence probability may be quite smaller than the individual ones, unless some technique is adopted to control this phenomenon.

For parametric MANOVA models, a variety of simultaneous inference procedures are available in the literature. A detailed account of this is given in Roy (1957). The developments in the nonparametric case are comparatively piecemeal and less general; this is mainly due to the lack of invariance of nonparametric statistics under nonsingular transformations on the observation vectors (the parametric procedures are mostly invariant in this respect). For some simple ANOVA models, some nonparametric procedures are discussed in Miller (1966); some later contributions are due to Sen (1966b, 1969c), Gabriel and Sen (1968), and Ghosh and Sen (1973), among others. Some of these, along with other relevant literature, are discussed in detail in Chapters 6 and 7 of Puri and Sen (1971). In these univariate problems, the nonparametric procedures remain invariant under monotone transformations on the variables, and hence things do work out reasonably well. In the multivariate case, though the coordinatewise rankings remain invariant under coordinatewise monotone transformations, the picture becomes very cumbrous when one considers the class of transformations $\mathbf{X} \to \mathbf{Y} = \mathbf{a} + \mathbf{BX}$, with nonsingular $\mathbf{B}$, not necessarily of the diagonal matrix form. Thus, in a nonparametric problem, in the multivariate case, one needs to preserve the identity of the coordinates—and this is mostly justified in practical problems, where a linear combination of the different coordinate variates may not make much sense. (There are some exceptions in some psychometric and educational testing problems, where of course, such linear combinations possess good physically interpretable properties.) Bearing in mind this restriction on the nonparametric MANOVA models, a general review of simultaneous statistical procedures based on ranks (for some common cases) has been made in Sen (1980b). Essentially, the results of Sen (1966b, 1969c) and Gabriel and Sen (1968) are generalized there in the multivariate setup and presented for the one-way and two-way MANOVA models. Though quite adaptable for such models, these procedures may not be directly applicable to the general subhypothesis-testing problems treated in this chapter. We consider here simultaneous inference procedures for general MANOVA models based on the rank-order estimators in Chapter 6 and Roy's (1953, 1957) largest-root criterion. Recently, for the normal-theory model, Wijsman (1979) has pointed out the distinctive advantage of using the largest-root criterion in simultaneous inference (over the likelihood-ratio and the Lawley–Hotelling trace criteria), and, in view of the asymptotic normality of the estimators in Chapter 6, we advocate the use of the same criterion for the nonparametric case too.

Consider the general MANOVA model in (6.4.1), and denote the rank-order estimators of $\boldsymbol{\beta}_0$ and $\boldsymbol{\beta}$ by $\hat{\boldsymbol{\beta}}_{0,N}$ and $\hat{\boldsymbol{\beta}}_N$, respectively (see Section 6.4.2 for details). Define the matrix $\mathbf{T}$ as in (6.5.16)–(6.4.18), $\mathbf{C}_N$ as in (6.3.3), and let

$$\mathbf{C}_N^* = \begin{pmatrix} \mathbf{N} & \mathbf{0} \\ \mathbf{0} & \mathbf{C}_N \end{pmatrix}. \tag{7.5.11}$$

Also, define $\hat{\mathbf{M}}_N$ as in (7.4.21)–(7.6.23). Further, define $\mathbf{B}$ as in (7.4.43) and (5.5.15). If we take $\hat{\boldsymbol{\beta}}_{N,j}$ in (6.4.33) and $\hat{\boldsymbol{\beta}}_{N,j} + N^{-1/2}\,\mathbf{a}$ for some $\mathbf{a} \neq \mathbf{0}$, then taking $L_N^{(j)}(\hat{\boldsymbol{\beta}}_{N,j})$ and $L_N^{(j)}(\hat{\boldsymbol{\beta}}_{N,j} + N^{-1/2}\mathbf{a})$ and using (6.4.65), we arrive at an estimator $\hat{\gamma}_{Nj}$ of γ_j, where γ_j is nothing but $B(F_{[j]}, \phi_j)$, defined by (5.5.15), $j = 1, \ldots, p$. Let us denote

$$\mathbf{B}_N = \operatorname{diag}(\hat{\gamma}_{N1}, \ldots, \hat{\gamma}_{Np}), \tag{7.5.12}$$

$$\hat{\mathbf{T}}_N = \hat{\mathbf{B}}_N^{-1}\hat{M}_N\hat{\mathbf{B}}_N^{-1} = \left(\left(\frac{\hat{m}_{Njl}}{\hat{\gamma}_{Nj}\hat{\gamma}_{Nl}}\right)\right). \tag{7.5.13}$$

We roll out the $p \times (q+1)$ matrix $(\hat{\boldsymbol{\beta}}_{0,N}, \hat{\boldsymbol{\beta}}_N)$ into a $p(q+1)$-vector and denote it by $\hat{\boldsymbol{\beta}}_N^*$. Similarly, the rollout form of $(\boldsymbol{\beta}_0, \boldsymbol{\beta})$ is denoted by $\boldsymbol{\beta}^*$. Let then

$$\mathbf{Q}_N = (\hat{\boldsymbol{\beta}}_N^* - \boldsymbol{\beta}^*)(\hat{\boldsymbol{\beta}}_N^* - \boldsymbol{\beta}^*)' \tag{7.5.14}$$

$$\mathbf{W}_N = \hat{\mathbf{T}}_N \otimes \mathbf{C}_N^{*-1}, \tag{7.5.15}$$

$$l_N = \operatorname{Tr}\{\mathbf{Q}_N \mathbf{W}_N^{-1}\} = (\hat{\boldsymbol{\beta}}_N^* - \boldsymbol{\beta}^*)' \mathbf{W}_N^{-1} (\hat{\boldsymbol{\beta}}_N^* - \boldsymbol{\beta}^*). \tag{7.5.16}$$

By Theorem 6.4.4, we conclude that under the model (6.4.1), l_N has asymptotically chi-square distribution with $p(q+1)$ DF, and hence, for large N,

$$P\{(\hat{\boldsymbol{\beta}}_N^* - \boldsymbol{\beta}^*)' \mathbf{W}_N^{-1} (\hat{\boldsymbol{\beta}}_N^* - \boldsymbol{\beta}^*) \le \chi^2_{p(q+1),\alpha} | \boldsymbol{\beta}^*\} \simeq 1 - \alpha, \tag{7.5.17}$$

where $1 - \alpha$ $(0 < \alpha < 1)$ is the desired coverage probability. On the other hand, for every real $p(q+1)$-vector $\mathbf{b}$ $(\neq \mathbf{0})$,

$$\sup\left\{ \frac{|\mathbf{b}'(\hat{\boldsymbol{\beta}}_N^* - \boldsymbol{\beta}^*)|}{(\mathbf{b}'\mathbf{W}_N\mathbf{b})^{1/2}} : \mathbf{b} \neq \mathbf{0} \right\} = l_N^{1/2}, \tag{7.5.18}$$

and hence, by (7.5.17) and (7.5.18), we obtain that asymptotically

$$P\{\mathbf{b}'\hat{\boldsymbol{\beta}}_N^* - \chi_{p(q+1),\alpha}(\mathbf{b}'\mathbf{W}_N\mathbf{b})^{1/2} \le \mathbf{b}'\boldsymbol{\beta}^*$$

$$\le \mathbf{b}'\hat{\boldsymbol{\beta}}_N^* + \chi_{p(q+1),\alpha}(\mathbf{b}'\mathbf{W}_N\mathbf{b})^{1/2} \ \forall \mathbf{b} \neq \mathbf{0}\} \cong 1 - \alpha. \tag{7.5.19}$$

Note that (7.5.19) provides a simultaneous confidence interval for all possible linear combinations of $\boldsymbol{\beta}^*$ with an asymptotic coverage probability $1 - \alpha$. Also, simultaneous tests for all possible linear combinations of $\boldsymbol{\beta}^*$ can be made by using (7.5.19) and identifying those $\mathbf{b}'\boldsymbol{\beta}^*$ as different from 0 for which both the upper and lower limits in (7.5.19) are of the same sign; this will insure that the overall significance level of this simultaneous test is asymptotically equal to α.

It may be noted that the class of all linear combinations of $\boldsymbol{\beta}^*$, which has been treated earlier, is somewhat more general than what we face in a MANOVA model. The canonical reduction of the subhypothesis-testing problem, treated earlier in this section, arises typically in situations where we restrict ourselves to the subclass of linear combination of the form

$$\mathbf{a}'(\boldsymbol{\beta}_0, \boldsymbol{\beta})\mathbf{l}, \qquad \mathbf{a}' = (a_1, \dots, a_p), \quad \mathbf{l}' = (l_1, \dots, l_{q+1}). \tag{7.5.20}$$

For this factorizable situation, the critical value $\chi_{p(q+1),\alpha}$ can be replaced by a smaller quantity, and this results in a simultaneous confidence region with smaller diameters. Towards this, let

$$\mathbf{Q}_N^* = (\hat{\boldsymbol{\beta}}_{0,N} - \boldsymbol{\beta}_0, \hat{\boldsymbol{\beta}}_N, \hat{\boldsymbol{\beta}}_N - \boldsymbol{\beta})\mathbf{C}_N^*(\hat{\boldsymbol{\beta}}_{0,N} - \boldsymbol{\beta}_0, \hat{\boldsymbol{\beta}}_N - \boldsymbol{\beta})' \tag{7.5.21}$$

and

$$l_N^* = \text{largest root of } \mathbf{Q}_N^* \hat{\mathbf{T}}_N^{-1}, \tag{7.5.22}$$

where all the notation is the same as before. Note that if we let $s = \min(p, q+1)$, then $\mathbf{Q}_N^*$ has asymptotically the Wishart structure with ∞ DF and parameters s and $\mathbf{T}$. Thus, l_N^* has asymptotically the same distribution as the largest characteristic root of $\mathbf{Z}\mathbf{Z}'$, where $\mathbf{Z}$ has the normal distribution with null mean and dispersion matrix $\mathbf{I}_p \otimes \mathbf{I}_s$. We denote the upper $100\alpha\%$ point of the latter distribution by $l(\alpha, s)$. Then we have asymptotically

$$P\{\, l_N^* \le l^2(\alpha, s) | H_0^* \} \cong 1 - \alpha. \tag{7.5.23}$$

On the other hand,

$$\sup\left\{ \frac{|\mathbf{a}'(\hat{\boldsymbol{\beta}}_{0,N} - \boldsymbol{\beta}_0, \hat{\boldsymbol{\beta}}_N - \boldsymbol{\beta})\mathbf{l}|}{[(\mathbf{a}'\hat{\mathbf{T}}_N\mathbf{a})(\mathbf{l}'\mathbf{C}_N^{*-1}\mathbf{l})]^{1/2}} : \mathbf{a} \ne \mathbf{0}, \mathbf{l} \ne \mathbf{0} \right\}$$

$$= \sup\left\{ \left[\frac{\mathbf{a}'(\hat{\boldsymbol{\beta}}_{0,N} - \boldsymbol{\beta}_0, \hat{\boldsymbol{\beta}}_N - \boldsymbol{\beta})\mathbf{C}_N^*(\hat{\boldsymbol{\beta}}_{0,N} - \boldsymbol{\beta}_0, \hat{\boldsymbol{\beta}}_N - \boldsymbol{\beta})'a}{\mathbf{a}'\hat{\mathbf{T}}_N\mathbf{a}} \right]^{1/2} : \mathbf{a} \ne \mathbf{0} \right\}$$

$$= [\text{largest root of } (\hat{\boldsymbol{\beta}}_{0,N} - \boldsymbol{\beta}_0, \hat{\boldsymbol{\beta}}_N - \boldsymbol{\beta})\mathbf{C}_N^*(\hat{\boldsymbol{\beta}}_{0,N} - \boldsymbol{\beta}_0, \hat{\boldsymbol{\beta}}_N - \boldsymbol{\beta})'\hat{\mathbf{T}}_N^{-1}]^{1/2}$$

$$= (l_N^*)^{1/2}. \tag{7.5.24}$$

Therefore, from (7.5.23) and (7.5.24), we conclude that asymptotically,

$$P\{\mathbf{a}'(\hat{\boldsymbol{\beta}}_{0,N}, \hat{\boldsymbol{\beta}}_N)\mathbf{l} - l(\alpha, s)(\mathbf{a}'\hat{\mathbf{T}}_N\mathbf{a})^{1/2}(\mathbf{l}'\mathbf{C}_N^{*-1}\mathbf{l})^{1/2} \le \mathbf{a}'(\boldsymbol{\beta}_0, \boldsymbol{\beta})\mathbf{l} \le$$

$$\mathbf{a}'(\hat{\boldsymbol{\beta}}_{0,N}, \hat{\boldsymbol{\beta}}_N)\mathbf{l} + l(\alpha, s)(\mathbf{a}'\hat{\mathbf{T}}_N\mathbf{a})^{1/2}(\mathbf{l}'\mathbf{C}_N^{*-1}\mathbf{l})^{1/2} \;\forall \mathbf{a} \ne \mathbf{0}, \mathbf{l} \ne \mathbf{0}\} \cong 1 - \alpha.$$

$$\tag{7.5.25}$$

This provides the desired simultaneous confidence region for the class of linear combinations in (7.5.20). It may be remarked that if in (7.5.20) we further restrict ourselves to the subclass $\mathbf{a}'\boldsymbol{\beta}\mathbf{l}$, where $\mathbf{a} \neq \mathbf{0}$ and $\mathbf{l} \neq \mathbf{0}$ are respectively p- and q-vectors, then in (7.5.21) we may simply work with $(\hat{\boldsymbol{\beta}}_N - \boldsymbol{\beta})\mathbf{C}_N(\hat{\boldsymbol{\beta}}_N - \boldsymbol{\beta})'$, and the resulting matrix in (7.5.22) will have asymptotically the Wishart structure with parameters $\min(p, q)$ and $\mathbf{I}_p$. Hence in this case we may obtain, parallel to (7.5.25), a simultaneous confidence interval for all $\mathbf{a}'\boldsymbol{\beta}\mathbf{l}$, where $l(\alpha, s)$ is to be replaced by a smaller quantity $l(\alpha, s')$, $s' = \min(p, q) \leq s$. In general, we may adapt this procedure for any subset of $(\boldsymbol{\beta}_0, \boldsymbol{\beta})$ (of dimension $p' \times q'$, for $p' \leq p$ and $q' \leq q + 1$), where we need to choose the corresponding minor of $\mathbf{C}_N^*$ in (7.5.21) and of $\hat{\mathbf{T}}_N$ in (7.5.22). The resulting critical value $l(\alpha, s)$ in (7.5.23) will depend on p' and q'. In any such case, a simultaneous test for the same subset of parameters can be made by identifying those linear combinations for which both the upper and lower limits in (7.5.25), or its analogue, are of the same sign. The overall level of significance will thus be asymptotically equal to α. Some generalizations of this procedure for the sequential case dealing with the univariate ANOVA models are due to Ghosh and Sen (1973). It may be noted that in the ANOVA model, the estimation of $\hat{\mathbf{T}}_N$ may be avoided by using the linearity results in (6.4.65) and considering aligned rank statistics. In MANOVA models too, such aligned rank statistics may be used to provide simultaneous inference in some specific models. But for the general model this may not work out well. In any case, one needs to estimate the rank covariance matrix. Generalizations of the proposed procedure to the case of multivariate analysis of covariance models (which will be introduced in Chapter 8) are due to Sen and Krishnaiah (1974).

In the developments sketched so far, we have considered infinite decomposition schemes [viz., (7.5.20)]. In many cases, one may be interested in a finite decomposition scheme, where $\mathbf{l}$ is allowed to take on only a finite number of distinct realizations. In such a case, we have a restricted maximization problem in (7.5.18) or (7.5.24), and the distribution theory of these restrained maxima depends on the joint distribution of several correlated quadratic forms (cf. Khatri, Krishnaiah, and Sen, 1977). These are discussed in Krishnaiah and Sen (1971) and Sen (1980b).

7.6. RANK STATISTICS FOR HYPOTHESES TESTING UNDER RESTRICTED ALTERNATIVES

In many problems involving linear models, though the null hypothesis is linear, the alternative hypotheses may be specified by nonlinear (or inequality) restraints. For example, in the multivariate one-sample location model, the null hypothesis that $\boldsymbol{\theta}$, the location vector, is equal to $\mathbf{0}$ may be tested against the *orthant-restricted* (one-sided) *alternatives*

$$H^{+}: \boldsymbol{\theta} \geq \mathbf{0} \qquad (\boldsymbol{\theta}'\boldsymbol{\theta} > 0). \tag{7.6.1}$$

Similarly, in a completely randomized (or randomized block) layout, one may test for the equality $\tau_1 = \cdots = \tau_k$ of the treatment effects $(\tau_1, \ldots, \tau_k)$, against *ordered alternatives*

$$H^{<}: \tau_1 \leq \cdots \leq \tau_k \qquad (\tau_k > \tau_1). \tag{7.6.2}$$

More generally, for a general multivariate linear model, as in (7.4.1), we may partition $\boldsymbol{\beta} = (\boldsymbol{\beta}_1, \boldsymbol{\beta}_2)$ as in (7.4.2) and want to test for $H_0: \boldsymbol{\beta}_2 = \mathbf{0}$ against ordered or orthant alternatives (treating $\boldsymbol{\beta}_1$ as a nuisance parameter).

In all these problems, the tests considered in earlier sections (for global alternatives) may turn out to be inefficient; the same criticism may be made against the global likelihood-ratio test for the parametric model. We may refer to Ghosh and Sen (1985) for some relevant asymptotic theory. For some specific parametric models, tests for suitable linear hypotheses against restricted alternatives have been considered by a host of workers; a detailed account (along with a useful bibliography) of these works is available with Barlow, Bartholomew, Bremner, and Brunk (1972). In this context, Roy's (1953) *union–intersection* (UI) *principle* can be employed to generate a rich class of tests, and specific solutions often depend on some basic theorems in *nonlinear programming*. In the nonparametric case, this approach has been systematically exploited by Chatterjee and De (1972, 1974), De (1976), Chinchilli and Sen (1981a, b, 1982), and Sen (1982a), among others.

Since by convenient reparametrization (and canonical reduction), ordered alternative problems may be reduced to an orthant restriction problem (see Exercises 7.6.1 and 7.6.2), we consider first a simple orthant restriction problem and then proceed to the general case.

Let $\mathbf{X}_i$, $1 \leq i \leq N$, be i.i.d.r.v.'s with a p-variate ($p \geq 1$) d.f. $F(\mathbf{x} - \boldsymbol{\theta})$, where F is diagonally symmetric about $\mathbf{0}$, and suppose that we want to test for

$$H_0: \boldsymbol{\theta} = \mathbf{0} \quad \text{vs.} \quad H^{+}: \boldsymbol{\theta} \geq \mathbf{0}. \tag{7.6.3}$$

Section 5.9.3 is devoted to rank tests for H_0 against $H: \boldsymbol{\theta} \neq \mathbf{0}$. In view of the restricted class of alternatives (H^{+}), we would like to incorporate Roy's (1953) UI principle to generate suitable test statistics, which have better (at least, in the asymptotic sense) performance when H^{+} holds. With the definition of $\mathbf{S}_N^{+}$ in (5.4.37) and (5.9.5), we are in a position to use the results in Chapter 4 and obtain that

$$N^{-1/2}(\mathbf{S}_N^{+} - N\boldsymbol{\mu}^{+}) \sim \mathcal{N}_p(\mathbf{0}, \mathbf{V}(\mathbf{S}_N^{+})), \tag{7.6.4}$$

where $\boldsymbol{\mu}^+ \geq \mathbf{0}$ iff $\boldsymbol{\theta} \geq \mathbf{0}$ ($\boldsymbol{\mu}^+ = \mathbf{0}$ iff $\boldsymbol{\theta} = \mathbf{0}$) and for $\boldsymbol{\theta}$ close to $\mathbf{0}$,

$$\boldsymbol{\mu}^+ \cong \mathbf{B}\boldsymbol{\theta}, \qquad \mathbf{B} = \operatorname{diag}(B_1, \ldots, B_p), \tag{7.6.5}$$

and the B_j, defined by (5.5.15), are positive finite constants (depending on F and ϕ). For testing $\boldsymbol{\theta} = \mathbf{0}$ (or equivalently, $\boldsymbol{\mu}^+ = \mathbf{0}$) against a specific alternative

$$H_{\boldsymbol{\gamma}}: \boldsymbol{\mu}^+ = \Delta\boldsymbol{\gamma}, \qquad \boldsymbol{\gamma} \text{ specified}, \quad \Delta > 0 \tag{7.6.6}$$

(where $\boldsymbol{\gamma} \geq \mathbf{0}$), we may use the linear combination

$$S_N^+(\mathbf{b}) = \mathbf{b}'S_N, \qquad \mathbf{b}' = (b_1, \ldots, b_p), \tag{7.6.7}$$

and choose $\mathbf{b}$ in such a way that $S_N^+(\mathbf{b})$ has the best power against (7.6.6), at least asymptotically. Let $\tilde{\mathbf{V}}_N$ be defined as in (5.4.49)–(5.4.51), and consider the test statistic

$$\frac{N^{-1/2}S_N^+(\mathbf{b})}{\{\mathbf{b}'\mathbf{V}_N\mathbf{b}\}^{1/2}} = \mathbf{S}_N^*(\mathbf{b}), \quad \text{say.} \tag{7.6.8}$$

Under $H_0: \boldsymbol{\theta} = \mathbf{0}$, for every $\mathbf{b} \in E^p$, $\mathbf{S}_N^*(\mathbf{b})$ has a distribution symmetric about $\mathbf{0}$ (asymptotically normal), and for $\boldsymbol{\theta}$ close to $\mathbf{0}$, by (7.6.5), (7.6.8), and the stochastic convergence of $\tilde{\mathbf{V}}_N$ to $\boldsymbol{\nu}$ [defined by (5.4.17)], the asymptotic power of $S_N^*(\mathbf{b})$ is maximized when $\mathbf{b}$ satisfies

$$\mathbf{b}'\boldsymbol{\gamma} \text{ is maximum}, \qquad \text{subject to} \quad \mathbf{b}'\boldsymbol{\nu}\mathbf{b} = \text{const.} \tag{7.6.9}$$

The solution is, of course, $\mathbf{b} = k\boldsymbol{\nu}^{-1}\boldsymbol{\gamma}$ and may well be approximated by $\mathbf{b}(\boldsymbol{\gamma}) = k\tilde{\mathbf{V}}_N^{-1}\boldsymbol{\gamma}$, where k ($\neq 0$) is an arbitrary constant, which we take (without any loss of generality) as 1. Thus, to test for $\boldsymbol{\theta} = \mathbf{0}$ against (7.6.6), $\boldsymbol{\gamma}$ specified, we would recommend the test with the critical region $\omega(\boldsymbol{\gamma}, \alpha_0)$ specified by

$$N^{-1/2}S_N^*\left(\tilde{\mathbf{V}}_N^{-1}\boldsymbol{\gamma}\right) \geq \tau_{\alpha^0}, \tag{7.6.10}$$

where τ_{α^0} is the upper $100\alpha^0\%$ point of the standard normal d.f. and α^0 ($0 < \alpha^0 < 1$) is the (asymptotic) significance level of this component test. If we let

$$H_{0\boldsymbol{\gamma}}: \boldsymbol{\gamma}'\boldsymbol{\theta} = 0, \qquad \boldsymbol{\gamma} \in \boldsymbol{\Gamma}^+ = \{\boldsymbol{\gamma}: \boldsymbol{\gamma} \geq \mathbf{0}\}, \tag{7.6.11}$$

then, by (7.6.3), (7.6.6), (7.6.11), and the positive homogeneity of $\boldsymbol{\Gamma}^+$, we conclude that

$$H_0 = \bigcap_{\boldsymbol{\gamma}\in\boldsymbol{\Gamma}^+} H_{0\boldsymbol{\gamma}} \quad \text{and} \quad H^+ = \bigcup_{\boldsymbol{\gamma}\in\boldsymbol{\Gamma}^+} H_{\boldsymbol{\gamma}}. \tag{7.6.12}$$

Thus, in accordance with the UI principle, the critical region for the overall hypotheses H_0 vs. H^+ will be the union of $\omega(\boldsymbol{\gamma}, \alpha^0)$, $\boldsymbol{\gamma} \in \boldsymbol{\Gamma}^+$, and the acceptance region will be the intersection of the complements of $\omega(\boldsymbol{\gamma}, \alpha^0)$. Hence, we let

$$\omega^*(\alpha) = \bigcup_{\boldsymbol{\gamma}\in\boldsymbol{\Gamma}^+} \omega(\boldsymbol{\gamma}, \alpha^0)$$

$$= \left\{(\mathbf{x}_1, \ldots, \mathbf{x}_N) : N^{-1/2} S_N^{*+} \geq \tau_{\alpha^0}\right\}, \tag{7.6.13}$$

where

$$S_N^{*+} = \sup\left\{ S_N^*\left(\tilde{\mathbf{V}}_N^{-1}\boldsymbol{\gamma}\right) : \boldsymbol{\gamma} \in \boldsymbol{\Gamma}^+\right\} \tag{7.6.14}$$

and α ($\geq \alpha^0$) is the significance level of the overall test. The crux of the problem is to find an explicit solution for S_N^* and to determine α^0, such that $\omega^*(\alpha)$ is of size α, for some given α ($0 < \alpha < 1$). It may be remarked that if $\boldsymbol{\Gamma}^0 = \{\boldsymbol{\gamma} : \boldsymbol{\gamma} \neq \mathbf{0}\}$, then $S_N^{*0} = \sup\{S_N^*(\tilde{\mathbf{V}}_N^{-1}\boldsymbol{\gamma}) : \boldsymbol{\gamma} \neq \mathbf{0}\} = \sup\{(\boldsymbol{\gamma}'\tilde{\mathbf{V}}_N^{-1}\mathbf{S}_N^+)/(\boldsymbol{\gamma}'\tilde{\mathbf{V}}_N^{-1}\boldsymbol{\gamma})^{1/2} : \boldsymbol{\gamma} \neq \mathbf{0}\} = [\text{largest root of } \tilde{\mathbf{V}}_N^{-1}\mathbf{S}_N^+(\mathbf{S}_N^+)']^{1/2} = [(\mathbf{S}_N^+)'\tilde{\mathbf{V}}_N^{-1}\mathbf{S}_N^+]^{1/2} = (N\mathscr{L}_N^+)^{1/2}$, where $\mathscr{L}_N^+$ is defined by (5.9.6). Hence the UI test for $\boldsymbol{\theta} = \mathbf{0}$ against $\boldsymbol{\theta} \neq \mathbf{0}$ is based on the statistic $\mathscr{L}_N^+$ in (5.9.6) and agrees with the conventional test considered in Section 5.9.3. Now, looking at (7.6.8) and (7.6.14), we gather that our basic problem is to

$$\text{maximize} \quad \mathbf{b}'\tilde{\mathbf{V}}_N^{-1}\mathbf{S}_N^+, \quad \text{subject to} \quad \mathbf{b}'\mathbf{V}_N\mathbf{b} = \text{const},$$

$$\text{when } \mathbf{b} \text{ belongs to the positive orthant } \{\mathbf{b} \geq \mathbf{0}\}. \tag{7.6.15}$$

Since, $S_N^{*+} = -\inf\{(-\mathbf{b}'\tilde{\mathbf{V}}_N^{-1}\mathbf{S}_N^+) : \mathbf{b}'\tilde{\mathbf{V}}_N\mathbf{b} = \text{const}, \mathbf{b} \geq 0\}$, (7.6.15) is exactly the constrained minimization problem in the theory of nonlinear programming, and the Kuhn–Tucker–Lagrange (point) theorem (Mangasarian, 1969) leads to the following:

Let J be any subset of $\mathscr{P} = \{1, \ldots, p\}$ and $\bar{J}$ be its complement. (There are 2^p such subsets.) For each $J \in \mathscr{P}$ ($\varnothing \subseteq J \subseteq \mathscr{P}$), we partition (and rearrange the components) of $\mathbf{S}_N^+$ and $\tilde{\mathbf{V}}_N$ as

$$(\mathbf{S}_N^+)' = \left((S_{N(J)}^+)', (S_{N(\bar{J})}^+)'\right), \tag{7.6.16}$$

$$\tilde{\mathbf{V}}_N = \begin{pmatrix} \tilde{\mathbf{V}}_{N(JJ)} & \tilde{\mathbf{V}}_{N(J\bar{J})} \\ \tilde{\mathbf{V}}_{N(\bar{J}J)} & \tilde{\mathbf{V}}_{N(\bar{J}\bar{J})} \end{pmatrix}. \tag{7.6.17}$$

Let then

$$\mathbf{S}^{+}_{N(J:\bar{J})} = \mathbf{S}^{+}_{N(J)} - \tilde{\mathbf{V}}_{N(J\bar{J})}\tilde{\mathbf{V}}^{-1}_{N(\bar{J}\bar{J})}\mathbf{S}^{+}_{N(\bar{J})}, \tag{7.6.18}$$

$$\tilde{\mathbf{V}}_{N(JJ:\bar{J})} = \tilde{\mathbf{V}}_{N(JJ)} - \tilde{\mathbf{V}}_{N(J\bar{J})}\tilde{\mathbf{V}}^{-1}_{N(\bar{J}\bar{J})}\tilde{\mathbf{V}}_{N(\bar{J}J)}, \tag{7.6.19}$$

and

$$N\mathscr{L}^{+}_{N(J:\bar{J})} = \left(\mathbf{S}^{+}_{N(J:\bar{J})}\right)'\tilde{\mathbf{V}}^{-1}_{N(JJ:\bar{J})}\left(\mathbf{S}^{+}_{N(J:\bar{J})}\right), \tag{7.6.20}$$

where for $J = \varnothing$, $\mathscr{L}^{+}_{N(0:\bar{0})} = 0$, and for $J = \mathscr{P}$, $\mathscr{L}^{+}_{N(J:\bar{J})} = \mathscr{L}^{+}_{N}$, defined by (5.9.6). Further, let $I(A)$ stand for the indicator function of the set A. Then we have

$$N^{-1/2}S^{*+}_{N} = \sum_{\varnothing \subseteq J \subseteq \mathscr{P}} \left\{\mathscr{L}^{+}_{N(J:\bar{J})}\right\}^{1/2} I\left(\mathbf{S}^{+}_{N(J:\bar{J})} > \mathbf{0}\right) I\left(\tilde{\mathbf{V}}^{-1}_{N(\bar{J}\bar{J})}\mathbf{S}_{N(\bar{J})} \leq \mathbf{0}\right). \tag{7.6.21}$$

In Chapter 5, we have noticed that $\mathscr{L}^{+}_{N}$ is a genuinely (conditionally) distribution-free statistic, whose distribution is generated by the 2^N (conditionally) equally likely sign inversions of $\mathbf{X}_1, \ldots, \mathbf{X}_N$. Note that $\tilde{\mathbf{V}}_N$ remains invariant under these sign inversions and $\mathscr{L}^{*+}_{N} = N^{-1/2}S^{*+}_{N}$ is also genuinely (conditionally) distribution-free under the same sign-invariance structure. Thus, for small values of N, an exact (conditional) test based $\mathscr{L}^{*+}_{N}$ can be constructed. For large N, we may note that under $H_0: \boldsymbol{\theta} = \mathbf{0}$, if for every J we denote

$$\rho_1(J:\bar{J}) = \lim_{N\to\infty} P_0\left\{N^{-1/2}\mathbf{S}^{+}_{N(J:\bar{J})} > 0\right\}, \tag{7.6.22}$$

$$\rho_2(\bar{J}) = \lim_{N\to\infty} P_0\left\{N^{-1/2}\tilde{\mathbf{V}}^{-1}_{N(\bar{J}\bar{J})}\mathbf{S}_{N(\bar{J})} \leq 0\right\}, \tag{7.6.23}$$

$$\rho(J:\bar{J}) = \rho_1(J:\bar{J})\rho_2(\bar{J}) \qquad \forall \phi \subseteq J \subseteq \mathscr{P}, \tag{7.6.24}$$

and let

$$\rho_r = \Sigma^{*}_{r}\rho(J:\bar{J}), \qquad r = 0, 1, \ldots, p, \tag{7.6.25}$$

where Σ^{*}_{r} extends over all J containing exactly r of the elements of $\mathscr{P}$, then for every $x \geq 0$,

$$\lim_{N\to\infty} P_0\left\{N^{-1/2}S^{*+}_{N} \leq x\right\} = \sum_{r=0}^{p} \rho_r P\{\chi_r \leq x\}, \tag{7.6.26}$$

where χ_r has the central chi distribution with r DF ($\chi_0 = 0$ with probability 1), for $0 \leq r \leq p$. In the literature, (7.6.26) is known as the chi-bar d.f., as it is the average of $p+1$ chi distributions with $0, 1, \ldots, p$ DF, where the ρ_r add up to 1. In view of the asymptotic normality (under the conditional model) of $N^{-1/2}\tilde{\mathbf{V}}_N^{-1/2}\mathbf{S}_N^+$ (under H_0), the probabilities in (7.6.22)–(7.6.25) can be computed by reference to the bivariate and multivariate probability integrals for the normal distributions. Tables for these are available with Gupta (1963) and Barlow et al. (1972), among others. Unfortunately, even for local alternatives, the asymptotic power function of the test in (7.6.13)–(7.6.14) does not appear as $\sum_{r=0}^{p}\rho_r P\{\chi_{r,\Delta_r} \leq x\}$ (where the χ_{r,Δ_r} are noncentral chi variates), but in more complicated forms, viz.

$$\sum_{\phi \subseteq J \subseteq \mathscr{P}} a(J, \boldsymbol{\lambda}) P\{\chi_{k(J), \Delta_J} \leq x\}, \tag{7.6.27}$$

where $k(J)$ stands for the cardinality of J, and the Δ_J (≥ 0) are appropriate noncentralities, while the $a(J, \boldsymbol{\lambda})$ depend on $\boldsymbol{\lambda}$ (where under $H_N^+ : \boldsymbol{\theta} = N^{-1/2}\boldsymbol{\lambda}$, $\boldsymbol{\lambda} \geq 0$) and are generally different from the $\rho(J : \bar{J})$ [$= a(J, \mathbf{0})$]. It may be very difficult to compare this restricted test with the unrestricted one ($\mathscr{L}_N^+$) in the light of the classical Pitman efficiency; for $p = 2$ this has been done by Chatterjee and De (1974) and Chinchilli (1979), while for $p \geq 3$, some numerical studies, due to Chinchilli (1979), reveal that generally the asymptotic power of the restricted test is greater than the global test when $\boldsymbol{\lambda}$ lies in the positive orthant. Note that if $\bar{\chi}(\alpha)$ is the upper $100\alpha\%$ point of (7.6.26), while $\chi_q^2(\alpha)$ is the parallel point for the chi-square distribution with q DF, then by (7.6.26) and the ordering of the chi distributions with different numbers of DF,

$$\sum_{r=0}^{p} \rho_r P\{\chi_r > x\} \leq P\{\chi_p > x\} \qquad \forall x \geq 0. \tag{7.6.28}$$

Thus $\bar{\chi}(\alpha) \leq \chi_p(\alpha)$, and if in fact $\boldsymbol{\theta} > \mathbf{0}$ (as under H^+), $\mathbf{S}_N^+$ will have positive components with high probability, and hence $(N^{-1/2}S_N^{*+})^2$ and $\mathscr{L}_N^+$ will agree, while by (7.6.28), they have different critical levels, and hence $\mathscr{L}_N^{*+}$ will have better power properties. This provides the intuitive reason for the use of $\mathscr{L}_N^{*+}$ instead of $\mathscr{L}_N^+$ for the positive-orthant problem.

Let us now consider the general regression model in (7.4.1), roll out the matrix $\boldsymbol{\beta}$ into a pq-vector, and write

$$\boldsymbol{\beta}' = (\underset{a \times 1}{\boldsymbol{\beta}'_{(1)}}, \underset{b \times 1}{\boldsymbol{\beta}'_{(2)}}), \qquad b = pq - a \geq 0, \quad a \geq 0. \tag{7.6.29}$$

The orthant problem consists of testing for

$$H_0 : \boldsymbol{\beta} = \mathbf{0} \quad \text{vs.} \quad H_1^+ : \boldsymbol{\beta} \neq \mathbf{0},\ \boldsymbol{\beta}_{(1)} \geq \mathbf{0}. \tag{7.6.30}$$

If we let $\mathbf{A} = [\mathbf{I}_a, \mathbf{0}_b]$, then the parameter space defined by H_1^+ in (7.6.30) is $\{\boldsymbol{\beta} : \mathbf{A}\boldsymbol{\beta} \geq \mathbf{0}\}$. We define the linear rank statistics $\mathbf{L}_N$ as in (7.4.6)–(7.6.7), $\mathbf{C}_N^*$ as in (7.4.11), and $\mathbf{V}_N$ as in (5.4.10). We roll out $\mathbf{L}_N$ into a pq-vector and define

$$\mathbf{U}_N = N^{-1/2}\left(\mathbf{C}_N^* \otimes \hat{\boldsymbol{\Gamma}}_N\right)^{-1}\mathbf{L}_N, \tag{7.6.31}$$

where $\hat{\boldsymbol{\Gamma}}_N = \operatorname{diag}(\hat{\gamma}_{11,N}, \ldots, \hat{\gamma}_{pp,N})$ and $\hat{\gamma}_{jj,N}$ is an estimator of B_j [defined by (5.5.15)], which we take as in (6.3.39). Similarly, we let

$$\hat{\boldsymbol{\Sigma}}_N = \left(\mathbf{C}_N^*\right)^{-1} \otimes \hat{\boldsymbol{\Gamma}}_N^{-1}\mathbf{V}_N\hat{\boldsymbol{\Gamma}}_N^{-1}. \tag{7.6.32}$$

We partition $\mathbf{U}_N' = [\mathbf{I}_{N(1)}', \mathbf{U}_{N(2)}']$ and $\hat{\boldsymbol{\Sigma}}_N$ as in (7.6.16)–(7.6.17), where $\mathbf{U}_{N(1)}$ is an a-vector, $\mathbf{U}_{N(2)}$ is a b-vector, and the component matrices of $\hat{\boldsymbol{\Sigma}}_N$ are $a \times a$, $a \times b$, $b \times a$, and $b \times b$, respectively. We also define $\mathbf{U}_{N(2:1)}$ and $\hat{\boldsymbol{\Sigma}}_{N(22:1)}$ as in (7.6.18) and (7.6.19), respectively. For convenience, we let $\mathbf{Z}_N = \mathbf{U}_{N(1)} = \mathbf{A}\mathbf{U}_N$ and $\hat{\boldsymbol{\Delta}}_N = \hat{\boldsymbol{\Sigma}}_{N(1)} = \mathbf{A}\hat{\boldsymbol{\Sigma}}_N\mathbf{A}'$, and define the partitions of $\mathbf{Z}_N$ and $\hat{\boldsymbol{\Delta}}_N$ as in (7.6.16)–(7.6.17). Here, we let J be any subset of $\mathscr{A} = \{1, \ldots, a\}$, and $\bar{J}$ be its complement, and for each J ($\varnothing \subseteq J \subseteq \mathscr{A}$), we define $\mathbf{Z}_{N(J)}$, $\mathbf{Z}_{N(\bar{J})}$, $\mathbf{Z}_{N(J:\bar{J})}$, $\hat{\boldsymbol{\Delta}}_{N(\bar{J}\bar{J})}$, and $\hat{\boldsymbol{\Delta}}_{N(JJ:\bar{J})}$ as in (7.6.16)–(7.6.19), and $N\mathscr{L}_{N(J:\bar{J})}^*$ as in (7.6.20).

The UI test statistic in this case turns out to be

$$\mathscr{L}_N^* = \sum_{\varnothing \subseteq J \subseteq \mathscr{A}} \left\{\mathbf{U}_{N(2:1)}'\hat{\boldsymbol{\Sigma}}_{N(22:1)}^{-1}\mathbf{U}_{N(2:1)} + \mathscr{L}_{N(J:\bar{J})}^*\right\}^{1/2}$$

$$\times I\left(\mathbf{Z}_{N(J:\bar{J})} > \mathbf{0}\right) I\left(\hat{\boldsymbol{\Delta}}_{N(\bar{J}\bar{J})}^{-1}\mathbf{Z}_{N(\bar{J})} \leq \mathbf{0}\right), \tag{7.6.33}$$

which apart from the first factor $\mathbf{U}_{N(2:1)}'\hat{\boldsymbol{\Sigma}}_N^{-1}\mathbf{U}_{N(2:1)}$ resembles (7.6.21); this additional factor is the contribution due to the unrestricted part $\boldsymbol{\beta}_{(2)}$ in H_1^+. Unlike $N^{-1/2}S_N^{*+}$ in (7.6.21), $\mathscr{L}_N^*$ is not genuinely distribution-free, but is ADF. Further, we may define the ρ_r as in (7.6.22)–(7.6.25), wherein $N^{-1/2}\mathbf{S}_{N(J:\bar{J})}^+$, $N^{-1/2}\mathbf{S}_{N(\bar{J})}^+$, and so on are replaced by $\mathbf{Z}_{N(J:\bar{J})}$, $Z_{N(\bar{J})}$, and so on; and then, parallel to (7.6.26), we have

$$\lim_{N \to \infty} P_0\left\{\mathscr{L}_N^* \leq x\right\} = \sum_{r=0}^{a} \rho_r P\{\chi_{b+r} \leq x\} \qquad \forall x \geq 0, \tag{7.6.34}$$

where $b = pq - a$. The discussion after (7.6.26) also applies to the general case in (7.6.33)–(7.6.34). Exercises 7.6.3 and 7.6.4 relate to the UI test for some related problems in subhypotheses testing under restricted alternatives.

Profile analysis is a special collection of testing problems for the p-response, q-sample model $\mathbf{X}_i = c_{Ni1}\boldsymbol{\beta}_1 + \cdots + c_{Niq}\boldsymbol{\beta}_q + \boldsymbol{\varepsilon}_i$, $i = 1, \ldots, N$, where the $\boldsymbol{\beta}_j$ are all p-vectors, c_{NiK} is equal to 1 or 0 according as $\mathbf{X}_i$ is from the kth sample or not ($1 \le k \le q$, $1 \le i \le N$), the $\boldsymbol{\varepsilon}_i$ are i.i.d.r.v.'s, and the p responses on each individual are comparable. The problem is to test that the q samples are equivalent, with the possible information that the response–sample interactions are null. The strategy is to conduct a preliminary test for the response–sample interaction, and then to test for an appropriate hypothesis of homogeneity. In this respect, the theory to be developed in Section 7.7 is useful, and the solution will be considered in Section 7.7. For this problem the UI principle leads to a conventional test statistic.

7.7. RANK STATISTICS FOR PRELIMINARY TEST INFERENCE

In Chapters 5, 6, and the current one, we have considered a variety of tests and estimates based on rank statistics for a variety of problems in linear models. In some problems of statistical inference, one encounters an incompletely specified model: From some extraneous considerations, some parametric restraints may suggest themselves for incorporation in the basic model under consideration, while there may not be enough evidence to adopt them without any reservation. If these constraints really hold and are adopted, the performance characteristics of the resultant procedures will be better than their counterparts with the constraints ignored. On the other hand, being based on the constraints, these procedures may not be robust against possible departures from them, and in this respect their unconstrained counterparts may perform better. Therefore, as a compromise, it may be appropriate to make a preliminary test of the validity of the constraints to be imposed on the model, and, depending on the outcome of this test, then to choose between the constrained and the unconstrained procedures. Such a statistical inference procedure following a preliminary test on some constraints is termed a *preliminary-test inference* (PTI) *procedure*. A great deal of work on PTI has been done in the parametric case during the past forty years; we may refer to Bancroft and Han (1977, 1980) for some good reviews of these developments. In this section, we shall mainly confine ourselves to the nonparametric case, treated in detail by Saleh and Sen (1978, 1982, 1983a, b, 1984b), Sen (1979e, 1982c, d), and Sen and Saleh (1979, 1985), among others.

We illustrate the PTI procedures for the multivariate simple regression model; the case of the univariate simple regression model will follow readily from this, and the case of general linear models will be treated briefly at the end. Let $\mathbf{X}_i$, $i = 1,\dots,N$, be independent random vectors with d.f.'s F_i, $i = 1,\dots,N$, all defined on E^p, for some $p \geq 1$, where [see (5.4.1)]

$$F_i(\mathbf{x}) = F(\mathbf{x} - \boldsymbol{\theta} - \boldsymbol{\beta} c_i), \qquad \mathbf{x} \in E^p, \quad \boldsymbol{\theta} = (\theta_1,\dots,\theta_p)',$$

$$\boldsymbol{\beta} = (\beta_1,\dots,\beta_p)', \tag{7.7.1}$$

F, $\boldsymbol{\theta}$, and $\boldsymbol{\beta}$ are unknown, and $c_1,\dots,c_N$ are known constants. We are primarily interested in drawing inferences on the parameter $\boldsymbol{\theta}$ when it is suspected but not certain that $\boldsymbol{\beta}$ is close to $\mathbf{0}$ (or any other specified value). Note that when $\boldsymbol{\beta} = \mathbf{0}$, the $\mathbf{X}_i$ are i.i.d.r.v.'s with the location parameter $\boldsymbol{\theta}$, and tests and estimates of $\boldsymbol{\theta}$ can be constructed in the same way as in Chapters 5 and 6. In particular, the tests are outlined in Section 5.4.2 and the estimates in Section 6.4.

For $\boldsymbol{\beta} = \mathbf{0}$, based on the signed rank statistics $\mathbf{S}_N$ in (6.4.11)–(6.4.12), the rank-order estimator $\hat{\boldsymbol{\theta}}_N$ of $\boldsymbol{\theta}$ is defined as it was after (6.4.12) (note the difference in notation). When $\boldsymbol{\beta}$ is treated as a nuisance parameter, as in (6.4.10), we denote the rank-order estimator of $\boldsymbol{\beta}$ by $\tilde{\boldsymbol{\beta}}_N$ and consider the residuals $\hat{\mathbf{X}}_i = \mathbf{X}_i - \tilde{\boldsymbol{\beta}}_N c_i$, $i = 1,\dots,N$. Based on $\hat{\mathbf{S}}_N$, the signed rank statistics on these residuals, proceeding as we did after (6.4.12), we obtain the rank-order estimator $\tilde{\boldsymbol{\theta}}_N$ of $\boldsymbol{\theta}$. Further, for testing $H_0 : \boldsymbol{\beta} = \mathbf{0}$ against $H_1 : \boldsymbol{\beta} \neq \mathbf{0}$, we proceed as in Section 5.4, define the test statistic $\mathscr{L}_N$ as in (5.4.13), and formulate the test procedure as we did there; (5.4.36) provides a large-sample simplification for this procedure. Let $\xi(\mathscr{L}_N)$ be 1 or 0 according as this preliminary test on $\boldsymbol{\beta}$ leads to the rejection or acceptance of $H_0 : \boldsymbol{\beta} = \mathbf{0}$. Then the preliminary-test estimator (PTE) $\boldsymbol{\theta}_N^*$ of $\boldsymbol{\theta}$ is defined by

$$\boldsymbol{\theta}_N^* = \xi(\mathscr{L}_N) \cdot \tilde{\boldsymbol{\theta}}_N + \{1 - \xi(\mathscr{L}_N)\} \cdot \hat{\boldsymbol{\theta}}_N. \tag{7.7.2}$$

Thus, the PTE is an empirical mixture of the unrestricted estimator ($\tilde{\boldsymbol{\theta}}_N$) and the constrained estimator ($\hat{\boldsymbol{\theta}}_N$) with the mixing coefficient derived from the preliminary test on $\boldsymbol{\beta}$, which is suspected to be close to $\mathbf{0}$. We may note that $\hat{\boldsymbol{\theta}}_N$ has generally smaller (generalized) variance than $\tilde{\boldsymbol{\theta}}_N$ when $\boldsymbol{\beta} = \mathbf{0}$, but may be biased when $\boldsymbol{\beta} \neq \mathbf{0}$ while $\tilde{\boldsymbol{\theta}}_N$ is not (at least asymptotically). The basic purpose of the preliminary test is to retain the robustness of the estimator against $\boldsymbol{\beta}$ away from $\mathbf{0}$ without much compromise of its efficiency. As is usually the case, the PTE $\boldsymbol{\theta}_N^*$ is generally neither an unbiased estimator of $\boldsymbol{\theta}$ nor asymptotically multinormal (though $\tilde{\boldsymbol{\theta}}_N$ and $\hat{\boldsymbol{\theta}}_N$ are).

Since the PTE is of interest when $\boldsymbol{\beta}$ is suspected to be close to $\mathbf{0}$, we confine ourselves to local alternatives $\{K_N\}$ where

$$K_N : \boldsymbol{\beta} = \boldsymbol{\beta}_{(N)} = N^{-1/2}\boldsymbol{\gamma}, \qquad \boldsymbol{\gamma} = (\gamma_1, \dots, \gamma_p)' \text{ (fixed)} \in E^p. \tag{7.7.3}$$

Also, we assume that as $N \to \infty$, $\bar{c}_N = N^{-1}\Sigma_{i=1}^N c_i \to \bar{c}$ and $N^{-1}\Sigma_{i=1}^N (c_i - \bar{c}_N)^2 \to C^*$, where $|\bar{c}| < \infty$ and $C^* < \infty$. Further, we define the matrix $\mathbf{T} = \mathbf{T}(F)$ as in (5.5.25). Then, using Theorem 5.5.2 and the asymptotic linearity results in Section 6.4, we arrive at the following theorem; the proof is left as an exercise.

Theorem 7.7.1. *Under $\{K_N\}$ in (7.7.3) and the regularity conditions of Theorems 6.6.1 and 6.4.2, as $N \to \infty$,*

$$\begin{aligned} G_p^*(\mathbf{x}; \boldsymbol{\gamma}) &= \lim_{N\to\infty} P_{K_N}\{N^{1/2}(\boldsymbol{\theta}_N^* - \boldsymbol{\theta}) \le \mathbf{x}\} \\ &= G_p(\mathbf{x} - \bar{c}\boldsymbol{\gamma}; \mathbf{0}, \mathbf{T}) H_p(\chi^2_{p,\alpha}; \Delta^*) \\ &\quad + \int_{E(\boldsymbol{\gamma})} G_p\left(\mathbf{x} + \frac{\bar{c}}{C^{*1/2}}\mathbf{z}; \mathbf{0}, \mathbf{T}\right) dG_p(\mathbf{z}; \mathbf{0}, \mathbf{T}), \mathbf{x} \in E^p, \end{aligned} \tag{7.7.4}$$

where $G_p(\mathbf{x}; \boldsymbol{\mu}, \boldsymbol{\Sigma})$ is the p-variate normal d.f. with mean vector $\boldsymbol{\mu}$ and dispersion matrix $\boldsymbol{\Sigma}$, $H_p(x,; \delta)$ is the noncentral chi-square d.f. with p DF and noncentrality parameter δ, $E(\boldsymbol{\gamma}) = \{\mathbf{y} \in E^p : (\mathbf{y} + C^{*1/2}\boldsymbol{\gamma})'\mathbf{T}^{-1}(\mathbf{y} + C^{*1/2}\boldsymbol{\gamma}) \ge \chi^2_{p,\alpha}\}$, and $\Delta^* = C^*(\boldsymbol{\gamma}'\mathbf{T}^{-1}\boldsymbol{\gamma})$, with α $(0 < \alpha < 1)$ the size of the preliminary test.

The mean vector and the dispersion matrix of the d.f. G_p^* in (7.7.4) provide the *asymptotic bias* and *asymptotic dispersion matrix* (a.d.m.) of $N^{1/2}(\boldsymbol{\theta}_N^* - \boldsymbol{\theta})$. Some standard computations (see Exercises 7.7.2 and 7.7.3) lead us to the following:

$$\left[\text{asymptotic bias of } N^{1/2}(\boldsymbol{\theta}_N^* - \boldsymbol{\theta})\right] = \bar{c}\boldsymbol{\gamma}H_{p+2}(\chi^2_{p,\alpha}; \Delta^*), \tag{7.7.5}$$

$$\begin{aligned} &\left[\text{a.d.m. of } N^{1/2}(\boldsymbol{\theta}_N^* - \boldsymbol{\theta})\right] \\ &= \left(1 + \frac{\bar{c}^2}{C^*}\right)\mathbf{T}\left(1 - H_{p+2}(\chi^2_{p,\alpha}; \Delta^*)\right) \\ &\quad + \bar{c}^2\left[2H_{p+2}(\chi^2_{p,\alpha}; \Delta^*) - H_{p+4}(\chi^2_{p,\alpha}; \Delta^*)\right]\boldsymbol{\gamma}\boldsymbol{\gamma}'. \end{aligned} \tag{7.7.6}$$

For the unconstrained and the constrained estimator $\tilde{\boldsymbol{\theta}}_N$ and $\hat{\boldsymbol{\theta}}_N$, we refer to the results in Section 6.4, and conclude that under $\{K_N\}$,

$$\left[\text{asymptotic bias of } N^{1/2}(\hat{\boldsymbol{\theta}}_N - \boldsymbol{\theta})\right] = \bar{c}\boldsymbol{\gamma}, \tag{7.7.7}$$

$$\left[\text{asymptotic bias of } N^{1/2}(\tilde{\boldsymbol{\theta}}_N - \boldsymbol{\theta})\right] = \mathbf{0}, \qquad \forall \boldsymbol{\gamma}, \tag{7.7.8}$$

$$\left[\text{a.d.m. of } N^{1/2}(\hat{\boldsymbol{\theta}}_N - \boldsymbol{\theta})\right] = \mathbf{T} + \bar{c}^2\boldsymbol{\gamma}\boldsymbol{\gamma}', \tag{7.7.9}$$

$$\left[\text{a.d.m. of } N^{1/2}(\tilde{\boldsymbol{\theta}}_N - \boldsymbol{\theta})\right] = \left(1 + \frac{\bar{c}^2}{C^*}\right)\mathbf{T}. \tag{7.7.10}$$

It appears that when $\bar{c} = 0$, all three estimators behave similarly with respect to both the asymptotic bias and the asymptotic dispersion matrix. However, when $\bar{c} \neq 0$ but $\boldsymbol{\gamma} = \mathbf{0}$, though they all have asymptotic bias $\mathbf{0}$, they differ in their asymptotic dispersion matrices. If, as in Section 6.4, we adopt the notion of generalized variance (as the pth root of the asymptotic dispersion matrix) and judge the asymptotic relative efficiency by the reciprocal of the generalized variance, then these estimators may be ordered as $\hat{\boldsymbol{\theta}}_N \succ \boldsymbol{\theta}_N^* \succ \tilde{\boldsymbol{\theta}}_N$. On the other hand, when $\boldsymbol{\gamma} \neq \mathbf{0}$, the asymptotic bias $\bar{c}\boldsymbol{\gamma}$ of $\hat{\boldsymbol{\theta}}_N$ may dominate the others, and its generalized variance, computed from (7.7.8), may exceed both those computed from (7.7.6) and (7.7.10), particularly when $\boldsymbol{\gamma}$ moves away from $\mathbf{0}$. Actually, (7.7.5) has the nice property that it not only is bounded by (7.7.7) but also converges to 0 as $\Delta^* \to \infty$, that is, $\boldsymbol{\gamma}$ moves away from $\mathbf{0}$, while (7.7.7) does not. Thus, excepting in a small neighborhood of $\boldsymbol{\gamma} = \mathbf{0}$, the PTE performs (at least asymptotically) better than the constrained estimator, and in this neighborhood it is better than the unconstrained estimator. Also, the unconstrained estimator performs better than the constrained one when $\boldsymbol{\gamma}$ is away from $\mathbf{0}$, though in such a case, (7.7.10) dominates (7.7.6). This explains the robustness of the PTE, in terms of both validity and efficiency, when $\boldsymbol{\gamma}$ need not be equal to $\mathbf{0}$. For some detailed studies of this, including some numerical results, we may refer to Saleh and Sen (1978) and Sen and Saleh (1979, 1985). The allied shrinkage rank estimation problem has also been discussed there.

Let us now consider the preliminary test testing (PTT) problem. Here, we desire to test for $H_0^{(1)}: \boldsymbol{\theta} = \mathbf{0}$ (or any other specified value, reducible to $\mathbf{0}$ by translation of the observations) against $\boldsymbol{\theta} \neq \mathbf{0}$, when it is suspected but not evident that $\boldsymbol{\beta}$ is close to $\mathbf{0}$ (or some other specified value). As before, the rank-order test for $H_0^{(2)}: \boldsymbol{\beta} = \mathbf{0}$ against $\boldsymbol{\beta} \neq \mathbf{0}$ is based on the statistic $\mathscr{L}_N$ in (5.4.13) and is genuinely distribution-free; (5.4.36) provides the large-sample procedure for this test. The size of this test is denoted by α_2

$(0 < \alpha_2 < 1)$. The rank-order test for $H_0^{(1)}$ when $\boldsymbol{\beta}$ is assumed to be equal to $\mathbf{0}$ is based on the statistic $\mathscr{L}_N^+$ in (5.9.6) and is also genuinely distribution-free when $\boldsymbol{\theta} = \boldsymbol{\beta} = \mathbf{0}$; we denote the size of this test by α_1 $(0 < \alpha_1 < 1)$. Finally, when $\boldsymbol{\beta}$ is treated as a nuisance parameter, an ADF rank-order test for $H_0^{(1)}$ is based on the aligned rank statistic $\hat{\mathscr{L}}_N^{(1)}$, defined after (7.4.72) [and is a special case on $\hat{\mathscr{L}}_N^{(3)}$ defined in (7.4.66)]; we denote the (asymptotic) size of this test by α_3 $(0 < \alpha_3 < 1)$. Note that the critical values $l_{N,\alpha_j}^{(j)}$, $j = 1, 2, 3$, of these three test statistics converge to χ^2_{p,α_j}, $j = 1, 2, 3$, as $N \to \infty$. Then the PTT for testing $H_0^{(1)}: \boldsymbol{\theta} = \mathbf{0}$ is based on the following test function ξ_N^*:

$$\xi_N^* = \begin{cases} 1 & \text{if} \quad \mathscr{L}_N < l_{N,\alpha_2}^{(2)} \text{ and } \mathscr{L}_N^+ \geq l_{N,\alpha_1}^{(1)} \\ & \text{or} \quad \mathscr{L}_N \geq l_{N,\alpha_2}^{(2)} \text{ and } \hat{\mathscr{L}}_N^{(1)} \geq l_{N,\alpha_3}^{(3)}, \\ 0 & \text{otherwise.} \end{cases} \tag{7.7.11}$$

Note that the size of the PTT is given by

$$\alpha_N^* = P\left\{ \mathscr{L}_N < l_{N,\alpha_2}^{(2)}, \mathscr{L}_N^+ \geq l_{N,\alpha_1}^{(1)} | \boldsymbol{\theta} = \mathbf{0} \right\}$$

$$+ P\left\{ \mathscr{L}_N \geq l_{N,\alpha_2}^{(2)}, \hat{\mathscr{L}}_N^{(1)} \geq l_{N,\alpha_3}^{(3)} | \boldsymbol{\theta} = \mathbf{0} \right\}, \tag{7.7.12}$$

and it depends on α_1, α_2, and α_3 as well as on the nuisance parameter $\boldsymbol{\beta}$. Since $\boldsymbol{\beta}$ is suspected to be close to $\mathbf{0}$, here also, we confine ourselves to the sequence $\{K_N\}$ of alternative in (7.7.3). Further, to study the asymptotic power properties of the tests, as in earlier sections, we consider a sequence of local shift alternatives to $H_0^{(1)}$. Combining this with (7.3.3), we formulate $\{K_N^*\}$ by letting

$$K_N^*: (\boldsymbol{\theta}, \boldsymbol{\beta}) = N^{-1/2}(\boldsymbol{\gamma}_0, \boldsymbol{\gamma}), \qquad \boldsymbol{\gamma}_0 \in E^p, \quad \boldsymbol{\gamma} \in E^p. \tag{7.7.13}$$

Further, in (7.7.12), we replace the condition $\boldsymbol{\theta} = \mathbf{0}$ by K_N^* and denote the resulting probability by $\Pi_N^*(\boldsymbol{\gamma}_0, \boldsymbol{\gamma})$, and this stands for the power of the PTT under K_N^* in (7.7.13). In particular, when $\boldsymbol{\gamma}_0 = \mathbf{0}$, $\pi_N^*(\mathbf{0}, \boldsymbol{\gamma})$ stands for the size of the PTT when $H_0^{(2)}$ need not hold but (7.3.3) holds.

Now, under $\boldsymbol{\theta} = \boldsymbol{\beta} = \mathbf{0}$, $\mathscr{L}_N$ and $\mathscr{L}_N^+$ are asymptotically independent, each distributed according to central chi-square distribution with p DF. As the assumed regularity conditions insure the contiguity of the sequence of probability measures under $\{K_N^*\}$ to that under $\boldsymbol{\theta} = \boldsymbol{\beta} = \mathbf{0}$, as in Section 6.4 and 7.4, we conclude that under $\{K_N^*\}$, $\mathscr{L}_N$, and $\mathscr{L}_N^+$ are asymptotically independent, and we have noncentral chi-square distribution with p DF and noncentrality parameters Δ^* and Δ^+, respectively, where Δ^* is defined in Theorem 7.7.1 and

$$\Delta^+ = (\boldsymbol{\gamma}_0 + \bar{c}\boldsymbol{\gamma})'\mathbf{T}^{-1}(\boldsymbol{\gamma}_0 + \bar{c}\boldsymbol{\gamma}). \tag{7.7.14}$$

Also, under $\{K_N^*\}$, $\hat{\mathscr{L}}_N^{(1)}$ has asymptotically a noncentral chi-square distribution with p DF and noncentrality parameter

$$\Delta^{(1)} = \left(1 + \bar{c}^2/C^*\right)^{-1}\left(\boldsymbol{\gamma}_0'\mathbf{T}^{-1}\boldsymbol{\gamma}_0\right). \tag{7.7.15}$$

However, under $H_0^{(1)}$ or $\{K_N^*\}$, the two statistics $\mathscr{L}_N$ and $\hat{\mathscr{L}}_N^{(1)}$ are not (even asymptotically) independent; their joint distribution is asymptotically of the form of correlated (noncentral) chi-squares, as studied by Khatri, Krishnaiah, and Sen (1977) and others. If we denote this joint d.f. by $G^{**}(x, y;\ \boldsymbol{\gamma}_0, \boldsymbol{\gamma})$, then from (7.4.12), (7.4.13)–(7.4.15), and some routine steps, we obtain that

$$\lim_{N\to\infty} \pi_N^*(\boldsymbol{\gamma}_0, \boldsymbol{\gamma}) = \pi^*(\boldsymbol{\gamma}_0, \boldsymbol{\gamma}) = H_p\left(\chi^2_{p,\alpha_2};\ \Delta^*\right)\left[1 - H_p\left(\chi^2_{p,\alpha_1};\ \Delta^+\right)\right]$$
$$+\left[H_p\left(\chi^2_{p,\alpha_3};\ \Delta^{(1)}\right) - G^{**}\left(\chi^2_{p,\alpha_2}, \chi^2_{p,\alpha_3};\ \boldsymbol{\gamma}_0, \boldsymbol{\gamma}\right)\right], \tag{7.7.16}$$

where the $H_p(\cdot)$ are defined as in Theorem 7.7.1. Note that the unconstrained test based on the aligned rank statistic $\hat{\mathscr{L}}_N^{(1)}$ has the asymptotic size α_3 for all $\boldsymbol{\gamma}$. On the other hand, for the constrained test based on $\mathscr{L}_N^+$, the asymptotic size is given by $1 - H_p(\chi^2_{p,\alpha_1};\ \Delta_0^+)$, where $\Delta_0^+ = \bar{c}^2(\boldsymbol{\gamma}'\mathbf{T}^{-1}\boldsymbol{\gamma})$. Hence, if $\bar{c}$ is different from 0 and $\boldsymbol{\gamma} \neq \mathbf{0}$, then the asymptotic size is greater than α_1, and this may go to 1 when $\boldsymbol{\gamma}$ moves away from $\mathbf{0}$ (so that $\Delta_0^+ \to \infty$). by (7.7.16) and the above discussion, we conclude that the PTT procedure is more robust than the constrained procedure but less so than the unconstrained one when $\boldsymbol{\theta} = \mathbf{0}$. When $\boldsymbol{\gamma}_0 \neq \mathbf{0}$ but $\boldsymbol{\gamma} = \mathbf{0}$, we have $\Delta^+ > \Delta^{(1)}$ and the constrained test performs better than the unconstrained one, while the PTT performs better than the unconstrained one. Finally, when both $\boldsymbol{\gamma}_0$ and $\boldsymbol{\gamma}$ are different from $\mathbf{0}$, the PTT fares better than the constrained test for a wide range of the parameters, excluding a neighborhood of $(\mathbf{0}, \mathbf{0})$. It also fares better than the unconstrained one when Δ^+ is large compared to $\Delta^{(1)}$. For some numerical studies relating to this, we may refer to Saleh and Sen (1983b). Thus, from the efficiency–robustness point of view, the PTT has an edge on the others.

We are in a position to extend the theory of PTI to the general linear models treated in Section 7.4. Consider the model (7.4.1) and the decomposition in (7.4.2). If it is suspected but not certain that $\boldsymbol{\beta}_2$ is equal to $\mathbf{0}$, we may perform a test for the hypothesis $H_0^{(2)}: \boldsymbol{\beta}_2 = \mathbf{0}$ against $\boldsymbol{\beta}_2 \neq \mathbf{0}$, treating $\boldsymbol{\beta}_0$ and $\boldsymbol{\beta}_1$ as nuisance parameters, and incorporate this preliminary test in drawing inference on the other parameters. For the PTE problem the estimators $\hat{\boldsymbol{\beta}}_0, \hat{\boldsymbol{\beta}}_1$ of $\boldsymbol{\beta}_0, \boldsymbol{\beta}_1$ when $\boldsymbol{\beta}_2 = 0$ can be obtained as in Section 6.4, where we need to consider the reduced model with $\boldsymbol{\beta}\mathbf{c}_i$ replaced by $\boldsymbol{\beta}_1\mathbf{c}_{i(1)}$, $i \geq 1$, in (7.4.1). Similarly, for the full model $\tilde{\boldsymbol{\beta}}_0, \tilde{\boldsymbol{\beta}}$ $[= (\tilde{\boldsymbol{\beta}}_1, \tilde{\boldsymbol{\beta}}_2)]$, the estimator $\boldsymbol{\beta}$ of $\boldsymbol{\beta}_0$ can be obtained as in Section 6.4. Further, the test for $H_0^{(2)}$ may be based on the aligned rank statistic $\hat{\mathscr{L}}_N^{(2)}$ in (7.4.25). The PTE $\boldsymbol{\beta}_0^*, \boldsymbol{\beta}_1^*$ may then be defined as in (7.7.2), and with (7.7.3) extended to a $p \times q_2$ matrix, Theorem 7.7.1 holds in this context with p replaced by $p(q_1 + 1)$ and some changes in the G_p's and H_p's: The first G_p should be $G_{p(q+1)}$, H_p should be H_{pq_2}, the next G_p should be $G_{p(q_1+1)}$, and the final one should be G_{pq_2}. Then $E(\boldsymbol{\gamma})$ will be a subspace of E^{pq_2}, and the covariance matrix of the asymptotic distribution in (6.4.74)–(6.4.75) will replace $\mathbf{T}^{-1}$. The results on the asymptotic bias and dispersion matrices are very parallel to (7.7.5)–(7.7.9), but more complicated in view of the $q \times q$ matrix $\mathbf{D}$ in (7.4.75) replacing C^*. The discussion of the robustness of the PTE given after (7.7.13) also pertains to this general case. As for the PTT, the unconstrained test for $H_0^*: \boldsymbol{\beta}_0 = \mathbf{0}$, $\boldsymbol{\beta}_1 = \mathbf{0}$ against at least one of $\boldsymbol{\beta}_0$ and $\boldsymbol{\beta}_1$ being non-null, treating $\boldsymbol{\beta}_2$ as a nuisance parameter, may be worked out as in Section 7.4 and is denoted by $\hat{\mathscr{L}}_N^{(1)}$. Similarly, the constrained test for H_0^*, given $\boldsymbol{\beta}_2 = \mathbf{0}$, may be based on the reduced model; let it be denoted by $\mathscr{L}_{\cdot N}^{(1)}$. Then the PTT may be formulated as in (7.7.11). Since $\hat{\mathscr{L}}_N^{(2)}$ and $\mathscr{L}_{\cdot N}^{(1)}$ are asymptotically independent under H_0^* and for $\boldsymbol{\beta} = N^{-1/2}\boldsymbol{\gamma}$, the robustness results on the power and size discussed after (7.7.11) also pertain to this general model. The details are therefore left as exercises.

We would like to apply the PTI theory to the profile-analysis problem mentioned at the end of Section 7.6. As we did there, consider the q-sample p-response model $\mathbf{X}_i = c_{Ni1}\boldsymbol{\beta}_1 + \cdots + c_{Niq}\boldsymbol{\beta}_q + \boldsymbol{\varepsilon}_i$, $i = 1, \ldots, N$, where the $\boldsymbol{\varepsilon}_i$ are i.i.d.r.v.'s, the $\boldsymbol{\beta}_j$ are all p-vectors, and c_{Nik} is equal to 1 or 0 according as the $\mathbf{X}_i$ is from the kth sample or not, for $1 \leq k \leq q$, $1 \leq i \leq N$. We wish to test that the q samples are equivalent, with the possible information that the response–sample interactions are null. If we define

$$\mathbf{G} = \left(\mathbf{I}_{p-1,p-1}, -\mathbf{J}_{p-1,1}\right) \quad \text{and} \quad \mathbf{M}' = \left(\mathbf{I}_{q-1,q-1}, -\mathbf{J}_{q-1,1}\right)$$

(where $\mathbf{J}$ has all elements equal to 1), then the sample-mean effects are mathematically represented by $\boldsymbol{\beta}\mathbf{M}$ and the response–sample interactions by $\mathbf{G}\boldsymbol{\beta}\mathbf{M}$, where $\boldsymbol{\beta} = (\boldsymbol{\beta}_1, \ldots, \boldsymbol{\beta}_q)$. The profile analysis involves the following hypotheses:

$$H_{RS}^0: \mathbf{G\beta M} = \mathbf{0} \quad \text{vs.} \quad H_{RS}^*: \mathbf{G\beta M} \neq \mathbf{0}, \tag{7.7.17}$$

$$H_S^0: \boldsymbol{\beta}\mathbf{M} = \mathbf{0} \quad \text{vs.} \quad H_S^*: \boldsymbol{\beta}\mathbf{M} \neq \mathbf{0}, \tag{7.7.18}$$

$$H_S^0: \boldsymbol{\beta}\mathbf{M} = \mathbf{0} \quad \text{vs.} \quad H_{RS}^*: \mathbf{G\beta M} \neq \mathbf{0}. \tag{7.7.19}$$

The strategy is to conduct a preliminary test for the response–sample interactions in (7.7.17), at significance level α_1. If H_{RS} is not rejected, then the testing problem of the sample equivalence in (7.7.18) is conducted at significance level α_2; otherwise, the testing problem in (7.7.19) is implemented at significance level α_3. The following procedure is adapted from Chinchilli and Sen (1982).

We let $n_1, \ldots, n_q$ denote the respective sizes of the q samples with $\Sigma_{k=1}^q n_k = N$. Then, for each k, $\bar{c}_{Nk} = n_k/N$, and we assume that there exist $\bar{c}_k$ $(0 < \bar{c}_k < 1)$ such that $\bar{c}_{Nk} \to \bar{c}_k$, for $k = 1, \ldots, q$. Let then $\mathbf{d}_{Ni} = N^{-1/2}(\mathbf{c}_{Ni} - \bar{\mathbf{c}}_N)$, $i = 1, \ldots, N$, so that $\mathbf{D}_N = \Sigma_{i=1}^N \mathbf{d}_{Ni}\mathbf{d}'_{Ni} = ((\bar{c}_{Nk}[\delta_{kk'} - \bar{c}_{Nk'}])) \to \mathbf{D} = ((\bar{c}_k[\delta_{kk'} - \bar{c}_{k'}]))$, where $\mathbf{D}$ is of rank $q-1$. We partition the $\mathbf{c}_{Ni}$, $\mathbf{d}_{Ni}$, $\mathbf{D}_N$, and $\mathbf{D}$ according to the first $q-1$ components (i.e.,

$$\mathbf{c}'_{Ni} = \left(\mathbf{c}'_{Ni(1)}, \mathbf{c}_{Niq}\right), \qquad \mathbf{D} = \begin{pmatrix} \mathbf{D}_{(11)} & \mathbf{D}_{(12)} \\ \mathbf{D}_{(21)} & \mathbf{D}_{(22)} \end{pmatrix}$$

etc.), and write $\boldsymbol{\lambda} = N^{1/2}\boldsymbol{\beta}$ and $\boldsymbol{\beta}_0^{(N)} = \boldsymbol{\beta}\bar{\mathbf{c}}_N$. Then, the model may be rewritten as $\mathbf{X}_i = \boldsymbol{\beta}_0^{(N)} + \boldsymbol{\lambda}\mathbf{M}\mathbf{d}_{Ni(1)} + \boldsymbol{\varepsilon}_i$, $i = 1, \ldots, N$, and the corresponding $\mathbf{D}_{N(11)}$ is of full rank $(= q-1)$. For this reduced model, we may apply the PTT theory developed earlier and carry out the desired tests. Let $\mathbf{U}_N$, $\mathbf{V}_N$, $\hat{\boldsymbol{\Gamma}}_N$ and $\hat{\boldsymbol{\Sigma}}_N$ be defined as in Section 7.6 [see (7.6.30)–(7.6.32)], and let

$$Q_{N,1}^2 = \mathbf{U}'_N\left[\mathbf{D}_{N(11)} \otimes \mathbf{G}\left(\mathbf{G}\hat{\boldsymbol{\Gamma}}_N^{-1}\mathbf{V}_N\hat{\boldsymbol{\Gamma}}_N^{-1}\mathbf{G}'\right)^{-1}\mathbf{G}\right]\mathbf{U}_N, \tag{7.7.20}$$

$$Q_{N,2}^2 = \mathbf{U}'_N\left[\mathbf{D}_{N(11)} \otimes \left(\hat{\boldsymbol{\Gamma}}_N^{-1}\mathbf{V}_N\hat{\boldsymbol{\Gamma}}_N^{-1}\right)^{-1}\right]\mathbf{U}_N. \tag{7.7.21}$$

Then $Q_{N,1}^2$ and $Q_{N,2}^2$ are respectively the UI test statistic for the testing problems in (7.7.17) and (7.7.18), while the parallel test statistic for (7.7.19) is

$$Q_{N,3}^2 = Q_{N,2}^2 - Q_{N,1}^2. \tag{7.7.22}$$

Under the sequence of alternative hypotheses $\{K_N\}$, where K_N relates to the model when $\boldsymbol{\lambda} = N^{1/2}\boldsymbol{\beta}$ is held fixed (as $N \to \infty$), we have from the results of Chapters 5 and 6 that as $N \to \infty$,

$$Q_{N,1}^2 \xrightarrow{\mathscr{D}} \chi^2_{(p-1)(q-1),\,\delta_1^2}, \qquad Q_{N,2}^2 \xrightarrow{\mathscr{D}} \chi^2_{p(q-1),\,\delta_2^2}, \quad \text{and} \quad Q_{N,3}^2 \xrightarrow{\mathscr{D}} \chi^2_{q-1,\,\delta_3^2}, \tag{7.7.23}$$

where the noncentrality parameters are found by substituting the corresponding values of $\boldsymbol{\lambda}\mathbf{M}$, $\mathbf{D}_{(11)}$, $\boldsymbol{\Gamma}$, and $\boldsymbol{\nu}$ for $\mathbf{U}_N$, $\mathbf{D}_{N(11)}$, $\hat{\boldsymbol{\Gamma}}_N$, and $\mathbf{V}_N$ in (7.7.20)–(7.7.22). Note that the overall significance level of this strategic decision rule is

$$\alpha_N^* = P_0\{Q_{N,1}^2 \ge s_1, Q_{N,2}^2 \ge s_2\} + P_0\{Q_{N,1}^2 < s_1, Q_{N,3}^2 \ge s_3\}, \tag{7.7.24}$$

where s_1, s_2, and s_3 are the critical levels for the $Q_{N,j}$ corresponding to the significance level α_j, $j = 1, 2, 3$. Since $Q_{N,2}^2 = Q_{N,1}^2 + Q_{N,3}^2$, and $Q_{N,1}^2$ and $Q_{N,3}^2$ are (asymptotically) independent, if we set $\alpha_2 = \alpha_3$ (a reasonable condition, since $Q_{N,2}^2$ and $Q_{N,3}^2$ test the same null hypothesis), we obtain, on using the fact that $s_1 + s_3 \ge s_2$, that

$$\begin{aligned} \alpha_N^* &= P_0\{Q_{N,1}^2 + Q_{N,3}^2 \ge s_2, Q_{N,1}^2 \ge s_1 + \alpha_3(1-\alpha_1) \\ &= \alpha_1 P_0\{Q_{N,1}^2 + Q_{N,3}^2 \ge s_2 | Q_{N,1}^2 \ge s_1\} + \alpha_2(1-\alpha_1) \\ &\ge \alpha_1\alpha_2 + \alpha_2(1-\alpha_1) = \alpha_2. \end{aligned} \tag{7.7.25}$$

This shows that the overall level of significance for the PTI rule is larger than α_2, the significance level for just testing (7.7.18). However, this difference is usually very small (see Saleh and Sen (1983a) for some special cases), and hence the PTI decision rule would still have a power advantage over $Q_{N,2}^2$. (This is due to the fact that $Q_{N,3}^2$ has a much smaller number of DF than $Q_{N,2}^2$.)

EXERCISES

7.2.1. Verify the asymptotic normality in (7.2.17).

7.2.2. For testing the null hypothesis H_0 in (7.2.2), where β is treated as a nuisance parameter, consider the usual likelihood-ratio statistic. Express it in terms of the maximum-likelihood estimators of the parameters, and, hence or otherwise, verify that the $-2\log$(likelihood ratio) criterion has asymptotically chi-square distribution with 1 DF. Under the alternatives in (7.2.9), show that it has the noncentral chi-square distribution with 1 DF and noncentrality parameter $\lambda I(f)$.

7.3.1. Invoke (2.7.39)–(2.7.40) to verify (7.2.34).

7.3.2. Along the lines of Exercise 7.2.2, verify (7.3.46).

7.3.3. Use (7.3.81), (7.3.82), and the joint (asymptotic) normality of the estimates to verify (7.3.83).

7.3.4. By arguments similar to those in Exercise 7.3.3, verify (7.3.85).

7.3.5. By arguments similar to those in the preceding exercise, verify that $\hat{\mathscr{L}}_N^0$ in (7.3.93) has asymptotically the chi-square distribution with $k-1$ DF under H_0^*.

7.3.6. For the normal-theory model, obtain the likelihood-ratio statistic in terms of some linear (least-squares) estimators, and, hence or otherwise, obtain the asymptotic distribution theory in the specific problem of the hypothesis in (7.3.71).

7.3.7. Obtain the noncentrality parameter of the statistic in the preceding problem and compare it with (7.3.95) to conclude that the ARE result agrees with (7.3.69).

7.3.8. Derive the noncentral chi-square distribution of $-2\log\lambda_N^{(2)}$, and show that the noncentrality parameter agrees with (7.3.101).

7.4.1. By the Laplace expansion of the determinants in (7.4.27), verify that (7.4.28) holds.

7.4.2. Verify the contiguity of the sequence of probability measures under $\{H_N^{(2)}\}$ in (7.4.40) to that under $H_0^{(2)}$.

7.6.1. Consider the simple linear model $X_i = c_{i1}\beta_1 + \cdots + c_{iq}\beta_q + e_i$, $i = 1,\ldots,N$, where the e_i are i.i.d.r.v.'s and we want to test for $H_0: \beta_1 = \cdots = \beta_q = 0$ against the ordered alternative that $\beta_1 \le \cdots \le \beta_q$, with at least one strict inequality. Let $d_{ij} = c_{ij} - c_{ij-1}$ for $j = 1,\ldots,q$, where $c_{i0} = 0$, and let $\gamma_j = \beta_j - \beta_{j-1}$ for $j = 1,\ldots,q$, where $\beta_0 = 0$. Then we have $X_i = d_{i1}\gamma_1 + \cdots + d_{iq}\gamma_q + e_i$, $i = 1,\ldots,N$, where $H_0: \gamma_1 = \cdots = \gamma_q = 0$ and the alternative hypotheses relate to $\gamma_j \ge 0$ for $1 \le j \le q$ with at least one strict inequality. Use this reparametrization to convert a general ordered alternative hypothesis to an orthant alternative hypothesis.

7.6.2. Consider the k-sample model ($k \ge 2$) with d.f.'s $F_1,\ldots,F_k$ and $F_j(x) = F((x-\mu_j)/\sigma_j)$, $j = 1,\ldots,k$. The null hypothesis is to test for the equality of the location parameters and the equality of the scale parameters, against the alternatives that the locations are ordered or the scale parameters are ordered. Reduce this alternative to that of an orthant alternative by suitable reparametrization, and construct suitable rank tests. (Sen, 1982a.)

7.6.3. Consider the partitioning in (7.6.29) and the null hypothesis that $\boldsymbol{\beta}_{(1)} = \mathbf{0}$ against $\boldsymbol{\beta}_{(1)} \geq \mathbf{0}$, treating $\boldsymbol{\beta}_{(2)}$ as a nuisance parameter. Instead of using $\mathbf{U}_N$ in (7.6.31), consider appropriate aligned rank statistics $\hat{\mathbf{L}}_{N(1)}$ for $\mathbf{L}_N$ and $\hat{\mathbf{U}}_{N(1)}$ for $\mathbf{U}_N$, and construct the UI test statistic. Obtain results parallel to (7.6.33) and (7.6.34), citing the correct number of DF in this case.

7.6.4. In the setup of the preceding exercise, instead of the orthant alternative $\boldsymbol{\beta}_{(1)} \geq \mathbf{0}$, consider an ordered alternative (similar to that in Exercises 7.6.1 and 7.6.2); construct the UI test statistic, and study its asymptotic distribution theory. (Chinchilli and Sen, 1981a, b.)

7.6.5. In the setup of Exercise 7.6.3, show that if the alternative hypothesis is specified by some equality constraints (instead of the inequalities), then the UI test statistic has asymptotically chi-square d.f. under the null hypothesis and noncentral chi-square under local alternatives. (Chinchilli and Sen, 1982.)

7.7.1. Provide a proof of Theorem 7.7.1. (Sen and Saleh, 1979.)

Hints. Use the linearity results in (6.4.26)–(6.4.27), the asymptotic normality result in (6.4.28) and the definitions of the constrained and unconstrained estimators.

7.7.2. With the notation in Theorem 7.7.1, show that for every $c > 0$ and $\delta = \mathbf{a}'\mathbf{B}^{-1}\mathbf{a}$ (> 0),

$$\int_{\mathbf{x}'\mathbf{B}^{-1}\mathbf{x} > c} dG_p(\mathbf{x}; \mathbf{a}, \mathbf{B}) = e^{-\delta/2} \sum_{r=0}^{\infty} \left(\frac{\delta}{2}\right)^r \frac{1 - H_{p+2r}(c; 0)}{r!}.$$

7.7.3. Use the identity in the preceding exercise, and by differentiation with respect to $\mathbf{a}$ (twice), show that

$$\int_{(\mathbf{x}+a)'\mathbf{B}^{-1}(\mathbf{x}+\mathbf{a}) > c} \mathbf{x}\, dG_p(\mathbf{x}; \mathbf{0}, \mathbf{B}) = \mathbf{a}\left[H_p(c;\delta) - H_{p+2}(c;\delta)\right]$$

and

$$\int_{(\mathbf{x}+\mathbf{a})'\mathbf{B}^{-1}(\mathbf{x}+\mathbf{a}) > c} \mathbf{x}\mathbf{x}'\, dG_p(\mathbf{x}; \mathbf{0}, \mathbf{B})$$

$$= \left[1 - H_{p+2}(c;\delta)\right]\mathbf{B} - \mathbf{a}\mathbf{a}'\left[H_p(c;\delta) - H_{p+2}(c;\delta) + H_{p+4}(c;\delta)\right].$$

Use these results along with (7.7.4) to derive (7.7.5) and (7.7.6). (Sen and Saleh, 1979.)

7.7.4. With the definitions of the $Q_{N,j}^2$, $j = 1, 2, 3$, in (7.7.20)–(7.7.22), verify the asymptotic distributional results in (7.7.23). (Chinchilli and Sen, 1982.)

Hints. Use the asymptotic linearity results in (6.4.21) and (6.4.44) along with the asymptotic multinormality results on the rank statistics in Chapter 5.

CHAPTER 8

Rank-Order Tests for Miscellaneous Problems in Linear Models

8.1. INTRODUCTION

For a variety of univariate as well as multivariate linear models, some (genuinely and asymptotically) distribution-free tests based on suitable rank statistics have been considered in Chapters 5 and 7. These linear models are characterized by *nonstochastic predictors* (regressors). In some problems of statistical inference, one may encounter *stochastic predictors*, and some conditionally distribution-free tests for such problems are available in the literature (Ghosh and Sen, 1971b). Nonparametric tests for regression with stochastic predictors are considered in Section 8.2. In the context of analysis of covariance and elsewhere, one may face *mixed models* involving both stochastic and nonstochastic predictors (Sen and Puri, 1970). Section 8.3 deals with nonparametric procedures for such mixed models. In many repeated-measurement designs, follow-up studies, and longitudinal studies, *growth-curve models* are quite popular in practice, and some nonparametric procedures are also available in the literature (Ghosh, Grizzle, and Sen, 1973; Sen, 1973b; and Woolson and Sen, 1974, among others). A general account of nonparametric growth-curve analysis is presented in Section 8.4.

In most practical applications, though the distributions of the underlying random variables may be continuous, due to roundoff errors or other practical limitations, collected data relate to a set of (finite or countable number of) class intervals (or ordered categories) along with their frequencies and other identifying factors. Ties among the observations are no longer negligible, and a somewhat different (conditional) approach may be more appropriate (Sen, 1967; Ghosh, 1973a, b, Padmanabhan and Puri, 1979, 1983, and others). Section 8.5 deals with nonparametric tests for some linear models relating to *grouped data* and examines the efficiency loss due to grouping. The last section is devoted to nonparametric testing under various types of censoring.

8.2. NONPARAMETRIC TESTS FOR REGRESSION WITH STOCHASTIC PREDICTORS

Let $\{\mathbf{Z}_i' = (\mathbf{Y}_i', \mathbf{X}_i')\ [= (Y_{1i}, \dots, Y_{pi}, X_{1i}, \dots, X_{qi})$ for some $p \geq 1,\ q \geq 1]$, $i \geq 1\}$ be a sequence of i.i.d.r.v.'s with a $(p+q)$-variate continuous d.f. $F(\mathbf{z})$, $\mathbf{z} \in R^{p+q}$. Let then $F_1(\mathbf{x})$ be the marginal d.f. of $\mathbf{X}_i$, and let $G(\mathbf{y}|\mathbf{x})$ be the conditional d.f. of $\mathbf{Y}_i$ given $\mathbf{X}_i = \mathbf{x}$. Usually, $G(\mathbf{y}|\mathbf{x})$ depends on $\mathbf{x}$ in some manner; for example, we may have

$$G(\mathbf{y}|\mathbf{x}) = G_0(\mathbf{y} - \boldsymbol{\beta}_0 - \boldsymbol{\beta}'\mathbf{x}), \tag{8.2.1}$$

where $\boldsymbol{\beta}_0$ $(p \times 1)$ and $\boldsymbol{\beta}'$ $(p \times q)$ are unknown parameters. In this sense, the $\mathbf{X}_i$ can be termed stochastic predictors. With respect to (8.2.1), one may be interested in testing the null hypothesis of no regression (i.e., $\boldsymbol{\beta} = \mathbf{0}$). Without imposing the linearity in (8.2.1), we frame the null hypothesis as

$$H_0: G(\mathbf{y}|\mathbf{x}) = G_0(\mathbf{y}) \qquad \forall \mathbf{x} \in R^q,\ \mathbf{y} \in R^p, \tag{8.2.2}$$

where G_0 is some continuous d.f. In this setup, we are really testing for the stochastic independence of the subvectors $\mathbf{Y}_i$ and $\mathbf{X}_i$. However, later on we shall also study the performance properties of the tests to be considered when (8.2.1) holds.

For testing H_0 in (8.2.2), both *pure-* and *mixed-rank statistics* have been suggested for the construction of suitable test statistics. The pure-rank statistics involve only the coordinatewise ranks of the $\mathbf{Z}_i$, and these are quite useful when the $\mathbf{Z}_i$ (or $\mathbf{X}_i$) are partially informed, that is, they are not observable, but only the ranks on them are available—a case that arises in many educational or psychometric problems involving ranked data. On the other hand, if the $\mathbf{X}_i$ are observable and a model like (8.2.1) can be conceived, mixed-rank statistics can be constructed which are natural analogues of linear rank statistics considered in Chapters 5 and 7.

Let R_{ji} (S_{ji}) be the rank of Y_{ji} (X_{ji}) among $Y_{j1}, \dots, Y_{jn}$ $(X_{j1}, \dots, X_{jn})$, for $i = 1, \dots, n$, $j = 1, \dots, p(q)$. Thus, we use separate ranking for the different rows of $\mathbf{Z} = (\mathbf{Z}_1, \dots, \mathbf{Z}_n)$. This leads us to the *rank-collection matrix*

$$\mathbf{R}^* = \begin{pmatrix} \mathbf{R} \\ \mathbf{S} \end{pmatrix}_{\substack{p\times n \\ q \times n}}, \qquad \mathbf{R} = ((R_{ji})), \quad \mathbf{S} = ((S_{ji})). \tag{8.2.3}$$

Consider now a set of *scores*

$$a_{nj}(i), \qquad 1 \leq i \leq n, \quad j = 1, \dots, p, \tag{8.2.4}$$

$$b_{nj}(i), \qquad 1 \leq i \leq n, \quad j = 1, \dots, q, \tag{8.2.5}$$

where the $a_{nj}(i)$ and $b_{nj}(i)$ are defined as in (4.2.4) with underlying *score functions* $\phi_j(u)$ and $\phi_j^*(u)$, respectively. Then the pure-rank statistics are defined by

$$\mathbf{M}_n = \left(\left(m_{njl} = \sum_{i=1}^{n} \left[a_{nj}(R_{ji}) - \bar{a}_{nj}\right]\left[b_{nl}(S_{li}) - \bar{b}_{nl}\right]\right)\right), \tag{8.2.6}$$

where

$$\bar{a}_{nj} = n^{-1}\sum_{i=1}^{n} a_{nj}(i), \quad 1 \le j \le p \quad \text{and} \quad \bar{b}_{nl} = n^{-1}\sum_{i=1}^{n} b_{nl}(i), \quad 1 \le l \le q. \tag{8.2.7}$$

On the other hand, the mixed-rank statistics are defined by

$$\mathbf{M}_n^* = \left(\left(m_{njl}^* = \sum_{i=1}^{n} \left[a_{nj}(R_{ji}) - \bar{a}_{nj}\right]\left[X_{li} - \bar{X}_{ln}\right]\right)\right), \tag{8.2.8}$$

where

$$\bar{X}_{ln} = n^{-1}\sum_{i=1}^{n} X_{li} \qquad \text{for} \quad l = 1, \dots, q. \tag{8.2.9}$$

In either case, the test statistic to be considered is a quadratic form in the elements of $\mathbf{M}_n$ or $\mathbf{M}_n^*$ and is based on the rank permutation principle of Chatterjee and Sen (1964) displayed in Section 5.4. For this, we define

$$\underset{p\times p}{\mathbf{V}_n^{(1)}} = \left(\left(n^{-1}\sum_{i=1}^{n} \left[a_{nj}(R_{ji}) - \bar{a}_{nj}\right]\left[a_{nl}(R_{li}) - \bar{a}_{nl}\right]\right)\right), \tag{8.2.10}$$

$$\underset{q\times q}{\mathbf{V}_n^{(2)}} = \left(\left(n^{-1}\sum_{i=1}^{n} \left[b_{nj}(S_{ji}) - \bar{b}_{nj}\right]\left[b_{nl}(S_{li}) - \bar{b}_{nl}\right]\right)\right) \tag{8.2.11}$$

and let

$$\underset{pq\times pq}{\mathbf{V}_n} = \mathbf{V}_n^{(1)} \otimes \mathbf{V}_n^{(2)} = \left(\left(v_{njj'}^{(1)} v_{nll'}^{(2)}\right)\right)_{j,j'=1,\dots,p,\; l,l'=1,\dots,q}. \tag{8.2.12}$$

Also, let

$$\mathbf{S}_n = (n-1)^{-1}\sum_{i=1}^{n}(\mathbf{X}_i - \overline{\mathbf{X}}_n)(\mathbf{X}_i - \overline{\mathbf{X}}_n)' = \left(\left(s_{njl}\right)\right), \quad (8.2.13)$$

$$\mathbf{V}_n^* = \mathbf{V}_n^{(1)} \otimes \mathbf{S}_n = \left(\left(v_{njj'}^{(1)} s_{nll'}\right)\right)_{j,j'=1,\ldots,p,\, l,l'=1,\ldots,q}. \quad (8.2.14)$$

Finally, we roll out $\mathbf{M}_n$ and $\mathbf{M}_n^*$ into $1 \times pq$ vectors and denote these by $\mathbf{m}_n$ and $\mathbf{m}_n^*$, respectively. Also, we rewrite (8.2.3) as

$$\mathbf{R}^* = \begin{pmatrix} \mathbf{R}_1 & \cdots & \mathbf{R}_n \\ \mathbf{S}_1 & \cdots & \mathbf{S}_n \end{pmatrix}, \quad (8.2.15)$$

where $\mathbf{R}_i = (R_{1i}, \ldots, R_{pi})'$ and $\mathbf{S}_i = (S_{1i}, \ldots, S_{qi})'$, $1 \le i \le n$.

8.2.1. Permutational Rank Tests for H_0 in (8.2.2)

Note that under H_0 in (8.2.2), $\mathbf{Y}_i$ and $\mathbf{X}_i$ are stochastically independent for every $1 \le i \le n$. Also, note that each row of $\mathbf{R}^*$ consists of the numbers $1, \ldots, n$, permuted in some order. Thus, there are $(n!)^{p+q}$ possible realizations of $\mathbf{R}^*$, and we denote the set of all these realizations by $\mathscr{R}^*$. For a given $\mathbf{R}^*$, we can generate a set of $n!$ realizations by letting

$$\mathbf{R}^*(\mathbf{i}) = \begin{pmatrix} \mathbf{R}_1 & \cdots & \mathbf{R}_n \\ \mathbf{S}_{i_1} & \cdots & \mathbf{S}_{i_n} \end{pmatrix}, \quad (8.2.16)$$

where $(i_1, \ldots, i_n)$ is any permutation of $(1, \ldots, n)$. Note that the distribution of $\mathbf{R}^*$ over $\mathscr{R}^*$ will in general depend on the unknown F, even when H_0 in (8.2.2) holds (exception: when $p = q = 1$). However, the conditional (permutational) distribution of $\mathbf{R}^*$ over the set of $n!$ realizations in (8.2.16) will be uniform under (8.2.2), and hence, if we denote this probability measure by $\mathscr{P}_n$, we have, by some standard steps (similar to those in Section 5.4),

$$E_{\mathscr{P}_n}(\mathbf{m}_n) = \mathbf{0}, \quad (8.2.17)$$

$$E_{\mathscr{P}_n}(\mathbf{m}_n' \mathbf{m}_n) = n^2(n-1)^{-1}\mathbf{V}_n, \quad (8.2.18)$$

and we use the following test statistic:

$$\mathscr{L}_n = n^{-2}(n-1)\left(\mathbf{m}_n \mathbf{V}_n^- \mathbf{m}_n'\right), \quad (8.2.19)$$

where $\mathbf{V}_n^-$ is a generalized inverse of $\mathbf{V}_n$. It may be noted that for $p = q = 1$, $\mathscr{L}_n$ in (8.2.19) reduces to

$$\mathscr{L}_n = n^{-2}(n-1)m_{n11}^2\left(v_{n11}^{(1)}v_{n11}^{(2)}\right)^{-1} \tag{8.2.20}$$

and is a genuinely distribution-free statistic when H_0 in (8.2.2) holds. For $p > 1$ or $q > 1$, $\mathscr{L}_n$ in (8.2.19) may not be genuinely distribution-free, but is a permutationally distribution-free statistic.

For the mixed-rank statistics, instead of (8.2.15), we use

$$\begin{pmatrix} \mathbf{R}_1 & \cdots & \mathbf{R}_n \\ \mathbf{X}_1 & \cdots & \mathbf{X}_n \end{pmatrix} \tag{8.2.21}$$

and in (8.2.16) we take

$$\begin{pmatrix} \mathbf{R}_1 & \cdots & \mathbf{R}_n \\ \mathbf{X}_{i_1} & \cdots & \mathbf{X}_{i_n} \end{pmatrix};$$

the rest of the permutational invariance remains the same, and hence, on parallel lines, we have

$$E_{\mathscr{P}_n}(\mathbf{m}_n^*) = \mathbf{0}, \qquad E_{\mathscr{P}_n}(\mathbf{m}_n^{*\prime}\mathbf{m}_n^*) = n\mathbf{V}_n^*, \tag{8.2.22}$$

and we consider the test statistic

$$\mathscr{L}_n^* = n^{-1}\mathbf{m}_n^*\mathbf{V}_n^{*-}\mathbf{m}_n^{*\prime}, \tag{8.2.23}$$

where $\mathbf{V}_n^{*-}$ is a generalized inverse of $\mathbf{V}_n^*$. Unlike (8.2.20), here, for $p = q = 1$, $\mathscr{L}_n^*$ is no longer genuinely distribution-free. Nevertheless, for general $p \geq 1$, $q \geq 1$, $\mathscr{L}_n^*$ is a permutationally distribution-free statistic.

For either $\mathscr{L}_n$ or $\mathscr{L}_n^*$, when n is small, the permutation distribution can be enumerated by considering all $n!$ possible equally likely permutations of $(i_1, \ldots, i_n)$ over $(1, \ldots, n)$, and thus a conditionally distribution-free test for H_0 in (8.2.2) can be constructed, rejecting H_0 for large values of $\mathscr{L}_n$ or $\mathscr{L}_n^*$. This task becomes prohibitively laborious as n increases, forcing us to use asymptotic distribution theory, which we present below.

First consider the case of $\mathscr{L}_n^*$. Note that under H_0 in (8.2.2), given $\mathbf{X}_1, \ldots, \mathbf{X}_n$, the $\mathbf{Y}_i$ are (conditionally) independent with the common d.f. $G_0(\mathbf{y})$, $\mathbf{y} \in R^p$, which does not depend on $\mathbf{X}_1, \ldots, \mathbf{X}_n$. Also, if we assume that

$$E(\mathbf{X}_i - E\mathbf{X}_i)(\mathbf{X}_i - E\mathbf{X}_i)' = \mathbf{\Sigma} \text{ is p.d. and finite,} \tag{8.2.24}$$

then writing $\mathbf{S}_n = [n/(n-1)]\{n^{-1}\Sigma_{i=1}^{n}(\mathbf{X}_i - E\mathbf{X}_i)(\mathbf{X}_i - E\mathbf{X}_i)' - (\overline{\mathbf{X}}_n - E\mathbf{X}_1)(\overline{\mathbf{X}}_n - E\mathbf{X}_1)'\}$ and using the Khinchin law of large numbers, we have

$$\mathbf{S}_n \to \mathbf{\Sigma} \quad \text{a.s.} \qquad \text{as} \quad n \to \infty. \tag{8.2.25}$$

On the other hand, under the conditions of Theorem 5.4.1,

$$\mathbf{V}_n^{(1)} \xrightarrow{P} \boldsymbol{\nu}(G) \qquad \text{as} \quad n \to \infty, \tag{8.2.26}$$

where $\boldsymbol{\nu}(G)$ is p.d. Hence, by (8.2.14), (8.2.25), and (8.2.26),

$$\mathbf{V}_n^* \xrightarrow{P} \boldsymbol{\nu}(G) \otimes \mathbf{\Sigma} \text{ and is p.d., in probability.} \tag{8.2.27}$$

Further, (8.2.24) insures that as $n \to \infty$,

$$\max_{1 \le k \le n} \left\{ n^{-1}\mathrm{Ch}_1\,(\mathbf{X}_k - \overline{\mathbf{X}}_n)(\mathbf{X}_k - \overline{\mathbf{X}}_n)' \right\} \to 0 \quad \text{a.s.} \tag{8.2.28}$$

Thus, we may virtually repeat the proof of Theorem 5.4.2, replacing the $\mathbf{c}_i$ by $\mathbf{X}_i$, and, conditional on $\mathbf{X}_1, \ldots, \mathbf{X}_n$ being given, we claim that the conditional distribution of $\mathscr{L}_n^*$ converges, in probability, to the central chi-square distribution with pq DF. Hence, under H_0 in (8.2.2),

$$\mathscr{L}_n^* \xrightarrow{\mathscr{L}} \chi_{pq}^2 \qquad \text{as} \quad n \to \infty. \tag{8.2.29}$$

Let now $F_{1[j]}(x)$ be the marginal d.f. of X_{ji} for $1 \le j \le q$, and let

$$X_{ji}^* = \phi_j^*\big(F_{1[j]}(X_{ji})\big), \qquad 1 \le j \le q, \quad 1 \le i \le n, \tag{8.2.30}$$

where ϕ_j^* is the score function underlying the scores $b_{nj}(i)$, $1 \le i \le n$. Suppose that in $\mathbf{m}_n^*$, we replace the X_{ji} by X_{ji}^* and denote the resulting vector by $\mathbf{m}_n^{**}$. In a similar manner, we define $\mathbf{S}_n^*$ analogous to $\mathbf{S}_n$ and let $\mathbf{V}_n^{**} = \mathbf{V}_n^{(1)} \otimes \mathbf{S}_n^*$ and $\mathscr{L}_n^{**} = n^{-1}\mathbf{m}_n^{**}\mathbf{V}_n^{**-}\mathbf{m}_n^{**\prime}$. Then, noting that under H_0 in (8.2.2) for every $1 \le j \le p$, $1 \le l \le q$, as $n \to \infty$,

$$n^{-1}E\left\{m_{njl} - m_{njl}^{**}\right\}^2 = v_{njj}^{(1)}\left\{\frac{n}{n-1}v_{nll}^{(2)} - S_{nll}^*\right\} \to 0, \tag{8.2.31}$$

we conclude that under H_0 in (8.2.2),

$$n^{-1}\|\mathbf{m}_n - \mathbf{m}_n^{**}\|^2 = \sum_{j=1}^{p}\sum_{l=1}^{q} n^{-1}\left\{m_{njl} - m_{njl}^{**}\right\}^2 \xrightarrow{P} 0. \tag{8.2.32}$$

Likewise, from our results in Section 5.4, $\mathbf{\Sigma}^*$ p.d. and finite insures that

$$\mathbf{S}_n^* - \mathbf{V}_n^{(2)} \xrightarrow{P} \mathbf{0} \quad \text{as} \quad n \to \infty. \tag{8.2.33}$$

Hence, we conclude that under H_0 in (8.2.2),

$$\mathscr{L}_n - \mathscr{L}_n^{**} \xrightarrow{P} 0 \quad \text{as} \quad n \to \infty. \tag{8.2.34}$$

On the other hand, for $\mathscr{L}_n^{**}$, the same proof as led to (8.2.29) holds, so that $\mathscr{L}_n^{**} \xrightarrow{\mathscr{L}} \chi_{pq}^2$ when H_0 in (8.2.2) and the regularity conditions of Theorem 5.4.1 and (8.2.24) hold. Thus, under H_0,

$$\mathscr{L}_n \xrightarrow{\mathscr{L}} \chi_{pq}^2 \quad \text{as} \quad n \to \infty. \tag{8.2.35}$$

Thus, for large n, $\chi_{pq,\alpha}^2$, the upper $100\alpha\%$ point of the central chi-square d.f. with pq DF, provides the critical point for the test for H_0 based on $\mathscr{L}_n$ or $\mathscr{L}_n^*$.

8.2.2. Asymptotic Non-null Distribution Theory of $\mathscr{L}_n$ and $\mathscr{L}_n^*$

More than one approach to the study of this asymptotic distribution theory may be conceived. First, the LeCam–Hájek contiguity approach, has been developed in the current context by Ghosh and Sen (1971b). Here, we do not need more restrictive regularity conditions on the score functions ϕ_j and ϕ_j^*, but need the existence of a finite Fisher information (matrix) for the d.f. G_0 (requiring the existence of an absolutely continuous pdf g_0 for G_0). We may also explore the classical Chernoff–Savage approach, which has been outlined in Chapter 8 of Puri and Sen (1971). This entails more restrictive conditions on the score functions, but does not require the existence of a finite Fisher information matrix for G_0. We develop the second approach here and refer to Ghosh and Sen (1971b) for the first approach.

Note that the stochastic convergence of $\mathbf{V}_n^{(1)}$ or $\mathbf{V}_n^{(2)}$ to some appropriate $\boldsymbol{\nu}^{(1)}$ or $\boldsymbol{\nu}^{(2)}$ follows along the lines of Theorem 5.4.1 [even when H_0 in (8.2.2) does not hold—as in any case $\mathbf{Y}_i$, $1 \le i \le n$, are i.i.d.r.v.'s, so also are the $\mathbf{X}_i$]. Also, the stochastic convergence of $\mathbf{S}_n$ to $\mathbf{\Sigma}$ (or of $\mathbf{S}_n^*$ to $\mathbf{\Sigma}^*$) follows as in (8.2.25), even when H_0 in (8.2.2) does not hold. Hence, the basic problem in the study of the asymptotic distribution theory of $\mathscr{L}_n$ or $\mathscr{L}_n^*$ is the asymptotic multinormality of the standardized form of $\mathbf{m}_n$ or $\mathbf{m}_n^*$. The asymptotic multinormality of the standardized form of $\mathbf{m}_n^*$ follows precisely as in Section 5.5 if we replace there the $\mathbf{c}_i$ by $\mathbf{X}_i$ and note that for the conditional d.f.'s of the $\mathbf{Y}_i$ given the $\mathbf{X}_i$, (5.5.9) holds a.s. as $n \to \infty$ [by (8.2.28)]. Hence, if one considers a sequence $\{H_n\}$ of alternative hypotheses, where under H_n, (8.2.1) holds for $\boldsymbol{\beta} = n^{-1/2}\boldsymbol{\gamma}$ and G_0 satisfies the conditions (a), (b), and (c) of Section 5.5 above (5.5.15), then the asymptotic multinormality of $n^{-1/2}\mathbf{m}_n^*$ [with asymptotic mean vector $\mathbf{B}\boldsymbol{\gamma}\boldsymbol{\Sigma}$ and asymptotic dispersion matrix $\boldsymbol{\nu}(G) \otimes \boldsymbol{\Sigma}$, where $\mathbf{B} = \operatorname{diag}(B_1, \ldots, B_p)$ and the B_j are defined by (5.5.15) with the $F_{[j]}$ replaced by $G_{0[j]}$, $1 \le j \le p$] follows as in Section 5.5. Hence, the asymptotic noncentral chi-square distribution of $\mathscr{L}_n^*$ (under $\{H_n\}$) with pq DF and noncentrality parameter

$$\Delta_{\mathscr{L}^*} = \sum_{j=1}^{p} \sum_{j'=1}^{p} \sum_{l=1}^{q} \sum_{l'=1}^{q} \nu^{jj'}(G)\sigma_{ll'}B_jB_{j'}\gamma_{jl}\gamma_{j'l'} \tag{8.2.36}$$

[where $\boldsymbol{\nu}^{-1}(G) = ((\nu^{jj'}(G)))$] follows from the above results.

Let us consider the case of $\mathscr{L}_n$ and study first the asymptotic (multi)normality of $n^{-1/2}\mathbf{m}_n$. For notational convenience, we let

$$G_{n[j]}^* = (n+1)^{-1} \sum_{i=1}^{n} u(x - Y_{ji}), \qquad 1 \le j \le p, \tag{8.2.37}$$

$$F_{1n[l]}^*(x) = (n+1)^{-1} \sum_{i=1}^{n} u(x - X_{li}), \qquad 1 \le l \le q, \tag{8.2.38}$$

[where $u(t) = 1$ or 0 according as $t \ge 0$ or < 0], and let $F_{1n[jj']}(x, y)$, $F_{2n[jl]}(x, y)$, and $F_{3n[ll']}(x, y)$ be respectively the bivariate marginal empirical d.f.'s of $(Y_{ji}, Y_{j'i})$, (Y_{ji}, X_{li}), and $(X_{li}, X_{l'i})$ for $j \ne j' = 1, \ldots, p$ and $l \ne l' = 1, \ldots, q$. For the scores in (8.2.4) and (8.2.5), for every j $(1 \le j \le p)$, l $(1 \le l \le q)$, and $i(= 1, \ldots, n)$, we let

$$a_{nj}(i) = \phi_{nj}\left(\frac{i}{n+1}\right), \quad b_{nl}(i) = \phi_n^*\left(\frac{i}{n+1}\right), \tag{8.2.39}$$

and extend the domain of ϕ_{nj} and ϕ_{nl}^* by letting them have constant values on each half-open interval $(i - 1/n, i/n]$, $1 \le i \le n$. Then we make the following assumptions:

(a) $\lim_{n\to\infty}\phi_{nj}(u) = \phi_j(u)$ and $\lim_{n\to\infty}\phi_{nl}^*(u) = \phi_l^*(u)$ exist for every $u \in (0,1)$, where $\phi_j(u)$ and $\phi_l^*(u)$ are not constant [inside $(0,1)$], $1 \le j \le p$, $1 \le l \le q$.

(b) $\phi_j(u)$ and $\phi_l^*(u)$ have continuous first derivatives $\phi_j^{(1)}(u)$ and $\phi_l^{*(1)}(u)$ for every $u \in (0,1)$, where for every j $(1 \le j \le p)$, l $(1 \le l \le q)$, and $r = 0, 1$ there exist positive constants K ($< \infty$) and δ (> 0) such that

$$\left| \frac{d^4 \phi_j(u)}{du^r} \right| \leq K[u(1-u)]^{-1/4+\delta-r},$$

$$\left| \frac{d^r \phi_l^*(u)}{du^r} \right| \leq K[u(1-u)]^{-1/4+\delta-r}, \qquad 0 < u < 1. \qquad (8.2.40)$$

[It is possible to replace the exponents in (8.2.40) by $-1/2a - r + \delta$ and $-1/2b - r + \delta$, respectively, where a and b are positive numbers such that $a^{-1} + b^{-1} = 1$. Also, it is possible to make both a and b dependent on j and l. However, for the sake of symmetry, we prefer the setup in (8.2.40).]

(c) For every j (= 1, ..., p) and l (= 1, ..., q), as $n \to \infty$,

$$n^{1/2} \int_{R^2}\!\!\int \Big[\phi_{nj}\big(G^*_{n[j]}(x)\big) \phi^*_{nl}\big(F^*_{1n[l]}(y)\big)$$

$$- \phi_j\big(G^*_{n[j]}(x)\big) \phi^*_l\big(F^*_{1n[l]}(y)\big)\Big] \, dF_{2n[jl]}(x, y) \xrightarrow{P} 0. \qquad (8.2.41)$$

Let us then define

$$\mu_{jl} = \int_{R^2}\!\!\int \phi_j\big(G_{[j]}(x)\big) \phi^*_l\big(F_{1[l]}(y)\big) \, dF_{2[jl]}(x, y), \qquad (8.2.42)$$

where $G_{[j]}(x) = P\{Y_{ji} \leq x\}$, $F_{1[l]}(y) = P\{X_{li} \leq y\}$, and $F_{2[jl]}$ is the bivariate d.f. of (Y_{ji}, X_{li}), for $j = 1, \ldots, p$, $l = 1, \ldots, q$. Further, let

$$U_{jl,i} = \phi_j\big(G_{[j]}(Y_{ji})\big) \phi^*_l\big(F_{1[l]}(X_{li})\big)$$

$$+ \int_{R^2}\!\!\int \Big\{ \big[u(x - Y_{ji}) - G_{[j]}(x)\big] \phi_j^{(1)}\big(G_{[j]}(x)\big) \phi^*_l\big(F_{1[l]}(y)\big)$$

$$+ \big[u(y - X_{li}) - F_{1[l]}(y)\big] \phi_l^{*(1)}\big(F_{1[l]}(y)\big) \phi_j\big(G_{[j]}(x)\big) \Big\} \, dF_{2[jl]}(x, y)$$

$$(8.2.43)$$

for $i = 1, \ldots, n$ and $j = 1, \ldots, p$, $l = 1, \ldots, q$. Finally, let

$$\mu_j^0 = \int_0^1 \phi_j(u) \, du, \quad 1 \leq j \leq p, \qquad \mu_l^{*0} = \int_0^1 \phi^*_l(u) \, du, \quad 1 \leq l \leq q.$$

$$(8.2.44)$$

Then, we have the following.

Theorem 8.2.1. *Under assumptions* (a), (b) *and* (c), *for arbitrary (continuous) F, for every* $j = 1, \ldots, p$ *and* $l = 1, \ldots, q$, *as* $n \to \infty$,

$$n^{1/2}\left\{n^{-1}m_{njl} - \left(\mu_{jl} - \mu_j^0\mu_l^{*0}\right) - n^{-1}\sum_{i=1}^{n} U_{jl,i}\right\} \xrightarrow{P} 0. \qquad (8.2.45)$$

The proof of this theorem is given in the Appendix. Now, $\{U_{jl,i}, 1 \le j \le p, 1 \le l \le q, i \ge 1\}$ are independent vectors, and the central limit theorem holds; this insures the asymptotic multinormality of $n^{-1/2}\{m_{njl} - n(\mu_{jl} - \mu_j^0\mu_l^{*0}), 1 \le j \le p, 1 \le l \le q\}$. Also, (8.2.45) remains true even when F is replaced by a sequence $\{F_{(n)}\}$ of d.f. such as in (8.2.1) with $\boldsymbol{\beta} = n^{-1/2}\boldsymbol{\gamma}$ (as is the case under $\{H_n\}$) and the multinormality also holds.

Note that we may rewrite (8.2.42) as

$$\mu_{jl} = \int_{R^{p+q}} \cdots \int \phi_j\left(G_{[j]}(y_j)\right)\phi_l^*\left(F_{1[l]}(x_l)\right) dF(\mathbf{y}, \mathbf{x}), \qquad (8.2.46)$$

where $\mathbf{y} = (y_1, \ldots, y_p)$ and $\mathbf{x} = (x_1, \ldots, x_q)$. Let us now denote

$$\psi_j(u) = -\frac{\partial}{\partial x}\log g_{[j]}(x)\Big|_{x = G_{[j]}^{-1}(u)}, \qquad 0 < u < 1, \qquad (8.2.47)$$

where $g_{[j]}(x) = G'_{[j]}(x)$, and we assume that

$$\int_0^1 \psi_j^2(u)\, du < \infty \qquad \forall 1 \le j \le p. \qquad (8.2.48)$$

Further, let $B_j = \int_0^1 \psi_j(u)\phi_j(u)\, du$, $1 \le j \le p$, be defined as in (5.5.15), and for every $l, l' = 1, \ldots, q$, let

$$B_{ll'}^* = \int_{R^2}\int x_l \phi_{l'}^*\left(F_{1[l']}(x_{l'})\right) dF_{3[ll']}(x_l, x_{l'}), \qquad (8.2.49)$$

$$\mathbf{B}^* = \left(\left(B_{ll'}^*\right)\right)_{l,l'=1,\ldots,q} = \left(\mathbf{B}_1^*, \ldots, \mathbf{B}_q^*\right), \quad \text{say}. \qquad (8.2.50)$$

Then, by (8.2.44), (8.2.46), (8.2.47), (8.2.49), and (8.2.50), we obtain that under $\{H_n\}$ [where H_n: (8.2.1) holds with $\boldsymbol{\beta} = n^{-1/2}\boldsymbol{\gamma}$ and $\boldsymbol{\gamma}' = (\boldsymbol{\gamma}_1', \ldots, \boldsymbol{\gamma}_p')$] and the assumed regularity conditions,

$$n^{1/2}\left\{\mu_{jl} - \mu_j^0\mu_l^{*0}\right\} \to B_j\boldsymbol{\gamma}_j\mathbf{B}_l^* \qquad \forall 1 \le j \le p, 1 \le l \le q. \quad (8.2.51)$$

Moreover, under $\{H_n\}$, the dispersion matrix of $n^{-1/2}\{\sum_{i=1}^n U_{jl,i}, 1 \le j \le p,$

$1 \le l \le q\}$ has the same limit as that under H_0. Hence, by virtue of Theorem 8.2.1, (8.2.51), and the above, we conclude that under $\{H_n\}$, $\mathscr{L}_n$ has asymptotically a noncentral chi-square d.f. with pq DF and noncentrality parameter

$$\Delta_{\mathscr{L}} = \sum_{j=1}^{p}\sum_{j'=1}^{p}\sum_{l=1}^{q}\sum_{l'=1}^{q} \nu^{jj'}(G_0)\nu^{ll'}(F_1)B_jB_{j'}(\boldsymbol{\gamma}_j\mathbf{B}_l^*)(\boldsymbol{\gamma}_{j'}\mathbf{B}_{l'}^*)$$

$$= \sum_{j=1}^{p}\sum_{j'=1}^{p} \nu^{jj'}(G_0)B_jB_{j'}\sum_{s=1}^{q}\sum_{s'=1}^{q}\gamma_{js}\gamma_{j's'}$$

$$\sum_{l=1}^{q}\sum_{l'=1}^{q} B_{sl}^*\nu^{ll'}(F_1)B_{s'l'}^*$$

$$= \sum_{j=1}^{p}\sum_{j'=1}^{p} \tau^{jj'}(G_0)\sum_{s=1}^{q}\sum_{s'=1}^{q}\gamma_{js}\gamma_{j's'}\tau_{ss'}^*, \tag{8.2.52}$$

where

$$\tau^{jj'}(G_0) = \nu^{jj'}(G_0)B_jB_{j'}, \qquad 1 \le j, j' \le p, \tag{8.2.53}$$

$$\tau_{ss'}^* = \sum_{l=1}^{q}\sum_{l'=1}^{q} B_{sl}^*B_{s'l'}^*\nu^{ll'}(F_1), \quad s, s' = 1, \dots, q. \tag{8.2.54}$$

[In particular, if $\phi_l^*(u) = F_{1[l]}^{-1}(u)$, $0 < u < 1$, $1 \le l \le q$, then $\mathbf{B}^* = \boldsymbol{\Sigma}$, so that on noting that $\boldsymbol{\nu}(F_1) = \boldsymbol{\Sigma}$, we have $\tau_{ss'}^* = \sigma_{ss'}$ $\forall 1 \le s, s' \le q$.]

8.2.3. Asymptotic Relative Efficiency of $\mathscr{L}_n$ and $\mathscr{L}_n^*$

Let $f(\mathbf{z})$ ($\mathbf{z} \in R^{p+q}$), $f_1(\mathbf{x})$ ($\mathbf{x} \in R^q$), and $g_0(\mathbf{y})$ ($\mathbf{y} \in R^p$) be respectively the p.d.f.'s corresponding to the d.f.'s F, F_1, and G_0. Then

$$f(\mathbf{z}) = g_0(\mathbf{y} - \boldsymbol{\beta}_0 - \boldsymbol{\beta}'\mathbf{x})f_1(\mathbf{x}), \qquad \mathbf{z}' = (\mathbf{y}', \mathbf{x}'). \tag{8.2.55}$$

Under $H_0: \boldsymbol{\beta} = \mathbf{0}$, we have

$$f(\mathbf{z}) = f_0(\mathbf{z}) = g_0(\mathbf{y} - \boldsymbol{\beta}_0)f_1(\mathbf{x}). \tag{8.2.56}$$

Thus, the likelihood function for the problem is

$$L_n(\mathbf{Z}_1,\ldots,\mathbf{Z}_n) = \prod_{i=1}^{n} f(\mathbf{Z}_i) = \prod_{i=1}^{n} g_0(\mathbf{Y}_i - \boldsymbol{\beta}_0 - \boldsymbol{\beta}'\mathbf{X}_i) f_1(\mathbf{X}_i), \tag{8.2.57}$$

and a similar expression holds when $\boldsymbol{\beta} = \mathbf{0}$. Assuming the form of g_0, f_1 to be known (and to satisfy the usual regularity conditions for the likelihood-ratio statistic to have the optimal properties—see Section 5.8) and denoting the MLE of $\boldsymbol{\beta}$ by $\hat{\boldsymbol{\beta}}$ (where we take $\boldsymbol{\beta}_0 = \mathbf{0}$), the likelihood-ratio criterion is

$$\lambda_n = \prod_{i=1}^{n} \frac{g_0(\mathbf{Y}_i)}{g_0(\mathbf{Y}_i - \hat{\boldsymbol{\beta}}\mathbf{X}_i)}. \tag{8.2.58}$$

By the same technique as in Section 5.8 or 7.3, we conclude that:

1. Under H_0 in (8.2.2) (i.e., $\boldsymbol{\beta} = \mathbf{0}$), $-2\log\lambda_n$ has asymptotically the central chi-square distribution with pq DF.
2. Under $\{H_n\}$, it has a noncentral chi-square distribution with pq DF and noncentrality parameter

$$\Delta^* = \sum_{j=1}^{p} \sum_{j'=1}^{p} \sum_{l=1}^{q} \sum_{l'=1}^{q} \gamma_{jl}\gamma_{j'l'}\sigma_{ll'}\mathscr{I}_{jj'}, \tag{8.2.59}$$

where

$$((\sigma_{ll'})) = E(\mathbf{X} - E\mathbf{X})(\mathbf{X} - E\mathbf{X})', \tag{8.2.60}$$

$$\mathscr{I}_{jj'} = E\left\{ \frac{\partial \log g_0(\mathbf{Y})}{\partial Y_j} \frac{\partial \log g_0(\mathbf{Y})}{\partial Y_{j'}} \right\}, \quad j, j' = 1,\ldots,p. \tag{8.2.61}$$

Thus, from (8.2.52) and (8.2.58), we conclude that the ARE of $\mathscr{L}_n$ with respect to λ_n is

$$e(\mathscr{L},\lambda) = \frac{\Delta_{\mathscr{L}}}{\Delta^*}, \tag{8.2.62}$$

and it depends on $\boldsymbol{\gamma}$, $\mathbf{T}$, $\mathbf{T}^*$, $\boldsymbol{\Sigma}$, and $\mathscr{I}$. Since both $\Delta_{\mathscr{L}}$ and Δ^* are quadratic forms in the rolled-out $\boldsymbol{\gamma}$ with discriminants $\mathbf{T}^{-1} \otimes \mathbf{T}^*$ and $\mathscr{I} \otimes \boldsymbol{\Sigma}$, respectively, by the Courant theorem,

$$\mathrm{Ch}_p(\mathbf{T}^{-1}\mathcal{J}^{-1})\mathrm{Ch}_p(\mathbf{\Sigma}^{-1}\mathbf{T}^*) = \inf_{\gamma} e(\mathcal{L}, \lambda) \le e(\mathcal{L}, \lambda)$$

$$\le \sup_{\gamma} e(\mathcal{L}, \lambda) = \mathrm{Ch}_1(\mathbf{T}^{-1}\mathcal{J}^{-1})\mathrm{Ch}_1(\mathbf{\Sigma}^{-1}\mathbf{T}^*), \quad (8.2.63)$$

so that the bounds are explicit functions of $\mathbf{T}$, $\mathbf{T}^*$, $\mathcal{J}$ and $\mathbf{\Sigma}$. In particular, if $\mathbf{T} = \mathcal{J}^{-1}$ and $\mathbf{T}^* = \mathbf{\Sigma}$, then the left- and right-hand sides of (8.2.63) are both equal to 1, so that $e(\mathcal{L}, \lambda) = 1$ for all γ. For $\mathbf{T}^* = \mathbf{\Sigma}$, we need $\phi_l^*(F_{1[l]}(x))$ linear in x (a.e.) for all $l = 1, \dots, q$ while for $T = \mathcal{J}^{-1}$, the same conditions as in (5.8.13)–(5.8.17) suffice. In particular, the above holds if F is multinormal and we use normal scores for all the $p + q$ variates; hence, for multinormal F, the normal-score test has asymptotically the same optimality properties as of the likelihood-ratio test.

In a similar way, for $\mathcal{L}_n^*$, the asymptotic distribution under $\{H_n\}$ has the noncentrality parameter $\Delta_{\mathcal{L}^*}$ in (8.2.36), so that

$$e(\mathcal{L}^*, \lambda) = \frac{\Delta_{\mathcal{L}^*}}{\Delta^*}, \quad (8.2.64)$$

and for all γ,

$$\mathrm{Ch}_p(\mathbf{T}^{-1}\mathcal{J}^{-1}) \le e(\mathcal{L}^*, \lambda^*) \le \mathrm{Ch}_1(\mathbf{T}^{-1}\mathcal{J}^{-1}), \quad (8.2.65)$$

so that for asymptotic optimality, we again need $\mathbf{T} = \mathcal{J}^{-1}$. Finally,

$$e(\mathcal{L}, \mathcal{L}^*) = \frac{\Delta_{\mathcal{L}}}{\Delta_{\mathcal{L}^*}} \quad (8.2.66)$$

and for all γ,

$$\mathrm{Ch}_p(\mathbf{\Sigma}^{-1}\mathbf{T}^*) \le e(\mathcal{L}, \mathcal{L}^*) \le \mathrm{Ch}_1(\mathbf{\Sigma}^{-1}\mathbf{T}^*), \quad (8.2.67)$$

so that the ARE bounds depend on $\mathbf{\Sigma}$ and $\mathbf{T}^*$.

8.3. RANK TESTS FOR MIXED MODELS

In Chapters 5 and 7, we have considered rank tests for linear models involving nonstochastic predictors; in Section 8.2 we considered the case of stochastic predictors. In a variety of problems (particularly in the context of analysis of covariance), one encounters a *mixed model* where some of the predictors are nonstochastic and some are stochastic. In this section, we shall consider such models and study appropriate rank tests.

As in Section 8.2, let $\{\mathbf{Z}_i = (\mathbf{Y}_i', \mathbf{X}_i')', \ i \geq 1\}$ be independent random vectors with continuous $(p + q)$-variate d.f.'s $\{F_i(\mathbf{z}), \mathbf{z} \in R^{p+q}, \ i \geq 1\}$. We assume that the $\mathbf{X}_i$ are i.i.d.r.v.'s with an unspecified (q-variate) continuous d.f. $F^*(\mathbf{x})$, $\mathbf{x} \in R^q$, and denote by $G_i(\mathbf{y}|\mathbf{x})$ the conditional d.f. of $\mathbf{Y}_i$ given $\mathbf{X}_i = \mathbf{x}$, for $\mathbf{y} \in R^p$, $\mathbf{x} \in R^q$, $i \geq 1$. Then, we assume that

$$G_i(\mathbf{y}|\mathbf{x}) \equiv G_0(\mathbf{y} - \boldsymbol{\beta}_0 - \boldsymbol{\beta}\mathbf{c}_i|\mathbf{x}), \qquad i \geq 1, \tag{8.3.1}$$

where $\boldsymbol{\beta}_0$ $(p \times 1)$ and $\boldsymbol{\beta}$ $(p \times t)$ are unknown parameters, G_0 is some unspecified d.f., and $\mathbf{c}_i = (c_{i1}, \dots, c_{it})'$, $i \geq 1$, are known constants (vectors). In the classical normal-theory model, we let

$$G_i(\mathbf{Y}_i|\mathbf{X}_i) = G_0(\mathbf{Y}_i - \boldsymbol{\beta}_0 - \boldsymbol{\beta}\mathbf{c}_i - \boldsymbol{\gamma}\mathbf{X}_i), \qquad i \geq 1, \tag{8.3.2}$$

where G_0 is a normal d.f. and $\boldsymbol{\gamma}$ is an unknown $p \times q$ matrix, and we desire to test suitable linear hypotheses concerning $\boldsymbol{\beta}$, treating $\boldsymbol{\gamma}$ as a nuisance parameter. In the nonparametric setup, we do not wish to impose the normality of G_0 or the linearity in (8.3.2), and for this reason we proceed through a multivariate approach given below.

For simplicity, we consider first the case of a simple null hypothesis, viz.

$$H_0: \boldsymbol{\beta} = \mathbf{0} \quad \text{vs.} \quad H_1: \boldsymbol{\beta} \neq \mathbf{0}. \tag{8.3.3}$$

We define the ranks R_{ji} and S_{li}, $1 \leq i \leq n$, $1 \leq j \leq p$, $1 \leq l \leq q$, as in (8.2.3) and let

$$L_{njs}^{(1)} = \sum_{i=1}^{n} (c_{is} - \bar{c}_{ns}) a_{nj}(R_{ji}), \qquad 1 \leq j \leq p, \quad 1 \leq s \leq t, \tag{8.3.4}$$

$$L_{nls}^{(2)} = \sum_{i=1}^{n} (c_{is} - \bar{c}_{ns}) b_{nl}(S_{li}), \qquad 1 \leq l \leq q, \quad 1 \leq s \leq t, \tag{8.3.5}$$

where $\bar{c}_{ns} = n^{-1}\sum_{i=1}^{n} c_{is}$, $1 \leq s \leq t$. Also, let m_{njl}, $\nu_{njj'}^{(1)}$ and $\nu_{nll'}^{(2)}$ be defined as in (8.2.6), (8.2.10)–(8.2.12), and let

$$\begin{aligned} \nu_{njl} &= n^{-1} \sum_{i=1}^{n} \left[a_{nj}(R_{ji}) - \bar{a}_{nj}\right]\left[b_{nl}(S_{li}) - \bar{b}_{nl}\right] \\ &= n^{-1} m_{njl}, \qquad 1 \leq j \leq p, \quad 1 \leq l \leq q; \end{aligned} \tag{8.3.6}$$

$$\mathbf{V}_n = \begin{pmatrix} \mathbf{V}_n^{(1)} & \frac{1}{n}\mathbf{M}_n \\ \frac{1}{n}\mathbf{M}_n' & \mathbf{V}_n^{(2)} \end{pmatrix} = \begin{pmatrix} \mathbf{V}_{n11} & \mathbf{V}_{n12} \\ \mathbf{V}_{n21} & \mathbf{V}_{n22} \end{pmatrix}, \quad \text{say.} \tag{8.3.7}$$

Further, let

$$\underset{t\times t}{\mathbf{C}_n} = \sum_{i=1}^{n} (\mathbf{c}_i - \bar{\mathbf{c}}_n)(\mathbf{c}_i - \bar{\mathbf{c}}_n)'. \tag{8.3.8}$$

Then, under H_0 in (8.3.3), the $\mathbf{Z}_i$ are i.i.d.r.v.'s, so that the permutational invariance structure, developed in Section 5.4 for the rank matrix $\mathbf{R}$ also holds for $\begin{pmatrix}\mathbf{R}\\ \mathbf{S}\end{pmatrix}$. Thus, if we roll out the elements in (8.3.4)–(8.3.5) in the form of a $(p+q)\times t$ vector $\mathbf{L}_n$, then under this permutational measure $\mathscr{P}_n$,

$$E_{\mathscr{P}_n}\mathbf{L}_n = \mathbf{0}, \tag{8.3.9}$$

$$E_{\mathscr{P}_n}\mathbf{L}_n\mathbf{L}_n' = (n-1)^{-1}n\mathbf{V}_n \otimes \mathbf{C}_n. \tag{8.3.10}$$

If we let

$$\mathbf{L}_n' = (\underset{pt\times 1}{L_n^{(1)\prime}}, \underset{qt\times 1}{L_n^{(2)\prime}}) \tag{8.3.11}$$

and fit a linear regression of $\mathbf{L}_n^{(1)}$ on $\mathbf{L}_n^{(2)}$ (justifiable by the asymptotic multinormality of $\mathbf{L}_n$ under $\mathscr{P}_n$), then the residual of this regression is

$$\mathbf{L}_n^{*\prime} = \mathbf{L}_n^{(1)} - \mathbf{V}_{n12}\mathbf{V}_{n22}^{-1}\mathbf{L}_n^{(2)}, \tag{8.3.12}$$

and then, rolling out $\mathbf{L}_n^*$ into a $pt \times 1$ vector,

$$E_{\mathscr{P}_n}\mathbf{L}_n^*\mathbf{L}_n^{*\prime} = (n-1)^{-1}n\mathbf{V}_n^* \otimes \mathbf{C}_n, \tag{8.3.13}$$

where

$$\mathbf{V}_n^* = \mathbf{V}_{n11} - \mathbf{V}_{n12}\mathbf{V}_{n22}^{-1}\mathbf{V}_{n21}. \tag{8.3.14}$$

Then, parallel to (5.4.13), we consider the test statistic

$$\mathscr{L}_n^* = \sum_{j=1}^{p}\sum_{j'=1}^{p}\sum_{s=1}^{t}\sum_{s'=1}^{t} \nu_n^{*jj'}c_n^{ss'}L_{njs}^*L_{nj's'}^*$$

$$= \mathbf{L}_n^{*\prime}(\mathbf{V}_n^{*-1} \otimes \mathbf{C}_n^{-1})\mathbf{L}_n^*, \tag{8.3.15}$$

where

$$\left(\left(\nu_n^{*jj'}\right)\right) = \mathbf{V}_n^{*-1} \quad \text{and} \quad \left(\left(c_n^{ss'}\right)\right) = \mathbf{C}_n^{-1}. \tag{8.3.16}$$

The main difference between $\mathscr{L}_n$ in (5.4.13) and $\mathscr{L}_n^*$ in (8.3.15) lies in the fact that whereas the L_{njs} ignore the concomitant variates, the L_{njs}^* in (8.3.12) are adjusted for the concomitant variates and this is also reflected in the replacement of $\mathbf{V}_n^{-1}$ in (5.4.13) by $\mathbf{V}_n^{*-1}$ in (8.3.15).

We may virtually repeat the permutation argument in Section 5.4 to show that $\mathscr{L}_n^*$ is a permutationally distribution-free statistic, and we use large values of $\mathscr{L}_n^*$ as the critical region for rejecting H_0 in (8.3.3). Further, Theorems 5.4.1 and 5.4.2 can be extended directly for the $(p+q)$-variate case (i.e., for $\mathbf{L}_n$ and $\mathbf{V}_n$), and hence the asymptotic multinormality of $\mathbf{L}_n^*$ (under $\mathscr{P}_n$) follows with mean vector $\mathbf{0}$ and dispersion matrix $\mathbf{V}_n^* \otimes \mathbf{C}_n$. Consequently, under $\mathscr{P}_n$, $\mathscr{L}_n^*$ has asymptotically the central chi-square distribution with pt DF. Thus, for large n, we have the following procedure:

$$\text{if} \quad \mathscr{L}_n^* \begin{cases} \geq \chi^2_{pt,\alpha}, & \text{reject } H_0 \text{ in (8.3.3)}, \\ < \chi^2_{pt,\alpha}, & \text{accept } H_0. \end{cases} \tag{8.3.17}$$

For the case where H_0 may not hold, we may proceed as in section 5.5 (replacing the p-variate case by the $(p+q)$-variate case everywhere), and hence, using Lemma 5.5.1 and Theorem 5.5.2, along with our (8.3.12)–(8.3.13) and (8.3.15), we may derive the distribution of $\mathscr{L}_n^*$. Let us consider a sequence $\{H_n\}$ of alternative hypotheses H_n, where

$$H_n\colon \text{(8.3.1) holds for } \boldsymbol{\beta} = n^{-1/2}\boldsymbol{\beta}^0, \tag{8.3.18}$$

and where we assume that

$$n^{-1}\mathbf{C}_n \to \mathbf{C}^* \quad \text{(p.d.)} \qquad \text{as} \quad n \to \infty. \tag{8.3.19}$$

Further, we assume that conditions (a), (b), (c) of Section 5.5 hold for $F_{[j]} = G_{[j]}$, $1 \leq j \leq p$, and define $B_j = B(G_{[j]}, \phi_j)$ as in (5.5.15) for $j = 1, \dots, p$. Also, note that

$$\mathbf{V}_n \xrightarrow{P} \boldsymbol{\nu} = \begin{pmatrix} \boldsymbol{\nu}_{11} & \boldsymbol{\nu}_{12} \\ \boldsymbol{\nu}_{21} & \boldsymbol{\nu}_{22} \end{pmatrix} \quad \text{(p.d.)}, \tag{8.3.20}$$

and let

$$\boldsymbol{\nu}^* = \boldsymbol{\nu}_{11} - \boldsymbol{\nu}_{12}\boldsymbol{\nu}_{22}^{-1}\boldsymbol{\nu}_{21}. \tag{8.3.21}$$

Let then

$$\mathbf{T}^* = ((\tau_{jj'}^*)) = \left(\left(\frac{\nu_{jj'}^*}{B_j B_{j'}}\right)\right)_{j,j'=1,\dots,p}, \tag{8.3.22}$$

$$\mathbf{T}^{*-1} = ((\tau^{*jj'})). \tag{8.3.23}$$

Then, as a direct generalization of Theorem 5.5.2, we obtain that under $\{H_n\}$ in (8.3.18) and the regularity conditions assumed above, $\mathscr{L}_n^*$ has asymptotically a noncentral chi-square distribution with pt DF and noncentrality parameter

$$\Delta_{\mathscr{L}^*} = \sum_{j=1}^{p} \sum_{j'=1}^{p} \sum_{s=1}^{t} \sum_{s'=1}^{t} \beta_{js}^0 \beta_{j's'}^0 c_{ss'}^* \tau^{*jj'}. \tag{8.3.24}$$

As a result, the results in Section 5.7 on asymptotic relative efficiency for $\mathscr{L}_n$ also apply to $\mathscr{L}_n^*$, provided we replace, in (5.7.4) and elsewhere, $\mathbf{T}$ by $\mathbf{T}^*$ and $\boldsymbol{\Sigma}$ by $\boldsymbol{\Sigma}^*$, where $\boldsymbol{\Sigma}^* = \boldsymbol{\Sigma}_{11} - \boldsymbol{\Sigma}_{12}\boldsymbol{\Sigma}_{22}^{-1}\boldsymbol{\Sigma}_{21}$.

For the special case of the one-way analysis-of-covariance problem [where the c_{ij} are 0 or 1 for $1 \le j \le t$ ($\Sigma_{j=1}^t c_{ij} = 1$), $i \ge 1$], $\mathscr{L}_n^*$ has been considered in detail by Puri and Sen (1969a, b), and a special case by Quade (1967); a more general case is treated in a subsequent paper by Sen and Puri (1970).

Let us now consider the case of a composite null hypothesis where we let

$\boldsymbol{\beta} = (\underset{p\times t_1}{\boldsymbol{\beta}_1}, \underset{p\times t_2}{\boldsymbol{\beta}_2})$ and frame

$$H_0: \boldsymbol{\beta}_2 = \mathbf{0} \quad \text{vs.} \quad H_1: \boldsymbol{\beta}_2 \neq \mathbf{0}, \tag{8.3.25}$$

treating $\boldsymbol{\beta}_1$ as a nuisance parameter. For this, we proceed as in Section 7.4 with modifications as in (8.3.12)–(8.3.16).

Let $\mathbf{B} = (\mathbf{b}_1', \dots, \mathbf{b}_{p+q}')'$ be a $(p+q) \times t$ matrix of real constants, and define

$$\mathbf{Z}_i(\mathbf{B}) = \mathbf{Z}_i - \mathbf{B}\mathbf{C}_i, \qquad 1 \le i \le n, \quad \mathbf{B} \in R^{(p+q)t}. \tag{8.3.26}$$

Also, let $R_{ji}(\mathbf{b}_j)$ [$= R_{ji}(\mathbf{B})$] be the rank of $Y_{ji} - \mathbf{b}_j\mathbf{c}_i$ among $Y_{j1} - \mathbf{b}_j\mathbf{c}_1, \dots, Y_{jn} - \mathbf{b}_j\mathbf{c}_n$ for $1 \le i \le n$, $1 \le j \le p$, and let $S_{li}(\mathbf{b}_{l+p})$ [$= \mathbf{S}_{li}(\mathbf{B})$] be the rank of $X_{li} - \mathbf{b}_{l+p}\mathbf{c}_i$ among $X_{l1} - \mathbf{b}_{l+p}\mathbf{c}_1, \dots, X_{ln} - \mathbf{b}_{l+p}\mathbf{c}_n$ for $1 \le i \le n$, $1 \le l \le q$. Then, in (8.3.4) and (8.3.5), we replace R_{ji} and S_{li} by $R_{ji}(\mathbf{b}_j)$ and

$S_{li}(\mathbf{b}_{l+p})$, respectively, and denote the resulting statistics by $\mathbf{L}_{nj}^{(1)}(\mathbf{b}_j)$ and $\mathbf{L}_{nl}^{(2)}(\mathbf{b}_{l+p})$, for $1 \le j \le p$, $1 \le l \le q$. Also, partition $\mathbf{B}$ as $(\mathbf{B}_1, \mathbf{B}_2)$ where $\mathbf{B}_j$ is $(p+q) \times t_j$, $j = 1, 2$, and similarly, let $\mathbf{b}_j = (\mathbf{b}_j^{(1)}, \mathbf{b}_j^{(2)})$, $1 \le j \le p+q$, where $b_j^{(r)}$ is $1 \times t_r$, $1 \le j \le p+q$, $r = 1, 2$. Let then

$$\mathbf{L}_{n(1)}(\mathbf{B}_1, \mathbf{0}) = \begin{pmatrix} \mathbf{L}_{n(1)}^{(1)}(\mathbf{B}_1, \mathbf{0}) \\ \mathbf{L}_{n(1)}^{(2)}(\mathbf{B}_1, \mathbf{0}) \end{pmatrix} \begin{matrix} {}_{p \times t_1} \\ {}_{q \times t_1} \end{matrix}, \tag{8.3.27}$$

where $L_{n(1)}^{(1)}(\mathbf{B}_1, \mathbf{0}) = ((\mathbf{L}_{nj}^{(1)}(\mathbf{b}_j^{(1)}, \mathbf{0}),\ 1 \le j \le p))$ and $L_{n(1)}^{(2)}(\mathbf{B}_1, \mathbf{0}) = ((L_{nl}^{(2)}(b_{l+p}^{(1)}, \mathbf{0}),\ 1 \le l \le q))$. For each j $(= 1, \dots, p)$, proceeding as in (7.3.13)–(7.3.14), we define $\hat{\boldsymbol{\beta}}_{j(1),n}^{(1)}$ and for each l $(= 1, \dots, q)$, in the same manner, we define $\hat{\boldsymbol{\beta}}_{l(1),n}^{(2)}$. Let then

$$\hat{\boldsymbol{\beta}}_{1,n}' = \left(\hat{\boldsymbol{\beta}}_{1(1),n}^{(1)\prime}, \dots, \hat{\beta}_{p(1),n}^{(1)\prime}, \hat{\beta}_{1(1),n}^{(2)\prime}, \dots, \hat{\beta}_{q(1),n}^{(2)\prime}\right) \tag{8.3.28}$$

Finally, let

$$\hat{\mathbf{L}}_{n(2)} = \begin{pmatrix} \mathbf{L}_{n(2)}^{(1)}(\hat{\boldsymbol{\beta}}_{1,n}, \mathbf{0}) \\ \mathbf{L}_{n(2)}^{(2)}(\hat{\boldsymbol{\beta}}_{1,n}, \mathbf{0}) \end{pmatrix} \begin{matrix} {}_{p \times t_2} \\ {}_{q \times t_2} \end{matrix} = \begin{pmatrix} \hat{\mathbf{L}}_{n(2)}^{(1)} \\ \hat{\mathbf{L}}_{n(2)}^{(2)} \end{pmatrix}, \tag{8.3.29}$$

where $\mathbf{L}_{n(2)}^{(1)}(\mathbf{B}) = ((L_{njs}^{(1)}(\mathbf{B})))_{j=1,\dots,p,\ s=t_1+1,\dots,t}$, $\mathbf{L}_{n(2)}^{(2)}(\mathbf{B}) = ((L_{nls}^{(2)}(\mathbf{B})))_{l=1,\dots,q,\ s=t_1+1,\dots,t}$, and $\mathbf{B} \in R^{(p+q)t}$. Then the test is based on the aligned rank statistics in (8.3.29). To formulate it, we let $\hat{R}_{ji} = R_{ji}(\hat{\boldsymbol{\beta}}_{j(1),n}^{(1)}, \mathbf{0})$, $1 \le j \le p$, and $\hat{S}_{li} = S_{li}(\hat{\boldsymbol{\beta}}_{(1),n}^{(2)}, \mathbf{0})$, $1 \le l \le q$, for $i = 1, \dots, n$, and in (8.3.6)–(8.3.7), on replacing the original ranks by these aligned ranks, we denote the resulting matrix by

$$\hat{\mathbf{V}}_n = \begin{pmatrix} \hat{\mathbf{V}}_{n11} & \hat{\mathbf{V}}_{n12} \\ \hat{\mathbf{V}}_{n21} & \hat{\mathbf{V}}_{n22} \end{pmatrix}.$$

Further, let

$$\hat{\mathbf{V}}_n^* = \hat{\mathbf{V}}_{n11} - \hat{\mathbf{V}}_{n12}\hat{\mathbf{V}}_{n22}^{-1}\hat{\mathbf{V}}_{n21}, \tag{8.3.30}$$

$$\hat{\mathbf{L}}_n^{*\prime} = \hat{\mathbf{L}}_{n(2)}^{(1)} - \hat{\mathbf{V}}_{n12}\hat{\mathbf{V}}_{22}^{-1}\hat{\mathbf{L}}_{n(2)}^{(2)}. \tag{8.3.31}$$

Then we proceed as in (7.4.15) through (7.4.26) and let

$$\hat{\mathbf{H}}_n^* = \left(\left(\hat{L}_{njs}^*\hat{L}_{nj's'}^*\right)\right)_{j,j'=1,\dots,p;\ s,s'=t_1+1,\dots,t}, \tag{8.3.32}$$

$$\hat{\mathbf{G}}_n^* = \hat{\mathbf{V}}_n^* \otimes \mathbf{C}_n^0, \tag{8.3.33}$$

where

$$\mathbf{C}_n^0 = \mathbf{C}_{n22} - \mathbf{C}_{n21}\mathbf{C}_{n11}^{-1}\mathbf{C}_{n12}, \qquad \mathbf{C}_n = \begin{pmatrix} \mathbf{C}_{n11} & \mathbf{C}_{n12} \\ \mathbf{C}_{n21} & \mathbf{C}_{n22} \end{pmatrix}. \tag{8.3.34}$$

Then the proposed test statistic is

$$\begin{aligned} \hat{\mathscr{L}}_n^* &= T_r\left(\hat{\mathbf{H}}_n^*\hat{\mathbf{G}}_n^{*-1}\right) \\ &= \sum_{j=1}^{p}\sum_{j'=1}^{p}\sum_{s=t_1+1}^{t}\sum_{s'=t_1+1}^{t} \hat{L}_{njs}^*\hat{L}_{nj's'}^*\hat{\nu}_n^{*jj'}C_n^{0ss'} \end{aligned} \tag{8.3.35}$$

where

$$\hat{\mathbf{V}}_n^{*-1} = \left(\left(\hat{\nu}_n^{*jj'}\right)\right) \quad \text{and} \quad \left(\mathbf{C}_n^0\right)^{-1} = \left(\left(C_n^{0ss'}\right)\right). \tag{8.3.36}$$

As in Section 7.4, $\hat{\mathscr{L}}_n^*$ will be an asymptotically distribution-free statistic. We may virtually repeat Lemmas 7.4.1, 7.4.2, and 7.4.3 for the $(p + q)$-variate case, and then using (8.3.30) through (8.3.36) obtain the following: Under $H^0: \boldsymbol{\beta}_2 = \mathbf{0}$ and the regularity conditions of Section 7.4, $\hat{\mathscr{L}}_n^*$ has asymptotically the central chi-square distribution with pt_2 DF. Hence, for testing $H_0: \boldsymbol{\beta}_2 = \mathbf{0}$ vs. $H_1: \boldsymbol{\beta}_2 \neq \mathbf{0}$, we may proceed as in (8.3.17) with $\chi^2_{pt,\alpha}$ being replaced by $\chi^2_{pt_2,\alpha}$.

For the study of the asymptotic power properties of the test based on $\hat{\mathscr{L}}_n^*$, we may proceed as in (7.4.40) through (7.4.44): parallel to (7.4.40), we assume that under $\{H_n\}$, (8.3.1) holds for $\boldsymbol{\beta}_2 = n^{-1/2}\boldsymbol{\delta}$ for some $\boldsymbol{\delta} \in R^{pt_2}$, proceed to prove the contiguity as in the discussion after (7.4.41), and thereby obtain the theory of the non-null distribution from that in the null case. Parallel to Theorem 7.4.5, we obtain that under $\{H_n\}$, $\mathscr{L}_n^*$ will have asymptotically a noncentral chi-square distribution with pt_2 DF and noncentrality parameter

$$\Delta_{\mathscr{L}^*} = \sum_{j=1}^{p}\sum_{j'=1}^{p}\sum_{s=t_1+1}^{t}\sum_{s'=t_1+1}^{t} \delta_{js}\delta_{j's'}\tau^{*jj'}c_{ss'}^{0*}, \tag{8.3.37}$$

where the $\tau^{*jj'}$ and $c_{ss'}^{0*}$ are defined in (8.3.19) and (8.3.22) with $\mathbf{C}^{0*} = \mathbf{C}_{22}^* - \mathbf{C}_{21}^*\mathbf{C}_{11}^{*-1}\mathbf{C}_{12}^*$.

The ARE results in (7.4.55) also extend to this case; the only changes are that $\mathbf{T}$ and $\boldsymbol{\Sigma}$ are replaced by $\mathbf{T}^*$ and $\boldsymbol{\Sigma}^*$ respectively. As a result, (7.4.56) holds also with $\mathbf{T}$ and $\boldsymbol{\Sigma}$ replaced by $\mathbf{T}^*$ and $\boldsymbol{\Sigma}^*$ respectively.

8.4. SOME NONPARAMETRIC PROCEDURES IN LONGITUDINAL STUDIES

In Chapters 5 and 7, we have considered general linear models where we have n independent p-vectors $\mathbf{X}_1, \ldots, \mathbf{X}_n$ and

$$\mathbf{X}_i = \boldsymbol{\beta}_0 + \boldsymbol{\beta}\mathbf{c}_i + \mathbf{e}_i, \qquad i \geq 1, \tag{8.4.1}$$

where $\boldsymbol{\beta}_0$, $\boldsymbol{\beta}$ are unknown parameters, the $\mathbf{c}_i$ are specified vectors, and the $\mathbf{e}_i$ are i.i.d.r.v.'s with a p-variate continuous d.f. F. In a typical longitudinal study, the p components of $\mathbf{X}_i$ are the measurements on the same individual at possibly different time points. Thus, these components are generally correlated. Nevertheless, their realization can be described by means of a deterministic part and a stochastic part—the latter mainly explains the dependence structure of the components of $\mathbf{e}_i$, while the former, in many cases, can be described by some simple model. For example, let $\mathbf{X}_i$ have p components $X_{i1}, \ldots, X_{ip}$, where X_{ij} stands for the blood pressure of the ith person at the end of j weeks of treatment for $j = 1, \ldots, p$, $i = 1, \ldots, n$. In such a case, it may be reasonable to assume that for every $i(= 1, \ldots, n)$,

$$X_{ij} = \alpha_0 + \alpha_1 b_j + \cdots + \alpha_q b_j^{q-1} + e_{ij}, \qquad 1 \leq j \leq p, \tag{8.4.2}$$

where q $(\leq p - 1)$ is usually small compared to p (in many situations, we may take $q = 2$ or 3). We may also consider a more general model where we allow the α_j's in (8.4.2) to depend on i (e.g., sex or age of the patient, occupational factor, etc.). In this manner, we may conceive of an index set $I = \{\mathbf{i} = (i_1, \ldots, i_m) : 1 \leq i_j \leq k_j\ (\geq 1), 1 \leq j \leq m\}$ and let $k = k_1 \cdots k_m$. Then the cardinality of I is equal to k. Also, let $T = \{(t_1, \ldots, t_p) : t_1 < \cdots < t_p\}$ be a set of p (≥ 1) distinct time points; for each $\mathbf{i} \in I$ we may conceive of $n(\mathbf{i})$ observations, which lead us to a totality of $n = \sum_{\mathbf{i} \in I} n(\mathbf{i})$ observations, for which (8.4.2) can be rewritten as

$$\mathbf{X}_i = \boldsymbol{\beta}_0 + \mathbf{G}\boldsymbol{\theta}\mathbf{c}_i + \mathbf{e}_i, \qquad 1 \leq i \leq n, \tag{8.4.3}$$

where $\boldsymbol{\beta}_0$ is a $p \times 1$ vector (unknown), $\mathbf{G}$ is a specified $p \times q$ matrix of known constants (depending on $\mathbf{T}$), $\boldsymbol{\theta}$ is a $q \times t$ matrix of unknown parameters, and the $\mathbf{c}_i$ are known t-vectors (depending on the factors in I). In this setup, we are primarily interested in testing linear hypotheses on $\boldsymbol{\theta}$. Here $q \leq p$, and without any loss of generality we may assume that $\mathbf{G}$ has the rank q.

In the parametric case, one assumes that the $\mathbf{e}_i$ are i.i.d.r.v.'s with a p-variate normal d.f. with mean $\mathbf{0}$ and a p.d. dispersion matrix $\boldsymbol{\Sigma}$. Linear statistical inference on $\boldsymbol{\theta}$ has then been studied by Rao (1958, 1967), Potthoff and Roy (1964), Grizzle and Allen (1969), and others. Nonparametric analogues of these have been considered by Ghosh, Grizzle, and Sen (1973) and Sen (1973b), and these will be presented here in a more unified way. The basic idea underlying these growth-curve models is the reduction of the dimensionality of $\boldsymbol{\beta}$ to $\boldsymbol{\theta}$, and this is accompanied by an increase in the precision of the statistical analysis based on such models.

Keeping (8.4.3) in mind, we conceive of transformations

$$\mathbf{X}_i \to \mathbf{Y}_i = (\mathbf{G}'\mathbf{G})^{-1}\mathbf{G}'\mathbf{X}_i, \qquad 1 \le i \le n, \tag{8.4.4}$$

so that the $\mathbf{Y}_i$ are q-vectors, following the model

$$\begin{aligned}\mathbf{Y}_i &= (\mathbf{G}'\mathbf{G})^{-1}\mathbf{G}'\boldsymbol{\beta}_0 + \boldsymbol{\theta}\mathbf{c}_i + (\mathbf{G}'\mathbf{G})^{-1}\mathbf{G}'\mathbf{e}_i \\ &= \boldsymbol{\theta}_0 + \boldsymbol{\theta}\mathbf{c}_i + \mathbf{e}_i^*, \quad \text{say}, \qquad 1 \le i \le n,\end{aligned} \tag{8.4.5}$$

where $\boldsymbol{\theta}_0$ is a q-vector and the $\mathbf{e}_i^*$ are also i.i.d.r.v.'s with a q-variate distribution F^*, derivable from that of $\mathbf{e}_i$. If, then, we want to test for

$$H_0 : \boldsymbol{\theta} = \mathbf{0} \quad \text{vs.} \quad H_1 : \boldsymbol{\theta} \ne \mathbf{0}, \tag{8.4.6}$$

or if we partition $\boldsymbol{\theta}$ as $(\boldsymbol{\theta}_1, \boldsymbol{\theta}_2)$ and want to test

$$H_0^{(1)} : \boldsymbol{\theta}_1 = \mathbf{0} \quad \text{vs.} \quad H_1^{(1)} : \boldsymbol{\theta}_1 \ne \mathbf{0}, \tag{8.4.7}$$

treating $\boldsymbol{\theta}_2$ as a nuisance parameter, then we have no problem in applying the theory developed in Chapters 5 and 7 and using rank tests based on the $\mathbf{Y}_i$, $1 \le i \le n$. By virtue of the dimension reduction in (8.4.4)–(8.4.5), our rank procedure will be based on q-dimensional vectors and usually will lead to some increase in the power of the test. This way, we have a nonparametric analogue of the parametric procedures suggested by Potthoff and Roy (1964).

In the parametric case, too, Rao (1965) has pointed out that an obvious drawback of the transformation in (8.4.4) is that it throws away some information in this dimension-reduction process, and that this loss can be recovered by employing a multivariate analysis-of-covariance approach. In the nonparametric case too, by virtue of our results in Section 8.3, we are in a position to apply the multivariate analysis-of-covariance model along with (8.4.4) to gain additional information, not acquired in the reduction process in (8.4.4). Consider the transformation

$$\mathbf{X}_i \to \mathbf{Z}_i = \begin{pmatrix} \mathbf{Y}_i \\ \mathbf{Y}_i^* \end{pmatrix} \begin{matrix} q\times 1 \\ (p-q)\times 1 \end{matrix}, \tag{8.4.8}$$

where

$$\mathbf{Y}_i = (\mathbf{G}'\mathbf{G})^{-1}\mathbf{G}'\mathbf{X}_i, \qquad \mathbf{Y}_i^* = \mathbf{H}\mathbf{X}_i, \tag{8.4.9}$$

and we choose $\mathbf{H}$ of order $(p-1) \times p$ in such a way that

$$\mathbf{H}\mathbf{G} = \mathbf{0}. \tag{8.4.10}$$

If we now denote

$$\begin{pmatrix} (\mathbf{G}'\mathbf{G})^{-1}G' \\ \mathbf{H} \end{pmatrix} \boldsymbol{\beta}_0 = \begin{pmatrix} \boldsymbol{\theta}_0 \\ \boldsymbol{\theta}_0^* \end{pmatrix}, \qquad \boldsymbol{\theta}^* = \begin{pmatrix} \boldsymbol{\theta} \\ \mathbf{0} \end{pmatrix}, \qquad \mathbf{e}_i^{**} = \begin{pmatrix} (\mathbf{G}'\mathbf{G})^{-1}\mathbf{G}' \\ \mathbf{H} \end{pmatrix} \mathbf{e}_i,$$

$$i \geq 1, \tag{8.4.11}$$

then, by virtue of (8.4.3), (8.4.8), (8.4.9), (8.4.10), and (8.4.11), we have

$$\mathbf{Z}_i = \begin{pmatrix} \boldsymbol{\theta}_0 \\ \boldsymbol{\theta}_0^* \end{pmatrix} + \boldsymbol{\theta}^*\mathbf{c}_i + \mathbf{e}_i^{**}, \qquad \mathbf{e}_i^{**} = \begin{pmatrix} \mathbf{e}_i^* \\ \mathbf{e}_i^{0*} \end{pmatrix}, \qquad i \geq 1, \tag{8.4.12}$$

where the lower block of $\boldsymbol{\theta}^*$ is a null matrix and the $\mathbf{e}_i^*$ are still i.i.d.r.v.'s. Further, $\mathbf{Z}_i' = (\mathbf{Y}_i', \mathbf{Y}_i^{*\prime})$, where

$$\mathbf{Y}_i^* = \boldsymbol{\theta}_0^* + \mathbf{e}_i^{0*}, \qquad i \geq 1. \tag{8.4.13}$$

We can now compare (8.4.12)–(8.4.13) with the basic model in (8.3.1) and and argue that the $\mathbf{Y}_i^*$ can be treated as concomitant vectors and the $\mathbf{Y}_i$ as the primary vectors. Thus the rank-based procedures developed in Section 8.3 are all applicable here.

One of the common drawbacks of both the analysis-of-variance and analysis-of-covariance approaches is the arbitrariness of the transformations $\mathbf{X}_i \to \mathbf{Y}_i$ and $\mathbf{X}_i \to \mathbf{Z}_i$. This drawback is particularly felt for the rank approach, where a coordinatewise ranking is made and different choices of $(\mathbf{G}'\mathbf{G})^{-1}\mathbf{G}'$ and $\mathbf{H}$ in (8.4.9) may result in somewhat different conclusions in some situations. However, in many typical problems, $\mathbf{G}$ can be defined in a natural way and thus the arbitrariness be removed to a greater extent.

The choice of q in (8.4.3) is important. An underestimation of q may result in an incorrect specification of the model, while an overestimation may increase the dimension of the reduced $\mathbf{Y}_i$ without contributing much to the statistical information. Comparison of two such rank tests based on two possible different values of q may involve central (under the null hypothesis) and noncentral chi-square variates with possibly different noncentrality parameters and different numbers of DF. For example, suppose we want to compare two groups of patients (men and women) with respect to their responses (repeated measurements) to some drugs. If we assume that the response can be described by a second-degree polynomial (in time) plus errors, we may have a 3-variate rank analysis-of-variance test, where the rank statistic has asymptotically a chi-square d.f. with 3 DF. On the other hand, if we fit a cubic polynomial, the resulting DF will be 4. Suppose now that the actual degree of the polynomial is 2. Then the two test statistics will have asymptotically chi-square distributions with 3 and 4 DF, while they may have different noncentrality parameters too. In such a case, the ratio of the two noncentrality parameters, as employed in the study of the Pitman ARE, fails to provide a true comparison of their relative performance, and some alternative measures of ARE are desired. Some of these are studied by Woolson and Sen (1974). We treat a few of these measures in exercises at the end of this chapter.

8.5. RANK-ORDER TESTS FOR GROUPED DATA

As in Chapters 5, 6, and 7, we consider a sequence of independent r.v.'s $\{X_i, i \geq 1\}$ where X_i has an absolutely continuous d.f. $F_i(x)$, $i \geq 1$, with

$$F_i(x) = F(x - \boldsymbol{\beta}'\mathbf{c}_i), \qquad x \in R, \quad i \geq 1, \tag{8.5.1}$$

$\boldsymbol{\beta}$ is an unknown parameter, the $\mathbf{c}_i$ are specified constants (vectors), and the form of F may not be known. However, in the current context, the X_i are not observable; the observable random variables X_i^* are defined in terms of a set $J = \{j = 0, \pm 1, \pm 2, \dots\}$ of non-overlapping class intervals

$$I_j = \{a_j < x \leq a_{j+1}\}, \qquad j \in J \tag{8.5.2}$$

(where $-\infty < \cdots < a_{-1} < a_0 < a_1 < \cdots < \infty$) in the following way:

$$X_i^* = \sum_{j \in J} I_j Z_{ij}, \qquad i \geq 1, \tag{8.5.3}$$

where for every $i \geq 1$ and $j \in J$,

$$Z_{ij} = \begin{cases} 1, & X_i \in I_j, \\ 0, & \text{otherwise.} \end{cases} \tag{8.5.4}$$

We may remark that in almost all real-life problems, data are essentially grouped. For example, even if the underlying variate is continuous (e.g., height or weight of a man), the process of recording involves a rounding off (e.g., to the nearest inch or pound), resulting in a set of class intervals. Also, in many problems, data relate to some ordered categories, where the end points $\{a_j\}$ may even be unspecified. In such a case, all we know is the particular class interval (or category) where X_i belongs, and this is exactly how X_i^* is defined in (8.5.3). Our primary concern is to develop suitable statistical inference procedures for testing $H_0: \boldsymbol{\beta} = \mathbf{0}$ (or other similar null hypotheses) when the X_i^* are available. In passing, we may remark that for the X_i^*, ties may no longer be neglected in probability, and the probabilities of the different cells $\{I_j\}$ depend on the underlying F. This makes the distributions of the usual rank statistics (even if adjusted for ties) dependent on the underlying F, and hence these statistics are not generally distribution-free. Nevertheless, the basic permutational invariance structure, discussed in Chapter 5, can be adapted in such a case (as has been done by Sen, 1967, and Ghosh, 1973a, b), and this enables us to develop some conditionally distribution-free tests based on ranks of the X_i^*. These will be studied here.

To motivate the rank tests, we consider first the simplest case, where in (8.5.1) both β and c_i are scalar quantities. Suppose that we want to test for

$$H_0: \beta = 0 \quad \text{vs.} \quad H_1: \beta > 0 \text{ or } H_2: \beta \neq 0. \tag{8.5.5}$$

If the d.f. F possesses an absolutely continuous pdf f with a finite Fisher information $I(f) = \int (f'/f)^2\, dF$, then one can develop a locally most powerful test for H_0 vs. H_1. Let

$$F_j = F(a_j), \quad P_j = F_{j+1} - F_j = P(I_j), \quad j \in J, \tag{8.5.6}$$

$$\psi(u) = -f'(F^{-1}(u))/f(F^{-1}(u)), \qquad 0 < u < 1 \tag{8.5.7}$$

$$\Delta_j^* = \frac{1}{P_j} \int_{F_j}^{F_{j+1}} \psi(u)\, du = P_j^{-1}\{f(a_j) - f(a_{j+1})\}, \quad j \in J. \tag{8.5.8}$$

Then the likelihood function for $X_1^*, \ldots, X_n^*$ [under (8.5.1)] is

$$\prod_{i=1}^{n} \left\{ \prod_{j \in J} \left[F(a_{j+1} - \beta c_i) - F(a_j - \beta c_i) \right]^{Z_{ij}} \right\}, \tag{8.5.9}$$

where the Z_{ij} are defined by (8.5.4). Thus, for testing $H_0: \beta = 0$ vs. $H_1: \beta > 0$, the log-likelihood ratio statistic is obtainable from (8.5.9) as

$$\log L_n = \sum_{i=1}^{n} \sum_{j \in J} Z_{ij} \log\left\{ P_j^{-1}\left[F(a_{j+1} - \beta c_i) - F(a_j - \beta c_i) \right] \right\}. \tag{8.5.10}$$

Consider now the family of alternatives $H_\gamma = \{H_1 : \beta > 0,\ 0 < \beta \le \gamma\}$. Then for small γ, by direct expansion of F in (8.5.10), we obtain that

$$\log L_n = \beta \sum_{i=1}^{n} \sum_{j \in J} Z_{ij} c_i \Delta_j^* + o(\beta) \qquad \forall 0 < \beta \le \gamma. \tag{8.5.11}$$

Thus, for testing $H_0 : \beta = 0$ vs. H_γ, a locally most powerful test is based on the test statistic

$$T_n = \sum_{i=1}^{n} c_i \left(\sum_{j \in J} \Delta_j^* Z_{ij} \right), \tag{8.5.12}$$

rejecting H_0 for large values of T_n.

In the nonparametric case, F and hence $\psi(u)$, as well as the Δ_j^*, are not known. However, keeping T_n in mind, we proceed to construct some similar statistics based on the sample d.f. Note that the observed cell frequencies are

$$n_j = \sum_{i=1}^{n} Z_{ij}, \quad F_{nj+1} = \frac{1}{n} \sum_{s \le j} n_s, \qquad j \in J. \tag{8.5.13}$$

Also, let

$$P_{nj} = \frac{n_j}{n} = F_{nj+1} - F_{nj}, \qquad j \in J. \tag{8.5.14}$$

Further, let $\phi(u)$, $0 < u < 1$, be a square-integrable function, and define

$$\Delta_{nj} = \begin{cases} P_{nj}^{-1} \int_{F_{nj}}^{F_{nj+1}} \phi(u)\, du, & P_{nj} > 0 \\ \phi(F_{nj}), & P_{nj} = 0 \end{cases}, \qquad j \in J. \tag{8.5.15}$$

Thus, Δ_{nj} is analogous to (8.5.8), where we replace F_j by F_{nj} and $\psi(u)$ by $\phi(u)$. Then, analogous to T_n in (8.5.12), we consider the statistic

$$S_n = \sum_{i=1}^{n} c_i \left(\sum_{j \in J} \Delta_{nj} Z_{ij} \right). \tag{8.5.16}$$

Note that the Δ_{nj} are also random variables, and hence, unlike T_n, S_n may no longer be a linear combination of independent r.v.'s. Moreover, the joint distribution of the Δ_{nj} and the Z_{ij} generally depends on the underlying F, and hence S_n is usually not genuinely distribution-free under $H_0: \beta = 0$. However, under H_0, the X_i^* are still i.i.d.r.v.'s, so that their joint distribution remains invariant under the permutation group. This permutation group has $n!$ elements and generates a (uniform) permutational probability measure $\mathscr{P}_n$ which attaches the equal conditional probability $1/n!$ to each of these permutations, whatever the underlying F may be. This enables us to construct permutational or conditionally distribution-free tests based on S_n. It follows that for every $i(\neq i') = 1, \ldots, n$,

$$E_{\mathscr{P}_n} Z_{ij} = P_{nj} = E_{\mathscr{P}_n} Z_{ij}^2, \quad j \in J, \tag{8.5.17}$$

$$E_{\mathscr{P}_n} Z_{ij} Z_{ij'} = 0 \qquad \text{for} \quad j \neq j' \, (\in J), \tag{8.5.18}$$

$$E_{\mathscr{P}_n} Z_{ij} Z_{i'j'} = \frac{n_j (n_{j'} - \delta_{jj'})}{n(n-1)}, \quad j, j' \in J, \tag{8.5.19}$$

where $\delta_{jj'}$ is the usual Kronecker delta. Thus, by (8.5.16) through (8.5.19), we obtain that

$$\begin{aligned} E_{\mathscr{P}_n} S_n &= \sum_{i=1}^n c_i \Big(\sum_{j \in J} \Delta_{nj} P_{nj} \Big) = \Big(\sum_{i=1}^n c_i \Big) \Big(\sum_{j \in J} \int_{F_{nj}}^{F_{nj+1}} \phi(u) \, du \Big) \\ &= n \bar{\phi} \bar{c}_n, \qquad \bar{\phi} = \int_0^1 \phi(u) \, du, \end{aligned} \tag{8.5.20}$$

$$V_{\mathscr{P}_n} S_n = \frac{n}{n-1} \Big\{ \sum_{i=1}^n (c_i - \bar{c}_n)^2 \Big\} A_n^2(J) = \frac{n}{n-1} C_n^2 A_n^2(J), \tag{8.5.21}$$

where $\bar{c}_n = n^{-1} \Sigma_{i=1}^n c_i$, $C_n^2 = \Sigma_{i=1}^n (c_i - \bar{c}_n)^2$, and

$$A_n^2(J) = \sum_{j \in J} \Delta_{nj}^2 P_{nj} - \bar{\phi}^2. \tag{8.5.22}$$

Then we may consider the standardized form

$$M_n = \frac{S_n - E_{\mathscr{P}_n} S_n}{\{ V_{\mathscr{P}_n} S_n \}^{1/2}} = \left(\frac{n-1}{n A_n^2(J)} \right)^{1/2} \sum_{i=1}^n c_{ni}^* \Big(\sum_{j \in J} \Delta_{nj} Z_{ij} \Big), \tag{8.5.23}$$

where $c_{ni}^* = (c_i - \bar{c}_n)/C_n$ (so that $\Sigma c_{ni}^* = 0$ and $\Sigma (c_{ni}^*)^2 = 1$).

For small values of n, the exact permutational distribution of M_n (under $\mathscr{P}_n$) can be evaluated by enumeration of the $n!$ possible realizations of M_n (generated by the permutations of $X_1^*, \dots, X_n^*$ among themselves), and a permutationally (conditionally) distribution-free test for $H_0: \beta = 0$ vs. $H_1: \beta > 0$ can be made, with the critical region formed by the larger values of M_n (constituting $[\alpha n!]$ of these realizations). For testing $H_0: \beta = 0$ vs. $H_2: \beta \neq 0$, one should use $|M_n|$ and base the test on its permutation distribution (instead of M_n's). The task of enumerating the permutation distribution of M_n (or $|M_n|$) becomes prohibitively laborious as n increases, and for this reason we consider the following large-sample solutions.

Note that by (8.5.22), for every J,

$$0 \le A_n^2(J) = \sum_{j \in J} P_{nj}^{-1} \left(\int_{F_{nj}}^{F_{nj+1}} \phi(u)\, du \right)^2 - \bar{\phi}^2$$

$$\le \int_0^1 \phi^2(u)\, du - \bar{\phi}^2 = A^2 \ (< \infty), \tag{8.5.24}$$

$$A_n^2(J) = A^2 - \sum_{j \in J} \int_{F_{nj}}^{F_{nj+1}} \left[\phi(u) - \Delta_{nj} \right]^2 du. \tag{8.5.25}$$

Let then

$$\Delta_j = P_j^{-1} \int_{F_j}^{F_{j+1}} \phi(u)\, du, \qquad j \in J, \tag{8.5.26}$$

$$A^2(J) = \sum_{j \in J} \Delta_j^2 P_j - \bar{\phi}^2 = A^2 - \sum_{j \in J} \int_{F_j}^{F_{j+1}} \left[\phi(u) - \Delta_j \right]^2 du. \tag{8.5.27}$$

Under $H_0: \beta = 0$, the X_i^* are i.i.d.r.v.'s, so that

$$\max_{j \in J} |F_{nj} - F_j| \to 0 \quad \text{a.s.} \qquad \text{as} \quad n \to \infty, \tag{8.5.28}$$

and this implies that

$$\max_{j \in J} |P_{nj} - P_j| \to 0 \quad \text{a.s.} \qquad \text{as} \quad n \to \infty, \tag{8.5.29}$$

$$\max_{j \in J} |\Delta_{nj} P_{nj} - \Delta_j P_j| \xrightarrow{P} 0 \qquad \text{as} \quad n \to \infty. \tag{8.5.30}$$

Therefore, starting with a finite number of class intervals and using

(8.5.28)–(8.5.30), it follows that

$$A_n^2(J) \xrightarrow{P} A^2(J) \quad \text{as} \quad n \to \infty. \tag{8.5.31}$$

Next, we can readily extend (8.5.31) to the case where J has a countable number of class intervals by approximating both $A_n^2(J)$ and $A^2(J)$ by $A_n^2(J^*)$ and $A^2(J^*)$, respectively, where J^* has only a finite number of cells. This is possible because of (8.5.25) and (8.5.27), where we allow the cells of J to be appropriately amalgamated to form the cells of J^*—the square-integrability of ϕ insures that J can always be replaced by such a J^* to any desired level of approximation.

We assume that

$$0 < A^2(J)\,(\leq A^2) < \infty. \tag{8.5.32}$$

Let now

$$S_n^* = \sum_{i=1}^{n} c_{ni}^*\Big(\sum_{j\in J} Z_{ij}\Delta_{nj}\Big), \qquad S_n^0 = \sum_{i=1}^{n} c_{ni}^*\Big(\sum_{j\in J} \Delta_j Z_{ij}\Big). \tag{8.5.33}$$

Then, by (8.5.17), (8.5.18), (8.5.19), and (8.5.33), we have

$$E_{\mathscr{P}_n}\big(S_n^* - S_n^0\big)^2 = \frac{n}{n-1}\sum_{j\in J}\big(\Delta_j - \Delta_{nj}\big)^2 P_{nj}, \tag{8.5.34}$$

and arguing on lines parallel to (8.5.28) through (8.5.31), we conclude that (8.5.34) converges to 0 in probability. Hence,

$$S_n^* - S_n^0 \xrightarrow{P} 0 \quad \text{under } \mathscr{P}_n \qquad \text{(a.e.)}. \tag{8.5.35}$$

On the other hand, if we let $W_i = \sum_{j\in J}\Delta_j Z_{ij}$, $i \geq 1$, then the W_i are i.i.d.r.v.'s with mean $\bar{\phi}$ and variance $A^2(J)$, where by (8.5.32), $0 < A(J) < \infty$. Thus, whenever

$$\max_{1\leq i\leq n} |c_{ni}^*| \to 0 \quad \text{as} \quad n \to \infty, \tag{8.5.36}$$

the central limit theorem holds for $\sum_{i=1}^{n} c_{ni}^* W_i$ $(= S_n^0)$, and hence, S_n^0 is asymptotically normal with mean 0 and variance $A^2(J)$. By (8.5.31), (8.5.35), and the above, we conclude that under H_0,

$$M_n \xrightarrow{\mathscr{L}} \mathscr{N}(0,1) \quad \text{as} \quad n \to \infty, \tag{8.5.37}$$

so that for large sample sizes, tests based on M_n can be constructed by using the normal-distribution percentage points.

Let us now consider the model (8.5.1) where $\boldsymbol{\beta}$ and $\mathbf{c}_i$, $i \geq 1$, are q-vectors for some $q \geq 1$. We define the n_j, P_{nj}, and Δ_{nj} as in (8.5.13), (8.5.14), and (8.5.15), respectively, and let

$$\mathbf{S}_n = \sum_{i=1}^{n} \mathbf{c}_i \left(\sum_{j \in J} \Delta_{nj} Z_{ij} \right). \tag{8.5.38}$$

Consider the null hypothesis [with respect to (8.5.1)] $H_0: \boldsymbol{\beta} = \mathbf{0}$ vs. $H_1: \boldsymbol{\beta} \neq \mathbf{0}$. Then, under H_0, the same permutational invariance structure discussed after (8.5.16) holds, and proceeding as in (8.5.17) through (8.5.22), we obtain that

$$E_{\mathscr{P}_n} \mathbf{S}_n = n\bar{\phi}\bar{\mathbf{c}}_n, \qquad \bar{\mathbf{c}}_n = n^{-1} \sum_{i=1}^{n} \mathbf{c}_i, \tag{8.5.39}$$

$$E_{\mathscr{P}_n} (\mathbf{S}_n - n\bar{\phi}\bar{\mathbf{c}}_n)(\mathbf{S}_n - n\bar{\phi}\bar{\mathbf{c}}_n)' = \frac{n}{n-1} A_n^2(J)\mathbf{C}_n, \tag{8.5.40}$$

where

$$\mathbf{C}_n = \sum_{i=1}^{n} (\mathbf{c}_i - \bar{\mathbf{c}}_n)(\mathbf{c}_i - \bar{\mathbf{c}}_n)'. \tag{8.5.41}$$

Thus, we proceed to use the quadratic form

$$\begin{aligned} \mathscr{L}_n &= (\mathbf{S}_n - n\bar{\phi}\bar{\mathbf{c}}_n)' \left[\frac{n}{n-1} A_n^2(J)\mathbf{C}_n \right]^{-1} (\mathbf{S}_n - n\bar{\phi}\bar{\mathbf{c}}_n) \\ &= \frac{n-1}{nA_n^2(J)} \sum_{j=1}^{q} \sum_{l=1}^{q} (S_{nj} - n\bar{\phi}\bar{c}_{nj})(S_{nl} - n\bar{\phi}\bar{c}_{nl}) C_n^{jl} \end{aligned} \tag{8.5.42}$$

as a test statistic, where $((C_n^{jl})) = \mathbf{C}_n^{-1}$, and we assume that $\mathbf{C}_n$ is a p.d. matrix. We reject $H_0: \boldsymbol{\beta} = \mathbf{0}$ when $\mathscr{L}_n$ is large. For the evaluation of the critical values of $\mathscr{L}_n$, again we may use, for small n, the exact permutation distribution of $\mathscr{L}_n$ [similarly to the case of M_n in (8.5.23)]; the task becomes prohibitively laborious as n increases, and for this reason we take recourse to the following asymptotic solution.

First, for each of the q coordinates of $\mathbf{S}_n$, we may proceed as in (8.5.33) through (8.5.35), and hence, on defining $\mathbf{S}_n^0 = \Sigma_{i=1}^n \mathbf{c}_i(\Sigma_{j\in J}\Delta_j Z_{ij})$, we obtain that under $H_0: \boldsymbol{\beta} = \mathbf{0}$,

$$(\mathbf{S}_n - \mathbf{S}_n^0)'\mathbf{C}_n^{-1}(\mathbf{S}_n - \mathbf{S}_n^0) \xrightarrow{P} 0. \tag{8.5.43}$$

On the other hand, $\mathbf{S}_n^0 = \Sigma_{i=1}^n \mathbf{c}_i W_i$, where the W_i are i.i.d.r.v.'s with mean 0 and variance $A^2(J)$. Thus, for every $\boldsymbol{\lambda} \neq \mathbf{0}$, $\boldsymbol{\lambda}'\mathbf{S}_n^0 = \Sigma_{i=1}^n(\boldsymbol{\lambda}'\mathbf{c}_i)W_i$ conforms to S_n^0 in (8.5.33), so that the central limit theorem holds. Thus, $\boldsymbol{\lambda}'\mathbf{S}_n^0$ will be asymptotically $\mathscr{N}((\boldsymbol{\lambda}'\bar{\mathbf{c}}_n)n\bar{\phi},\ A^2(J)\cdot\boldsymbol{\lambda}'\mathbf{C}_n\boldsymbol{\lambda})$. Hence,

$$\mathscr{L}_n^* = A^{-2}(J)(\mathbf{S}_n^0 - n\bar{\phi}\bar{\mathbf{c}}_n)'\mathbf{C}_n^{-1}(\mathbf{S}_n^0 - n\bar{\phi}\bar{\mathbf{c}}_n) \tag{8.5.44}$$

will have asymptotically the central chi-square distribution with q DF. By (8.5.42), (8.5.43), and (8.5.44), under H_0,

$$\mathscr{L}_n - \mathscr{L}_n^* \xrightarrow{P} 0 \quad \text{as} \quad n \to \infty, \tag{8.5.45}$$

so that $\mathscr{L}_n$ has also asymptotically the central chi-square distribution with q DF. Thus, we may proceed as follows:

$$\text{If} \quad \mathscr{L}_n \begin{cases} \geq \chi_{q,\alpha}^2, & \text{reject } H_0: \boldsymbol{\beta} = \mathbf{0}, \\ < \chi_{q,\alpha}^2, & \text{accept } H_0. \end{cases} \tag{8.5.46}$$

Let us now proceed to study the asymptotic power properties of the tests based on M_n and $\mathscr{L}_n$. First, we consider the case of M_n and consider a sequence $\{H_n\}$ of alternative hypotheses, where

H_n: (8.5.1) holds with $\beta' c_i$ replaced by γc_{ni}^*, where the c_{ni}^* are defined after (8.5.23), γ is a constant, and (8.5.36) holds. (8.5.47)

Also, assume that F satisfies the conditions stated after (8.5.5). Then, proceeding as in (8.5.9)–(8.5.11), but working with (8.5.47), it follows that for testing $H_0: \gamma = 0$ vs. H_n, the log-likelihood ratio statistic is

$$\log L_n = \gamma \sum_{i=1}^n c_{ni}^* \sum_{j\in J} \Delta_j^* Z_{ij} - \tfrac{1}{2}\gamma^2 \sum_{i=1}^n c_{ni}^{*2} \sum_{j\in J} \Delta^{*2} Z_{ij} + o_p(1). \tag{8.5.48}$$

Under $H_0 : \gamma = 0$,

$$\sum_{i=1}^{n} c_{ni}^{*2} \sum_{j \in J} \Delta_j^{*2} Z_{ij} \xrightarrow{P} \sum_{j \in J} \Delta_j^{*2} P_j,$$

so that proceeding as in (8.5.36)–(8.5.37), we obtain that under H_0,

$$\log L_n \sim \mathcal{N}\left(-\tfrac{1}{2}\gamma^2 A^{*2}(J), \gamma^2 A^{*2}(J)\right), \tag{8.5.49}$$

where

$$A^{*2}(J) = \sum_{j \in J} \Delta_j^{*2} P_j = \int_0^1 \psi^2(u)\, du - \sum_{j \in J} \int_{F_j}^{F_{j+1}} \{\psi(u) - \Delta_j^*\}^2\, du$$

$$\left(\leq \int_0^1 \psi^2(u)\, du = \int_{-\infty}^{\infty} \left[\frac{f'(x)}{f(x)} \right]^2 dF(x) \right). \tag{8.5.50}$$

Now, (8.5.49) insures the contiguity of the joint distribution of the X_i^* under $\{H_n\}$ to that under H_0. As a result of this contiguity and (8.5.35), we conclude that

$$S_n^* - S_n^0 \xrightarrow{P} 0 \qquad \text{under } \{H_n\} \text{ as well.} \tag{8.5.51}$$

On the other hand, $E(Z_{ij}|H_n) = F(a_{j+1} - \gamma c_{ni}^*) - F(a_j - \gamma c_{ni}^*) = P_j\{1 + \gamma c_{ni}^* \Delta_j^* + o(c_{ni}^*)\}$, $j \in J$, $1 \leq i \leq n$, where the Δ_n^* are defined by (8.5.8). Hence,

$$E(S_n^0 | H_n) = \gamma \sum_{j \in J} \Delta_j \Delta_j^* P_j + o(1). \tag{8.5.52}$$

Further, under $\{H_n\}$, S_n^0 is still a linear combination of independent r.v.'s (forming a triangular array), and hence the central limit theorem holds. Thus, noting that by the contiguity and (8.5.36) we have

$$A_n^2(J) \xrightarrow{P} A^2(J) \qquad \text{under } \{H_n\} \text{ as well,}$$

we conclude that under $\{H_n\}$,

$$M_n \sim \mathcal{N}\left(\frac{\gamma\left(\sum_{j \in J} \Delta_j \Delta_j^* P_j\right)}{A(J)}, 1 \right). \tag{8.5.53}$$

Thus, for testing $H_0: \gamma = 0$ vs. $\{H_n\}$, the asymptotic power of the one-sided test based on M_n is given by

$$1 - \Phi\left(\tau_\alpha - \frac{\gamma \sum_{j \in J} \Delta_j \Delta_j^* P_j}{A(J)}\right), \tag{8.5.54}$$

where Φ is the standard normal d.f. and $\Phi(\tau_\alpha) = 1 - \alpha$. On the other hand, working with L_n in (8.5.48) and using the contiguity, we obtain on parallel lines that the asymptotic power of the test based on L_n is given by

$$1 - \Phi(\tau_\alpha - \gamma A^*(J)). \tag{8.5.55}$$

Since by (8.5.12) and (8.5.48), under H_0 (as well as $\{H_n\}$), $\log L_n = \gamma T_n - \frac{1}{2}\gamma^2 A^{*2}(J) + o_p(1)$, the locally most powerful test (based on T_n) has the same asymptotic power in (8.5.55), and thus (8.5.55) represents the envelope power function for local alternatives $\{H_n\}$.

Several useful conclusions can be derived from the last two formulae. First, if the X_i were observable, then the locally most powerful test (according to our result in Chapter 5) would have the asymptotic power function (for H_0 vs. H_n)

$$1 - \Phi(\tau_\alpha - \gamma I^{1/2}(f)), \tag{8.5.56}$$

where $I(f) = \int_0^1 \psi^2(u)\,du$. By (8.5.50), (8.5.55), and (8.5.56), we conclude that the *intrinsic loss in efficiency due to grouping* is

$$\frac{A^{*2}(J)}{I(f)} = 1 - \frac{1}{I(f)}\left\{\sum_{j \in J} \int_{F_j}^{F_{j+1}} [\psi(u) - \Delta_j^*]^2\,du\right\}. \tag{8.5.57}$$

It is also clear from (8.5.57) that

$$\text{if} \quad \max_{j \in J} P_j \text{ is small} \quad \text{then} \quad \frac{A^*(J)}{I^{1/2}(f)} \simeq 1, \tag{8.5.58}$$

indicating that the loss is also negligible. On the other hand, if the P_j are not small, then (8.5.57) may be quite below 1. Second, if the X_i were observable, as in Chapter 5, we could have used a linear rank statistic for testing H_0 vs. $\{H_n\}$, and the asymptotic power function of this test would have been

$$1 - \Phi(\tau_\alpha - \gamma I^{1/2}(f)\rho(\phi, \psi)), \tag{8.5.59}$$

where

$$\rho^2(\phi, \psi) = \frac{\left(\int_0^1 \phi(u)\psi(u)\,du\right)^{-2}}{A^2 I(f)} \;(\leq 1). \tag{8.5.60}$$

Thus, by (8.5.53), (8.5.59), and (8.5.60), we conclude that in this situation, the loss of efficiency due to grouping is

$$[I(f)\rho^2(\phi, \psi)]^{-1} \frac{\left(\sum_{j \in J} \Delta_j \Delta_j^* P_j\right)^2}{A^2(J)} = \frac{A^2}{A^2(J)} \left[\frac{\int_0^1 \phi(u)\psi(u)\,du}{\sum_{j \in J} \Delta_j \Delta_j^* P_j}\right]^{-2}, \tag{8.5.61}$$

which depends on both ϕ and ψ as well as on the P_j. Third, the Pitman ARE of the test based on M_n with respect to the one based on L_n is given by

$$e(M, L) = \frac{\left(\sum_{j \in J} \Delta_j \Delta_j^* P_j\right)^2}{A^2(J) A^{*2}(J)} = \frac{\left(\sum_{j \in J} \Delta_j \Delta_j^* P_j\right)^2}{\left(\sum_{j \in J} \Delta_j^2 P_j\right)^2 \left(\sum_{j \in J} \Delta_j^{*2} P_j\right)}$$

$$= \rho^2(\phi, \psi; J) \leq 1 \tag{8.5.62}$$

with equality holding only when $\quad \Delta_j = \Delta_j^* \quad \forall j \in J.$ (8.5.63)

Thus, as in the case of ungrouped data, the optimal score function is still $\phi = \psi$.

So far, we have considered the case of one-sided tests based on T_n, M_n, and L_n. The results go through for the two-sided tests based on these statistics. Instead of (8.5.54), we have the asymptotic power function for this case as

$$1 - \Phi\left(\tau_{\alpha/2} - \frac{\gamma \sum_{j \in J} \Delta_j \Delta_j^* P_j}{A(J)}\right) + \Phi\left(-\tau_{\alpha/2} - \frac{\gamma \sum_{j \in J} \Delta_j \Delta_j^* P_j}{A(J)}\right), \tag{8.5.64}$$

and similar changes are to be made in (8.5.55), (8.5.56), and (8.5.59). But (8.5.57), (8.5.61), and (8.5.62)–(8.5.63) remain intact.

Let us now consider the general case of $\mathscr{L}_n$ when in (8.5.1) $q \geq 1$. Since $\mathbf{C}_n$ in (8.5.41) is assumed to be p.d. for every n ($\geq n_0$), we use the following canonical reduction of the sequence of alternative hypotheses. Let $\mathbf{D}_n$ be such that $\mathbf{D}_n\mathbf{C}_n\mathbf{D}_n' = \mathbf{I}_q$; we write $\mathbf{D}_n = \mathbf{C}_n^{-1/2}$. Also, let

$$\mathbf{c}_{ni}^* = \mathbf{D}_n(\mathbf{c}_i - \bar{\mathbf{c}}_n), \qquad i = 1, \ldots, n; \tag{8.5.65}$$

then $\Sigma\mathbf{c}_{ni}^* = \mathbf{0}$ and $\Sigma\mathbf{c}_{ni}^*\mathbf{c}_{ni}^{*\prime} = \mathbf{D}_n\mathbf{C}_n\mathbf{D}_n' = \mathbf{I}_q$. We consider the sequence $\{H_n\}$ of alternative hypotheses

$$H_n: P\{X_i \leq x\} = F(x - \boldsymbol{\gamma}'\mathbf{c}_{ni}^*), \qquad 1 \leq i \leq n, \quad \boldsymbol{\gamma} \in R^q. \tag{8.5.66}$$

For testing $H_0: \boldsymbol{\gamma} = \mathbf{0}$ vs. H_n in (8.5.66), we may consider the usual likelihood-ratio statistic. Parallel to (8.5.48), we have

$$\log L_n = \boldsymbol{\gamma}' \sum_{i=1}^{n} \mathbf{c}_{ni}^* \left(\sum_{j \in J} \Delta_j^* Z_{ij} \right)$$
$$- \tfrac{1}{2}\boldsymbol{\gamma}' \left(\sum_{i=1}^{n} \mathbf{c}_{ni}^*\mathbf{c}_{ni}^{*\prime} \sum_{j \in J} \Delta_j^{*2} Z_{ij} \right) \boldsymbol{\gamma} + o_p(1) \tag{8.5.67}$$

and parallel to (8.5.49)–(8.5.50), we have under $H_0: \boldsymbol{\gamma} = \mathbf{0}$

$$\log L_n \sim \mathscr{N}\left(-\tfrac{1}{2}\boldsymbol{\gamma}'\boldsymbol{\gamma}A^{*2}(J), \boldsymbol{\gamma}'\boldsymbol{\gamma}A^{*2}(J)\right). \tag{8.5.68}$$

Thus, the contiguity of the probability measure under $\{H_n\}$ to that under $\{H_0\}$ is insured, and therefore, using direct vector generalizations of (8.5.51)–(8.5.52), we conclude that under $\{H_n\}$,

$$\mathbf{D}_n(\mathbf{S}_n - n\bar{\phi}\bar{\mathbf{c}}_n) \sim \mathscr{N}\left(\boldsymbol{\gamma}\left(\sum_{j \in J} \Delta_j \Delta_j^* P_j\right), A^2(J) \cdot \mathbf{I}_q\right), \tag{8.5.69}$$

so that $\mathscr{L}_n$ has asymptotically a noncentral chi-square distribution with q DF and the noncentrality parameter

$$\Delta_{\mathscr{L}} = (\boldsymbol{\gamma}'\boldsymbol{\gamma})A^{-2}(J)\left(\sum_{j \in J} \Delta_j \Delta_j^* P_j\right)^2. \tag{8.5.70}$$

If we proceed to maximize (8.5.67) with respect to $\boldsymbol{\gamma}$ and then construct the likelihood-ratio test statistic $-\log\lambda_n$, then it follows from (8.5.67)–(8.5.68) that (by an appeal to contiguity again) under $\{H_n\}$, $-\log\lambda_n$ has asymptotically a noncentral chi-square distribution with q DF and the noncentrality parameter

$$\Delta_\lambda = (\boldsymbol{\gamma}'\boldsymbol{\gamma})A^{*2}(J). \tag{8.5.71}$$

Thus, the Pitman ARE of the rank test based on $\mathscr{L}_n$ with respect to the likelihood-ratio test based on λ_n is

$$e(M,\lambda) = \frac{\left(\sum_{j\in J}\Delta_j\Delta_j^*P_j\right)^2}{A^2(J)A^{*2}(J)} = \rho^2(\psi,\phi;J), \tag{8.5.72}$$

defined by (8.5.62). This agrees with the case of $q = 1$, discussed earlier, and hence all the conclusions made after (8.5.63) hold for general $q \geq 1$. In passing, we may remark that the likelihood-ratio test based on $-\log\lambda_n$ has the Wald optimality (asymptotically), as discussed in Section 5.8. Hence, for $\Delta_j = \Delta_j^*$, $j \in J$ (or $\phi = \psi$), the same asymptotic optimality applies to the rank test based on M_n.

Let us finally sketch the multivariate case, where in (8.5.1) we have

$$F_i(\mathbf{x}) = F(\mathbf{x} - \boldsymbol{\beta}'\mathbf{c}_i), \qquad i = 1,\dots,n, \tag{8.5.73}$$

the $\mathbf{X}_i$ are p-vectors ($p \geq 1$), $\boldsymbol{\beta}$ is a $p \times q$ matrix of unknown parameters, and the $\mathbf{c}_i$ are specified q-vectors. Here the $I_{\mathbf{j}}$ in (8.5.2) are p-dimensional blocks

$$I_{\mathbf{j}} = \{\mathbf{a}_{\mathbf{j}} \leq \mathbf{x} < \mathbf{a}_{\mathbf{j}+\mathbf{1}}\}, \qquad \mathbf{j} \geq \mathbf{0}; \tag{8.5.74}$$

similarly, the Z_{ij} in (8.5.4) are defined for each $\mathbf{j} \geq \mathbf{0}$, and

$$\mathbf{X}_i^* = \sum_{\mathbf{j}\geq\mathbf{0}} I_{\mathbf{j}} Z_{i\mathbf{j}}, \qquad i \geq 1. \tag{8.5.75}$$

For each of the p marginal variates in $\mathbf{X}_i$ (or $\mathbf{X}_i^*$), we defined the scores $\Delta_{n\mathbf{j}}^{(k)}$ $(k = 1,\dots,p)$ as in (8.5.15), where we use $\phi_k(u)$ for the kth coordinate and where $F_{n\mathbf{j}}$ is replaced by $F_{n[k]\mathbf{j}}$, the marginal empirical d.f. for the kth coordinate. In this way, we get for every $k(=1,\dots,p)$,

$$S_{nkl} = \sum_{i=1}^{n} c_{il}\left(\sum_{j\in J}\Delta_{\mathbf{j}k}^* Z_{i\mathbf{j}k}\right), \quad 1 \leq l \leq q, \tag{8.5.76}$$

where

$$Z_{ijk} = \begin{cases} 1, & X_{ik} \in I_j \\ 0, & \text{otherwise} \end{cases}, \qquad 1 \le k \le p, \quad j \in J. \tag{8.5.77}$$

The permutation distribution theory developed after (8.5.15) also holds in this multivariate case. It follows that

$$E_{\mathscr{P}_n} S_{nkl} = n\bar{c}_{nl}\bar{\phi}_k, \qquad 1 \le k \le p, \ 1 \le l \le q, \tag{8.5.78}$$

where

$$\bar{\phi}_k = \int_0^1 \phi_k(u)\,du \quad \text{and} \quad \bar{c}_{nl} = n^{-1}\sum_{i=1}^{n} c_{il}. \tag{8.5.79}$$

Further, if $F_{n[kk']}$ is the bivariate empirical d.f. of the (k, k')th coordinate of the $\mathbf{X}_i$, and if $I_{\mathbf{j}(kk')}$ is the two-dimensional cell obtained from $I_{\mathbf{j}}$ by considering only the (k, k')th coordinates, then we let

$$P_{n\mathbf{j}(kk')} = \text{empirical probability of } I_{\mathbf{j}(kk')}. \tag{8.5.80}$$

Let then $A_{nkk}(J)$ be defined by $A_n^2(J)$ in (8.5.22) with ϕ replaced by ϕ_k, and for $k \neq k'$, let

$$A_{nkk'}(J) = \sum_{j \in J} \Delta_{n\mathbf{j}(k)}\Delta_{n\mathbf{j}(k')}P_{n\mathbf{j}(kk')}. \tag{8.5.81}$$

Then we have

$$E_{\mathscr{P}_n}(S_{nkl}S_{nk'l'}) = \frac{n}{n-1}(\mathbf{A}_n(J)) \otimes \mathbf{C}_n, \tag{8.5.82}$$

where $\mathbf{C}_n$ is defined by (8.5.41) and

$$\mathbf{A}_n(J) = ((A_{nkk'}(J)))_{k,k'=1,\ldots,p}. \tag{8.5.83}$$

Consider now the $pq \times pq$ matrix

$$\mathbf{H}_n = (([S_{nkl} - n\bar{c}_{nl}\bar{\phi}_k][S_{nk'l'} - n\bar{c}_{nl'}\bar{\phi}_{k'}])), \tag{8.5.84}$$

and as in Chapter 5, consider the test statistic

$$\mathscr{L}_n = \frac{n-1}{n}\operatorname{Tr}\left(\mathbf{H}_n[\mathbf{A}_n(J) \otimes \mathbf{C}_n]^{-1}\right). \tag{8.5.85}$$

For small n, the exact permutation-distribution theory of $\mathscr{L}_n$ can be incorporated to find the critical values—while for large n, by arguments similar to those in (8.5.43) through (8.5.45), we claim that $\mathscr{L}_n$ has asymptotically the central chi-square distribution with pq DF. Further, the concept of contiguity extends readily to this multivariate case, and the noncentral-distribution theory follows as a direct multivariate extension of that of $\mathscr{L}_n$ in (8.5.70).

8.6. NONPARAMETRIC TESTING UNDER CENSORING

Mostly, in clinical trials, life-testing experiments, and problems of survival analysis, the *response* (dependent) variable (Y) is a nonnegative random variable (failure time, span of life, etc.), while the concomitant variates (**X**) relate to various assignable factors in such experiments. Because of practical limitations (of time and cost), the experiment cannot continue indefinitely, and is usually curtailed after a preplanned duration or when a prefixed number of responses have occurred. Typically, this is termed a *right-censored model*; a *left-censored model* may also be defined in an analogous manner. In either case, one has to curtail experimentation at an intermediate stage and draw valid statistical conclusions from the (incomplete) data at hand; here, the observations pertaining to the tenure of the experimentation are *recorded*, while the others are *censored*. Censoring may be of different types.

(I) Truncation or Type I Censoring: Suppose that $Y_1, \ldots, Y_N$ are the independent random variables relating to the response (failure times), and $\mathbf{X}_1, \ldots, \mathbf{X}_N$ the respective covariates. Also, suppose that for some pre-fixed $T\ (< \infty)$, experimentation is curtailed at time T. Define

$$Y_i^* = Y_i \wedge T = \min(Y_i, T), \qquad 1 \leq i \leq N. \tag{8.6.1}$$

Then the Y_i^* are the truncated r.v.'s, and the model relates to a single point truncation (type I censoring) scheme. Let $r(T) =$ (number of $Y_i \leq T$, $i \leq N$), so that $r(T)$ is a nonnegative and integer-valued r.v. Thus, in a truncation scheme, T is fixed but $r(T)$ is stochastic.

(II) Type II Censoring: Let $Y_{N,1} \leq \cdots \leq Y_{N,N}$ be the order statistics corresponding to $Y_1, \ldots, Y_N$, and for some $p \in (0, 1)$, let $r = [Np] + 1$ be a pre-fixed positive integer. If the experiment is stopped after the rth failure $Y_{N,r}$ has occurred, then the model relates to a type II (single point) censoring scheme. Here r is pre-fixed, but $Y_{N,r}$ is stochastic.

(III) Random Censoring: We imagine a set of r.v.'s $T_1, \ldots, T_N$, and define

$$Y_i^0 = Y_i \wedge T_i \text{ and } \delta_i = \begin{cases} 1, & Y_i^0 = Y_i \\ 0, & \text{otherwise} \end{cases}, \quad i = 1, \ldots, n. \qquad (8.6.2)$$

This model typically arises when the T_i refer to the withdrawal times or when the subjects enter into the schemes at random points but are censored at a fixed point of time. Usually, the T_i are assumed to be i.i.d.r.v.'s with some (unknown) d.f., and this enables nonparametric procedures to be applicable in some testing problems. Equation (8.6.2) may also arise in the context of random withdrawal *cum* staggered entry of the individuals into the scheme.

The theory of rank-order tests, developed in Chapters 5, 7, and 8 (earlier sections), can be extended to cover such censoring schemes. We shall present a brief review of this in this section. For the specific two-sample and multisample problems, such tests have been developed by Halperin (1960), Gehan (1965), Basu (1967a, b), Breslow (1970), and others, while for general linear rank statistics, more general results have been obtained by Chatterjee and Sen (1973), Majumdar and Sen (1977, 1978a, b), Gardiner and Sen (1978), Sen (1976b, 1979b), and others.

To start with, we consider the case of nonstochastic $\mathbf{X}_i$, and as in (5.3.1), we assume that

$$F_i(x) = P\{Y_i \leq x\} = F(x - \alpha - \boldsymbol{\beta}'\mathbf{c}_i), \qquad 1 \leq i \leq N, \quad (8.6.3)$$

where the $\mathbf{c}_i$ are known vectors of regression constants, and suppose that as in (5.3.2), we want to test for $H_0: \boldsymbol{\beta} = \mathbf{0}$ against $H_1: \boldsymbol{\beta} \neq \mathbf{0}$. Consider first the case of type II censoring, so that r $(= [Np] + 1)$ is specified in advance. If $Y_1, \ldots, Y_N$ were all observed and $R_{N1}, \ldots, R_{NN}$ were their ranks (among themselves), then we would have considered the test statistic $\mathscr{L}_N$ in (5.3.7) based on the vector of linear rank statistics $\mathbf{L}_N$ in (5.3.4). But in this case, if $R_{Ni} \leq r$, we have an exact idea of the rank, while for the $Y_i > Y_{N,r}$, all we know is that the corresponding R_{Ni} are $> r$. This calls for some adjustments, which we consider below.

Let $S_{N1}, \ldots, S_{NN}$ be the antiranks of $Y_1, \ldots, Y_N$, that is, we have

$$Y_i = Y_{N, R_{Ni}}, \quad Y_{N,i} = Y_{S_{Ni}}, \qquad 1 \leq i \leq N, \qquad (8.6.4)$$

$$R_{NS_{Ni}} = S_{NR_{Ni}} = i, \qquad 1 \leq i \leq N. \qquad (8.6.5)$$

With this, we may rewrite $\mathbf{L}_N$ in (5.3.4) as

$$\mathbf{L}_N = \sum_{i=1}^{N} (\mathbf{c}_{S_{Ni}} - \bar{\mathbf{c}}_N) a_N(i). \tag{8.6.6}$$

Also, let $\mathbf{S}_N^{(k)} = (S_{N1}, \ldots, S_{Nk})$, $1 \le k \le N$, $\mathbf{S}_N^{(0)} = \{0\}$. Then in a type II censoring scheme, we are given the sets $(Y_{N,1}, \ldots, Y_{N,r})$ and $\mathbf{S}_N^{(r)}$, and would like to construct a test based on these sets. Note that under $H_0: \boldsymbol{\beta} = \mathbf{0}$, $(S_{N1}, \ldots, S_{NN})$ assumes all possible permutations of $(1, \ldots, N)$ with the equal probability $(N!)^{-1}$. As such,

$$P_0\{\mathbf{S}_N^{(r)} = (i_1, \ldots, i_r)\} = (N^{[r]})^{-1} \tag{8.6.7}$$

for every $(i_1, \ldots, i_r): 1 \le i_1 \ne \cdots \ne i_r \le N$. Moreover, given $\mathbf{S}_N^{(r)}$, S_{Ni} (for any $i: r+1 \le i \le N$) can assume any one of the remaining $N - r$ values $(\{1, \ldots, N\} \setminus \{i_1, \ldots, i_r\})$ with the equal probability $(N-r)^{-1}$. Hence,

$$\begin{aligned}
E_0\{\mathbf{L}_N | \mathbf{S}_N^{(r)}\} &= \sum_{i=1}^{r} (\mathbf{c}_{S_{Ni}} - \bar{\mathbf{c}}_N) a_N(i) + \sum_{i=r+1}^{N} a_N(i) E_0\{\mathbf{c}_{S_{Ni}} - \bar{\mathbf{c}}_N | \mathbf{S}_N^{(r)}\} \\
&= \sum_{i=1}^{r} (\mathbf{c}_{S_{Ni}} - \bar{\mathbf{c}}_N) \left(a_N(i) + \sum_{i=r+1}^{N} a_N(i) \frac{1}{N-r} \sum_{j=r+1}^{N} (\mathbf{c}_{S_{Nj}} - \bar{\mathbf{c}}_N) \right) \\
&= \sum_{i=1}^{r} (\mathbf{c}_{S_{Ni}} - \bar{\mathbf{c}}_N) [a_N(i) - a_N^*(r)] = \mathbf{L}_{Nr}, \quad \text{say},
\end{aligned} \tag{8.6.8}$$

where

$$a_N^*(r) = \begin{cases} (N-r)^{-1} \displaystyle\sum_{j=r+1}^{N} a_N(j), & r \le N-1, \\ 0, & r = N. \end{cases} \tag{8.6.9}$$

$\mathbf{L}_{Nr}$ is termed a censored linear rank statistic, where censoring takes place at the rank r. Note that if we let

$$a_{Nr}(i) = \begin{cases} a_N(i), & 1 \le i \le r, \\ a_N^*(r), & r+1 \le i \le N, \end{cases} \tag{8.6.10}$$

then, by definition, $\mathbf{L}_{Nr} = \sum_{i=1}^{N} (\mathbf{c}_i - \bar{\mathbf{c}}_N) a_{Nr}(R_{Ni})$ is a linear rank statistic [vector with scores given by (8.6.10)]. Thus, if we replace A_N^2, defined by (5.2.11), by

$$A_{Nr}^2 = \frac{1}{N-1}\left\{\sum_{i=1}^{r} a_N^2(i) + (N-r)\left[a_N^*(r)\right]^2 - N\bar{a}_N^2\right\}, \tag{8.6.11}$$

then, as in (5.3.5), we have

$$E\{\mathbf{L}_{Nr}|H_0\} = \mathbf{0} \quad \text{and} \quad E\{\mathbf{L}_{Nr}\mathbf{L}'_{Nr}|H_0\} = A_{Nr}^2\mathbf{C}_N, \tag{8.6.12}$$

where $\mathbf{C}_N$ is defined by (5.3.6). This leads us to the following test statistic:

$$\mathscr{L}_{Nr} = A_{Nr}^{-2}\left(\mathbf{L}'_{Nr}\mathbf{C}_N^{-}\mathbf{L}_{Nr}\right). \tag{8.6.13}$$

Note that $\mathbf{L}_{Nr}$ is a measurable function of $S_N^{(r)}$, where, by (8.6.7), under $H_0: \boldsymbol{\beta} = \mathbf{0}$, the (joint) probability law of $\mathbf{S}_N^{(r)}$ does not depend on the underlying F. Hence $\mathbf{L}_{Nr}$ (and, consequently, $\mathscr{L}_{Nr}$) are genuinely distribution-free statistics. For small values of N, the exact distribution of $\mathscr{L}_{Nr}$ can be obtained by using (8.6.7), while for N large, we may directly appeal to Theorem 5.3.1 [where the $a_N(i)$ are to be replaced by $b_N(i)$, $1 \le i \le N$] and conclude that when rank$(\mathbf{C}_N) = q$, under H_0, $\mathscr{L}_{Nr}$ has asymptotically chi-square distribution with q DF.

For every $p: 0 < p < 1$, we let

$$\phi_p(u) = \begin{cases} \phi(u), & 0 < u \le p, \\ \phi_p^* = \dfrac{1}{1-p}\displaystyle\int_p^1 \phi(t)\,dt, & p < u < 1, \end{cases} \tag{8.6.14}$$

$$A_p^2 = \int_0^1 \phi_p^2(u)\,du - \bar{\phi}^2 \tag{8.6.15}$$

and

$$\begin{aligned} \gamma_p &= \int_{-\infty}^{\infty} \frac{d}{dx}\phi_p(F(x))\,dF(x) \\ &= \int_0^1 \phi_p(u)\psi(u)\,du = \int_0^1 \phi_p(u)\psi_p(u)\,du \\ &= \int_0^1 \phi(u)\psi_p(u)\,du, \end{aligned} \tag{8.6.16}$$

where $\psi(u) = -f'(F^{-1}(u))/f(F^{-1}(u))$, $0 < u < 1$, and $\psi_p(u)$ is defined as in (8.6.14) with ϕ replaced by ψ. Note that $\int_0^1 \psi^2(u)\,du = \mathscr{I}(f)$, the Fisher information. Let then

$$\mathscr{I}_p(f) = \int_0^1 \psi_p^2(u)\,du, \qquad 0 < p < 1, \tag{8.6.17}$$

so that $\mathscr{I}_p(f) \to \mathscr{I}(f)$ as $p \to 1$, and let

$$\rho(\phi_p, \psi_p) = \frac{\gamma_p}{\left\{A_p^2 \mathscr{I}_p(f)\right\}^{1/2}}. \tag{8.6.18}$$

Note that by (8.6.11), (8.6.14), and (8.6.15), as in Chapter 5,

$$A_{Nr}^2 \to A_p^2 \qquad \text{as} \quad N \to \infty. \tag{8.6.19}$$

Also, if we consider the sequence $\{H_N\}$ of alternative hypotheses in (5.5.1)–(5.5.2) (for the univariate case), then precisely on the same line as in the proof of Theorem 5.5.2, we arrive at the following: Under $\{H_N\}$ and the regularity conditions of Theorem 5.5.2, $\mathscr{L}_{Nr}$ has asymptotically a noncentral chi-square distribution with q DF and noncentrality parameter

$$\Delta_{\mathscr{L},p} = A_p^{-2}\gamma_p^2 \sum_{j=1}^{q} \sum_{l=1}^{q} \tilde{C}_{jl}\beta_j^0\beta_l^0. \tag{8.6.20}$$

Note that

$$\frac{\gamma_p^2}{A_p^2} = \mathscr{I}_p(f)\rho^2(\phi_p, \psi_p), \tag{8.6.21}$$

so that

$$\sup_{\{\phi\}} A_p^{-2}\gamma_p^2 = \mathscr{I}_p(f) \qquad (\text{increasing for } p \in (0,1]), \tag{8.6.22}$$

and the equality is attained when $\psi_p(u) \equiv \phi_p(u)$. The efficiency results of Chapter 5 remain applicable to this censored case, with $\mathscr{I}(f)$ replaced by $\mathscr{I}_p(f)$. In the particular case of $q = 1$, tests can also be based on $A_{Nr}^{-1}C_N^{-1}L_{Nr}$.

Consider next the case of stochastic predictors, as outlined in Section 8.2, when the Y_i (univariate case) are subject to type II censoring [only on the dependent (Y) variable]. Define the $a_N(i)$ and $b_{Nj}(i)$, $1 \le i \le N$, $1 \le j \le q$, as in (8.2.4)–(8.2.5), and the ranks as in (8.2.3). Note that the S_{ji} are all observable, while the R_{1i} $(= R_i)$ are subject to the censoring. In the same way, L_{Nr} is a projection of L_N, given $\mathbf{S}_N^{(r)}$, and we have the following statistic, which is a projection of $\mathbf{M}_N$ in (8.2.6):

$$\mathbf{M}_{Nr} = E_0\left(\mathbf{M}_N | \mathbf{S}_N^{(r)}\right)$$

$$= \left(\sum_{i=1}^{N} \left[a_{Nr}(R_i) - \bar{a}_N\right]\left[b_{Nj}(S_{ji}) - \bar{b}_{Nj}\right], 1 \leq j \leq q \right), \qquad (8.6.23)$$

where the $a_{Nr}(i)$ are defined by (8.6.10). (Note that the *antiranks* $S_{N1}, \ldots, S_{Nr}$ for the Y-variables are different from the ranks S_{ji}, $1 \leq j \leq q$, $1 \leq i \leq N$ of the predictors.)

Let us define A_{Nr}^2 as in (8.6.11), and $\mathbf{V}_N^{(2)}$ as in (8.2.11). Then, for testing H_0 in (8.2.2), based on the censored data, we consider the test statistic

$$\mathscr{L}_{Nr}^0 = \left(nA_{Nr}^2\right)^{-1}\left\{\mathbf{M}_{Nr}'\left(\mathbf{V}_N^{(2)}\right)^{-}\mathbf{M}_{Nr}\right\}. \qquad (8.6.24)$$

Unlike the case of $\mathscr{L}_{Nr}$ in (8.6.13), for $q \geq 2$, $\mathscr{L}_{Nr}^0$ is not genuinely distribution-free, but it is conditionally (given the reduced rank collection matrix) distribution-free. For $q = 1$, however, $\mathscr{L}_{Nr}^0$ is genuinely distribution-free. For small N, the (conditional) null distribution of $\mathscr{L}_{Nr}^0$ may be obtained by direct enumeration, while for large N, by an appeal to (8.2.29) [where the $a_N(i)$ are replaced by $a_{Nr}(i)$, $1 \leq i \leq N$, and the Y_i are real-valued r.v.], we conclude that under H_0 in (8.2.2), $\mathscr{L}_{Nr}^0$ has asymptotically a chi-square distribution with q DF.

Side by side, we may also introduce the mixed-rank statistic $\mathbf{M}_{Nr}^*$, defined by

$$\mathbf{M}_{Nr}^* = E_0\left\{\mathbf{M}_N^* | \mathbf{S}_N^{(r)}\right\}, \qquad (8.6.25)$$

where $\mathbf{M}_{Nr}^*$ is defined by (8.2.8), and this reduces to

$$\mathbf{M}_{Nr}^* = \left(\left(\sum_{i=1}^{N} \left[a_{Nr}(i) - \bar{a}_N\right]\left(X_{ji} - \bar{X}_{jn}\right), 1 \leq j \leq q \right)\right). \qquad (8.6.26)$$

If we define the covariance matrix $\mathbf{S}_N$ as in (8.2.13), then the test statistic reduces to

$$\mathscr{L}_{Nr}^* = \frac{N-1}{N^2 A_{Nr}^2} \mathbf{M}_{Nr}^{*\prime} \mathbf{S}_N^{-1} \mathbf{M}_{Nr}^*. \qquad (8.6.27)$$

$\mathscr{L}_{Nr}^{*}$ is a conditionally distribution-free test statistic, and as in (8.2.29), for large N the null distribution of $\mathscr{L}_{Nr}^{*}$ can be approximated by the chi-square distribution with q DF.

We define the covariance matrices $\boldsymbol{\nu}^{(2)}$ and $\boldsymbol{\Sigma}$ as in Section 8.2.2 [before (8.2.26)]. Then under local alternatives $\{H_N\}$, relating to (8.2.1) with $\boldsymbol{\beta} = N^{-1/2}\boldsymbol{\lambda}$, $\mathscr{L}_{Nr}^{0}$ has asymptotically a noncentral chi-square distribution with q DF and noncentrality parameter

$$\Delta_{\mathscr{L}^0, p} = \gamma_p^2 A_p^{-2}(\boldsymbol{\lambda}'\boldsymbol{\nu}^{(2)-}\boldsymbol{\lambda}), \tag{8.6.28}$$

and $\mathscr{L}_{Nr}^{*}$ has asymptotically a noncentral chi-square distribution with q DF and noncentrality parameter

$$\Delta_{\mathscr{L}^*, p} = \gamma_p^2 A_p^{-2}(\boldsymbol{\lambda}'\boldsymbol{\Sigma}^{-}\boldsymbol{\lambda}), \tag{8.6.29}$$

where γ_p^2 and A_p^2 are defined as in (8.6.15)–(8.6.16). Note that the discussion following (8.6.20) also holds for (8.6.28)–(8.6.29).

Let us now consider the case of type I censoring (truncation) schemes. As has been described in (8.6.1), the situation relates to a combination of grouped and ungrouped data; the observed responses on $(-\infty, T]$ relate to the ungrouped part, and the number of censored responses [belonging to (T, ∞)] relates to the group part. Thus, here one needs to combine the results of Section 8.5 with those in Chapter 5. We follow Chatterjee and Sen (1973) and Majumdar and Sen (1978a) and consider the alternative approach via type II censoring schemes. As in the discussion after (8.6.1), let $r(T) =$ (number of $Y_i \le T$, $1 \le i \le N$). Then, under the null hypothesis H_0: $\boldsymbol{\beta} = \mathbf{0}$ [relating to (8.6.3)], the probability law of $r(T)$ is given by

$$P_0\{r(T) = r\} = \binom{N}{r} P^r(1-P)^{N-r}, \quad 0 \le r \le N, \tag{8.6.30}$$

where $P = F(T - \alpha) \in (0,1)$. We define $\mathbf{L}_{Nr(T)}$ as in (8.6.8) when $r(T) = r$ [for $r(T) = 0$, $\mathbf{L}_{N0} = \mathbf{0}$]. Then $\mathbf{L}_{Nr(T)}$ is employed (instead of $\mathbf{L}_{Nr}$) for the testing.

Note that (8.6.30) depends on the unknown F through $F(T-\alpha)$, and hence $\mathbf{L}_{Nr(T)}$ is not genuinely distribution-free (even under H_0). On the other hand, under H_0, the set of antiranks $S_{N1}, \dots, S_{NN}$ and the set of order statistics $Y_{N,1} < \cdots < Y_{N,N}$ are stochastically independent, while $r(T)$ depends only on the order statistics [i.e., $r(T) = \max\{k: Y_{N,k} \le T\}$]. Hence, given $r(T) = r$, conditionally $\mathbf{L}_{Nr(T)}$ is distribution-free with mean and covariance matrix given by (8.6.12). Thus, (8.6.13) provides a conditionally distribution-free test [for $r(T) = r$]; we denote the test statistic by

$$\mathscr{L}_{Nr(T)} = \begin{cases} \dfrac{\mathbf{L}'_{Nr(T)}\mathbf{C}_N^{-}\mathbf{L}_{Nr(T)}}{A^2_{Nr(T)}}, & r(T) > 0, \\ 0, & r(T) = 0. \end{cases} \tag{8.6.31}$$

Note that by (8.6.30), under H_0,

$$N^{-1}r(T) \to P = F(T - \alpha) \quad \text{a.s.} \quad \text{as} \quad N \to \infty. \tag{8.6.32}$$

Further, if $\beta_{Nk} = \beta(\mathbf{S}_N^{(k)})$ denotes the σ-field generated by the subvector $\mathbf{S}_N^{(k)}$ of antiranks, $0 \le k \le N$, then β_{Nk} is nondecreasing in k, and by (8.6.8), $\{\mathbf{L}_{Nk}, \beta_{Nk} : 0 \le k \le N\}$ is a martingale sequence, closed on the right by $\mathbf{L}_N$. Using this martingale property along with (8.6.32), we obtain by the Kolmogorov maximal inequality that under H_0, for every $\varepsilon > 0$ and $\eta > 0$, there exist a $\delta : 0 < \delta < 1$ and an N_0 such that for every $N \ge N_0$,

$$P\Big\{ \max_{r : |r/N - P| \le \delta} |\mathscr{L}_{Nr} - \mathscr{L}_{N[NP]}| > \varepsilon \Big\} < \eta. \tag{8.6.33}$$

From (8.6.31)–(8.6.33) and the asymptotic chi-square distribution of $\mathscr{L}_{N[NP]}$ [following from (8.6.13)–(8.6.20)], we conclude that under H_0, $\mathscr{L}_{Nr(T)}$ has asymptotically a chi-square distribution with q DF. Thus, for type I censoring too, the statistic $\mathscr{L}_{Nr(t)}$ can be used as in a type II censoring case. Note that under $\{H_N\}$, and granted the "contiguity"-insuring regularity conditions [as assumed for (8.6.20)], (8.6.32)–(8.6.33) remain true for $\{H_N\}$ too, and hence, by (8.6.20) and (8.6.31), we conclude that under $\{H_N\}$, $\mathscr{L}_{Nr(T)}$ has asymptotically a noncentral chi-square distribution with q DF and noncentrality parameter $\Delta_{\mathscr{L}, P}$ defined by (8.6.20) with $P = F(T - \alpha)$. Therefore the discussions following (8.6.20) also applying to type I censoring.

Since (8.6.30) and (8.6.32) hold for both the cases where $\mathbf{M}_{Nr(T)}$ and $\mathbf{M}^*_{Nr(T)}$ are used instead of $\mathscr{L}_{Nr(T)}$, and both $\{\mathbf{M}_{Nk}, \beta_{Nk} : 0 \le k \le N\}$ and $\{M^*_{Nk}, \beta_{Nk} : 0 \le k \le N\}$ are (by definition) martingales under H_0, (8.6.33) also holds for $\mathscr{L}_{Nr}$ and $\mathscr{L}_{N[NP]}$ replaced by $\mathscr{L}^0_{Nr}$ and $\mathscr{L}^0_{N[NP]}$ (or by $\mathscr{L}^*_{Nr}$ and $\mathscr{L}^*_{N[NP]}$), respectively, and hence, for a type I censoring scheme, we are also in a position to use $\mathscr{L}^0_{Nr(T)}$ and $\mathscr{L}^*_{Nr(T)}$ for testing H_0 when the predictors are stochastic.

In passing, we may remark that for the two-sample or several-sample case, various attempts have been made to construct locally most powerful rank (LMPR) tests for censored data, when some specific density is conceived for the model (8.6.1). A recent paper by Basu, Ghosh, and Sen (1983) relates to a unified way of deriving an LMPR test for censored data. We pose their results as exercises at the end of this chapter.

Let us now consider the random censoring scheme considered after (8.6.2). In such a case, tests may be based either solely on the Y_i^0 (ignoring the tagging variable δ_i) or on the vectors (Y_i^0, δ_i), $i \geq 1$. Note that the joint distribution of (Y_i^0, δ_i) depends explicitly on that of (Y_i, T_i). If the T_i can have different distributions, then even if the Y_i have the same distribution, the Y_i^0 may not be i.i.d.r.v.'s. Thus, usually a simplifying assumption is made: the T_i are i.i.d.r.v.'s with a distribution function G. This may quite well be justified in a random censoring scheme. It is further assumed that for each i, Y_i and T_i are stochastically independent. Thus, if $F_i^*(y)$ is the d.f. of Y_i^0, then we have

$$1 - F_i^*(y) = [1 - G(y)][1 - F_i(y)], \qquad i \geq 1, \tag{8.6.34}$$

where F_i is the d.f. of Y_i, $i \geq 1$. Under the model (8.6.3) with $\boldsymbol{\beta} = \mathbf{0}$, or in general, when the F_i are all equal, the F_i^* are also all equal. Hence, if under any null hypothesis the Y_i are i.i.d.r.v.'s and (8.6.34) is assumed, then the Y_i^0 are also. Thus the rank tests developed in Sections 5.2–5.3 for the Y_i can readily be adapted for the Y_i^0, with the underlying d.f. F replaced by F^*. This justifies the validity and distribution-freeness of the rank tests based on the Y_i^0 alone. However, these rank tests may not be very informative (since the information on the δ_i has not been incorporated in the construction of the test statistics). To make this point clear, note that by (8.6.34), for every $i \neq i' = 1, \ldots, N$,

$$F_i^*(x) - F_{i'}^*(x) = [1 - G(x)][F_i(x) - F_{i'}(x)], \tag{8.6.35}$$

where $1 - G(x) \in (0,1)$. This damping factor diminishes the distance between F_i^* and $F_{i'}^*$ (compared to F_i and $F_{i'}$) and leads to a smaller noncentrality parameter in Theorem 5.5.2. If we look at (5.5.15)–(5.5.16) and replace the F_i by F_i^*, then under $\{H_N^{(1)}\}$ in (5.5.2), (5.5.17) holds with B replaced by B^* (in the univariate case), where

$$B^* = \int_{-\infty}^{\infty} \phi(F^*(x))[1 - G(x)] f(x)\, dF^*(x). \tag{8.6.36}$$

Therefore, in (5.5.25), we need to replace B by B^* (in the univariate case), so that (5.5.24) reduces to

$$\Delta_{\mathscr{L}} = \frac{(B^*)^2}{A^2} \sum_{j=1}^{q} \sum_{l=1}^{q} \tilde{C}_{jl} \beta_j^0 \beta_l^0. \tag{8.6.37}$$

Since $B^* \leq B$, (3.6.37) is numerically smaller than (5.5.24), and this accounts for the loss of efficiency of the censored rank procedure. The relative efficiency of the random censoring scheme with respect to the uncensored case is equal to

$$\left(\frac{B^*}{B}\right)^2, \tag{8.6.38}$$

and this depends on the average damping factor in (8.6.35). If the d.f. G is shifted farther to the right of F, then (8.6.38) will be close to 1, while it may go to zero as G is shifted to the left.

In the above development, information on the δ_i has not been incorporated. Incorporating this information, an estimator of F^*, known as the Kaplan–Meier (1958) (or the product-limit) estimator, is available in the literature. Procedures based on this estimator are, however, not solely rank-based ones, and we shall not discuss them. We refer to Breslow and Crowley (1974) and Susarla, van Ryzin, and Koul (1981) for some nice account of developments in this direction.

So far, we have considered the case where the dependent variables Y_i are univariate, so that censoring is defined quite unambiguously. However, even if the response Y_i is multivariate, sometimes the censoring is done only on one of the characteristics, while the others are all observable. In such a case, the results extend directly to the multivariate case. We need to incorporate the theory in Sections 5.4–5.5 and 8.2. These are, of course, more complicated to present. A very especial case of this multivariate model is the analysis of covariance (ANOCOVA) model, where besides the primary response (subject to possible type I or II or random censoring), there are concomitant variates (covariates) which are not subject to censoring. In such a case, the results treated earlier in this section extend readily and are dealt with in detail in Sen (1979b, 1981b, 1984a, b). We pose some of these as exercises and refer to Chapter 11 of Sen (1981a) for detailed discussions. Also, throughout this section, we have considered the case where under H_0, the Y_i are i.i.d.r.v.'s. In Chapter 7, for the uncensored case, the general theory of rank-order tests for subhypotheses have been studied. It is a natural question how far the theory goes over to the censored case. The main barrier is the estimation of the nuisance parameters in a censored model based on rank statistics and the linearity theorems pertaining to the aligned censored rank-order statistics. While some fragmentary work has been done in this direction, the general theory needs to be developed in a more unified manner before it can be adapted for the subhypotheses-testing problem. Finally, the concept of fixed-point (type I or II) censoring has led to a more general concept of progressive censoring, where the experiment is monitored from the beginning and a possible early termination (depending on the outcome) is planned. Nonparametric testing under progressive censoring based on linear rank statistics has been systematically developed by Chatterjee and Sen (1973), Majumdar and Sen (1977, 1978a, b), Gardiner and Sen (1978), Koziol and Petkau (1978), and Sen (1976b, 1978a, 1979a, b, 1981b, 1984), and others. Most of these developments include the single-point censoring schemes as particular cases, and rest on some invariance principles for censored linear rank statistics. These are discussed in detail in Chapter 11 of Sen (1981a) and Chapter 2 of Sen (1985). We shall not enter into this discussion.

EXERCISES

8.2.1. Provide a formal proof of the assertion (8.2.29).

8.2.2. Provide a formal proof of the assertion (8.2.32).

8.2.3. Provide a formal proof of the assertion (8.2.34).

8.2.4. Provide that under H_0, L_n^{**} defined in Section 8.2.1 has asymptotically the central chi-square distribution with pq DF.

8.2.5. Prove that under H_n, L_n^* has asymptotically the noncentral chi-square distribution with pq DF and noncentrality parameter Δ_L^* given by (8.2.36).

8.3.1. Prove that under H_0, L_n^* defined in (8.3.15) has asymptotically the central chi-square distribution with pt DF.

8.3.2. Prove that under H_n defined in (8.3.18), L_n^* has asymptotically the noncentral chi-square distribution with pt DF and noncentrality parameter Δ_L^* given by (8.3.24).

8.3.3. Prove that under $H_0: \boldsymbol{\beta}_2 = \mathbf{0}$ and the regularity conditions of Section 7.4, $\hat{L}_n^*$ has asymptotically the central chi-square distribution with pt_2 DF.

8.3.4. Prove that under $H_n: \boldsymbol{\beta}_2 = n^{-1/2}\boldsymbol{\delta}$, $\hat{L}_n^*$ has asymptotically the noncentral chi-square distribution with pt_2 DF and noncentrality parameter Δ_{L^*} given by (8.3.37).

8.5.1. Verify the assertions (8.5.20) and (8.5.21).

8.5.2. Verify the assertions (8.5.34) and (3.5.35).

8.5.3. Provide a formal proof of the assertion (8.5.37).

8.5.4. Provide a formal proof of the assertion (8.5.43).

8.5.5. Prove that under H_0, S_N^0 defined in (8.5.33) has asymptotically the normal distribution with mean 0 and variance $A^2(J)$ given by (8.5.27).

8.5.6. Verify the assertion (8.5.45).

8.5.7. Prove that under H_0, L_n defined in (8.5.42) has asymptotically the central chi-square distribution with q DF.

8.5.8. Provide a formal proof of the assertion (8.5.53).

8.5.9. Provide a formal proof of the assertions (8.5.68) and (8.5.69).

8.5.10. Prove that under H_n, L_n has asymptotically the noncentral chi-square distribution with q DF and noncentrality parameter Δ_L given by (8.5.70).

8.5.11. Verify the assertions (8.5.78) and (8.5.82).

8.5.12. Prove that under H_0, L_n defined in (8.5.85) has asymptotically the central chi-square distribution with pq DF, and under a sequence of contiguous alternatives, it has a noncentral chi-square distribution.

8.6.1. For the model (8.6.3), show that under $H_0: \boldsymbol{\beta} = \mathbf{0}$, for every k $(1 \leq k \leq N)$, $(S_{N1}, \ldots, S_{Nk})$ and $(Y_{N,1}, \ldots, Y_{N,k})$ are stochastically independent. Hence, or otherwise, verify (8.6.7).

8.6.2. Verify (8.6.22).

8.6.3. Define $r(T)$ as was done after (8.6.1). Use the result of Exercise 8.6.1 to obtain the exact distribution of $r(T)$ under $H_0: \boldsymbol{\beta} = \mathbf{0}$. Hence, or otherwise, verify (8.6.32).

8.6.4. Verify (8.6.33). (Chatterjee and Sen, 1973.)

8.6.5. Let T_1, T_2 be two (possibly vector-valued) statistics. For each θ, let $L_\theta(T_1, T_2)$, $L_{1\theta}(T_1)$, and $L_{2\theta}(T_2)$ be the joint and marginal densities of T_1 and T_2. Then show that $L_{1\theta}(t_1)/L_{1\theta_0}(t_1) = E_{\theta_0}\{L_\theta(T_1, T_2)/L_{\theta_0}(T_1, T_2) | T_1 = t_1\}$. (Basu, Ghosh, and Sen 1983.)

8.6.6. Specify suitable regularity conditions on L_θ, under which for all $\theta = \theta_0 + \lambda$, $0 \leq \lambda \leq \Delta$, as $\Delta \to 0$, $L_\theta(T_1, T_2)/L_{\theta_0}(T_1, T_2) = 1 + \lambda\psi(T_1, T_2) + o(\lambda)$ a.e. Then show that under the same regularity conditions, $L_{1\theta}(T_1)/L_{1\theta_0}(T_1) = 1 + \lambda\psi_1(T_1) + o(\lambda)$ a.e., where $\psi_1(t_1) = E_{\theta_0}\{\psi(T_1, T_2) | T_1 = t_1\}$. (Basu, Ghosh, and Sen 1983.)

8.6.7. Considering the vectors T_1 and T_2 as $(S_{N1}, \ldots, S_{Nr})$ and $(S_{Nr+1}, \ldots, S_{NN})$ and using the result in Exercise 8.6.6, derive the locally most powerful rank test statistics for the simple regression model under type

II censoring. Further, use (8.6.33) to extend this result to type I censoring schemes. (Basu, Ghosh and Sen 1983.)

8.6.8. Use the result of Exercise 8.6.6 to derive locally most powerful rank tests for independence in the bivariate case under (a) type II censoring and (b) type I censoring schemes. (Shirahata, 1975; Basu, Ghosh, and Sen, 1983.)

8.6.9. Verify that in (8.6.38), $(B^*/B)^2$ is bounded from above by 1. (Sen, 1983a, b.)

8.6.10. Consider the ANOCOVA model in (8.3.1), and define $\mathbf{L}_n$, $\mathbf{V}_n$, $\mathbf{C}_n$, and so on as in (8.3.4)–(8.3.11). In this setup, take $p = 1$, so that there is only one primary variate and the rest are all covariates. The primary variate is subject to censoring, while the covariates are not. For the primary variate, define the antiranks as in (i.6.4)–(8.6.5).

(a) Obtain the expression for $\mathbf{L}_{nr} = E_{\mathscr{P}_n}(\mathbf{L}_n | S_{n1}, \ldots, S_{nr})$, $E_{\mathscr{P}_n}\mathbf{L}_{nr}$, and the permutational dispersion matrix of $\mathbf{L}_{nr}$. Show that under the null hypothesis H_0 in (8.3.3), the $\mathbf{L}_{nr}$ form a martingale sequence $(0 \leq r \leq n)$ for every n (≥ 1).

(b) Parallel to (8.3.12), obtain the residuals of regression of $\mathbf{L}_{nr}^{(1)}$ on $\mathbf{L}_{nr}^{(2)}$, and parallel to (8.3.13)–(8.3.14), obtain its permutational covariance matrix.

(c) Deduce the form of the test statistic [parallel to (8.3.15)] for testing the null hypothesis H_0 in (8.3.3) when one has a type II censoring scheme, the censoring being done at the rth failure on the primary variate.

(d) Use (8.6.33) to extend the results in (a)–(c) to type I censoring schemes, and justify these under the usual conditional arguments. (Sen, 1979b, 1981b, 1984a, c.)

Appendix

A.1. INTRODUCTION

Throughout the text, for smooth reading, deviations of some of the results (stated, mostly, in the form of lemmas and propositions) were postponed; these will be presented collectively here. For readers interested mainly in the final results and applications, this appendix will not be of much concern, while for others interested in proofs as well, these may provide some useful tools. The proofs are presented in the order of the results proved.

A.2. PROOF OF LEMMAS 2.6.2 AND 2.6.3

The proof is due to Hoeffding (1973). Note that

$$\sum_{i=1}^{N}\left|\phi\left(\frac{i}{N+1}+\right)-\phi\left(\frac{i}{N+1}\right)\right| \le \sum_{i=1}^{N}\int_{t=i/(N+1)} d\phi(t)$$

$$\le \left\{\max_{1\le i\le N}\left(\frac{i}{N+1}\right)^{-1/2}\left(\frac{N-i+1}{N+1}\right)^{-1/2}\right\}$$

$$\times \sum_{i=1}^{N}\int_{t=i/(N+1)} t^{1/2}(1-t)^{1/2}\,d\phi(t)$$

$$\le 2N^{1/2}J(\phi), \tag{A.2.1}$$

where $J(\phi)$ is defined by (2.6.10). Hence, to prove (2.6.13), it suffices to show that

$$\sum_{i=1}^{N}\left|E\phi\left(U_N^{(i)}\right)-\phi\left(\frac{i}{N+1}+\right)\right| \le c_1^* J(\phi)N^{1/2}, \tag{A.2.2}$$

where c_1^* $(= C_1 + 2)$ is a finite constant. Let $G_{N,i}$ be the d.f. of $U_N^{(i)}$, and let

$$H_{N,i}(t) = I\left(t \le \frac{i}{N+1}\right)G_{N,i}(t) + I\left(t > \frac{i}{N+1}\right)[1 - G_{N,i}(t)], \tag{A.2.3}$$

so that, noting that

$$E\phi(U_N^{(i)}) - \phi\left(\frac{i}{N+1}+\right) = \int_0^1 \left\{\phi(u) - \phi\left(\frac{i}{N+1}+\right)\right\} dG_{N,i}(u)$$

$$= \int_{i/(N+1)+} H_{N,i}(t)\, d\phi(t) - \int_0^{i/(N+1)} H_{N,i}(t)\, d\phi(t), \tag{A.2.4}$$

we have

$$\sum_{i=1}^{N} \left| E\phi(U_N^{(i)}) - \phi\left(\frac{i}{N+1}+\right)\right| \le \int_0^1 \sum_{i=1}^{N} H_{N,i}(t)\, d\phi(t). \tag{A.2.5}$$

Hence, it suffices to show that

$$\sum_{i=1}^{N} H_{N,i}(t) \le C_1^* N^{1/2}\{t(1-t)\}^{1/2} \qquad \forall 0 < t < 1. \tag{A.2.6}$$

Since $G_{N,i}(t) = P\{W_N(t) \ge i\}$, where $W_N(t)$ has the binomial distribution with parameters (N, t), by (A.2.3),

$$\sum_{i=1}^{N} H_{N,i}(t) = E|W_N(t) - j| \quad \text{if } j < (N+1)t \le j+1, \qquad 0 \le j \le N. \tag{A.2.7}$$

Finally, $E|W_N(t) - j| \le \{E(W_N(t) - j)^2\}^{1/2} = \{Nt(1-t) + (Nt - j)^2\}^{1/2}$ and $(Nt - j)^2 \le Nt(1-t)$, $\forall j < (N+1)t \le j+1$, so (A.2.6) holds with $C_1^* = \sqrt{2}$. Hence, (2.6.13) holds.

To prove (2.6.14), we note that

$$E\left\{\phi\left(\frac{R_{Ni}}{N+1}\right)\middle| X_i = x\right\} = E\phi\left(\frac{V_i(x)}{N+1}\right), \tag{A.2.8}$$

where $V_i(x) = 1 + \sum_{j=1}^{N} u(x - X_j) - u(x - X_j) = 1 + V(x) - u(x - X_i)$ for $i = 1, \dots, N$. [Note that $V(x) \le V_i(x) \le 1 + V(x)$ and $V_i(x) \le N$ for all x, $i \ge 1$.] Hence, for every $i(= 1, \dots, N)$,

$$\left|E\phi\left(\frac{R_{Ni}}{N+1}\right) - \int \phi(H(x))\, dF_i(x)\right| \le \int \left|E\phi\left(\frac{V_i(x)}{N+1}\right) - \phi(H(x))\right| dF_i(x)$$

$$\le \int E\left|\phi\left(\frac{V_i(x)}{N+1}\right) - \phi(H(x))\right| dF_i(x), \tag{A.2.9}$$

where

$$\phi\left(\frac{V_i(x)}{N+1}\right) \le g(V(x)), \qquad g(u) = \min\left\{\phi\left(\frac{u+1}{N+1}\right), \phi\left(\frac{N}{N+1}\right)\right\}. \tag{A.2.10}$$

Further, for nondecreasing ϕ, $\phi(V_i(x)/(N+1)) \ge \phi(V(x)/(N+1))\ \forall i \ge 1$, so that by (A.2.9), (A.2.10), and some standard steps, the lhs of (2.6.14) is bounded from above by

$$N\int E|g(V(x)) - \phi(H(x))|\,dH(x) + \phi\left(\frac{N}{N+1}\right) - \phi\left(\frac{1}{N+1}\right)$$

$$\le N\int E|g(V(x)) - g([NH(x)])|\,dF(x)$$

$$+N\int_0^1 |g([Nt]) - \phi(t)|\,dt + \phi\left(\frac{N}{N+1}\right) - \phi\left(\frac{1}{N+1}\right), \tag{A.2.11}$$

where

$$E|g(V(x)) - g(\lfloor NH(x)\rfloor)| = \sum_{i=1}^{[NH(x)]} \{g(i) - g(i-1)\}P\{V(x) \le i-1\}$$

$$+ \sum_{i=\lfloor NH(x)\rfloor+1}^{N} \{g(i) - g(i-1)\}P\{V(x) \ge i\}, \tag{A.2.12}$$

and defining $W_N(t)$ as in (A.2.7), we obtain, by using the tail-dominated property of the binomial distribution for heterogeneous p's, that

$$P\{V(x) \le j\} \le P\{W_N(H(x)) \le j\} \qquad \forall j \le NH(x) - 1,$$

$$P\{V(x) \ge j\} \le P\{W_N(H(x)) \ge j\} \qquad \forall j \ge NH(x) + 1, \tag{A.2.13}$$

so that by (A.2.12) and (A.2.13),

$$E|g(V(x)) - g(\lfloor NH(x)\rfloor)| \le E|g(W_N(H(x)) - g(\lfloor NH(x)\rfloor)|. \tag{A.2.14}$$

Thus,

$$\int E|g(V(x)) - g(\lfloor NH(x)\rfloor)|\,dH(x)$$

$$\le \int_0^1 E|g(W_N(t)) - g(\lfloor Nt\rfloor)|\,dt = \sum_{i=1}^{N-1} \{g(i) - g(i-1)\}$$

$$\times\left(\int_{i/N}^1 P\{W_N(t) \le i-1\}\,dt + \int_0^{i/N} P\{W_N(t) \ge i\}\,dt\right). \tag{A.2.15}$$

Further, note that

$$\int_{i/N}^{1} P\{W_N(t) \le i-1\}\,dt + \int_0^{i/N} P\{W_N(t)B \ge i\}\,dt$$

$$= \int_{i/N}^{1}\{1 - G_{N,i}(t)\}\,dt + \int_0^{i/N} G_{N,i}(t)\,dt$$

$$= \int_0^1 \left|u - \frac{i}{N}\right| dG_{N,i}(u) \le \left\{\int_0^1 \left(u - \frac{1}{N}\right)^2 dG_{N,i}(u)\right\}^{1/2}$$

$$\le 2N^{-1/2}\left\{\frac{i}{N}\left(1 - \frac{i}{N}\right)\right\}^{1/2}, \qquad \text{(A.2.16)}$$

and $g(i) = \phi((i+1)/(N+1))\ \forall 1 \le i \le N-1$. Thus, by (A.2.15) and (A.2.16),

$$\int_0^1 E|g(W_N(t)) - g(\lfloor Nt \rfloor)|\,dt$$

$$\le 2N^{-1/2}\sum_{i=1}^{N-1}\left\{\phi\left(\frac{i+1}{N+1}\right) - \phi\left(\frac{i}{N+1}\right)\right\}\left(\frac{i}{N}\right)^{1/2}\left(1 - \frac{i}{N}\right)^{1/2}$$

$$= 2N^{-1/2}J(\phi_N^*) \qquad \left[\phi_N^*(t) = \phi\left(\frac{[Nt]+1}{N+1}\right),\ 0 \le t \le 1\right]$$

$$\le 4N^{-1/2}J(\phi), \qquad \text{as } \phi \text{ is increasing.} \qquad \text{(A.2.17)}$$

Thus, by (A.2.11), (A.2.15), and (A.2.17), we have the left-hand side of (2.6.14) bounded by

$$4N^{1/2}J(\phi) + N\int_0^1 |g(\lfloor Nt \rfloor) - \phi(t)|\,dt + \left\{\phi\left(\frac{N}{N+1}\right) - \phi\left(\frac{1}{N+1}\right)\right\}. \qquad \text{(A.2.18)}$$

Finally.

$$N\int_0^1 |g(\lfloor Nt \rfloor) - \phi(t)|\,dt = N\int_0^1 \left|\phi\left(\frac{[Nt]+1}{N+1}\right) - \phi(t)\right| dt$$

$$\le N\int_0^1 \left\{\phi\left(\frac{Nt+1}{N+1}\right) - \phi\left(\frac{Nt}{N+1}\right)\right\} dt$$

$$= (N+1)\left\{\int_{N/(N+1)}^{1} \phi(t)\,dt - \int_0^{1/(N+1)} \phi(t)\,dt\right\}, \qquad \text{(A.2.19)}$$

$$\phi\left(\frac{N}{N+1}\right) - \phi\left(\frac{1}{N+1}\right) \le (N+1)\left\{\int_{N/(N+1)}^{1} \phi(t)\,dt - \int_0^{1/(N+1)} \phi(t)\,dt\right\}, \qquad \text{(A.2.20)}$$

and

$$\int_{N/(N+1)}^{1} \phi(t)\,dt - \int_0^{1/(N+1)} \phi(t)\,dt = \int_0^1 \min\left(u, (N+1)^{-1}, 1-u\right) d\phi(u)$$

$$\leq N^{-1/2} \int_0^1 u^{1/2}(1-u)^{1/2}\, d\phi(u) = N^{-1/2} J(\phi), \qquad \text{(A.2.21)}$$

so that by (A.2.18)–(A.2.21), we conclude that (2.6.14) holds with $C_2 = 8$, Q.E.D.

A.3. PROOFS OF THEOREMS 5.2.1, 5.2.2, 5.2.4, 5.2.5, AND 5.4.2

For L_N in Theorem 5.2.1, the permutation distribution is unaltered by an *a priori* ordering of the $a_N(i)$ or $b_N(i)$, and hence we may assume without any loss of generality that $a_N(1) \leq \cdots \leq a_N(N)$ and $b_N(1) \leq \cdots \leq b_N(N)$. Then, by an appeal to Theorem 5.2.2, it suffices to verify the asymptotic normality of $(T_N - N\bar{a}_N\bar{b}_N)/A_N B_N$, where $B_N^2 = \sum_{i=1}^N [b_N(i) - \bar{b}_N]^2$. If we let

$$Z_{Ni} = \frac{[b_N(i) - \bar{b}_N][a_N^0(U_i) - \bar{a}_N]}{A_N B_N}, \qquad 1 \leq i \leq N, \qquad \text{(A.3.1)}$$

then by (5.2.21), $EZ_{Ni} = 0$; $EZ_{Ni}^2 = \{[b_N(i) - \bar{b}_N]^2/B_N^2\}(N-1)/N$, $1 \leq i \leq N$; and $\sum_{i=1}^N EZ_{Ni}^2 = (N-1)/N \to 1$ as $N \to \infty$. Further, (5.2.17) insures that the Z_{Ni} are uniformly asymptotically negligible. Hence, to verify the asymptotic normality of $\sum_{i=1}^N Z_{Ni}$, it suffices to show that for every $\varepsilon > 0$, as $N \to \infty$,

$$\sum_{i=1}^N P\{|Z_{Ni}| > \varepsilon\} \to 0, \qquad \text{(A.3.2a)}$$

$$\sum_{i=1}^N \left\{ EZ_{Ni}^2 I(|Z_{Ni}| \leq \varepsilon) - [EZ_{Ni} I(|Z_{Ni}| \leq \varepsilon)]^2 \right\} \to 1. \qquad \text{(A.3.2b)}$$

Since the U_i have the uniform (0, 1) d.f., using (5.2.21), it is easy to verify that (5.2.18) insures both (A.3.2a) and (A.3.2b).

To prove Theorem 5.2.2, we first write

$$E(L_N - T_N)^2 = E\left[E\left\{ (L_N - T_N)^2 \middle| U_N^{(1)}, \ldots, U_N^{(N)} \right\}\right], \qquad \text{(A.3.3)}$$

where the $U_N^{(i)}$ stand for the ordered U_i. Given the $U_N^{(i)}$, $T_N = \Sigma_{i=1}^N[b_N(i) - \bar{b}_N][a_N^0(U_N^{(R_{Ni})}) - \bar{a}_N] + N\bar{a}_N\bar{b}_N$, so that $L_N - T_N$ is a linear rank statistic, for which (5.2.10) yields the variance. Hence, (A.3.3) is bounded from above by

$$B_N^2(N-1)^{-1}\sum_{i=1}^{N} E\left[a_N^0(U_i) - a_N(R_{Ni})\right]^2$$

$$= \frac{N}{N-1}B_N^2 E\left[a_N^0(U_1) - a_N(R_{n1})\right]^2, \tag{A.3.4}$$

where $a_N(R_{N1}) = a_N^0(N^{-1}R_{N1})$. Thus, to show that $B_N^{-2}A_N^{-2}E(L_N - T_N)^2 \to 0$, it suffices to show that as $N \to \infty$,

$$A_N^{-2}E\left[a_N^0(U_1) - a_N^0\left(\frac{1}{N}R_{N1}\right)\right]^2 \to 0. \tag{A.3.5}$$

Towards this, we let $u(t)$ be equal to 1 or 0 according as $t >$ or ≤ 0, and write for every $t \in (0,1]$

$$a_N^0(t) = \sum_{k=1}^{N-1}\{a_N(k+1) - a_N(k)\}u\left(t - \frac{k}{N}\right). \tag{A.3.6}$$

Then, note that $u(U_1 - k/N) - u(N^{-1}R_{N1} - k/N)$ is equal to 0 if (1) $U_1 \leq k/N$, $R_{N1} \leq k$ or (2) $U_1 > k/N$ and $R_{N1} \geq k+1$, while it is equal to $+1$ if $U_1 > k/N$, $R_{N1} \leq k$, and -1 if $U_1 \leq k/N$, $R_{N1} \geq k+1$, for $1 \leq k \leq N-1$. Further, R_{N1} assumes the values $1, \ldots, N$ with equal probability $1/N$, while for $R_{N1} = q$, $U_1 = U_N^{(q)}$ has the density

$$N\binom{N-1}{q-1}u^{q-1}(1-u)^{N-q}, \quad 0 < u < 1, \qquad \text{for} \quad q = 1, \ldots, N.$$

Hence, using (A.3.5) for each of $a_N^0(U_1)$ and $a_N^0(N^{-1}R_{N1})$, it follows by some standard steps (see Hájek, 1961, for details) that

$$E\left[a_N^0(U_1) - a_N(R_{N1})\right]^2 \leq \frac{2}{N}\max_{1\leq k\leq N}|a_N(k) - \bar{a}_N|$$

$$\times\left\{\sum_{i=1}^{N}[a_N(i) - \bar{a}_N]^2\right\}^{1/2}, \tag{A.3.7}$$

and hence, (5.2.17) and (A.3.7) insure (A.3.5).

The proof of Theorem 5.2.4 runs parallel to that of Theorem 5.2.1. By virtue of the asymptotic equivalence result in Theorem 5.2.5, it suffices to show that T_N in (5.2.40) is asymptotically normal. For this purpose, we define $Z_{Ni} = A_N^{-1}B_N^{-1}b_N(i)a_N^0(U_i)\text{sgn } X_i$, $i = 1, \ldots, N$, where $B_N^2 = \sum_{i=1}^N b_N^2(i)$ and $A_N^2 = \sum_{i=1}^N a_N^2(i)$. Then, as in (A.3.1), it is easy to verify that (5.2.38) insures (A.3.3), so that the asymptotic normality follows by an appeal to the classical central limit theorem.

The proof of Theorem 5.2.5 also runs almost parallel to that of Theorem 5.2.2. Additionally, here the r.v. U_i and sgn X_i are stochastically independent, and hence, verification of (A.3.3)–(A.3.7), with obvious modifications, follows precisely on the same lines. Therefore, the details are omitted.

Finally, we proceed to the proof of Theorem 5.4.2. In the general multivariate case, there is a technical point against using the coordinatewise Hájek-projection result, as the permutation distribution (being a conditional one) differs from the unconditional null distribution. For this reason, we adapt the following martingale approach, systematically explored in Chatterjee and Sen (1973) (in the univariate case) and Sen (1979b, 1983a).

We define $\mathbf{L}_N$ as in (5.4.4)–(5.4.5), $\mathbf{R}_N$ as in (5.4.6), and $\mathbf{R}_N^*$ as in the discussion after (5.4.6). Define then the vector (of antiranks) $\mathbf{S}_N = (S_{N1}, \ldots, S_{NN})$ by letting

$$R_{NS_{Ni}}^{(1)} = i \qquad \text{for} \quad i = 1, \ldots, N, \tag{A.3.8}$$

so that $\mathbf{S}_N$ relates to the column permutation of $\mathbf{R}_N$, reducing it to $\mathbf{R}_N^*$. To prove the theorem, it suffices to show that for an arbitrary $p \times q$ matrix $\Lambda = ((\lambda_{ij}))$ ($\neq \mathbf{0}$), the permutational distribution of $\text{Tr}(\Lambda'\mathbf{L}_N)$ is asymptotically (in probability) normal. We let $Z_N = \text{Tr}(\Lambda'\mathbf{L}_N)$ and let $\mathbf{S}_{Nr} = (S_{N1}, \ldots, S_{Nr})$, $1 \leq r \leq N$. Further, let $\mathscr{B}_{Nr}$ be the σ-field generated by $\mathbf{S}_{Nr}$, $0 \leq r \leq N$, where $\mathscr{B}_{N0}$ is the trivial σ-field. Finally, let

$$Z_{Nr} = E_{P_N}\{Z_N | \mathscr{B}_{Nr}\}, \qquad 0 \leq r \leq N, \tag{A.3.9}$$

so that $\{Z_{Nr}, \mathscr{B}_{Nr}, 0 \leq r \leq N\}$ is a martingale, closed on the right by Z_N (under $\mathscr{P}_N$). The basic idea is to verify a general martingale central limit theorem for the Z_{Nr}, where for the (martingale) differences $Y_{Nr} = Z_{Nr} - Z_{Nr-1}$, $1 \leq r \leq N$ ($Y_{N0} = 0$), we need to verify that under $\mathscr{P}_N$,

$$\left(E_{\mathscr{P}_N}(Z_N^2)\right)^{-1} \sum_{r=1}^{N} E_{\mathscr{P}_N}\{Y_{Nr}^2 | \mathscr{B}_{Nr-1}\} \xrightarrow{P} 1, \tag{A.3.10}$$

and the Y_{Nr} satisfy the classical Lindeberg conditional (under $\mathscr{P}_N$) in probability. If we define for every $1 \le j \le p$, $0 \le k \le N$,

$$a_{Nj}^*(k) = (N-k)^{-1}\left\{N\bar{a}_{Nj} - \sum_{i=1}^{k} a_{Nj}\left(R_{ij}^*\right)\right\}, \tag{A.3.11}$$

(where $a_{Nj}^*(N) = 0$, $1 \le j \le p$), then we have

$$Y_{Nr} = \sum_{j=1}^{p}\sum_{l=1}^{q}\lambda_{jl}\left\{a_{Nj}\left(R_{rj}^*\right) - a_{Nj}^*(r)\right\}$$
$$\times\left\{\left(c_{S_{Nr}l} - \bar{c}_{Nl}\right) - \frac{1}{N-r+1}\sum_{m=r}^{N}\left(c_{S_{Nm}l} - \bar{c}_{Nl}\right)\right\}, \tag{A.3.12}$$

where under $\mathscr{P}_N$, given $\mathbf{S}_{Nr-1}$, S_{Nm} can assume any one value from the set $\{1,\dots,N\}\setminus\mathbf{S}_{Nr-1}$ with the equal (conditional) probability $(N-r+1)^{-1}$, while R_{rj}^* and $a_{Nj}^*(r)$ are held fixed $(1 \le r \le N)$. Hence

$$E_{P_N}\left(Y_{rr}^2|\mathscr{B}_{Nr-1}\right) = \sum_{j=1}^{p}\sum_{j'=1}^{p}\sum_{l=1}^{q}\sum_{l'=1}^{q}\lambda_{jl}\lambda_{j'l'}\left\{a_{Nj}\left(R_{rj}^*\right) - a_{Nj}^*(r)\right\}$$
$$\times\left\{a_{Nj'}\left(R_{rj'}^*\right) - a_{Nj'}^*(r)\right\}$$
$$\times\frac{1}{N-r+1}\sum_{i=r}^{N}\left[\left(c_{S_{Ni}l} - \bar{c}_{Nl}\right) - \frac{1}{N-r+1}\sum_{m=r}^{N}\left(c_{S_{Nm}l} - \bar{c}_{Nl}\right)\right]$$
$$\times\left[\left(c_{S_{Ni}l'} - \bar{c}_{Nl'}\right) - \frac{1}{N-r+1}\sum_{m=r}^{N}\left(c_{S_{Nm}l'} - \bar{c}_{Nl'}\right)\right] \tag{A.3.13}$$

for $r = 1,\dots,N$. By virtue of the martingale property and Theorem 5.4.1,

$$\sum_{r=1}^{N}\frac{E_{\mathscr{P}_N}\left(Y_{Nr}^2\right)}{\sum_{j=1}^{p}\sum_{j'=1}^{p}\sum_{l=1}^{q}\sum_{l'=1}^{q}\lambda_{jl}\lambda_{j'l'}\nu_{jj'}C_{Nll'}} \xrightarrow{P} 1 \tag{A.3.14}$$

as $N \to \infty$ (where the denominator goes to $+\infty$ as $N \to \infty$). From this, using the polynomial approximation (Lemma 2.5.4) for the score functions ϕ_j $(1 \le j \le p)$ (under the hypothesis of Theorem 5.4.1), it follows that (5.3.8)–(5.3.10) and (A.3.13) insure the conditional Lindeberg condition. On the other hand, using the same polynomial approximation [thereby bounding the $a_{Nj}(i)$, $1 \le i \le N$, $1 \le j \le p$], we may express the lhs of (A.3.10) as a ratio of quadratic forms in the $\mathbf{c}_i$, each of which converges in L_1 norm (under $\mathscr{P}_N$) to a common value, and this insures (A.3.10). (For details, we may refer to Sen, 1979b, 1981b.) This complete the proof of Theorem 5.4.2.

A.4. ASYMPTOTIC LINEARITY RESULTS ON RANK STATISTICS

The asymptotic linearity results in (6.2.44), (6.2.45), (6.4.21), and (6.3.44) will be systematically studied here. These results are mainly adapted from Jurečková (1969, 1971a, b) and van Eeden (1972), with further simplifications in Sen (1981a, Theorems 4.5.2 and 5.5.1). For simplicity of presentation, we present only the proofs of the results in (6.2.44), (6.2.45), (6.3.21), and (6.3.44); more general results are appended with only brief indications of their derivation.

A.4.1. Proof of (6.2.44)

By virtue of Theorem 6.2.1, for every $\varepsilon > 0$ and $K < \infty$, there exists a positive integer M $(= M_{\varepsilon K})$ such that, on letting $b_j = -K + jM^{-1}K$, $j = 0, \ldots, 2M$, the lhs of (6.2.44) is bounded from above by

$$\max_{0 \le j \le 2M} \left| L_N\left(N^{-1/2}b_j\right) - L_N(0) + N^{1/2}bjC_0^2\rho(\phi, \psi)A_\phi A_\psi \right| + \tfrac{1}{2}\varepsilon\sqrt{N}\,. \tag{A.4.1}$$

Hence, it suffices to show that under the assumed regularity conditions, when β in (6.2.1) is equal to 0,

$$N^{-1/2}\left| L_N\left(N^{-1/2}b_j\right) - L_N(0) + N^{1/2}b_jC_0^2\rho(\phi, \psi)A_\phi A_\psi \right| \xrightarrow{P} 0 \tag{A.4.2}$$

uniformly in j $(0 \le j \le 2M)$; in this context, we may without any loss of generality assume that ϕ is nondecreasing, and a very similar proof works for ϕ the difference of two nondecreasing functions. We consider a sequence $\{\phi^{(k)}(u), 0 \le u \le 1, k \ge 1\}$ of bounded and nondecreasing score functions for which $\phi^{(k)} \to \phi$ as $k \to \infty$, in the sense that

$$\lim_{k \to \infty} \int_0^1 \{\phi^{(k)}(u) - \phi(u)\}^2\, du = 0. \tag{A.4.3}$$

If, in the definition of the $L_N(N^{-1/2}b)$, we replace the ϕ by $\phi^{(k)}$ and denote the resulting quantity by $L_N^{(k)}(N^{-1/2}b)$, $k \ge 1$, $b \in [-K, K]$, then by

(A.4.3), under $H_0: \beta = 0$, $N^{-1/2}|L_N^{(k)}(0) - L_N(0)|$ converges in mean square (and hence, in probability) to 0 as $k \to \infty$. Also, under $H_0: \beta = 0$, $L_N(N^{-1/2}b)$ [or $L_N^{(k)}(N^{-1/2}b)$] has the same distribution as $L_N(0)$ [or $L_N^{(k)}(0)$] under $K_N: \beta = -N^{-1/2}b$. Further, by (6.2.35), (6.2.36), and the finite-Fisher-information assumption following (6.2.39), the contiguity of the sequence of probability measures under K_N to that under H_0 follows as in Section 2.7 [see (2.7.39)]. Hence, for every j $(= 0, 1, \ldots, 2M)$, as $k \to \infty$,

$$N^{-1/2}\left|L_N\left(N^{-1/2}b_j\right) - L_N^{(k)}\left(N^{-1/2}b_j\right)\right| \xrightarrow{P} 0. \tag{A.4.4}$$

At this stage, we may note that by (A.4.3),

$$A_{\phi^{(k)}}\rho\left(\phi^{(k)}, \psi\right) \to A_{\phi}\rho\left(\phi, \psi\right) \quad \text{as} \quad k \to \infty, \tag{A.4.5}$$

while for any k (fixed), $\phi^{(k)}$ being bounded, we may appeal to Theorem 2.4.5 and claim that under $H_0: \beta = 0$ and the assumed regularity conditions,

$$N^{-1/2}\left|L_N^{(k)}\left(N^{-1/2}b_j\right) - L_N^{(k)}(0) + N^{1/2}b_jC_0^2\rho\left(\phi^{(k)}, \psi\right)A_{\phi^{(k)}}A_{\psi}\right| \xrightarrow{P} 0, \tag{A.4.6}$$

so that the desired result follows by choosing k adequately large and appealing to (A.4.4), (A.4.5), and (A.4.6).

It may be noted that if in (6.2.2), we define $X_i(b) = X_i - bd_{Ni}$, $1 \le i \le N$, where the d_{Ni} are real specified constants, and if we assume the following concordance condition (due to Jurečková, 1969):

$$\left(c_i - c_j\right)\left(d_{Ni} - d_{Nj}\right) \ge 0 \qquad \forall i \ne j = 1, \ldots, N, \tag{A.4.7}$$

then the monotonicity result in Theorem 6.2.1 holds when for the $Y_i(b)$ one takes $Y_i - bd_{Ni}$, $1 \le i \le N$. Thus, if we assume that

$$\lim_{N\to\infty} N^{-1/2}\sum_{i=1}^{N} d_{Ni}\left(c_i - \bar{c}_N\right) = C^* \text{ exists}, \tag{A.4.8}$$

$$\sup_N \sum_{i=1}^{N} d_{Ni}^2 < \infty, \tag{A.4.9}$$

then (6.2.44) also holds for this more general model with C_0^2 replaced by C^*. The line of the proof in (A.4.1) through (A.4.6) remains unaltered in this case.

A.4.2. Proof of (6.2.45)

By virtue of Lemma 6.2.3, for every fixed b, $S_N(N^{-1/2}a, N^{-1/2}b)$ is decreasing in a. Also, for every fixed a, $X_i - N^{-1/2}a - N^{-1/2}b(c_i - \bar{c}_N)$ is monotonically nondecreasing or nonincreasing in b according as $c_i <$ or $> \bar{c}_N$. It will be more convenient to write for every $i(= 1, \ldots, N)$,

$$c_{Ni}^{+} = \max\{c_i - \bar{c}_N, 0\} \text{ and } c_{Ni}^{-} = \min\{0, c_{Ni} - \bar{c}_N\}, \quad \text{(A.4.10)}$$

so that $X_i - N^{-1/2}[a + b(c_i - \bar{c}_N)] = X_i - N^{-1/2}(a + bc_{Ni}^{+} + bc_{Ni}^{-})$, $1 \le i \le N$. Thus, considering the residuals $X_i - N^{-1/2}(a + b_1 c_{Ni}^{+} + bc_{Ni}^{+})$, $1 \le i \le N$, we may construct a signed rank statistic, denoted by $S_N(N^{-1/2}(a, b_1, b_2))$, and consider the resulting three-parameter process where $(a, b_1, b_2) \in J = [-K, K]^3$. Note that for every $(a, b_1, b_2) \in J$,

$$\begin{aligned} S_N&\big(N^{-1/2}(a, b_1, b_2)\big) - S_N(\mathbf{0}) \\ &= \big\{S_N\big(N^{-1/2}(a, b_1, b_2)\big) - S_N\big(N^{-1/2}(0, b_1, b_2)\big)\big\} \\ &\quad + \big\{S_N\big(N^{-1/2}(0, b_1, b_2)\big) - S_N\big(N^{-1/2}(0, 0, b_2)\big)\big\} \\ &\quad + \big\{S_N\big(N^{-1/2}(0, 0, b_2)\big) - S_N\big(N^{-1/2}(0, 0, 0)\big)\big\}. \qquad \text{(A.4.11)} \end{aligned}$$

The first term on the rhs of (A.4.11) is decreasing in a (for given b_1, b_2), while the second term is also decreasing in b_1 (for given b_2), as the c_{Ni}^{+} for all ≥ 0. Finally, $c_{Ni}^{-} \le 0$, and hence, the last term in increasing in b_2. This coordinatewise monotonicity property enables us to bound $\sup\{N^{-1/2}|S_N(N^{-1/2}(a, b_1, b_2)) - S_N(\mathbf{0}) + N^{1/2}(a + b_1\bar{c}_N^{+} + b_2\bar{c}_N^{-}) A_\phi A_\psi \rho(\phi, \psi)| : (a, b_1, b_2) \in J\}$ by $\max\{N^{-1/2}|S_N(N^{-1/2}(a_j, b_{1k}, b_2)) - S_N(\mathbf{0}) + N^{-1/2}(a_j + b_{1k}\bar{c}_N + b_2\bar{c}_N^{-})A_\phi A_\psi \rho(\phi, \psi)| : 0 \le j \le 2M,\ 0 \le k \le 2M,\ 0 \le l \le 2M\} + \frac{1}{2}\varepsilon$, where $\varepsilon > 0$, M depends on ε and K, and $\bar{c}_N^{+} = N^{-1}\sum_{i=1}^{N} c_{Ni}^{+}$, $\bar{c}_N^{-} = N^{-1}\sum_{i=1}^{N} c_{Ni}^{-}$ (so that $\bar{c}_N^{+} + \bar{c}_N^{-} = 0$). [Viewed from this angle, for $b_1 = b_2$, $b_1\bar{c}_N^{+} + b_2\bar{c}_N^{-} = 0$, so (6.2.45) will correspond to a restricted case of $b_1 = b_2$.] Once this is done, we may virtually repeat the steps in (A.4.2)–(A.4.6), with the contiguity property borrowed from Chapter 3, and complete the proof. Also, as in (A.4.7)–(A.4.9), we may replace the $c_i - \bar{c}_N$ by d_{Ni}, $1 \le i \le N$, and (A.4.7) will insure the directional monotonicity in (A.4.11), so that the rest of the proof will be the same.

A.4.3. Proof of (6.3.21)

Without any loss of generality, we may take $\|\mathbf{u}\|$ as the Euclidean norm, and also take $N^{-1}(\mathbf{c}_{(j)} - \bar{c}_j \mathbf{1}_N)(\mathbf{c}_{(j)} - \bar{c}_j \mathbf{1}_N)' = 1$, $j = 1, \ldots, q$. Since the

Euclidean norm of a vector is bounded by the sum of the coordinatewise distances, to prove (6.3.21) it suffices to show that for each j $(= 1, \dots, q)$,

$$P_0\left\{ \sup_{\mathbf{u}:\|\mathbf{u}\| \le K} \left| L_{Nj}(N^{-1/2}\mathbf{u}) - L_{Nj}(\mathbf{0}) + N^{1/2}\rho(\phi,\psi)A_\phi A_\psi \boldsymbol{\lambda}_j \mathbf{u} \right| > N^{1/2}\varepsilon \right\} \to 0, \quad \text{as } N \to \infty, \tag{A.4.12}$$

where $\boldsymbol{\lambda}_j$ is the jth row of $\boldsymbol{\Lambda}$, and $\varepsilon > 0$ is an arbitrary number. Let $R_{Ni}(\mathbf{b}_1, \mathbf{b}_2)$ be the rank of $X_i - \mathbf{b}_1'(\mathbf{c}_i^{(1)} - \bar{\mathbf{c}}_N^{(1)}) - \mathbf{b}_2'(\mathbf{c}_i^{(2)} - \bar{\mathbf{c}}_N^{(2)})$ among the N residuals of the same type, where the $\mathbf{c}_i$ defined in (6.3.1) are written as $\mathbf{c}_i^{(1)} + \mathbf{c}_i^{(2)}$ and $\bar{\mathbf{c}}_N^{(k)} = N^{-1}\Sigma_{i=1}^N \mathbf{c}_i^{(k)}$ for $k = 1, 2$. With this decomposition, we consider the statistics $N^{-1/2}L_{Nj}(N^{-1/2}\mathbf{u}^{(1)}, N^{-1/2}\mathbf{u}^{(2)}) = N^{-1/2}\Sigma_{i=1}^N(c_{ij} - \bar{c}_j)a_N(R_{Ni}(N^{-1/2}\mathbf{u}^{(1)}, N^{-1/2}\mathbf{u}^{(2)})) = N^{-1/2}\Sigma_{k=1}^2 L_{Nj}^{(k)}(N^{-1/2}\mathbf{u}^{(1)}, N^{-1/2}\mathbf{u}^{(2)})$, where $L_{Nj}^{(k)} = \Sigma_{i=1}^N(c_{ij}^{(k)} - \bar{c}_j^{(k)})a_N(N^{-1/2}\mathbf{u}^{(1)}, N^{-1/2}\mathbf{u}^{(2)})$ for $k = 1, 2$. Now, in view of the concordance–discordance condition in (6.3.13)–(6.3.16), and proceeding as in the proof of Theorem 6.2.1 (see also Theorem 2.1 of Jurečková, 1969), it follows that $L_{Nj}^{(1)}(N^{-1/2}\mathbf{u}^{(1)}, N^{-1/2}\mathbf{u}^{(2)})$ is nonincreasing (nondecreasing) in the elements of $\mathbf{u}^{(1)}$ $(\mathbf{u}^{(2)})$, and $L_{Nj}^{(2)}(N^{-1/2}\mathbf{u}^{(1)}, N^{-1/2}\mathbf{u}^{(2)})$ is nondecreasing (nonincreasing) in the elements of $\mathbf{u}^{(1)}$ $(\mathbf{u}^{(2)})$. Now, consider any j $(= 1, \dots, q)$ such that $\Sigma_{i=1}^N(c_{ij}^{(1)} - \bar{c}_j^{(1)})^2 > 0$ for all but a finite number of N. Denote by $J = \{\mathbf{x} : \|\mathbf{x}\| \le K\}$ and $\gamma = \rho(\phi,\psi)A_\phi A_\psi$. We then prove that as $N \to \infty$,

$$P\left\{ \sup_{\mathbf{u}^{(1)},\mathbf{u}^{(2)} \in J} N^{-1/2}\left| L_{Nj}^{(1)}(N^{-1/2}\mathbf{u}^{(1)}, N^{-1/2}\mathbf{u}^{(2)}) - L_{Nj}^{(1)}(\mathbf{0},\mathbf{0}) + N^{-1/2}\gamma \sum_{l=1}^q u_l^{(1)}\left(\mathbf{c}_{(j)}^{(1)} - \bar{\mathbf{c}}_{(j)}^{(1)}\right)\left(\mathbf{c}_{(l)}^{(1)} - \bar{\mathbf{c}}_{(l)}^{(1)}\right)' + N^{-1/2}\gamma \sum_{l=1}^q u_l^{(2)}\left(\mathbf{c}_{(j)}^{(1)} - \bar{\mathbf{c}}_{(j)}^{(1)}\right)\left(\mathbf{c}_{(l)}^{(2)} - \bar{\mathbf{c}}_{(l)}^{(2)}\right)' \right| \ge N^{-1/2}\varepsilon\left\| \mathbf{c}_{(j)}^{(1)} - \bar{\mathbf{c}}_{(j)}^{(1)} \right\| \right\} \to 0. \tag{A.4.13}$$

Note that if in (A.4.13) we delete the restriction on the sup over J, then for every fixed $\mathbf{u}^{(1)}, \mathbf{u}^{(2)}$ in J, the probability will converge to 0 by virtue of the contiguity of the probability measures (see, for example, Theorem 3.1 of

Jurečková, 1969). To prove the stronger version, we proceed as in Jurečková (1971a) and for each j $(= 1, \ldots, q)$, $\mathbf{x} = (x_1, \ldots, x_q)'$ in J, consider a partition of $[-K, K]$, $-K = a_{(0)} < a_{(1)} < \cdots < a_{(r)} = K$, such that

$$\left|\gamma\left(a_{(k+1)} - a_{(k)}\right)\right| \leq \varepsilon(2qM)^{-1/2} \quad \text{for} \quad 0 \leq k \leq r - 1, \tag{A.4.14}$$

where M is a constant, given in (6.3.19). Then the following inequality is a consequence of the monotonicity of the $L_{Nj}^{(1)}$ in the components of $\mathbf{u}^{(1)}$ and $\mathbf{u}^{(2)}$ and of (A.4.14) (we leave the proof as an exercise):

$$\begin{aligned}
\sup\Bigg\{ & N^{-1/2}\Big| L_{Nj}^{(1)}\big(N^{-1/2}(\mathbf{u}^{(1)}, \mathbf{u}^{(2)})\big) - L_{Nj}^{(1)}(\mathbf{0}, \mathbf{0}) \\
& + \gamma N^{-1/2} \sum_{l=1}^{q} u_l^{(1)}\left(\mathbf{c}_{(j)}^{(1)} - \bar{\mathbf{c}}_{(j)}^{(1)}\right)\left(\mathbf{c}_{(l)}^{(1)} - \bar{\mathbf{c}}_{(l)}^{(1)}\right)' \\
& + N^{-1/2}\gamma \sum_{l=1}^{q} u_l^{(2)}\left(\mathbf{c}_{(j)}^{(2)} - \bar{\mathbf{c}}_{(j)}^{(1)}\right)\left(\mathbf{c}_{(l)}^{(2)} - \bar{\mathbf{c}}_{(l)}^{(2)}\right)'\Big| : \mathbf{u}^{(1)}, \mathbf{u}^{(2)} \in J \Bigg\} \\
\leq\, & N^{-1/2}\frac{\varepsilon}{2}\left\|\mathbf{c}_{(j)}^{(1)} - \bar{\mathbf{c}}_{(j)}^{(1)}\right\| \\
& + 2\max\Bigg\{ N^{-1/2}\Big| L_{Nj}^{(1)}\big(N^{-1/2}(\mathbf{u}_{(\mathbf{p}')}, \mathbf{u}_{(\mathbf{p}'')})\big) - L_{Nj}^{(1)}(\mathbf{0}, \mathbf{0}) \\
& + \gamma N^{-1/2} \sum_{l=1}^{q} u_{(p_l')}\left(\mathbf{c}_{(j)}^{(1)} - \bar{\mathbf{c}}_{(j)}^{(1)}\right)\left(\mathbf{c}_{(l)}^{(1)} - \bar{\mathbf{c}}_{(l)}^{(1)}\right)' \\
& + N^{-1/2}\gamma \sum_{l=1}^{q} u_{(p_l'')}\left(\mathbf{c}_{(j)}^{(1)} - \bar{\mathbf{c}}_{(j)}^{(1)}\right)\left(\mathbf{c}_{(l)}^{(2)} - \bar{\mathbf{c}}_{(l)}^{(2)}\right)'\Big| : \mathbf{p}', \mathbf{p}'' \in \mathbf{P} \Bigg\},
\end{aligned} \tag{A.4.15}$$

where $\mathbf{P} = \{(p_1, \ldots, p_q) : 0 \leq p_j \leq r,\ 1 \leq j \leq q\}$ and the $u_{(p_j')}$ (or $u_{(p_j'')}$) refer to the points a_j in (A.4.14). Since there are a finite number (q^{2r}) of points in $\mathbf{P}$, expressing the maximum in (A.4.15) as less than the sum over each of these terms, and proceeding for each term as in the proof of (6.2.44), we conclude that the rhs of (A.4.15) converges in probability to 0 as $N \to \infty$. An analogous result holds for $L_{Nj}^{(2)}(N^{-1/2}(\mathbf{u}^{(1)}, \mathbf{u}^{(2)}))$ with such j that $\Sigma_{i=1}^{N}(c_{ij}^{(2)} - \bar{c}_j^{(2)})^2 > 0$ for all but a finite number of N. This completes the proof for the case where $\Sigma_{i=1}^{N}(c_{ij}^{(k)} - \bar{c}_j^{(k)})^2 > 0$ for all but a finite number of N, $k = 1, 2$.

There remains two other cases when one of these sums may converge to 0 while the other does not; if both converge to 0, we have a degenerate case, excluded by the assumption that $N^{-1}\Sigma_{i=1}^{N}(c_{ij} - \bar{c}_{N(j)})^2 = 1$. If $\Sigma_{i=1}^{N}(c_{ij}^{(1)} - \bar{c}_{N(j)}^{(1)})^2 > 0$ and $\Sigma_{i=1}^{N}(c_{ij}^{(2)} - \bar{c}_{N(j)}^{(2)})^2 = 0$ for $N \geq N_0$, then $N^{-1}\Sigma_{i=1}^{N}(c_{ij} - \bar{c}_{(j)})^2 = N^{-1}\|\mathbf{c}_{(j)}^{(1)} - \bar{\mathbf{c}}_{(j)}^{(1)}\|^2$ for all $N \geq N_0$ and $L_{Nj}(N^{-1/2}\mathbf{u}^{(1)}) = L_{Nj}^{(1)}(N^{-1/2}(\mathbf{u}^{(1)}, \mathbf{u}^{(2)}))$ for all $\mathbf{u}^{(2)}$. Thus, the dependence on $\mathbf{u}^{(2)}$ can be neglected and the proof follows more simply by treating the process in (A.4.15) as a function of $\mathbf{u}^{(1)}$ alone. A similar case holds for the other situation where $\Sigma_{i=1}^{N}(c_{ij}^{(1)} - \bar{c}_{N(j)}^{(1)})^2 = 0$ but $\Sigma_{i=1}^{N}(c_{ij}^{(2)} - \bar{c}_{N(j)}^{(2)})^2 > 0$ for $N \geq N_0$. Hence, the proof of the result is complete.

A.4.4. Proof of (6.3.44)

Let $\mathbf{c}_i^+ = (c_{i1}^+, \ldots, c_{iq}^+)'$ and $\mathbf{c}_i^- = (c_{i1}^-, \ldots, c_{iq}^-)'$ be defined by $c_{ij}^+ = \max\{c_{ij} - \bar{c}_{N(j)}, 0\}$ and $c_{ij}^- = \min\{c_{ij} - \bar{c}_{N(j)}, 0\}$, $j = 1, \ldots, q$, $i = 1, \ldots, N$. Then $X_i - N^{-1/2}[a + \mathbf{b}'(\mathbf{c}_i - \bar{\mathbf{c}}_N)] = X_i - N^{-1/2}(a + \mathbf{b}'\mathbf{c}_i^+ + \mathbf{b}'\mathbf{c}_i^-)$ for $i = 1, \ldots, N$. As in Section A.4.3, we increase the dimension of the residual process by considering the residuals $X_i - N^{-1/2}(a + \mathbf{b}^{(1)\prime}\mathbf{c}_i^+ + \mathbf{b}^{(2)\prime}\mathbf{c}_i^-)$, $i = 1, \ldots, N$, and forming a signed rank statistic $S_N(N^{-1/2}(a, \mathbf{b}^{(1)}, \mathbf{b}^{(2)})$ with the parameters $(a, \mathbf{b}^{(1)}, \mathbf{b}^{(2)}) \in J^* = [-K, K] \times \{\mathbf{x} : \mathbf{x} \in R^q, \|\mathbf{x}\| \leq K\}^2$. Now, for every $(a, \mathbf{b}^{(1)}, \mathbf{b}^{(2)})$ in J^*, we may write

$$
\begin{aligned}
&S_N\big(N^{-1/2}(a, \mathbf{b}^{(1)}, \mathbf{b}^{(2)})\big) - S_N(0, \mathbf{0}, \mathbf{0}) \\
&\quad = \big\{S_N\big(N^{-1/2}(a, \mathbf{b}^{(1)}, \mathbf{b}^{(2)})\big) - S_N\big(N^{-1/2}(0, \mathbf{b}^{(1)}, \mathbf{b}^{(2)})\big)\big\} \\
&\qquad + \big\{S_N\big(N^{-1/2}(0, \mathbf{b}^{(1)}, \mathbf{b}^{(2)})\big) - S_N\big(N^{-1/2}(0, \mathbf{0}, \mathbf{b}^{(2)})\big)\big\} \\
&\qquad + \big\{S_N\big(N^{-1/2}(0, \mathbf{0}, \mathbf{b}^{(2)})\big) - S_N\big(N^{-1/2}(0, \mathbf{0}, \mathbf{0})\big)\big\}. \qquad \text{(A.4.16)}
\end{aligned}
$$

By arguments similar to those in (A.4.1)–(A.4.6) (also see Jurečková, 1971, Lemma 1), we may prove the following results by some standard steps (and we leave the proofs as exercises; see Exercises 6.3.11–6.3.13): as $N \to \infty$,

$$N^{-1/2}\{S_N(N^{-1/2}(a,\mathbf{0},\mathbf{0})) - S_N(N^{-1/2}(0,\mathbf{0},\mathbf{0}))\} + a\rho(\phi,\psi)A_\phi A_\psi \xrightarrow{P} 0, \tag{A.4.17}$$

$$N^{-1/2}\Bigg\{S_N(N^{-1/2}(0,\mathbf{b}^{(1)},\mathbf{0})) - S_N(N^{-1/2}(0,\mathbf{0},\mathbf{0})) + N^{-1/2}\rho(\psi,\phi)A_\phi A_\psi \sum_{l=1}^{q} b_l^{(1)} \sum_{i=1}^{N} c_{il}^{+}\Bigg\} \xrightarrow{P} 0, \tag{A.4.18}$$

$$N^{-1/2}\Bigg\{S_N(N^{-1/2}(0,\mathbf{0},\mathbf{b}^{(2)})) - S_N(N^{-1/2}(0,\mathbf{0},\mathbf{0})) + N^{-1/2}\rho(\phi,\psi)A_\phi A_\psi \sum_{l=1}^{q} b_l^{(2)} \sum_{i=1}^{N} c_{il}^{-}\Bigg\} \xrightarrow{P} 0. \tag{A.4.19}$$

Note that by the contiguity results in Section 2.7, for every $(a,\mathbf{b}^{(1)},\mathbf{b}^{(2)})$ in J^*, the contiguity of $\prod_{i=1}^{N} f(X_i + N^{-1/2}(a + \mathbf{b}^{(1)\prime}\mathbf{c}_i^{+} + \mathbf{b}^{(2)\prime}\mathbf{c}_i^{-}))$ to $\prod_{i=1}^{N} f(X_i)$ is insured. Hence, by the same method as in (A.4.18) and (A.4.19), we arrive at the following:

$$N^{-1/2}\Bigg\{S_N(N^{-1/2}(a,\mathbf{b}^{(1)},\mathbf{b}^{(2)})) - S_N(0,\mathbf{0},\mathbf{0}) + N^{-1/2}\rho(\phi,\psi)A_\phi A_\psi \sum_{l=1}^{q}\Bigg(b_l^{(1)}\sum_{i=1}^{N} c_{il}^{+} + b_l^{(2)}\sum_{i=1}^{N} c_{il}^{-}\Bigg)\Bigg\} + a\rho(\phi,\psi)A_\phi A_\psi \xrightarrow{P} 0. \tag{A.4.20}$$

Now, the $\mathbf{c}_i^{+}$ have all nonnegative elements while the $\mathbf{c}_i^{-}$ have all nonpositive elements. Hence, by virtue of Lemma 6.2.3, we conclude that $S_N(N^{-1/2}(a,\mathbf{b}^{(1)},\mathbf{b}^{(2)}))$ is nonincreasing in $a, b_1^{(1)},\ldots,b_q^{(1)}$ and is nondecreasing in $b_1^{(2)},\ldots,b_q^{(2)}$. Having this (coordinatewise) monotonicity property, we may then consider a partitioning of J^* into q^{2r+1} grid points [as in (A.4.14)] and bound the supremum of $|N^{-1/2}\{S_N(N^{-1/2}(a,\mathbf{b}^{(1)},\mathbf{b}^{(2)})) - S_N(0,\mathbf{0},\mathbf{0}) + N^{-1/2}\rho(\phi,\psi)A_\phi A_\psi \Sigma_{l=1}^{q}(b_l^{(1)}\Sigma_{i=1}^{N} c_{il}^{+} + b_l^{(2)}\Sigma_{i=1}^{N} c_{il}^{-})\} + a\rho(\phi,\psi)A_\phi A_\psi|$ (over J^*) by that of a maximum of the process over the q^{2r+1} grid points plus a residual term which can be made arbitrarily small by proper choice of r; this parallels (A.4.15), and the proof is left as an exercise (see Exercise 6.3.14). Thus, the rest of the proof remains virtually the same as in the last part of Section A.4.3. Hence the details are omitted.

In passing we may remark that in both (6.3.21) and (6.3.44), the assumption of concordance–discordance of the c_i has been tacitly utilized in making a particular representation of the rank statistics for which the coordinatewise monotonicity property holds, and this enables one to reduce the supremum over a compact region by that of a maximum over a finite set of grid points. Once this is done, the weak pointwise convergence insures the same property over the entire compact region.

A.5. PROOF OF THEOREM 8.2.1

The proof of Theorem 8.2.1 can be worked out in several ways under parallel regularity conditions. A proof based on the Chernoff–Savage (1958) decomposition for the m_{njl} has been chalked out in detail in Puri and Sen (1971, pp. 387–390). In their proof, the treatment of the higher-order terms was practically omitted [see the top of p. 389 of Puri and Sen (1971)] by reference to another theorem; unfortunately, there appears to be a technical error in the proof of the latter. Since we are dealing here with i.i.d. random vectors, this technical error can easily be avoided by invoking the Pyke–Shorack (1968a, b) weak-convergence approach: a proof of (8.2.45) has been provided by Ruymgaart, Shorack, and van Zwet (1972) and elaborated further by Ruymgaart (1973). For contiguous alternatives, as treated in Chapters 2, 3, 5, 7, and 8, a simpler proof is given in Ghosh and Sen (1971b). However, the elegant approach of Hájek (1968), which has been fully explored in Chapters 2, 3, and 4, has not been successfully incorporated in the proof of the basic result in Theorem 8.2.1. We fully endorse the basic goal that the proof of Theorem 8.2.1 should be available in full generality as in Hájek (1968), but for now we can only leave this as an open problem.

Given the state of this problem, what we would like to show here is how to combine the Chernoff–Savage approach in Puri and Sen (1971) with the weak-convergence approach in Ruymgaart (1973) to provide a justification which is highly intuitive, and provide the basic steps, so that one may complete the proof by working out the details in between. Define the empirical d.f.'s $G^*_{n[j]}$ and $F^*_{in[b]}$ as in (8.2.37)–(8.2.38). Also, the bivariate empirical d.f.'s are defined as in the discussion after (8.2.38). Further, the score functions are introduced as in (8.2.39). Then, we may write $n^{-1}m_{njl}$ as

$$\int\int \phi_{nj}\left(G^*_{n[j]}(x)\right)\phi^*_{nl}\left(F^*_{1n[l]}(y)\right)\,dF_{2n[jl]}(x,y)$$
$$-\left(\int_0^1 \phi_{nj}(u)\,du\right)\left(\int_0^1 \phi^*_{nl}(v)\,dv\right). \tag{A.5.1}$$

The last term differs from $\mu_j^0\mu_l^{*0}$ by $o(n^{-1/2})$, and hence, using (8.2.41) [Assumption (c)] and (8.2.42), it suffices to show that as $n \to \infty$,

$$n^{1/2}\left\{\int\int\left[\phi_j\left(G^*_{n[j]}(x)\right)\phi^*_l\left(F^*_{1n[l]}(y)\right)dF_{2n[jl]}(x,y)\right.\right.$$

$$\left.-\mu_{jl}-n^{-1}\sum_{i=1}^{n}U_{jl,i}\right\}\xrightarrow{P}0, \qquad (A.5.2)$$

where the $U_{jl,i}$ are defined in (8.2.43). We now write

$$\phi_j\left(G^*_{n[j]}\right)=\phi_j\left(G_{[j]}\right)+\left(G^*_{n[j]}-G_{[j]}\right)\phi'_j\left(G_{[j]}\right)+\xi_j, \qquad (A.5.3)$$

$$\phi^*_l\left(F^*_{1n[l]}\right)=\phi^*_l\left(F_{1[l]}\right)+\left(F^*_{in[l]}-F_{1[l]}\right)\phi^{*\prime}_l\left(F_{1[l]}\right)+\xi^*_l, \qquad (A.5.4)$$

and $dF_{2n[jl]}=dF_{2[jl]}+d(F_{2n[jl]}-F_{2[jl]})$. Then the integral in (A.5.2) may be decomposed in to 18 terms, of which the first four will cancel with $\mu_{jl}+n^{-1}\Sigma_{i=1}^nU_{jl,i}$, while the rest can be rearranged, and (A.5.2) reduces to

$$n^{1/2}\left\{\int\int\phi_j\left(G_{[j]}(x)\right)\left[F^*_{1n[l]}(y)-F_{1[l]}(y)\right]\right.$$

$$\times\phi^{*\prime}_l\left(F_{1[l]}(y)\right)d\left[F_{2n[jl]}(x,y)-F_{2[jl]}(x,y)\right]$$

$$+\int\int\phi^*_l\left(F_{1[l]}(y)\right)\left[G^*_{n[j]}(x)-G_{[j]}(x)\right]$$

$$\times\phi'_j\left(G_{[j]}(x)\right)d\left[F_{2n[jl]}(x,y)-F_{2[jl]}(x,y)\right]$$

$$+\int\int\left[G^*_{n[j]}-G_{[j]}\right]\phi'_j\left(G_{[j]}\right)\left[F^*_{1n[l]}-F_{1[l]}\right]\phi^{*\prime}_l\left(F_{1[l]}\right)dF_{2n[jl]}$$

$$+\int\int\xi_j(x)\phi^*_l\left(F_{1[l]}\right)dF_{2n[jl]}(\cdot)$$

$$+\int\int\xi_j(x)\left(F^*_{1n[l]}-F_{1[l]}\right)\phi^{*\prime}_l\left(F_{1[l]}\right)dF_{2n[jl]}(\cdot)$$

$$+\int\int\xi^*_l(y)\phi_j\left(G_{[j]}\right)dF_{2n[jl]}(\cdot)$$

$$+\int\int\xi^*_l(y)\left(G^*_{n[j]}-G_{[j]}\right)\phi^*_j\left(G_{[j]}\right)dF_{2n[jl]}(\cdot)$$

$$\left.+\int\int\xi_j(x)\xi^*(y)\,dF_{2n[jl]}(x,y)\right\}$$

$$=I_{n1}+\cdots+I_{n8},\quad\text{say.} \qquad (A.5.5)$$

At this stage, we make use of Theorem 2.11.9 of Puri and Sen (1971) and claim that for every $\eta > 0$, there exists a positive K_ν $(< \infty)$ such that

$$\sup\left\{\frac{n^{1/2}\left|G^*_{n[j]}(x) - G_{[j]}(x)\right|}{\left\{G_{[j]}(x)\left[1 - G_{[j]}(x)\right]\right\}^{1/2-\eta}} : x \in R\right\} \le K_\eta, \tag{A.5.6}$$

with a probability converging to 1, as $n \to \infty$, and a similar result holds for the empirical d.f.'s $F_{1n[l]}$. Also, for every $\varepsilon > 0$, there exist an $\eta > 0$ and a positive integer n_0 such that for every $n \ge n_0$,

$$P\left\{G_{[j]}(Y_{ji}) \notin \left[\frac{\eta}{n}, 1 - \frac{\eta}{n}\right] \text{ for some } i: 1 \le i \le n\right\} < \varepsilon, \tag{A.5.7}$$

and a similar result holds for the $F_{1[l]}(X_{li})$. By (8.2.40) and (A.5.6), we have

$$\sup_x\left\{n^{1/2}\frac{\left[G_{n[j]}(x) - G_{[j]}(x)\right]\phi_j'\left(G_{[j]}(x)\right)}{K_\eta \cdot K\left\{G_{[j]}(x)\left[1 - G_{[j]}(x)\right]\right\}^{-3/4+\delta-\eta}}\right\} \le 1, \tag{A.5.8}$$

with a probability converging to unity as $n \to \infty$, and a similar result holds for the $F_{1n[l]}(y)$. Also, over the set $\eta/n \le G_{[j]}(x) \le 1 - \eta/n$, we have

$$n^{-a} \le \eta^{-a}\left\{G_{[j]}(x)\left[1 - G_{[j]}(x)\right]\right\}^a \quad \text{for every} \quad a > 0. \tag{A.5.9}$$

In (A.5.8) below we take $\eta = \delta/4$, and in (A.5.9), $a = 3(1 - \delta)/4$. Now, note that $\phi_j(u)$ is a continuous function of $u \in (0,1)$, while $[F_{1n[l]}(y) - F_{1[l]}(y)]$. Furthermore, $\phi_l^{*\prime}(F_{1[l]}(y))$ has only n points of discontinuity (namely, at the observed X_{li}), where jumps of the magnitude $n^{-1}\phi_l^{*\prime}(F_{1[l]}(X_{li}))$ occur. For an arbitrary $\varepsilon > 0$, consider a compact set $C_\varepsilon = \{(x, y) : G_{[j]}(x)[1 - G_{[j]}(x)] \ge \varepsilon,\ F_{1[l]}(y)[1 - F_{1[l]}(y)] \ge \varepsilon\} = C_{\varepsilon j} \times C^*_{\varepsilon l}$. Inside $C_{\varepsilon l}$, by linearly interpolating $F^*_{1n[l]}(y)$ between the successive order statistics (of the X_{li}), we obtain a continuous function which differ from $F^*_{1n[l]}$ by n^{-1} at most. Thus, we obtain by using (A.5.8) that

$$\begin{aligned}\iint_{C_\varepsilon} & n^{1/2}\phi_j\left(G_{[j]}(x)\right)\left[F_{1n[l]}(y) - F_{1[l]}(y)\right] \\ & \times\phi_l^{*\prime}\left(F_{1[l]}(y)\right) d\left[F_{2n[jl]}(\cdot) - F_{2[jl]}(\cdot)\right] \\ = & \iint_{C_\varepsilon} g_n(x, y)\, d\left[F_{1n[jl]}(x, y) - F_{1[j]}(x, y)\right] \\ & + O_p\left(n^{-1/2}\varepsilon^{-3(1-\delta)/4}\right),\end{aligned} \tag{A.5.10}$$

where by (8.2.40), $g_n(x, y)$ is continuous and bounded (in probability), while $F_{2n[jl]}$ converges weakly to $F_{2[jl]}$, as $n \to \infty$. Note that (A.5.8) holds uniformly in $n \geq n_0$, and hence, using a version of the Helly–Bray Lemma [cf. Puri and Sen, 1971, p. 19) on the compact set C_ε, we conclude that (A.5.10) converges to 0, in probability, as $n \to \infty$. For the complementary set C_ε^c, we note that

$$C_\varepsilon^c \subseteq \left(C_{\varepsilon j}^c \times C_l^*\right) \cup \left(C_j \times C_{\varepsilon l}^{*c}\right), \qquad C_j = C_{\varepsilon j} \cup C_{\varepsilon j}^c \text{ and } C_l^* = C_{\varepsilon l}^* \cup C_{\varepsilon l}^{*c}. \tag{A.5.11}$$

Consider next

$$\begin{aligned}
&\Bigg| \iint_{C_{\varepsilon j}^c \times C_l^*} n^{1/2} \phi_j\left(G_{[j]}(x)\right)\left[F_{1n[l]}^*(y) - F_{1[l]}(y)\right] \\
&\qquad \times \phi_l^{*\prime}\left(F_{1[l]}(y)\right) d\left[F_{2n[jl]}(\cdot) - F_{2[jl]}(\cdot)\right] \\
&\qquad \leq \left\{\int_{C_{\varepsilon j}^c} \phi_j^4\left(G_{[j]}(x)\right) d\left[G_{n[j]}^*(x) + G_{[j]}(x)\right]\right\}^{1/4} \\
&\qquad \times \left\{\int_{C_l^*} \left| n^{1/2}\left[F_{1n[l]}^*(y) - F_{1[l]}(y)\right]\right.\right. \\
&\qquad \left.\left. \times\, \phi_l^{*\prime}\left(F_{1[l]}(y)\right)\right|^{4/3} d\left(F_{1n[l]}^* + F_{1[l]}\right)\right\}^{3/4}.
\end{aligned} \tag{A.5.12}$$

Now, by (8.2.40),

$$\int_{C_{\varepsilon j}^c} \phi_j^4\left(G_{[j]}(x)\right) dG_{[j]}(x) = O(\varepsilon^{4\delta}), \tag{A.5.13}$$

while, by the Khinchine law of large numbers, as $n \to \infty$,

$$\int_{C_{\varepsilon j}^c} \phi_j^4\left(G_{[j]}(x)\right) dG_{n[j]}^*(x) \to \int_{C_{\varepsilon j}^c} \phi_j^4\left(G_{[j]}(x)\right) dG_{[j]}(x) \quad \text{a.s.} \tag{A.5.14}$$

By (A.5.8), the second integral on the rhs of (A.5.12) is bounded in probability by

$$K^{**} \int_{-\infty}^{\infty} \left\{F_{1[l]}(y)\left[1 - F_{1[l]}(y)\right]\right\}^{-1+\delta} d\left[F_{1n[l]}^*(y) + F_{1[l]}(y)\right], \quad \text{where } K^{**} < \infty. \tag{A.5.15}$$

Again, by the Khinchine law of large numbers, as $n \to \infty$,

$$\int_{-\infty}^{\infty} \{F_{1[l]}(y)[1 - F_{1[l]}(y)]\}^{-1+\delta} dF^*_{1n[l]}(y)$$

$$\to \int_{-\infty}^{\infty} \{F_{1[l]}(y)[1 - F_{1[l]}(y)]\}^{-1+\delta} dF_{1[l]}(y) \quad \text{a.s.,} \quad \text{(A.5.16)}$$

where the rhs is $O(1)$. Hence, (A.5.15) is $O_p(1)$, so that by (A.5.12) through (A.5.16), we conclude that the lhs of (A.5.12) is $o_p(1)$ as $n \to \infty$. A similar case holds with $C_j \times C^{*c}_{\varepsilon l}$, so that by (A.5.10) through (A.5.16), we conclude that $I_{n1} = o_p(1)$ as $n \to \infty$. The treatment of I_{n2} is exactly the same (with the interchange of j and l), and hence is omitted.

For I_{n3}, by virtue of (A.5.7), we confine ourselves to the region where both the sets of $G_{[j]}$ and $F_{1[l]}$ satisfy (A.5.7). Note that by virtue of (A.5.8) (and a similar bound for the $F^*_{1n[l]}$), within the set satisfying (A.5.7) (for both the marginals), the integral in I_{n3} is bounded (in absolute value), in probability, by

$$K^{***}n^{-\delta/2}\int\int\{G_{[j]}(x)[1 - G_{[j]}(x)]\}^{-(1-\delta)/2}$$

$$\times\{F_{1[l]}(y)[1 - F_{1[l]}(y)]\}^{-(1-\delta)/2} dF_{2n[jl]}(x, y)$$

$$\leq K^{***}n^{-\delta/2}\left\{\int\{G_{[j]}[1 - G_{[j]}]\}^{-(1-\delta)} dG^*_{n[j]}\right.$$

$$\left.\times\int\{F_{1[l]}[1 - F_{1[l]}]\}^{-(1-\delta)} dF^*_{1n[l]}\right\}^{1/2},$$

where K^{***} ($< \infty$) is finite. By the application of the Khinchin law of large numbers to each of the integrals (above), we conclude that they are $O(1)$ a.s. as $n \to \infty$, so that $I_{n3} = O_p(n^{-\delta/2}) = o_p(1)$ as $n \to \infty$.

For the treatment of I_{n4}, I_{n5}, I_{n6}, I_{n7}, and I_{n8}, we make use of the following result due to Csáki (1977): With $\varepsilon_n = dn^{-1}\log\log n$ and $d \geq 0.236\ldots$,

$$\sup\left\{\frac{n^{1/2}|G^*_{n[j]}(x) - G_{[j]}(x)|}{\{G_{[j]}(x)[1 - G_{[j]}(x)]\}^{1/2}} : G_{[j]}(x)[1 - G_{[j]}(x)] \geq \varepsilon_n\right\}$$

$$= 2(\log\log n)^{1/2} \quad \text{a.s.} \quad \text{as} \quad n \to \infty. \quad \text{(A.5.17)}$$

A similar result holds for the $F^*_{1n[l]}$ too. As in the case of I_{n1}, we first consider (for I_{n4}) the integral over the region C_ε and proceed as in (A.5.10). Within the compact set C_ε, $\xi_j(x)$ converges to 0 (uniformly in x) in probability, so that the integral over this domain [by using the Hölder inequality as in (A.5.12), and the Khinchin law of large numbers as in (A.5.14)] converges to 0, in probability, as $n \to \infty$. For the complementary part, we break up (for each j and l) the range $\eta/n \le G_{[j]}(1 - G_{[j]} \le \varepsilon$ into two parts: $\eta/n \le G_{[j]}(1 - G_{[j]}) < \varepsilon_n$ and $\varepsilon_n \le G_{[j]}(1 - G_{[j]}) \le \varepsilon$, and for the second part we make use of (A.5.17), instead of (A.5.8), while for the first part, we make use of (8.2.40). By virtue of (8.2.40) and (A.2.17), for every $x : \varepsilon_n \le G_{[j]}(x)[1 - G_{[j]}(x)] \le \varepsilon$, we have $|\xi_j(x)| \le K_0^*\{G_{[j]}(x)[1 - G_{[j]}(x)]\}^{-3/4+\delta}$ (where $K_0^* < \infty$) in probability, so that we may proceed as in (A.5.12) through (A.5.16) and conclude that the contribution of this part to I_{n4} converges to 0 in probability as $n \to \infty$. For every $x: \eta/n \le G_{[j]}(x)[1 - G_{[j]}(x)] \le \varepsilon_n$, we make use of (A.5.3) and (8.2.40) and conclude that $|\xi_j(x)| \le K_{00}n^{3/4-\delta}$ in probability, where $K_{00} < \infty$. Since the number of points belonging to this strip is $O[\log\log n)$ a.s., while $dF_{2n[jl]}$ attaches the probability mass n^{-1} to each such point, we conclude that the contribution of this part to I_{n4} is $O(n^{3/4-\delta} \cdot \log\log n \cdot n^{-1}) = o(1)$ in probability as $n \to \infty$. Therefore $I_{n4} = o_p(1)$ as $n \to \infty$. The treatment of I_{n6} is identical, and hence is omitted. For the treatment of I_{n5} and I_{n7}, we make use of the same decomposition of the range as in I_{n4}, and in addition we use (A.5.8). The result follows on parallel lines. Finally, for I_{n8}, we make use of the same decomposition for each j and l, so that we have nine terms. For each of these subsets, we proceed as in the case of I_{n4}, making use of (A.5.8) and (A.5.17) as necessary. This leads us to $I_{n8} = o_p(1)$ as $n \to \infty$. This concludes the proof of Theorem 8.2.1.

We may remark in passing that in the above proof, we have made a detailed decomposition of the remainder terms, and for each term, we have used some appropriate theorem on the sample d.f.'s to incorporate suitable rates of convergence insuring the desired convergence result. It is possible to replace this proof by a neater one, depicted in Ruymgaart (1973, Chapter 2, pp. 30–36). The novelty of that approach lies in the incorporation of the weak convergence of the empirical d.f. (with respect to the ρ_q metric topology, for integrable, continuous, and U-shaped $q^{-2}(t)$, $0 < t < 1$), so that much of the detail can be avoided by reference to this stronger mode of weak convergence. We suggest that the reader go through that proof for additional insight in this direction.

References

The set of references given here is by no means exhaustive but is closely related to the topics covered in the various chapters of the book. We use the following abbreviations for the journals most frequently referred to:

AAM	*Advances in Applied Mathematics*
AISM	*Annals of the Institute of Statistical Mathematics*
AMS	*Annals of Mathematical Statistics*
AP	*Annals of Probability*
AS	*Annals of Statistics*
BC	*Biometrics*
BK	*Biometrika*
CJS	*Canadian Journal of Statistics*
CSAB	*Calcutta Statistical Association Bulletin*
CS,A	*Communications in Statistics. A. Theory and Methods*
ISR	*International Statistical Review*
JASA	*Journal of the American Statistical Association*
JMA	*Journal of Multivariate Analysis*
JRSS,B	*Journal of the Royal Statistical Society. Series B*
JSPI	*Journal of Statistical Planning and Inference*
MOS	*Mathematische Operationsforschung und Statistik, Series Statistics*
S,A	*Sankhyā, Series A*
S,B	*Sankhyā, Series B*
SD	*Statistics and Decisions*
SSMH	*Studia Scientarium Mathematicarum Hungarica*
TC	*Technometrics*
TVP	*Teoriya Veroyatnosteĭ i ee Primeneniya*
TAMS	*Transactions of the American Mathematical Society*
ZW	*Zeitschrift für Wahrscheinlichkeitstheorie und Verwandte Gebiete*

Adichie, J. N. (1967a). Asymptotic efficiency of a class of nonparametric tests for regression parameters. *AMS* **38**, 884–893.

——— (1967b). Estimation of regression parameters based on rank tests. *AMS* **38**, 894–904.

——— (1974). Rank score comparison of several regression parameters. *AS* **2**, 396–402.

——— (1978). Rank tests of sub-hypotheses in the general linear regression. *AS* 1012–1026.

Akritas, M., Saleh, A. K. Md. E., and Sen, P. K. (1984). Nonparametric Estimation of Intercepts after a Preliminary Test on Parallelism of Several Regression Lines In Biostatistics: B. G. Greenberg Volume (Ed. P. K. Sen), Amsterdam: North Holland, pp. 220–234.

Albers, W., Bickel, P. J., and van Zwet, W. R. (1976). Asymptotic expansions for the power of distribution-free tests in the one-sample problem. *AS* **4**, 108–156.

Anderson, T. W. (1959). *An Introduction to Multivariate Statistical Analysis*. New York: Wiley.

Bahadur, R. R. (1967). "An optimal property of the likelihood ratio statistic". In *Proc. Fifth Berkeley Symp. Math. Statist. Probab.*, Vol. 1 (Eds.: J. Neyman and L. LeCam) Berkeley: Univ. of California Press, pp. 13–26.

Bancroft, T. A. and Han, C. P. (1977). Inference based on conditional specification: A note and a bibliography. *ISR* **45**, 117–127.

——— (1980). On the reduction of erroneous statistical inferences due to incorrect specification of the model under analysis. *ISR* **48**, 309–316.

Barlow, R. E., Bartholomew, D. J., Bremner, J. M., and Brunk, H. D. (1972). *Statistical Inference under Order Restrictions. The Theory and Application of Isotonic Regression*. New York: Wiley.

Basu, A. P. (1967a). On the large sample properties of a generalized Wilcoxon–Mann–Whitney statistic. *AMS* **38**, 905–915.

——— (1967b). On a generalized Savage statistic with applications to life testing. *AMS* **38**, 1591–1604.

Basu, A. P., Ghosh, J. K., and Sen, P. K. (1983). A unified way of deriving LMP rank tests from censored data. *JRSS,B* **45**, 384–390.

Behnen, K. and Neuhaus, G. (1975). A central limit theorem under contiguous alternatives. *AS* **3**, 1349–1354.

——— (1983). Galton's test as a linear rank test with estimated scores and its local asymptotic efficiency. *AS* **11**, 588–599.

Behnen, K., Neuhaus, G., and Ruymgaart, F. (1983). Two-sample rank estimators of optimal nonparametric score functions and corresponding adaptive rank statistics. *AS* **11**, 1175–1189.

Bell, C. B. (1982). Signal Detection for Spherically Exchangeable Stochastic Processes (Tech. Rept. 4-82, Ser. Statist. & Biostatist.), San Diego State Univ.

Bell, C. B. and Donoghue, J. (1969). Distribution-free tests for randomness. *S,A* **31**, 157–176.

Bell, C. B. and Kurotschka, V. (1971). "Einige prinzipien zur behandlung nichtparameterischer hypothesen". In *Studi di Probabilita, Statistica e Ricera Operative in Onore di Giuseppe Pompilj*. Oderrisi-Buggio, pp. 164–186.

Bell, C. B. and Sen, P. K. (1984). "Randomization procedures". In *Handbook of Statistics, Vol. 4: Nonparametric Methods* (Eds.: P. R. Krishnaiah and P. K. Sen). Amsterdam: North Holland, pp. 1–29.

Bergström, H. and Puri, M. L. (1977). Convergence and remainder terms in linear rank statistics. *AS* **5**, 671–680.

Bhapkar, V. P. and Patterson, K. W. (1977). On some nonparametric tests for profile analysis of several multivariate samples. *JMA* **7**, 265–277.

Bhattacharya, G. K. and Johnson, R. A. (1970). A layer rank test for ordered bivariate alternatives. *AMS* **41**, 1296–1310.

Bhattacharya, P. K., Chernoff, H., and Yang, S. S. (1983). Nonparametric estimation of the slope of a truncated regression. *AS* **11**, 505–514.

Bhattacharya, R. N. (1985). Some recent results on Cramér–Edgeworth expansions with applications. In *Multivariate Analysis*, Vol. VI (Ed.: P. R. Krishnaiah). Amsterdam: North Holland, pp. 57–75.

Bhattacharya, R. N. and Ghosh, J. K. (1978). On the validity of the formal Edgeworth expansion. *AS* **6**, 434–451.

Bhattacharya, R. N. and Puri, M. L. (1983). On the order of magnitude of cumulants of von Mises functionals and related statistics. *AP* **11**, 346–354.

Bhattacharya, R. N. and Ranga Rao, R. (1976). *Normal Approximation and Asymptotic Expansions*. Wiley, New York.

Bickel, P. J., and van Zwet, W. R. (1978). Asymptotic expansions for the power of distribution-free tests in the two-sample problem. *AS* **6**, 937–1004.

Billingsley, P. (1968). *Convergence of Probability Measures*. Wiley, New York.

Bönner, N., Müller-Funk, U., and Witting, H., (1980). "A Chernoff-Savage theorem for correlation rank statistics with applications to sequential tests". In *Asymptotic Theory of Statistical Tests and Estimates* (Ed.: I. M. Chakravarti). Academic, New York.

Boyd, M. N. and Sen, P. K. (1983). Union–intersection rank tests for ordered alternatives in some linear models. *CS,A* **12**, 1737–1754.

——— (1984). Union–intersection rank tests for ordered alternatives in a complete block design. *CS,A* **13**, 285–303.

Breslow, N. (1970). A generalized Kruskal–Wallis test for comparing *K* samples subject to unequal pattern of censorship. *BK* **57**, 579–594.

Breslow, N. and Crowley, J. (1974). A large sample study of the life table and product limit estimates under random censoring. *AS* **2**, 437–453.

Brown, G. W. and Mood, A. M. (1950). "On median tests for linear hypotheses". In *Proc. Second Berkeley Symp. Math. Statist. Probab.*, Vol. 1 (Ed.: J. Neyman). Berkeley: Univ. of California, pp. 159–166.

Callaert, H. and Janssen, P. (1978). The Berry–Esséen theorem for *U*-statistics. *AS* **6**, 417–421.

Callaert, H. and Veraverbeke, N. (1981). The order of normal approximation for a studentized *U*-statistic. *AS* **9**, 194–200.

Carlson, M. A. and Puri, M. L. (1985). "Central limit theorem for linear rank statistics process". In *Recent Developments in Statistical Theory and Data Analysis* (Ed.: K. Matusita). Amsterdam: North Holland, pp. 103–145.

Chan, Y. and Wierman, J. (1977). On the Berry–Esséen theorem for *U*-statistics. *AS* **5**, 136–139.

Chatterjee, S. K. and De, N. K. (1972). Bivariate nonparametric location tests against restricted alternatives. *CSAB* **21**, 1–20.

——— (1974). On the power superiority of certain bivariate location tests against restricted alternatives. *CSAB* **23**, 73–84.

Chatterjee, S. K. and Sen, P. K. (1964). "Nonparametric tests for the bivariate two-sample location problem". *CSAB* **13**, 18–58.

——— (1973). Nonparametric testing under progressive censoring. *CSAB* **22**, 13–50.

Chernoff, H. and Savage, I. R. (1958). Asymptotic normality and efficiency of certain nonparametric test statistics. *AMS* **29**, 972–994.

Chiang, Ching-Yuan and Puri, M. L. (1984a). Rank order tests for the parallelism of several regression surfaces. *JSPI* **10**, 43–57.

——— (1984b). Rank procedures for testing subhypotheses in linear regression. *AISM* **36**, 35–50.

——— (1985). Tests of subhypotheses in linear regression based on rank order estimates. *SSMH* **20**, (To appear).

Chinchilli, V. M. (1979). Rank Tests for Restricted Alternative Problems in Multivariate Analysis. Unpublished Ph.D. Dissertation, Chapel Hill: Univ. of North Carolina.

Chinchilli, V. M. and Sen, P. K. (1981a). Multivariate linear rank statistic and the union-intersection principle for hypothesis testing under restricted alternatives. *S,B* **43**, 135–151.

——— (1981b). Multivariate linear rank statistics and the union–intersection principle for the orthant restriction problem. *S,B* **43**, 152–171.

——— (1982). Multivariate linear rank statistics for Profile Analysis. *JMA* **12**, 219–229.

Cox, D. R. (1972). Regression models and life tables. *JRSS,B* **34**, 187–220.

Cramér, H. (1946). *Mathematical Methods of Statistics*. Princeton Univ. Press.

Csáki, E. (1977). The law of the iterated logarithm for normalized empirical distribution function. *ZW* **38**, 147–167.

Csörgő, M. and Révész, P. (1980). *Strong Approximations in Probability and Statistics*. New York: Academic.

Davidson, R. R. and Bradley, R. A. (1969). Multivariate paired comparisons: The extension of a univariate model and associated estimation and test procedures. *BK* **56**, 81–95.

——— (1970). "Multivariate paired comparisons: Some large-sample results on estimation and test of equality of preference". In *Nonparametric Techniques in Statistical Inference* (Ed.: M. L. Puri). Cambridge U.P., pp. 111–125.

De, N. K. (1976). Rank tests for randomized blocks against ordered alternatives. *CSAB* **25**, 97–100.

DeLong, E. R. and Sen, P. K. (1981). Estimation of $P\{X > Y\}$ based on progressively truncated versions of the Wilcoxon–Mann–Whitney statistics. *CS,A* **10**, 963–981.

——— (1983). The extended two-sample problem: Progressively truncated estimation of $P\{X > Y\}$. *SD* **1**, 147–170.

Denker, M. and Puri, M. L. (1985a). Asymptotic Behavior of Multi-response Permutation Procedures. (Technical Report 1985-11, Dept. of Math.) Bloomington: Indiana Univ.

——— (1985b). Invariance principles for two sample linear rank statistics. (Technical Report 1985-15, Dept. of Math.) Bloomington: Indiana Univ.

Denker, M., Puri, M. L., and Rösler, U. (1985). A sharpening of the remainder term in the higher dimensional central limit theorem for multilinear rank statistics. *JMA* **15**, (To appear).

Does, R. J. M. M. (1982). *Higher Order Asymptotics for Simple Linear Rank Statistics*. Amsterdam: Mathematisch Centrum.

Dupač, V. (1970). "A contribution to the asymptotic normality of simple linear rank statistics". In *Nonparametric Techniques in Statistical Inference* (Ed.: M. L. Puri). Cambridge U.P., pp. 75–88.

Dupač, V. and Hájek, J. (1969). Asymptotic normality of simple linear rank statistics under alternatives II. *AMS* **40**, 1992–2017.

Feder, P. I. (1968). On the distribution of the log likelihood ratio test statistic when the true parameter is "near" the boundaries of the hypothesis regions. *AMS* **39**, 2044–2055.

Gabriel, K. R. and Sen, P. K. (1968). Simultaneous test procedures for one-way ANOVA and MANOVA based on rank scores. *S,A* **30**, 303–322.

Gardiner, J. C. and Sen, P. K. (1978). Asymptotic normality of a class of time-sequential statistics and applications. *CS,A* **7**, 373–388.

Gastwirth, J. L. (1965). Asymptotically most powerful rank tests for the two-sample problem with censored data. *AMS* **36**, 1243–1247.

Gehan, E. A. (1965). A generalized Wilcoxon test for comparing arbitrarily singly-censored samples. *BK* **52**, 203–223.

Ghosh, J. K. and Sen, P. K. (1985). On the asymptotic properties of the log-likelihood ratio statistics for the mixture model and related results. *Proc. Berkeley Symp. in Honor of J. Neyman and J. Kiefer* (eds: L. LeCam and R. A. Olshen), Wadsworth, Belmont, Calif. (In press).

Ghosh, M. (1971), On the Wald-optimality of rank order tests for paired comparisons. *AMS* **42**, 1970–1976.

——— (1973a). On a class of asymptotically optimal nonparametric tests for grouped data, I. *AISM* **25**, 91–108.

——— (1973b). On a class of asymptotically optimal nonparametric tests for grouped data, II. *AISM* **25**, 109–122.

——— (1975). On some properties of a class of Spearman rank statistics with applications. *AISM* **27**, 157–168.

Ghosh, M. and Dasgupta, R. (1982). "Berry–Esséen theorems for *U*-statistics in the non-i.i.d. case". In *Nonparametric Statistical Inference*, Vol. I (Eds.: B. V. Gnedenko, M. L. Puri, and I. Vincze). Amsterdam: North Holland, pp. 293–314.

Ghosh, M., Grizzle, J. E., and Sen, P. K. (1973). Nonparametric methods in longitudinal studies. *JASA* **68**, 29–36.

Ghosh, M. and Sen, P. K. (1970). On the almost sure convergence of von Mises' differentiable statistical functions. *CSAB* **19**, 41–44.

——— (1971a). On bounded length sequential confidence intervals based on one-sample rank order statistics. *AMS* **42**, 189–203.

——— (1971b). On a class of rank order tests for regression with partially informed stochastic predictors. *AMS* **42**, 650–661.

——— (1972). On bounded length confidence intervals for the regression coefficient based on a class of rank statistics. *S,A* **34**, 33–52.

——— (1973). On some sequential simultaneous confidence interval procedures. *AISM* **25**, 123–133.

——— (1977). Sequential rank tests for regression. *S,A* **39**, 45–62.

——— (1984). On Asymptotically Risk-Efficient Sequential Versions of Generalized *U*-Statistics. *Sequen. Anal* **3**, 233–252.

Ghosh, M. and Stores, D. (1977). Strong convergence of linear rank statistics for mixing processes. *S,B* **39**, 1–11.

Grams, W. F. and Serfling, R. J. (1973). Convergence rates for *U*-statistics and related statistics. *AS* **1**, 153–160.

Grizzle, J. E. and Allen, D. M. (1969). Analysis of growth and dose response curves. *BC* **25**, 307–318.

Gupta, S. S. (1963). Probability integrals of multivariate normal and multivariate *t*. *AMS* **34**, 792–838.

Hâjek, J. (1961). Some extensions of the Wald–Wolfowitz–Noether theorem. *AMS* **32**, 506–523.

——— (1962). Asymptotically most powerful rank order tests. *AMS* **33**, 1124–1147.

——— (1968). Asymptotic normality of simple linear rank statistics under alternatives. AMS **39**, 325–346.

——— (1974). Asymptotic sufficiency of the vector of ranks in the Bahadur sense. *AS* **2**, 75–83.

Hâjek, J. and Šidák Z. (1967). *Theory of Rank Tests*. Prague: Academia.

Hallin, M., Ingenbleek, J-F., and Puri, M. L. (1985a). Linear Serial Rank Tests for Independence Against ARMA Alternatives (Technical Report 1985-12, Department of Mathematics), Indiana University, Bloomington.

——— (1985b). Linear and Quadratic Serial Rank Tests for Randomness Against Serial Dependence (Technical Report 1985-13, Dept. of Math.) Bloomington: Indiana Univ.

Halperin, M. (1960). Extension of the Wilcoxon–Mann–Whitney test to samples censored at the same fixed point. *JASA* **55**, 125–138.

Hayakawa, T. and Puri, M. L. (1985). Asymptotic expansions of the distributions of some test statistics. *AISM* 37,.

Helmers, R. and Janssen, P. (1982). *On the Berry–Esséen Theorem for Multivariate U-Statistics.* (Tech. Rept. SW 90/82). Amsterdam: Math. Centrum.

Helmers, R. and van Zwet, W. R. (1982). "The Berry–Esséen bound for *U*-statistics". In *Statistical Decision Theory and Related Topics*, Vol. III (Eds.: S. S. Gupta and J. O. Berger). New York: Academic, pp. 497–512.

Hettmansperger, T. P. (1984). *Statistical Inference Based on Ranks*. New York: Wiley.

Hettmansperger, T. P. and Schrader, R. M. (1980). Robust analysis of variance based upon a likelihood ratio criterion. *BK* **67**, 93–101.

Hill, B. M. (1962). A test of linearity versus convexity of a median regression curve. *AMS* **33**, 1096–1123.

Hill, N. J., Padmanabhan, A. R., and Puri, M. L. (1985). Adaptive Nonparametric Procedures and Applications (Technical Report 1985-6, Dept. of Math.) Bloomington: Indiana Univ.

Hodges, J. L., Jr. and Lehmann, E. L. (1963). Estimates of location based on rank tests. *AMS* **34**, 598–611.

Hoeffding, W. (1948). A class of statistics with asymptotically normal distributions. *AMS*, **19**, 293–325.

——— (1951a). A combinatorial central limit theorem. *AMS* **22**, 558–566.

——— (1951b). "Optimum nonparametric tests". In *Proc. Second Berkeley Symp. Math. Statist. Probab.*, Vol. 1 (Ed.: J. Neyman). Univ. of California Press, pp. 203–220.

——— (1965). Asymptotically optimal tests for multinomial distributions. *AMS* **36**, 369–400.

——— (1973). On the centering of a simple linear rank statistic. *AS* **1**, 54–66.

Hušková, M. (1970). Asymptotic distribution of simple linear rank statistics for testing symmetry. *ZW* **12**, 308–322.

——— (1971). Asymptotic distribution of rank statistics used for multivariate testing symmetry. *JMA*, **1**, 461–484.

——— (1977a). The rate of convergence of simple linear rank statistics under hypothesis and alternatives. *AS* **5**, 658–670.

——— (1977b). Rates of convergence of quadratic rank statistics. *JMA* **7**, 63–73.

——— (1979). "The rate of convergence of simple linear rank statistics under alternatives". In *Contributions to Statistics: J. Hájek Memorial Volume* (Ed.: J. Jurečková). Prague: Academia, pp. 99–108.

——— (1980). "Some asymptotic results on the multivariate rank statistics". In *Multivariate Analysis*, Vol. V (Ed.: P. R. Krishnaiah). New York: Academic, pp. 223–237.

——— (1982). "On bounded length sequential confidence interval for parameter in regression model based on ranks". In *Nonparametric Statistical Inference*, Vol. I (Eds.: B. V. Gnedenko, M. L. Puri, and I. Vincze). Amsterdam: North Holland, pp. 435–463.

Hušková, M. and Jurečková, J. (1981). Second order asymptotic relations of *M*-estimators and *R*-estimators in two-sample location model. *JSPI* **5**, 309–328.

——— (1984). "Asymptotic representation of *R*-estimators of location". In *Proc. 4th Pannonian Symp. Math. Statist.*

Hušková, M. and Sen, P. K. (1984). On sequentially adaptive asymptotically efficient rank statistics. *Inst. Statist., Univ. N. Car. Mimeo Rept.* NO 1475.

Jaeckel, L. A. (1972). Estimating regression coefficients by minimizing the dispersion of the residuals. *AMS* **43**, 1449–1458.

Jurečková, J. (1969). Asymptotic linearity of a rank statistic in regression parameter. *AMS* **40**, 1889–1900.

——— (1971a). Nonparametric estimate of regression coefficients. *AMS* **42**, 1328–1338.

——— (1971b). Asymptotic independence of rank test statistic for testing symmetry on regression. *S,A*, **33**, 1–18.

——— (1973). Central limit theorem for Wilcoxon rank statistics process. *AMS* **6**, 1046–1060.

——— (1977a). Asymptotic relations of *M*-estimates and *R*-estimates in linear regression models. *AS* **5**, 664–672.

——— (1977b). "Asymptotic relations of least squares estimates and two robust estimates of regression parameter vector". In *Trans. 7th Prague Conf. Inform. Theory Statist. Dec. Funct. Random Proc.*

——— (1980a). Asymptotic representations of *M*-estimators of location. *MOS* **11**, 61–73.

——— (1980b). Rate of consistency of one sample tests of location. *JSPI* **4**, 248–257.

——— (1983a). "Robust estimators of location and regression parameters and their second order asymptotic relations". In *Trans. 9th Prague Conf. Inform. Theory Statist. Dec. Funct. Random Proc.*, pp. 79–92.

——— (1983b). Robust estimators and their relations. *Acta Univ. Carolin.—Math. Phys.* **24**, 49–59.

——— (1984/85). "Robust estimators of location and their second order asymptotic relations". In *ISR* (Centennial Vol.), in press.

Jurečková, J. and Puri, M. L. (1975). Order of normal approximation for rank test statistics distribution. *AP* **3**, 526–533.

Jurečková, J. and Sen, P. K. (1981). Invariance principles for some stochastic processes related to *M*-estimators and their role in sequential statistical inference. *S,A* **43**, 190–210.

——— (1982a). *M*-estimators and *R*-estimators of location: Uniform integrability and asymptotically risk-efficient sequential versions. *Sequen. Anal.* **1**, 27–56.

——— (1982b). Simultaneous *M*-estimator of the common location and the scale-ratio in the two-sample problem. *MOS* **13**, 263–269.

——— (1984). On adaptive scale-equivariant *M*-estimation in linear models. *SD* **2**, *Suppl.* 31–46.

Kalbfleisch, J. D. and Prentice, R. (1980). *The Statistical Analysis of Failure Time Data*. New York: Wiley.

Kallenberg, W. C. M. (1982). Cramér type large deviations for simple linear statistics. *ZW* **60**, 403–409.

Kaplan, E. L. and Meier, P. (1958). Nonparametric estimation from incomplete observations. *JASA* **53**, 457–481.

Khatri, C. G., Krishnaiah, P. R., and Sen, P. K. (1977). A note on the joint distribution of correlated quadratic forms. *JSPI* **1**, 175–186.

Koul, H. L. (1969). Asymptotic behavior of Wilcoxon type confidence regions in multiple regression. *AMS* **40**, 1950–1979.

——— (1970). Some convergence theorems for ranks and weighted empirical cumulatives. *AMS* **41**, 1768–1773.

Koul, H. L. and Staudte, R. G. (1972). Asymptotic normality of signed rank statistics. *ZW* **22**, 295–300.

Koziol, J. A. (1978). Multivariate signed rank statistics for shift alternatives. *MOS* **9**, 549–562.

Koziol, J. A., Maxwell, D. A., Fukushima, M., Colmerauer, M. E. M., and Pilch, Y. H. (1981). A distribution-free approach to tumor growth curve analyses with application to an animal tumor immunotherapy experiment. *BC* **37**, 383–390.

Koziol, J. A. and Petkau, A. J. (1978). Sequential testing of equality of two survival distributions using the modified Savage statistics. *BK* **65**, 615–623.

Kraft, C. H. and van Eeden, C. (1972). Linearized rank estimates and signed rank estimates for the general linear hypothesis. *AMS* **43**, 42–57.

Krishnaiah, P. R. and Sen, P. K. (1971). Some asymptotic simultaneous tests for multivariate moving averages processes. *S,A* **33**, 81–90.

Lai, T. L. (1975). On Chernoff–Savage statistics and sequential rank tests. *AS* **3**, 825–845.

Lea, C. D. and Puri, M. L. (1985a). Asymptotic Properties of Perturbed Empirical Distribution Functions Evaluated at a Random Point (Technical Report 1985-14, Dept. of Math.) Bloomington: Indiana Univ.

——— (1985b). Asymptotic Properties of Linear Functions of Order Statistics (Technical Report 1985-15, Dept. of Math.) Bloomington: Indiana Univ.

LeCam, L. (1960). Locally asymptotically normal families of distributions. *Univ. Calif. Publ. Statist.* **3**, 37–98.

Lehmann, E. L. (1966). Some concepts of dependence. *AMS* **37**, 1137–1153.

——— (1975). *Nonparametrics: Statistical Methods Based on Ranks*. San Francisco: Holden-Day.

Madow, W. G. (1948). On the limiting distributions of estimates based on samples from finite universes. *AMS* **19**, 535–545.

Majumdar, H. and Sen, P. K. (1977). Rank order tests for grouped data under progressive censoring. *CS,A* **6**, 507–524.

——— (1978a). Nonparametric tests for multiple regression under progressive censoring. *JMA* **8**, 73–95.

——— (1978b). Nonparametric testing for simple regression under progressive censoring with staggering entry and random withdrawal. *CS,A* **7**, 349–371.

Mangasarian, O. L. (1969). *Nonlinear Programming*. New York: McGraw-Hill.

McKean, J. W. and Hettmansperger, T. P. (1976). Tests of hypotheses based on ranks in the general linear model. *CS,A* **5**(8), 693–709.

Mehra, K. L. and Sen, P. K. (1969). On a class of conditionally distribution-free tests for interactions in factorial experiments. *AMS* **40**, 658–666.

Mielke, P. W. and Sen, P. K. (1981). On asymptotic nonnormal null distributions for locally most powerful rank test statistics. *CS,A* **10**, 1079–1094.

Miller, R. G., Jr. (1966). *Simultaneous Statistical Inference*. New York: McGraw-Hill.

Miller, R. G., Jr. and Sen, P. K. (1972). Weak convergence of *U*-statistics and von Mises' differentiable statistical functions. *AMS* **43**, 31–41.

Moses, L. E. (1953). "Nonparametric methods". In *Statistical Inference* (Eds.: Walker and Lev). New York: Henry Holt, Chap. 18, pp. 426–450.

Motoo, M. (1957). On Hoeffding's combinatorial central limit theorem. *AISM* **8**, 145–154.

Muller-Funk, U. and Witting, H. (1982). "On the rate of convergence in the CLT for signed linear rank statistics". In *Nonparametric Statistical Inference*, Vol. II (Eds.: B. V. Gnedenko, M. L. Puri, and I. Vincze), Amsterdam: North Holland, pp. 637–652.

Noether, G. E. (1949). On a theorem by Wald and Wolfowitz. *AMS* **20**, 455–458.

Padmanabhan, A. R. and Puri, M. L. (1979). Rank order estimates in the case of grouped data (linear regression). *S,B* **41**, 239–259.

——— (1983). Theory of nonparametric statistics for rounded-off data with applications. *MOS* **14**, 301–349.

——— (1984). Estimating the Ratio of Scale Parameters (Technical Report 1983-16, Dept. of Math.) Bloomington: Indiana Univ.

——— (1985a). Robust Estimation of Ratio of Scale Parameters and Applications (Technical Report 1985-7, Dept. of Math.) Bloomington: Indiana Univ.

——— (1985b). Adaptive nonparametric procedures and applications I. (Technical Report 1985-16, Dept. of Math.) Bloomington: Indiana Univ.

——— (1985c). Adaptive nonparametric procedures and applications II. (Technical Report 1985-17, Dept. of Math.) Bloomington: Indiana Univ.

Padmanabhan, A. R., Puri, M. L., and Saleh, A. K. Md. E. (1981). "A nonparametric test of ordered alternatives in the case of skewed data with a biomedical application". In *Statistics and Related Topics* (Eds.: M. Csörgő, D. A. Dawson, J. N. K. Rao, and A. K. Md. E. Saleh). Amsterdam: North Holland, pp. 279–283.

Petrov, V. V. (1975). *Sums of Independent Random Variables*. Berlin: Springer.

Potthoff R. F. and Roy, S. N. (1964). A generalized multivariate analysis of variance model useful especially for growth curve problems. *BK* **51**, 313–326.

Prášková, Z. (1982). "A local limit theorem and an asymptotic expansion for a two-sample rank test". In *Nonparametric Statistical Inference*, Vol. II (Eds.: B. V. Gnedenko, M. L. Puri, and I. Vincze). Amsterdam: North Holland, pp. 713–728.

Puri, M. L. and Rajaram, N. S. (1979). Asymptotic behavior of stochastic linear rank statistics. *Bull. Internat. Statist. Inst.* **47**(4), 292–297.

——— (1980). Asymptotic normality and convergence rates of linear rank statistics under alternatives, *Banach Center Publ.* **6**, 267–277.

——— (1982). "Stochastic integrals and rank statistics". In *Nonparametric Statistical Inference*,

Vol. II (Eds.: I. Vincze, B. V. Gnedenko, and M. L. Puri), Amsterdam: North Holland, pp. 729–747.

Puri, M. L. and Ralescu, S. (1982a). "On the degeneration of the variance in the asymptotic normality of signed rank statistics". In *Statistics and Probability: Essays in Honor of C. R. Rao*. (Eds.: G. Kallianpur, P. R. Krishnaiah, and J. K. Ghosh). Amsterdam: North Holland, pp. 591–608.

——— (1982b). "The asymptotic distribution theory of one sample signed rank statistic". In *Statistical Decision Theory and Related Topics III*, Vol. 2 (Eds.: Shanti S. Gupta and James O. Berger), New York: Academic, pp. 213–232.

——— (1984a). Centering of signed rank statistic with continuous score generating function. *TVP* **29**, 580–584.

——— (1984b). On Berry–Esséen rates, a law of the iterated logarithm and an invariance principle for the proportion of the sample below the sample mean. *JMA* **14**, 231–247.

——— (1985a). Limit Theorems for Random Central Order Statistics. (Technical Report 1985-10, Dept. of Math.) Bloomington: Indiana Univ.

——— (1985b). An asymptotic expansion of extremes. (Technical Report 1985-18, Dept. of Math.) Bloomington: Indiana Univ.

——— (1985c). Almost Sure Linearity for Signed Rank Statistics in the Non-I.I.D. Case. *Acta Mathematica Hungarica*, 45, (To appear).

——— (1986). Central limit theorem for perturbed empirical distribution functions evaluated at a random point. *JMA* **15**, in press.

Puri, M. L., Ralescu, S., and Seoh, M. (1983). Large Deviation Probabilities for Signed Linear Rank Statistics (Technical Report 1983-18, Dept. of Math.). Bloomington: Indiana Univ.

Puri, M. L. and Sen, P. K. (1969a). A class of rank order tests for a general linear hypothesis. *AMS* **40**, 1325–1343.

——— (1969b). Analysis of covariance based on general rank scores. *AMS* **40**, 610–618.

——— (1971). *Nonparametric Methods in Multivariate Analysis*. New York: Wiley.

——— (1973). A note on asymptotically distribution free tests for subhypotheses in multiple linear regression. *AS* **1**, 553–556.

Puri, M. L. and Seoh, M. (1984a). Edgeworth expansion for signed linear rank statistics. *JSPI*, **10**, 137–149.

——— (1984b). "Berry–Esseen theorems for signed linear rank statistics with regression constants". In *Limit Theorems in Probability and Statistics*. (Ed.: P. Révész). Amsterdam: North Holland, pp. 875–906.

——— (1984c). Berry–Esseen theorems for signed linear rank statistics under near location alternatives. *SSMH* **19**, in press.

——— (1984d). Edgeworth expansions for signed linear rank statistics under near location alternatives. *JSPI* **10**, 289–309.

——— (1985a). On the rate of convergence to asymptotic normality for generalized linear rank statistics *AISM* **37**, 25–43.

——— (1985b). Some Large Deviation Results for a Large Class of Statistics (Technical Report 1985-8, Dept. of Math.) Bloomington: Indiana Univ.

——— (1985c). On the Rate of Convergence in Normal Approximation for a Class of Statistics (Technical Report 1985-9, Dept. of Math.) Bloomington: Indiana Univ.

Puri, M. L. and Shane H. (1970). "Statistical inference in incomplete block designs". In *Nonparametric Techniques in Statistical Inference*. (Ed.: M. L. Puri). Cambridge U.P., pp.

131–153.

Puri, M. L. and Tran, L. (1980). Empirical distribution functions and functions of order statistics for mixing random variables. *JMA* **10**, 405–425.

——— (1981). "Invariance principles for rank statistics for testing independence". In *Contributions to Probability Theory* (Eds.: J. Gani and V. K. Rohatgi). New York: Academic, pp. 267–282.

Puri, M. L. and Wu, T. J. (1984). Asymptotic Normality of the Lengths of a Class of Nonparametric Confidence Intervals for Regression Parameters *CJS* **12**, 217–228.

——— (1985a). Gaussian Approximation of Signed Linear Rank Statistics Process *JSPI* **11**, (To appear).

——— (1985b). The order of normal approximation for signed linear rank statistics. *TVP* **30**, (To appear).

Pyke, R. and Shorack, G. R. (1968a). Weak convergence of a two-sample empirical process and a new approach to Chernoff–Savage theorems. *AMS* **39**, 755–771.

——— (1968b). Weak convergence and a Chernoff–Savage theorem for random sample sizes. *AMS* **39**, 1675–1685.

Quade, D. (1967). Rank analysis of covariance. *JASA* **62**, 1187–1200.

Ralescu, S. S. and Puri, M. L. (1984). On Berry–Esséen rates, a law of the iterated logarithm and an invariance principle for the proportion of the sample below the sample mean. *JMA* **14**, 231–247.

Ralescu, S. S. and Puri, M. L. (1985). On the rate of convergence in the central limit theorem for signed rank statistics. *AAM* **6**, 23–51.

Rao, C. R. (1958). Some statistical methods for comparison of growth curves. *BC* **14**, 1–17.

——— (1965). *Linear Statistical Inference and its Applications*. New York: Wiley.

——— (1967). "Least squares theory using an estimated dispersion matrix and its application to measurement of signals". In *Proc. Fifth Berkeley Symp. Math. Statist. Probab.*, Vol. 1 (Eds.: Lucien LeCam and Jerzey Neyman), Univ. of California Press, pp. 355–372.

Roussas, G. G. (1972). *Contiguity of Probability Measures: Some Applications in Statistics*. London: Cambridge U.P.

Roy, S. N. (1953). On a heuristic method of test construction and its use in multivariate analysis. *AMS* **24**, 220–238.

——— (1957). *Some Aspects of Multivariate Analysis*. Calcutta: Asia Publ. House.

Russell, C. T. and Puri, M. L. (1974). Joint asymptotic multinormality for a class of rank statistics in multivariate paired comparisons. *JMA* **4**, 88–105.

Ruymgaart, F. H. (1973). *Asymptotic Theory of Rank Tests for Independence* (Tract No. 43). Amsterdam: Mathematical Centre.

Ruymgaart, F. H., Shorack, G. R., and van Zwet, W. R. (1972). Asymptotic normality of nonparametric tests for independence. *AMS* **43**, 1122–1135.

Saleh, A. K. Md. E. and Puri, M. L. (1984). Locally most powerful rank test for the two-sample problem with censored samples. *Journal of Organizational Behaviour and Statistics*, **1**, 189–196.

Saleh, A. K. Md. E. and Sen, P. K. (1978). Nonparametric estimation of the location parameter after a preliminary test on regression. *AS* **6**, 154–168.

——— (1982). Nonparametric tests for location after a preliminary test on regression. *CS,A* **11**, 639–652.

——— (1983a). Nonparametric tests for location after a preliminary test on regression in the multivariate case. *CS,A* **12**, 1855–1872.

——— (1983b). Asymptotic properties of tests of hypotheses following a preliminary test. *SD* **1** 455–477.

——— (1984a). "Least squares and rank order preliminary test estimation in general multivariate linear models". In *Golden Jubilee Conference*. Calcutta: Indian Statistical Institute, pp. 237–253

——— (1984b). "Preliminary test prediction in general linear models". In *Recent Developments in Statistical Theory and Data Analysis* (Ed.: K. Matusita). Amsterdam: North Holland.

——— (1985). Nonparametric Shrinkage Estimation in a Parallelism Problem *S, A* 47. In press.

Sen, P. K. (1963). On the estimation of relative potency in dilution (direct) assays by distribution-free methods. *BC* **19**, 532–552.

——— (1966a). On a distribution-free method of estimating asymptotic efficiency of a class of nonparametric tests. *AMS* **37**, 1759–1770.

——— (1966b). On nonparametric simultaneous confidence regions and tests in the one-criterion analysis of variance problem. *AISM* **18**, 319–336.

——— (1967). Asymptotically most powerful rank order tests for grouped data. *AMS* **38**, 1229–1239.

——— (1968). Estimates of regression based on Kendall's tau. *JASA* **63**, 1379–1389.

——— (1969a). On a class of rank order tests for parallelism of several regression lines. *AMS* **40** 1668–1683.

——— (1969b). Nonparametric tests for multivariate interchangeability, Part two: The problem of MANOVA in two-way layouts. *S,A* **31**, 145–156.

——— (1969c). On nonparametric T-method of multiple comparisons in randomized blocks. *AISM* **21**, 329–333.

——— (1970a). Nonparametric inference in n replicated 2^m factorial experiment. *AISM* **20**, 281–294.

——— (1970b). On some convergence properties of one-sample rank order statistics. *AMS* **41**, 2140–2143.

——— (1970c). "On the distribution of one-sample rank order statistics". In *Nonparametric Techniques in Statistical Inference* (Ed.: M. L. Puri). New York: Cambridge U.P., pp. 53–72.

——— (1971a). Asymptotic efficiency of a class of aligned rank order tests for multiresponse experiments in some incomplete block designs. *AMS* **42**, 1104–1112.

——— (1971b). Robust statistical procedures in problems of linear regression with special reference to quantitative bio-assays, I. *ISR* **39**, 21–38.

——— (1972a). On a class of aligned rank order tests for the identity of the intercepts of several regression lines. *AMS* **43**, 2004–2012.

——— (1972b). Robust statistical procedures in problems of linear regression with special reference to quantitative bio-assays, II. *ISR* **40**, 161–172.

——— (1972c). Weak convergence and relative compactness of martingale processes with applications to nonparametric statistics. *JMA* **2**, 345–361.

——— (1973a). An almost sure invariance principle for multivariate Kolmogorov–Smirnov statistics. *AP* **1**, 488–496.

——— (1973b). "Some aspects of nonparametric procedures in multivariate statistical analysis". In *Multivariate Statistical Inference* (Eds.: D. G. Kabe and R. P. Gupta). Amsterdam:

North Holland, pp. 230–240.

——— (1974a). Almost sure behavior of *U*-statistics and von Mises' differentiable statistical functions. *AS* **2**, 387–395.

——— (1974b). On L_p-convergence of *U*-statistics. *AISM* **26**, 55–60.

——— (1975). "Rank statistics, martingales and limit theorems". In *Statistical Inference and Related Topics* (Ed.: M. L. Puri), New York: Academic, pp. 129–158.

——— (1976a). A two-dimensional functional permutational central limit theorem for linear rank statistics. *AP* **4**, 13–26.

——— (1976b). Asymptotically optimal rank order tests for progressive censoring. *CSAB* **25**, 65–78.

——— (1977). Tied-down Wiener process approximations for aligned rank order statistics and some applications. *AS* **5**, 1107–1123.

——— (1978a). "Nonparametric repeated significance tests". In *Developments in Statistics*, Vol. 1 (Ed.: P. R. Krishnaiah), New York: Academic, pp. 227–264.

——— (1978b). On some distribution-free tests for affine symmetry. *CSAB* **27**, 59–79.

——— (1979a). Nonparametric repeated significance tests for some analysis of covariance models. *CS,A* **9**, 819–841.

——— (1979b). Rank analysis of covariance under progressive censoring. *S,A* **41**, 147–169.

——— (1979c). The extended two-sample problem: Nonparametric case. *JSPI* **3**, 287–298.

——— (1979d). "Nonparametric tests for bivariate interchangeability under competing risks". In *Contributions to Statistics: Jaroslav Hájek Memorial Volume* (Ed.: J. Jurečkova). Prague: Academia, pp. 211–228.

——— (1979e). Asymptotic properties of maximum likelihood estimators based on conditional specifications. *AS* **7**, 1019–1033.

——— (1980a). On almost sure linearity theorems for signed rank statistics. *AS* **8**, 313–321.

——— (1980b). "Nonparametric simultaneous inference for some MANOVA models". In *Handbook of Statistics*, Vol. 1 (Ed.: P. R. Krishnaiah). Amsterdam: North Holland, pp. 673–702.

——— (1980c). Asymptotic theory of some tests for a possible change in the regression slope occurring at an unknown time point. *ZW* **52**, 203–218.

——— (1981a). *Sequential Nonparametrics: Invariance Principles and Statistical Inference*. New York: Wiley.

——— (1981b). "Rank analysis of covariance under progressive censoring. II". In *Statistics and Related Topics* (Eds.: M. Csörgő, D. A. Dawson, J. N. K. Rao, and A. K. Md. E. Saleh). Amsterdam: North Holland, pp. 285–295.

——— (1981c). Some invariance principles for mixed rank statistics and induced order statistics. *CS,A* **10**, 1691–1718.

——— (1982a). "The UI-principle and LMP rank tests". In *Nonparametric Statistical Inference*, Vol. II (Eds.: B. V. Gnedenko, M. L. Puri, and I. Vincze). Amsterdam: North Holland, pp. 843–858.

——— (1982b). On *M*-tests in linear models. *BK* **69**, 245–248.

——— (1982c). Asymptotic theory of some tests for constancy of regression relationship over time". *MOS* **13**, 21–32.

——— (1982d). Asymptotic properties of likelihood ratio tests under conditional specifications. *SD* **1**, 81–106.

——— (1983a). On permutational central limit theorems for general multivariate linear rank statistics. *S,A* **45**, 141–149.

——— (1983b). "Some recursive residual rank tests for change-points." In *Recent Advances in Statistics: Papers in Honor of Herman Chernoff's Sixtieth Birthday* (Eds.: M. H. Rizvi, D. Siegmund, and J. S. Rustagi), New York: Academic, 371–391.

——— (1983c). Sequential R-estimation of location in the general Behrens-Fisher model. *Sequen. Anal.* **2**, 311–335.

——— (1983d). Recursive M-tests for the change-points. *Bull. Internat. Statist. Inst.* (*Madrid Conf.*) **1**, 206–209.

——— (1984a). The Cox regression model, random censoring and locally optimal rank tests, *JSPI* **9**, 355–366.

——— (1984b). Subhypotheses testing against restricted alternatives for the Cox regression model. *JSPI* **10**, 31–42.

——— (1984c). Jackknifing L-estimators: Affine structure and asymptotics. *S, A* **46**, 207–219.

——— (1984d). On a Kolmogorov-Smirnov type aligned test. *Statist. Prob. Lett.* **2**, 193–196.

——— (1984e). Invariance principles for *U*-statistics and von Mises' functionals in the non-i.d. case. *S,A* **46** 416–425.

——— (1984f). On sequential nonparametric estimation of multivariate location. *Proc. 3rd Prague Confer. Asympt. Meth.* (Eds: P. Mandl & M. Hušková), 119–130.

——— (1984g). On James-Stein detour of U-Statistics. *CS, A* **13**, 2725–2747.

——— (1984h). Nonparametric testing against restricted alternatives under progressive consoring. *Inst. Statist., Univ. N. Car. Mimeo Rept.* NO 1473.

——— (1985). *Theory and Applications of Sequential Nonparametrics*. Philadelphia, SIAM.

Sen, P. K. and David, H. A. (1968). Paired comparisons for paired characteristics. *AMS* **39**, 200–208.

Sen, P. K. and Ghosh, M. (1971). On bounded length sequential confidence intervals based on one-sample rank order statistics. *AMS* **42**, 189–203.

——— (1972). On strong convergence of regression rank statistics. *S,A* **34**, 335–348.

——— (1973a). A Chernoff–Savage representation of rank order statistics for stationary ϕ-mixing processes. *S,A* **35**, 153–172.

——— (1973b). A law of iterated logarithm for one sample rank order statistics and some applications. *AS* **1**, 568–576.

——— (1974). Some invariance principles for rank statistics for testing independence. *ZW* **29**, 93–108.

Sen, P. K. and Krishnaiah, P. R. (1974). On a class of simultaneous rank tests in MANOCOVA. *AISM* **26**, 135–145.

Sen, P. K. and Puri, M. L. (1969). "On robust nonparametric estimation in some multivariate linear models". In *Multivariate Analysis*, Vol. II (Ed.: P. R. Krishnaiah). New York: Academic, pp. 33–52.

——— (1970). Asymptotic theory of likelihood ratio and rank order tests in some multivariate linear models. *AMS* **41**, 87–100.

——— (1972a). On the robustness of rank order tests and estimates in the generalized one-sample location problem. *ZW* **22**, 226–241.

——— (1972b). On some selection procedures in two-way layouts. *ZW* **22**, 242–250.

——— (1977). Asymptotically distribution-free aligned rank order tests for composite hypotheses for general linear models. *ZW* **39**, 175–186.

Sen, P. K. and Saleh, A. K. Md. E. (1979). Nonparametric estimation of location parameter after a preliminary test on regression in the multivariate case. *JMA* **9**, 322–331.

——— (1984a). "Nonparametric shrinkage estimators of location in a multivariate simple regression model". In *Proc. 4th Pannonian Conf. Math. Statist.* (to appear).

——— (1985). On some shrinkage estimators of multivariate location. *AS* **13** (in press).

Seoh, M., Ralescu, S. S., and Puri, M. L. (1985). Cramér-type large deviations for generalized linear rank statistics. *AP* **13**, in press.

Serfling, R. J. (1980). *Approximation Theorems of Mathematical Statistics.* New York: Wiley.

Shane, H. and Puri, M. L. (1969). Rank order tests for multivariate paired comparisons. *AMS* **40**, 2101–2117.

Shirahata, S. (1975). Locally most powerful rank tests for independence with censored data, *AS* **3**, 241–245.

Shorack, G. R. (1973). Convergence of reduced empirical and quantile processes with application to functions of order statistics in the non-i.i.d. case. *AS* **1**, 146–152.

Singer, J. M. and Sen, P. K. (1985a). *M*-methods in multivariate linear models. *JMA* **15** (in press).

——— (1985b). Asymptotic relative efficiency of multivariate *M*-estimators. *CS,A* **14** (to appear).

Sinha, A. N. and Sen, P. K. (1982). Tests based on empirical processes for progressive censoring schemes with staggering entry and random withdrawal. *S,B* **44**, 1–18.

——— (1984). "Staggering entry, random withdrawal and progressive censoring schemes: Some nonparametric procedures". In *Golden Jubilee Conference*. Calcutta: Indian Statistical Institute, pp. 531–547.

So, Y. C. and Sen, P. K. (1982a). *M*-estimators based repeated significance tests for one-way ANOVA with adaptations to multiple comparisons. *Sequen. Anal.* **1**, 101–119.

——— (1982b). Nonparametric repeated significance tests for one-way ANOVA with adaptations to multiple comparisons. *JSPI* **7**, 83–96.

Susarla, V., van Ryzin, J., and Koul, H. L. (1981). Regression analysis with random rightcensored data. *AS* **9**, 1276–1288.

Terry, M. E. (1952). Some rank order tests which are most powerful against specific parametric alternatives. *AMS* **23**, 346–366.

van Eeden, C. (1972). An analogue, for signed rank statistics, for Jurečková's asymptotic linearity theorem for rank statistics. *AMS* **43**, 791–802.

——— (1983). On the asymptotic relation between *L*-estimators and *M*-estimators and their asymptotic efficiency relative to the Cramér–Rao lower bound. *AS* **11**, 674–690.

van Zwet, W. R. (1982). "On the Edgeworth expansion for the simple linear rank statistic". In *Nonparametric Statistical Inference*, Vol. II (Eds.: B. V. Gnedenko, M. L. Puri, and I. Vincze). Amsterdam: North Holland, pp. 889–909.

——— (1984). A Berry-Esséen bound for symmetric statistics. *ZW* **66**, 425–440.

Vorličková, D. (1970). Asymptotic properties of rank tests under discrete distributions. *ZW* **14**, 275–289.

——— (1972). Asymptotic properties of rank tests of symmetry under discrete distributions. *AMS* **43**, 2013–2018.

Wald, A. (1943). Tests of statistical hypotheses concerning several parameters when the number of observations is large. *TAMS* **54**, 426–482.

Wald, A. and Wolfowitz, J. (1944). Statistical tests based on permutations of the observations. *AMS* **15**, 358–372.

Wijsman, R. A. (1979). Constructing all smallest simultaneous confidence sets in a given class, with applications to MANOVA. *AS* **7**, 1003–1018.

Withers, C. S. (1983). Expansions for the distribution and quantiles of a regular functional of the empirical distribution with applications to nonparametric confidence intervals. *AS* **11**, 577–587.

Wolfowitz, J. (1953). Estimation by the minimum distance method. *AISM* **5**, 9–23.

Woolson, R. F. and Sen, P. K. (1974). Asymptotic comparison of a class of multivariate multiparameter tests. *CS* **3**, 813–828.

Yohai, V. J. (1974). Robust estimation in the linear model. *AS* **2**, 562–567.

Yohai, V. J. and Maronna, R. A. (1979). Asymptotic behavior of *M*-estimators for the linear model. *AS* **7**, 258–268.

Author Index

Subject Index

Applied Probability and Statistics (Continued)

HOEL and JESSEN • Basic Statistics for Business and Economics, *Third Edition*
HOGG and KLUGMAN • Loss Distributions
HOLLANDER and WOLFE • Nonparametric Statistical Methods
IMAN and CONOVER • Modern Business Statistics
JAGERS • Branching Processes with Biological Applications
JESSEN • Statistical Survey Techniques
JOHNSON and KOTZ • Distributions in Statistics
Discrete Distributions
Continuous Univariate Distributions—1
Continuous Univariate Distributions—2
Continuous Multivariate Distributions
JOHNSON and KOTZ • Urn Models and Their Application: An Approach to Modern Discrete Probability Theory
JOHNSON and LEONE • Statistics and Experimental Design in Engineering and the Physical Sciences, Volumes I and II, *Second Edition*
JUDGE, HILL, GRIFFITHS, LÜTKEPOHL and LEE • Introduction to the Theory and Practice of Econometrics
JUDGE, GRIFFITHS, HILL, LÜTKEPOHL and LEE • The Theory and Practice of Econometrics, *Second Edition*
KALBFLEISCH and PRENTICE • The Statistical Analysis of Failure Time Data
KISH • Survey Sampling
KUH, NEESE, and HOLLINGER • Structural Sensitivity in Econometric Models
KEENEY and RAIFFA • Decisions with Multiple Objectives
LAWLESS • Statistical Models and Methods for Lifetime Data
LEAMER • Specification Searches: Ad Hoc Inference with Nonexperimental Data
LEBART, MORINEAU, and WARWICK • Multivariate Descriptive Statistical Analysis: Correspondence Analysis and Related Techniques for Large Matrices
McNEIL • Interactive Data Analysis
MAINDONALD • Statistical Computation
MANN, SCHAFER and SINGPURWALLA • Methods for Statistical Analysis of Reliability and Life Data
MARTZ and WALLER • Bayesian Reliability Analysis
MIKÉ and STANLEY • Statistics in Medical Research: Methods and Issues with Applications in Cancer Research
MILLER • Beyond ANOVA, Basics of Applied Statistics
MILLER • Survival Analysis
MILLER, EFRON, BROWN, and MOSES • Biostatistics Casebook
MONTGOMERY and PECK • Introduction to Linear Regression Analysis
NELSON • Applied Life Data Analysis
OSBORNE • Finite Algorithms in Optimization and Data Analysis
OTNES and ENOCHSON • Applied Time Series Analysis: Volume I, Basic Techniques
OTNES and ENOCHSON • Digital Time Series Analysis
PANKRATZ • Forecasting with Univariate Box-Jenkins Models: Concepts and Cases
PIELOU • Interpretation of Ecological Data: A Primer on Classification and Ordination
POLLOCK • The Algebra of Econometrics
PRENTER • Splines and Variational Methods
RAO and MITRA • Generalized Inverse of Matrices and Its Applications
RIPLEY • Spatial Statistics
SCHUSS • Theory and Applications of Stochastic Differential Equations
SEAL • Survival Probabilities: The Goal of Risk Theory
SEARLE • Linear Models

(continued from front)